Hammad Ullah, Abdur Rauf, Maria Daglia (Eds.)
**Nutraceuticals**

## Also of interest

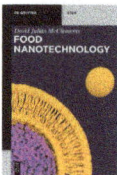

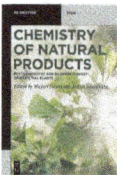

# Nutraceuticals

A Holistic Approach to Disease Prevention

Edited by
Hammad Ullah, Abdur Rauf, and Maria Daglia

**DE GRUYTER**

**Editors**
**Hammad Ullah, PhD**
Department of Pharmacy
University of Napoli Federico II
Via D. Montesano
80131 Naples
e-mail: hammad.ullah@unina.it

**Abdur Rauf, PhD**
Department of Chemistry
University of Swabi
94640 Khyber Pakhtunkhwa Swabi
Pakistan
e-mail: abdurrauf@uoswabi.edu.pk

**Maria Daglia, PhD**
Department of Pharmacy
University of Napoli Federico II
Via D. Montesano
80131 Naples
And
International Research Center for Food Nutrition and Safety
Jiangsu University
Zhenjiang 212013
China
e-mail: maria.daglia@unina.it

ISBN 978-3-11-131730-4
e-ISBN (PDF) 978-3-11-131760-1
e-ISBN (EPUB) 978-3-11-131787-8

**Library of Congress Control Number: 2023952308**

**Bibliographic information published by the Deutsche Nationalbibliothek**
The Deutsche Nationalbibliothek lists this publication in the Deutsche Nationalbibliografie; detailed
bibliographic data are available on the Internet at http://dnb.dnb.de.

© 2024 Walter de Gruyter GmbH, Berlin/Boston
Cover image: Gettyimages/Nikcoa
Typesetting: Integra Software Services Pvt. Ltd.
Printing and binding: CPI books GmbH, Leck

www.degruyter.com

# Preface

In the dynamic landscape of healthcare and wellness, the pursuit of optimal health has taken center stage, prompting individuals to explore diverse avenues for maintaining and enhancing their well-being. Among the myriad approaches, nutraceuticals have emerged as a fascinating intersection between nutrition and pharmaceuticals, offering a unique blend of science and nature to support health.

This book entitled *Nutraceuticals: A Holistic Approach to Disease Prevention* endeavors to unravel the intricate tapestry of nutraceuticals, providing readers with a comprehensive understanding of these compounds and their potential impact on health and disease. As we delve into this realm, we embark on a journey that transcends conventional boundaries, where food becomes medicine and science merges seamlessly with holistic well-being.

The term "nutraceutical" itself embodies the fusion of "nutrition" and "pharmaceuticals," encapsulating a diverse array of bioactive compounds found in foods and dietary supplements. From antioxidants to probiotics, omega-3 fatty acids to herbal extracts, each nutraceutical harbors the promise of promoting health and preventing illness. This book aims to demystify these compounds, shedding light on their origins, mechanisms of action, and evidence-based benefits.

As we navigate the chapters ahead, we will examine the cutting-edge research that underpins their efficacy, offering readers a glimpse into the exciting advancements that continue to shape this field. Moreover, practical insights on incorporating nutraceuticals into daily life, alongside considerations of safety and potential risks, will empower readers to make informed decisions about their health and wellness.

It is important to note that this book does not replace professional medical advice. Instead, it serves as a valuable resource, equipping readers with knowledge to engage in informed discussions with healthcare professionals and make personalized choices aligned with their unique health goals.

We extend our gratitude to the experts and researchers whose work has paved the way for this exploration, and to the readers who embark on this journey with curiosity and a commitment to their well-being. May this book serve as a beacon, guiding you through the vast landscape of nutraceuticals and empowering you to make choices that nurture a healthier, more vibrant life.

December, 2023

Hammad Ullah
University of Naples Federico II, Italy

Abdur Rauf
University of Swabi, Pakistan

Maria Daglia
University of Naples Federico II, Italy

https://doi.org/10.1515/9783111317601-202

# Contents

# About the editors

**Hammad Ullah** obtained his PhD in "Nutraceuticals, Functional Foods, and Human Health" at the Department of Pharmacy, University of Naples Federico II in liaison with the Department of Analytical and Food Chemistry, University of Vigo, Spain. The research activity focuses mainly on the comprehensive investigation of natural products, with the final aim of the development of new food supplement and functional food ingredients. The scientific activity is documented by more than 50 peer-reviewed articles, 4 book chapters, and 20 communications to national and international congresses. He is currently an editorial board member of *BMC Complementary Medicine and Therapies* and *Journal of Medicinal Food*. His scientific contributions are recognized by several research awards including MEDWELL Award for Best PhD Thesis (2023), PSE Dra. Mariola Macías Award (2023), and Young Researcher Award at MONASH INITIATE 2022. In addition, he is a regular member of American Chemical Society, Society for Medicinal Plant and Natural Product Research (GA), Royal Society of Chemistry (MRSC), International Natural Product Sciences Taskforce (INPST), Phytochemical Society of Europe (PSE), and IUPHAR Mediterranean Group of Natural Products Pharmacology.

**Abdur Rauf** works at the Department of Chemistry, University of Swabi, Khyber Pakhtunkhwa, Pakistan. He completed his PhD at the Institute of Chemical Sciences, University of Peshawar, Pakistan, in 2015. His research work focuses on phytochemistry, pharmacology, and nanotechnology. Abdur Rauf is the author and coauthor of more than 366 research papers published in peer-reviewed journals. He has edited 2 books and contributed 33 book chapters for international publishers. He also published one US patent, and submitted one national and three international patents for the discovery of novel drugs. Furthermore, he is the Associate Editor/Editorial Board Member of several journals including *Frontier in Pharmacology*, *Medicinal Chemistry*; *Green Processing and Synthesis*; *Open Chemistry*; and *Medicinal Chemistry*, among others. Abdur Rauf won the Young Scientist Award from the Directorate of Science and Technology in 2018 and the Research Productivity Award for 2016–17, from the Pakistan Science Foundation. His name is also in World's Top 1% Highly Cited Researchers 2022 list released by Clarivate. He has won 9 national and 10 international projects. As a scientist in the past few years, he contributed a lot to the fields of medicinal chemistry, pharmacology, and nanotechnology. His research interests also include exploring the local natural resources (i.e., Pakistani flora) for efficient treatment of different health disorders, which in turn would significantly impact the national economy.

**Maria Daglia** obtained her PhD degree in pharmaceutical chemistry and technology at the University of Pavia, Italy, and is currently Full Professor of Food Chemistry at the University of Naples Federico II, Italy, and a visiting professor at the International Research Center for Food Nutrition and Safety, Jiangsu University, Zhenjiang China. The scientific activity has developed along two lines of research: (i) the study of biological properties of components that are either naturally present or induced following thermo/technological treatments in foods useful in food and pharmaceutical fields; (ii) the development of analytical spectrophotometric and chromatographic methods useful in the identification and determination of biological active compounds occurring in foods. More recently, further research has been carried out, including (i) the investigation of the mechanism at the molecular level through which foods with nutraceutical activity and/or their components perform beneficial effects in humans, with particular reference to polyphenols, secondary metabolites of plants widely distributed in all categories of foods, from current foods to food supplements and (ii) the study of components with nutraceutical activity to be used in the health product industry, with *in vitro*, *in vivo* model systems and clinical studies. Her research activity is documented by more than 270 scientific papers, 2 books, 4 book chapters, and more than 120 communications to national and international congresses. As per Clarivate analytics, Prof. Daglia remained a highly cited researcher in the field of agricultural sciences in 2021. She is currently Editor-in-Chief of *Food Safety and Health* (Wiley), and editorial board member of *Phytomedicine* (Elsevier) and *Food Frontiers* (Wiley).

https://doi.org/10.1515/9783111317601-204

Hammad Ullah, Maria Daglia*

# Chapter 1
# Nutraceuticals and food supplements: basic concepts and regulatory aspects

**Abstract:** Nutraceuticals, a hybrid term of nutrient and pharmaceutical, are derived from food sources providing health effects beyond nutrition. Food supplements are manufactured products intended to supplement the human diet, whereas functional foods refer to foods containing bioactive components (in addition to basic nutrients) with potential health benefits. Nutraceuticals can be further classified based on their dietary sources. Dietary fibers are one of the critical components of food supplies that may promote human health via regulation of gastrointestinal digestion and gut microbiota. Prebiotics support the growth and activity of the gut flora. Probiotics are live microorganisms, providing benefits to host health via regulation of gut microbiota. Polyunsaturated fatty acids (PUFAs) possess potential to ameliorate systemic inflammation and the pathogenesis of chronic disorders (mainly of cardiac and metabolic origin). Vitamins and minerals are abundantly found in food sources, fighting against oxidative stress and regulate the cellular and biochemical processes of the body. Dietary polyphenols including flavonoids and nonflavonoids are one of the potent antioxidants with a wide array of biological effects. Dietary spices, known for their use as food additives and traditional medicine, possess a great impact on human health when consumed regularly in adequate amount. This chapter is focusing on the basic definitions and classification, efficacy, safety and regulation of nutraceuticals, functional foods, and food supplements.

# 1 Nutraceuticals: definition and classification

Foods contain various dietary factors with a range of biological effects, offering an opportunity to improve human health, while providing a protection against chronic diseases [1, 2]. The medicinal benefits of foods were recognized in 500 BC by a Greek physician Hippocrates, who quoted "Let food be thy medicine and medicine be thy food." Stephen DeFelice used the term "nutraceutical" in 1989 for the first time as a hybrid of "nutrition" and "pharmaceutical" to describe the health-promoting products

*Corresponding author: Maria Daglia**, Department of Pharmacy, University of Napoli Federico II, Via D. Montesano 49, 80131 Naples, NA, Italy; International Research Center for Food Nutrition and Safety, Jiangsu University, Zhenjiang 212013, China, e-mail: maria.daglia@unina.it
**Hammad Ullah,** Department of Pharmacy, University of Napoli Federico II, Via D. Montesano 49, 80131 Naples, NA, Italy

https://doi.org/10.1515/9783111317601-001

based on food-derived nutritional factors. DeFelice defined it as "a food (or part of a food) that provides medical or health benefits, including the prevention and/or treatment of a disease." Nutraceuticals may include whole food, isolated nutrients, food supplements, herbal products, or processed foods (i.e., beverages, cereals, and soups) that can also be used as medicine other than nutrition. In other words, nutraceuticals are food-derived products that possess physiological benefits and/or provide protection against chronic disorders. They may be used to delay the aging processes, enhance the life expectancy, prevent the chronic ailments, and support the anatomy or physiology of the body [3].

Functional foods refer to foods, containing bioactive components beyond basic nutrition. The Food and Agriculture Organization (FAO) defines functional foods as "a foodstuff containing other components, in addition to the basic nutrients, that may possess specific health effects, such as preventing and/or treating disease" [4]. However, there are certain issues with the commonly used definition such as (i) inclusion of commonly used healthy foods, that contain phytochemicals with a potential to enhance health and prevent disease, that is, peanuts, beetroot, strawberries, pomegranate juice, and sweet potato; (ii) many foods are contributing to healthy life, and it is not clear which food components (beyond basic nutrients) possess health-enhancing and disease-preventing potential. To address these issues, new definition for functional foods was proposed as "the novel foods, formulated to contain substances or live microorganisms, which may possess a potential health-improving or disease-preventing value. The added ingredients will be at safe and effective concentration to achieve the desired biological effects, and that may include nutrients, phytochemicals, dietary fiber, other components or probiotics" [5]. Some of the key examples of functional foods include margarine with enhanced omega-3 fatty acid levels, orange juice with added calcium, fermented foods containing probiotic strains, prebiotics supplemented foods, and margarine with added phytosterols and stanols. The subclasses of functional foods including fortified products, altered products, enriched products, and enhanced commodities are summarized in Figure 1.

In scientific community, the terms "nutraceuticals" and "functional foods" are usually used interchangeably, as both the definitions often overlap. Kalra (2003) proposed to redefine these terms as "a food being cooked or prepared using a scientific intelligence, with or without having knowledge of how and why it is being used is known as functional food" while "nutraceuticals can be defined as the functional foods with a potential to prevent and/or treat diseases' [6]. Other related products include dietary supplements, fortified foods, and herbal medicine. Dietary supplements refer to the pharmaceutical formulations containing food-derived components, indented to prevent or treat a deficiency, for example, micronutrients, amino acids, dietary fibers, and plant extract. Herbal medicines are defined by World Health Organization (WHO) as "herbs, herbal materials, herbal preparations and finished herbal products that contain as active ingredients parts of plants, or other plant materials, or combinations."

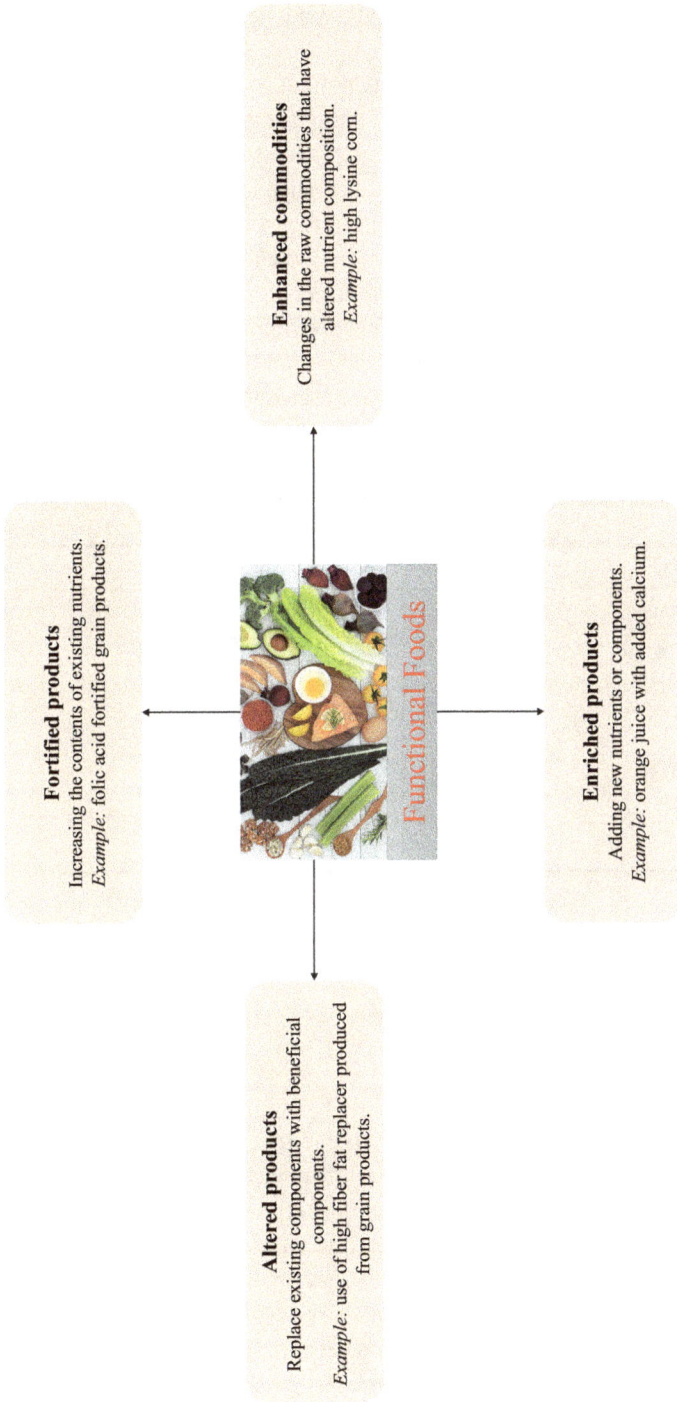

**Enhanced commodities**
Changes in the raw commodities that have altered nutrient composition.
*Example:* high lysine corn.

**Fortified products**
Increasing the contents of existing nutrients.
*Example:* folic acid fortified grain products.

**Enriched products**
Adding new nutrients or components.
*Example:* orange juice with added calcium.

**Altered products**
Replace existing components with beneficial components.
*Example:* use of high fiber fat replacer produced from grain products.

Functional Foods

**Figure 1:** Subclasses of functional foods and their examples.

Fortified foods are the foodstuffs to which substances with preventive or therapeutic efficacy be added [7].

Generally, nutraceuticals can be classified in two main categories including traditional and nontraditional nutraceuticals (Table 1). Traditional nutraceuticals are typically sourced from nature without any modification being made to the original structure. They are subgrouped divided in three classes including chemical constituents (i.e., nutrients, herbals, and phytochemicals), probiotic strains (i.e., yogurt and kefir), and nutraceutical enzymes (i.e., papain and lactase enzymes) [8]. Nontraditional nutraceuticals are enriched food supplements made with biotechnological techniques to maximize nutrient absorption. These include fortified nutraceuticals that are supplemented foods or drink products with added nutrients (such as orange juice with added calcium or milk with cholecalciferol) or recombinant nutraceuticals that are prepared via genetic recombination or biotechnology (such as fermented starch) [9].

**Table 1:** Key differences among traditional and nontraditional nutraceuticals.

| Traditional nutraceuticals | Nontraditional nutraceuticals |
|---|---|
| No change to the food is made. | Product of agricultural breeding or added ingredients and/or nutrients. |
| Whole foods with new information about their potential health qualities. | Made with biotechnological techniques to improve the nutritional content of foods. |
| Examples<br>– resveratrol in grapes<br>– omega-3 fatty acids in salmon<br>– lycopene in tomatoes | Examples<br>– orange juice fortified with calcium<br>– cereals with added vitamins or minerals<br>– flour with added folic acid |

Based on food sources, nutraceuticals can be classified in dietary fibers, prebiotics, probiotics, PUFAs, micronutrients, dietary polyphenols, and spices (Table 2).

## 1.1 Dietary fibers

Dietary fiber is a critical component of human health and food supply that contributes to the proper digestion and excretion of food. Dietary fibers include insoluble (i.e., celluloses, lignins, and some hemicelluloses) and soluble fibers (pectin, inulin, guar gum, β-glucan, mucilage, and hemicellulose). Insoluble fibers add bulk to the stool and speed up the passage of food through the gastrointestinal tract. They are found abundantly in vegetables, whole grains, and wheat bran. Soluble fibers turn to gel during digestion while attracting water, which may slow down the digestion process. They are found mostly in barley, nuts, oat brans, lentils, beans, and peas. Soluble fibers have increased water holding capacity and are subjected to colonic fermentation to produce short chain fatty acids

**Table 2:** Classification of nutraceuticals based on food sources.

| Nutraceutical class | Definition | Examples |
|---|---|---|
| Dietary fibers | Food-derived components that are not completely breakdown by digestive enzymes in the gastrointestinal tract. | Arabinoxylans, inulin, dextrins, chitins, lignin, pectins, β-glucans, cellulose, and oligosaccharides. |
| Prebiotics | Foods or supplements containing indigestible fibers that selectively support the growth and activity of indigenous probiotic strains. | Fructo-oligosaccharides (FOS), galacto-oligosaccharides (GOS), xylo-oligosaccharides (XOS), pectin, lactulose, inulin, resistant starch, and β-glucans. |
| Probiotics | Mono- or mixed cultures of living microorganisms that modulate gut microbiota, conferring beneficial effects to the host beyond nutrition. | *Lactobacillus*, *Bifidobacterium*, and *Saccharomyces* species. |
| Polyunsaturated fatty acids | Fatty acids containing more than one double bond in their carbon backbone and are precursors for eicosanoids. | Omega-3 fatty acids, omega-6 fatty acids |
| Micronutrients | Essential dietary elements required by body in trace amount for normal growth and development. | Vitamins and minerals. |
| Polyphenols | Group of secondary metabolites found in common food sources that possess wide range of biological effects and health benefits. | Flavonoids, phenolic acids, coumarins, xanthones, stilbenes, lignans, and curcuminoids. |
| Spices | Food adjuncts that enhance sensory quality of foods, minute quantity of which may possess potential biological effects such as antioxidant, anti-inflammatory, chemo-preventive, and immune modulatory effects. | Ginger, garlic, turmeric, cinnamon, cumin, pepper, fenugreek, lemon grass, and peppermint. |

(SCFAs) while insoluble fibers are fermented to a lesser extent and have minimal water holding capacity [10].

Intake of dietary fibers on regular basis confers range of beneficial effects to host health. Soluble fibers may improve functional gastrointestinal disorders via accelerating oro-anal transit. These effects are mainly attributed to the alteration of gut microbiome by SCFAs, that is, increased abundance of firmicutes, bacteroidetes, and actinobacteria. Butyrate has been shown effective in the management of ulcerative rectocolitis. More-over, butyrate is also found to act as epigenetic switch in regulation of immune system as it induces the regulatory T cells production in intestine and suppresses the interferon-γ-mediated inflammation, which suggest it as one of the potential therapies against the chronic inflammatory conditions of gastrointestinal tract that is, inflammatory bowel disease [11–15]. In addition, regular intake of dietary fibers (such as inulin and arabinoxylan oligosaccharides) may also improve insulin sensitivity, regulate glucose and lipids homeostasis, reduce body weight, increase fatty acid oxidation, and improve endothelial dysfunction [16, 17].

## 1.2 Prebiotics

Prebiotics is a subclass of dietary fibers, supporting the growth and activity of gut microbial species and probiotic strains. The most commonly used prebiotics include inulin, fructo-oligosaccharide, galacto-oligosaccharide, lactulose, and psyllium [18, 19]. The dietary sources of prebiotics include bananas, asparagus, onions, garlic, artichoke, tomatoes, leeks, oats, wheat, and soy beans. Intake of prebiotics on regular basis may increase SCFAs production, control the growth and/or activity of pathogens and toxins, and regulate the intestinal immune system [20]. It has been demonstrated that feeding infants with prebiotics such as inulin reduces their vulnerability to colds, diarrhea, and vomiting [21]. Prebiotics are primarily fermented by gut bacteria into SCFAs that may affect the intestinal immune system by enhancing the absorption of some nutrients and improving gut microbiota composition [22].

## 1.3 Probiotics

Probiotics are live microorganisms modulating gut microbiota and their consumption in adequate amount as food supplements or functional foods may provide health benefits [23]. Some common examples include *Lactobacillus, Bifidobacterium,* or *Saccharomyces* species, some strains of *Escherichia coli,* and some Gram-positive cocci. Probiotics should be nonpathogenic, having nontoxic properties and are able to produce SCFAs and to adhere to the gut epithelium. Beyond nutrition, they confer numerous beneficial health effects to the host including increasing integrity of the intestinal epithelium, regulating intestinal immunity, protecting the gut barrier disruption, and inhibiting pathogenic microbes [24–26]. The daily probiotic dose of $10^6$–$10^9$ colony forming units (CFU) is recommended for their health effects; thus probiotic strain in available formulations should be in concentration above the threshold, that is, $10^6$ CFU. However, some formulations may also contain 1,012 CFU of individual bacterial strain [27, 28].

## 1.4 Polyunsaturated fatty acids

In recent years, PUFAs have gained prominence as functional foods in ameliorating inflammatory processes and preventing the pathogenesis of chronic ailments (mainly cardio-metabolic risk factors). Considering that the human body does not produce them, plants and marine sources are vital to fulfill the dietary need of PUFAs [29]. Food sources rich in PUFAs include edible oils (corn oil, sunflower oil, soybean oil), salmon, and some seeds and nuts (walnuts, sunflower seeds, soybeans, and tofu). Based on the carbon chain length, they are sometimes classified as short chain PUFAs (18 C-atoms) and long-chain polyunsaturated fatty acids (≥20 C-atoms). Omega-3 polyunsaturated fatty acids (*n*-3 PUFAs), that is, α-linolenic, eicosapentanoic, and docosahexanoic acids are good for cardiac function in several ways like reduction of triglycerides levels, slowing up plaque formation to attenuate the risk for atherosclerosis, lowering blood pressure, and risk reduction for cardiac arrhythmias [30]. The *n*-3 PUFAs are now being incorporated into a variety of products in the food and pharmaceutical industries. Omega-6 polyunsaturated fatty acids (*n*-6 PUFAs), that is, linoleic acid may help in lowering blood pressure and reducing blood glucose levels, thus protecting against diabetes risk factors. Linoleic acid is also reported to protect against mammary and prostate cancers [31].

On the other hand, increased risk of chronic inflammatory conditions (i.e., gastrointestinal, metabolic, cardiovascular, hepatic, and neuroinflammatory diseases) has also been reported with high consumption of *n*-6 PUFAs enriched foods and supplements. Interestingly, eicosanoids derived from *n*-3 and *n*-6 PUFAs possess anti-inflammatory and proinflammatory actions, respectively. The anti-inflammatory effects of *n*-3 PUFAs may be antagonized by the concomitant intake of *n*-6 PUFAs [32,33]. Increasing the *n*-3 to *n*-6 ratio in diet may lessen the risk for chronic inflammation.

## 1.5 Micronutrients

Vitamins and minerals occupy a large part of the global nutraceutical market. Vitamins, through their ability to prevent oxidative reactions, act individually and synergistically to prevent several diseases. They are abundantly found in food sources that function by scavenging free radicals and providing protection against oxidative stress and pathogenic processes [34]. Visual health, immune function, and bone health are promoted by vitamins. Minerals are found primarily in vegetables and fruits that play a key role in cells metabolism and normal functioning of the organisms [35]. Their applications include strengthening bones and teeth, improving nerve and muscle function, and in the treatment of several diseased conditions [31].

A comprehensive discussion on vitamins and minerals is provided in Chapter 4.

## 1.6 Dietary polyphenols

Polyphenols are secondary metabolites, found abundantly in food sources such as vegetables, fruits, beverages, seeds, and in lesser amount in cereals and legumes. Chemically they contain a benzene ring with hydroxyl (OH) moieties. Based on the number of phenolic rings and structural components connecting these rings provided, polyphenols are further classified in flavonoids (Figure 2) and nonflavonoids (Figure 3) [36]. They possess potent antioxidant actions and are one of the highly consumed dietary antioxidants. Studies showed about 1 –g/day consumption of polyphenols by Western population from dietary sources. Over the last decades, they were thoroughly assessed for their biological properties and health benefits that include antidiabetic, cardioprotective, antimicrobials, antiasthma, antidepressant, anxiolytic, neuroprotective, and anticarcinogenic [37–43]. Their potent antioxidant actions make them vulnerable agents to delay the aging process and to prevent against the age-related disorders [44, 45].

However, despite the promising health effects in preclinical studies, the use of polyphenols in food supplements and functional foods has not been optimally realized due to their low bioavailability. The possible causes of which may include limited release from the food matrix, poor solubility in gastrointestinal fluids, low permeability across the epithelial cells, polyphenol–protein interaction, and/or molecular transformations in the gastrointestinal tract. Polyphenol's interaction with the intestinal flora may also lead to deep modification of molecules. Some of the scientific processes are suggested to effectively improve the bioavailability of polyphenols including food processing (i.e., fruits juicing or vegetables homogenization), encapsulation, nanoformulations, enzymatic treatments (i.e., enhanced bioavailability of isoflavones as a result of soy drink treatment with β-glucosidase), and food-based fermentation (breaking down complex polyphenols into simpler ones) [46].

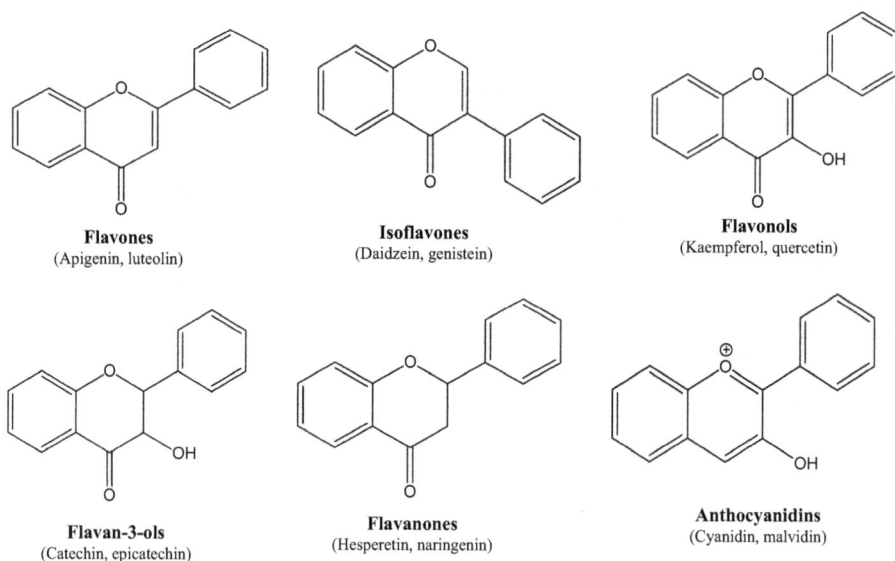

**Flavones**
(Apigenin, luteolin)

**Isoflavones**
(Daidzein, genistein)

**Flavonols**
(Kaempferol, quercetin)

**Flavan-3-ols**
(Catechin, epicatechin)

**Flavanones**
(Hesperetin, naringenin)

**Anthocyanidins**
(Cyanidin, malvidin)

**Figure 2:** Major classes of flavonoids.

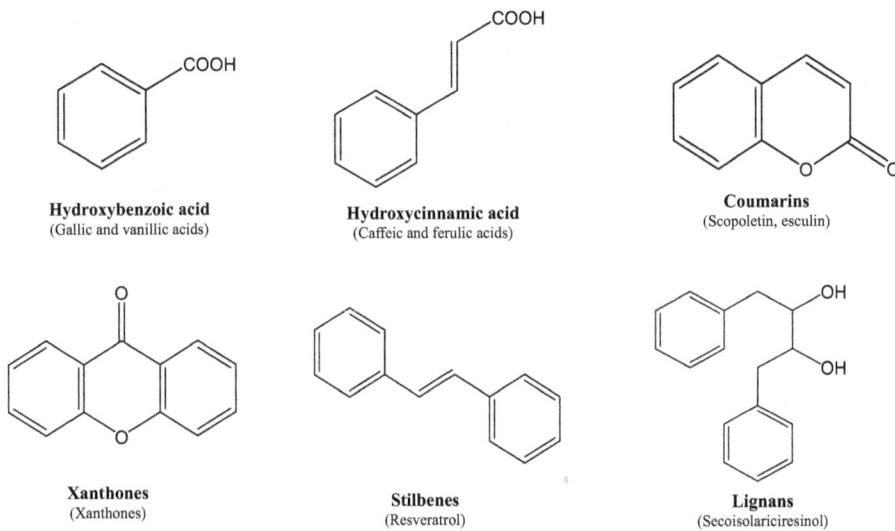

**Figure 3:** Major classes of nonflavonoids.

## 1.7 Spices

Dietary spices are collection of volatile and nonvolatile staple food additives that are known for their wide uses as food additive and in traditional medicine. Research trends over the last decades showed that they have diverse array of phytochemicals with complementary and overlapping actions, and their regular intake in minute quantities may possess a significant impact on health because of their potent antioxidant, anti-inflammatory, and chemopreventive actions [47–49]. In addition, spices and derived phytochemicals have the ability to limit the oxidative and microbial spoilage of foods while maintaining their organoleptic properties; thus their application in foods industry as biopreservatives is increasing [50].

## 2 Efficacy and safety concerns

The worldwide market share of nutraceuticals and dietary supplements is expanding, mainly due to increasing tendency of consumers toward preventive care, as increasing tendency of world population is seen preferring vegetable extracts for their primary healthcare [51]. These products are providing multiple health benefits beside basic nutritional support and possess ability to prevent the most challenging disorders. They are documented effective in subjects with overweight and obesity, hyperglycemia, hyperinsulinemia, and hypercholesterolemia as well as in delaying the aging process and age-related disorders. Consumption of these products may maintain good health and longevity in healthy people and improves the health condition in diseased subjects. They may help in increasing the absorption of nutrients, supporting the normal flora of the human body, regulating the immune system, and decreasing the reactive oxygen and nitrogen species and inflammatory cytokines [52, 53]. However, the response of consumers to nutraceuticals may vary from person to person, as susceptibility of an individual to any illness largely depends on the lifestyle factors (such as eating habits, smoking, and chronic alcohol consumption) and genetic predisposition.

Nutraceuticals and food supplements are generally considered safe ad nontoxic. Though these products are introduced in the market without any prior regulation and are available for use over-the-counter, safety remains a prime concern or else they may lead to undesirable effects. As these supplements are brought to the market without any solid evidence from human studies, there is scarcity of systematic studies reporting the associated adverse events. Usually, case report of side effects occurring after the intake of the supplement hints that there may be some sort of supplement-associated undesirable effect. Cluster of cases may establish the possibility of adverse effects with the intake of specific supplement [54]. Polyphenols are commonly described as dietary antioxidants but some of the commonly used phenolic compounds like quercetin and epigallocatechin gallate are suspected to cause oxidative stress-induced liver injury [55, 56]. Likewise, soy isoflavones are shown to possess estrogenic properties in preclinical studies, which may result in reproductive tract malformations or uterine hypertrophy, reduced fertility, reduced testis size, and stimulate the estrogen-dependent cancers [57–64]. In addition, patients usually do not disclose the history of the food supplements use to their healthcare providers, which increase the likelihood of supplement–drug interactions.

# 3 Regulation of nutraceuticals and food supplements

Nutraceuticals inhabit a gray line in between drugs and foods, with no agreement on the regulation of these products worldwide. They do not qualify for the available regulatory frameworks, that is, Food and Drug Authority (FDA) and European Medicines Agency (EMA), and are regulated differently in different countries. However, the global market of food-based therapeutics is considerably increasing with annual value of approximately US$180 billion in 2020, with regional shares of North America, Europe, and Asia-Pacific at 25%, 30%, and 31%, respectively, and it is estimated to reach at US$278 billion by 2024 [65]. The market size of dietary supplements in Europe in 2022 was estimated at US$68.94 billion and is expected at a compound annual growth rate of 6% by 2030. In 2014, ~18.8% of the European population was taking at least one supplement based on vegetable extract [66, 67]. The expanding market of nutraceuticals and food supplements calls for increased regulation to ensure their efficacy, safety, and quality.

## 3.1 United States of America (USA)

Unlike micronutrients, herbal products, and protein supplements, nutraceutical products (such as prebiotics, probiotics, and special foods for medical use) require clinical assessment of efficacy and safety prior to human use in USA [68]. In 1938, Federal Food, Drug and Cosmetic Act (FFDCA) allowed FDA, regulating the safety of drugs, foods, and cosmetics. FFDCA introduced two different regulatory frameworks for drugs and foods. Later in 1994, Dietary Supplement Health and Education Act (DSHEA) introduced dietary supplements as novel stream of foods [69], with a recommendation that nutraceuticals should be regulated as foods and should meet all the provisions for conventional foods as specified in the FFDCA. The DSHEA became the fundamental body regulating dietary supplements in the USA, aimed to distinguish between drugs and food regulatory framework. This act supposed manufacturers to assure the safety and quality of their products by clearly displaying the ingredients used in the formulation of product or any possible allergies. The act established a clear definition of dietary supplement and outlined their potential qualities. Moreover, it enables FDA to state jurisdiction over establishing good manufacturing practices specific to food supplement industry. In addition, DSHEA promoted research studies on dietary supplements assessing their potential benefits to human health. Clinical trials and safety studies are required in scenarios where the expected exposure to dietary supplement exceeds the historical consumption. The office of Dietary Supplement Regulation was established within the National Institutes of Health to ensure the survey requirements set by DSHEA [70] though it does not implement the DSHEA regulations. FDA is responsible to monitor the marketplace for misleading labeling and advertising claims or unsafe practices, in addition to premarketing evaluation of new ingredients for safety and toxicity. In general, premarketing authorization is not mandatory except for novel ingredients marketed in USA after 1994, as historical dietary supplement ingredients marketed in USA before DSHEA implementation are considered safe [70, 71].

## 3.2 Europe Union (EU)

In EU, drug and food safety is regulated by different agencies. According to the European Commission Concerted Action on Functional Food Science in Europe (FUFOSE), a food that beneficially targets one or more functions in the body, beyond the acceptable nutritional effects, either enhancing the general physical health and/or reducing the risk of disease development or progression can be regarded as functional food [72, 73]. The legal status of nutraceuticals in EU is based on their health promoting effects. A product can be considered as functional food ingredient if it contributes to maintain the normal health and wellness; but at the same time, it should be rendered as a medical substance if it modifies or tends to modify the pharmacological process [74]. The EMA is basically responsible to evaluate, supervise, and monitor the pharmacovigilance of the medicines across the European states. In 2002, the European Parliament and the Council adopted Regulation (EC) No. 178/2002 framed the General Food Law Regulation (requirements and principles of the food law), that laid down the foundation for general requirements, guidelines, principles, and procedures for the policymaking process related to food production, food distribution, and safety. An independent agency, EU Food Safety Authority (EFSA) was also established under these regulations that are in authority for scientific advice and safety assessment of food products. EU legislation of nutraceuticals is very complex, as the umbrella of these products in Europe is very wide, and different regulatory procedures are available for each of class of nutraceuticals and dietary supplements. To obtain marketing authority, nutraceuticals with medical claim, or food for special medical purposes should comply with the Directive 2001/82/EC and should be dealt with EU regulation No. 609/2013 and No. 1924/2006 to ensure safety and efficacy [67]. Following the directives of the Scientific Committees and panels of the EFSA, biological and safety data are required for foods with medical claims.

# 4 Summary

Nutraceuticals are considerably growing sector, using widely in pharmaceutical and food industries for their potential health benefits. They are extensively studied in recent years in line with the advancement of biochemistry and food chemistry research fields, underlying analytical techniques and their clinical relevance. Nutraceuticals and food supplements are quickly replacing pharmaceuticals in the prevention and/or management of chronic disorders, where it showed a huge success with great monetary benefits. However, it needs more in-depth assessment for their efficacy and safety to prove their preference over conventional therapeutic options. In addition, they are regulated differently in different countries, and gaps in the regulation processes as well as challenges in technical analysis can increase the significant health risks and economic burden. Development of a holistic regulating and monitoring system is essential to support their safe and effective use.

# References

[1] Gul, K., Singh, A.K. and Jabeen, R. 2016. Nutraceuticals and functional foods: The foods for the future world. Critical Reviews in Food Science and Nutrition, 56, 2617–2627.

[2] Ullah, H., de Filippis, A., Khan, H., Xiao, J. and Daglia, M. 2020. An overview of the health benefits of Prunus species with special reference to metabolic syndrome risk factors. Food and Chemical Toxicology, 144, 111574.

[3] Puri, V., Nagpal, M., Singh, I., Singh, M., Dhingra, G.A., Huanbutta, K., Dheer, D., Sharma, A. and Sangnim, T. 2022. A comprehensive review on nutraceuticals: Therapy support and formulation challenges. Nutrients, 14(21), 4637.

[4]  FAO. FAO term portal. Available online at https://www.fao.org/faoterm/viewentry/en/?entryId= 170967 (accessed June 11, 2023).

[5]  Temple, N.J. 2022. A rational definition for functional foods: A perspective. Frontiers in Nutrition, 9, 957516.

[6]  Kalra, E.K. 2003. Nutraceutical-definition and introduction. AAPS Pharmaceutical Sciences, 5(3), 25.

[7]  Andrew, R. and Izzo, A.A. 2017. Principles of pharmacological research of nutraceuticals. British Journal of Pharmacology, 174(11), 1177.

[8]  Chanda, S., Tiwari, R.K., Kumar, A. and Singh, K. 2019. Nutraceuticals inspiring the current therapy for lifestyle diseases. Advances in Pharmacological and Pharmaceutical Sciences, 2019.

[9]  Singh, J. and Sinha, S. 2012. Classification, regulatory acts and applications of nutraceuticals for health. International Journal of Pharmacy and Biological Sciences, 2, 177–187.

[10]  De Filippis, A., Ullah, H., Baldi, A., Dacrema, M., Esposito, C., Garzarella, E.U., Santarcangelo, C., Tantipongpiradet, A. and Daglia, M. 2020. Gastrointestinal disorders and metabolic syndrome: Dysbiosis as a key link and common bioactive dietary components useful for their treatment. International Journal of Molecular Sciences, 21(14), 4929.

[11]  Parada Venegas, D., De la Fuente, M.K., Landskron, G., González, M.J., Quera, R., Dijkstra, G., Harmsen, H.J., Faber, K.N. and Hermoso, M.A. 2019. Short chain fatty acids (SCFAs)-mediated gut epithelial and immune regulation and its relevance for inflammatory bowel diseases. Frontiers in Immunology, 277.

[12]  Den Besten, G., Van Eunen, K., Groen, A.K., Venema, K., Reijngoud, D.J. and Bakker, B.M. 2013. The role of short-chain fatty acids in the interplay between diet, gut microbiota, and host energy metabolism. Journal of Lipid Research, 54(9), 2325–2340.

[13]  El-Salhy, M., Ystad, S.O., Mazzawi, T. and Gundersen, D. 2017. Dietary fiber in irritable bowel syndrome. International Journal of Molecular Medicine, 40(3), 607–613.

[14]  Zhang, M., Zhou, Q., Dorfman, R.G., Huang, X., Fan, T., Zhang, H., Zhang, J. and Yu, C. 2016. Butyrate inhibits interleukin-17 and generates Tregs to ameliorate colorectal colitis in rats. BMC Gastroenterology, 16(1), 1–9.

[15]  Velazquez, O.C., Lederer, H.M. and Rombeau, J.L. 1997. Butyrate and the colonocyte: Production, absorption, metabolism, and therapeutic implications. In *Dietary Fiber in Health and Disease*. Vol. 427, Wimberley, TX, USA: Humana Press, 123–134.

[16]  Kjølbæk, L., Benítez-Páez, A., Del Pulgar, E.M.G., Brahe, L.K., Liebisch, G., Matysik, S., Rampelli, S., Vermeiren, J., Brigidi, P., Larsen, L.H. and Astrup, A. 2020. Arabinoxylan oligosaccharides and polyunsaturated fatty acid effects on gut microbiota and metabolic markers in overweight individuals with signs of metabolic syndrome: A randomized cross-over trial. Clinical Nutrition, 39(1), 67–79.

[17]  Li, K., Zhang, L., Xue, J., Yang, X., Dong, X., Sha, L., Lei, H., Zhang, X., Zhu, L., Wang, Z. and Li, X. 2019. Dietary inulin alleviates diverse stages of type 2 diabetes mellitus via anti-inflammation and modulating gut microbiota in db/db mice. Food & Function, 10(4), 1915–1927.

[18]  Roberfroid, M., Gibson, G.R., Hoyles, L., et al. 2010. Prebiotic effects: Metabolic and health benefits. British Journal of Nutrition, 104, S1–S63.

[19]  Zheng, H.J., Guo, J., Jia, Q., Huang, Y.S., Huang, W.J., Zhang, W., Zhang, F., Liu, W.J. and Wang, Y., 2019. The effect of probiotic and synbiotic supplementation on biomarkers of inflammation and oxidative stress in diabetic patients: a systematic review and meta-analysis of randomized controlled trials. Pharmacological Research, 142, 303–313.

[20]  Al-Sheraji, S.H., Ismail, A., Manap, M.Y., Mustafa, S., Yusof, R.M. and Hassan, F.A. 2013. Prebiotics as functional foods: A review. Journal of Functional Foods, 5(4), 1542–1553.

[21]  Thammarutwasik, P., Hongpattarakere, T., Chantachum, S., et al. 2009. Prebiotics – A Review. Songklanakarin Journal of Science and Technology, 4, 401–408.

[22] Feng, W., Ao, H. and Peng, C. 2018. Gut microbiota, short-chain fatty acids, and herbal medicines. Front Pharmacol, 9, 1–12.

[23] AlAli, M., Alqubaisy, M., Aljaafari, M.N., AlAli, A.O., Baqais, L., Molouki, A., Abushelaibi, A., Lai, K.S. and Lim, S.H.E. 2021. Nutraceuticals: Transformation of conventional foods into health promoters/disease preventers and safety considerations. Molecules, 26(9), 2540.

[24] Bron, P.A., Kleerebezem, M., Brummer, R.-J., Cani, P.D., Mercenier, A., MacDonald, T.T., Garcia-Ródenas, C.L. and Wells, J.M. 2017. Can probiotics modulate human disease by impacting intestinal barrier function? British Journal of Nutrition, 117, 93–107.

[25] Kang, H.-J. and Im, S.-H. 2015. Probiotics as an immune modulator. Journal of Nutritional Science and Vitaminology, 61, S103–S105.

[26] Krishna Rao, R. and Samak, G. 2013. Protection and restitution of gut barrier by probiotics: Nutritional and clinical implications. Current Research in Nutrition and Food Science, 9, 99–107.

[27] Sreeja, V. and Prajapati, J.B. 2013. Probiotic formulations: Application and status as pharmaceuticals – A review. Probiotics Antimicrob Proteins, 5, 81–91.

[28] Jesus, A.L.T., Fernandes, M.S., Kamimura, B.A., Prado-Silva, L., Silva, R., Esmerino, E.A., Cruz, A.G. and Sant'Ana, A.S. 2016. Growth potential of Listeria monocytogenes in probiotic cottage cheese formulations with reduced sodium content. Food Research International, 81, 180–187.

[29] De Deckere, E.A. 2001. Health aspects of fish and n-3 polyunsaturated fatty acids from plant and marine origin. Nutritional health: Strategies for disease prevention. 195–206.

[30] Freeman, L.M. 2010. Beneficial effects of omega-3 fatty acids in cardiovascular disease. Journal of Small Animal Practice, 51(9), 462–470.

[31] Heinze, V.M. and Actis, A.B. 2012. Dietary conjugated linoleic acid and long-chain n-3 fatty acids in mammary and prostate cancer protection: A review. International Journal of Food Sciences and Nutrition, 63(1), 66–78.

[32] Innes, J.K. and Calder, P.C. 2018. Omega-6 fatty acids and inflammation. Prostaglandins, Leukotrienes and Essential Fatty Acids, 132, 41–48.

[33] Patterson, E., Wall, R., Fitzgerald, G.F., Ross, R.P. and Stanton, C. 2012. Health implications of high dietary omega-6 polyunsaturated fatty acids. Journal of Nutrition and Metabolism, 2012.

[34] Das, L., Bhaumik, E., Raychaudhuri, U. and Chakraborty, R. 2012. Role of nutraceuticals in human health. Journal of Food Science and Technology, 49, 173–183.

[35] Killilea, D.W. and Killilea, A.N. 2022. Mineral requirements for mitochondrial function: A connection to redox balance and cellular differentiation. Free Radical Biology and Medicine, 182, 182–191.

[36] Ullah, H., De Filippis, A., Santarcangelo, C. and Daglia, M. 2020. Epigenetic regulation by polyphenols in diabetes and related complications. Mediterranean Journal of Nutrition and Metabolism, 13(4), 289–310.

[37] Khan, H., Reale, M., Ullah, H., Sureda, A., Tejada, S., Wang, Y., Zhang, Z.J. and Xiao, J. 2020. Anti-cancer effects of polyphenols via targeting p53 signaling pathway: Updates and future directions. Biotechnology Advances, 38, 107385.

[38] Kishimoto, Y., Tani, M. and Kondo, K. 2013. Pleiotropic preventive effects of dietary polyphenols in cardiovascular diseases. European Journal of Clinical Nutrition, 67(5), 532–535.

[39] Muceniece, R., Klavins, L., Kviesis, J., Jekabsons, K., Rembergs, R., Saleniece, K., Dzirkale, Z., Saulite, L., Riekstina, U. and Klavins, M. 2019. Antioxidative, hypoglycaemic and hepatoprotective properties of five *Vaccinium* spp. berry pomace extracts. Journal of Berry Research, 9(2), 267–282.

[40] Losada-Barreiro, S. and Bravo-Diaz, C. 2017. Free radicals and polyphenols: The redox chemistry of neurodegenerative diseases. European Journal of Medicinal Chemistry, 133, 379–402.

[41] Alvarez-Suarez, J.M., Giampieri, F., Cordero, M., Gasparrini, M., Forbes-Hernández, T.Y., Mazzoni, L., Afrin, S., Beltrán-Ayala, P., González-Paramás, A.M., Santos-Buelga, C. and Varela-Lopez, A. 2016. Activation of AMPK/Nrf2 signalling by Manuka honey protects human dermal fibroblasts against

oxidative damage by improving antioxidant response and mitochondrial function promoting wound healing. Journal of Functional Foods, 25, 38–49.

[42] Čanadanović-Brunet, J., Tumbas Šaponjac, V., Stajčić, S., Ćetković, G., Čanadanović, V., Ćebović, T. and Vulić, J. 2019. Polyphenolic composition, antiradical and hepatoprotective activities of bilberry and blackberry pomace extracts. Journal of Berry Research, 9(2), 349–362.

[43] Mazzoni, L., Perez-Lopez, P., Giampieri, F., Alvarez-Suarez, J.M., Gasparrini, M., Forbes-Hernandez, T.Y., Quiles, J.L., Mezzetti, B. and Battino, M. 2016. The genetic aspects of berries: From field to health. Journal of the Science of Food and Agriculture, 96(2), 365–371.

[44] Khan, H., Sureda, A., Belwal, T., Çetinkaya, S., Süntar, İ., Tejada, S., Devkota, H.P., Ullah, H. and Aschner, M., 2019. Polyphenols in the treatment of autoimmune diseases. Autoimmunity Reviews, 18(7), 647–657.

[45] Ullah, H., Khan, A. and Daglia, M. 2022. The focus on foods for special medical purposes and food supplements in age-related disorders. Food Frontiers, 3(3), 353–357.

[46] Polia, F., Pastor-Belda, M., Martínez-Blázquez, A., Horcajada, M.N., Tomás-Barberán, F.A. and García-Villalba, R. 2022. Technological and biotechnological processes to enhance the bioavailability of dietary (poly) phenols in humans. Journal of Agricultural and Food Chemistry, 70(7), 2092–2107.

[47] Gupta, M. 2010. Pharmacological properties and traditional therapeutic uses of important Indian spices: A review. International Journal of Food Properties, 13, 1092–1116.

[48] Singh, V.K., Yadav, P. and Tadigoppula, N. 2014. Recent advances in the synthesis, chemical transformations and pharmacological studies of some important dietary spice's constituents. Chemistry & Biology Interface, 4, 66–99.

[49] Kochhar, K. 2008. Dietary spices in health and diseases: I. Indian Journal of Physiology and Pharmacology, 52, 106–122.

[50] Mandal, D., Sarkar, T. and Chakraborty, R. 2023. Critical review on nutritional, bioactive, and medicinal potential of spices and herbs and their application in food fortification and nanotechnology. Applied Biochemistry and Biotechnology, 195(2), 1319–1513.

[51] Ekor, M. 2014. The growing use of herbal medicines: Issues relating to adverse reactions and challenges in monitoring safety. Frontiers in Pharmacology, 4, 177.

[52] Dillard, C.J. and German, J.B. 2000. Phytochemicals: Nutraceuticals and human health. Journal of the Science of Food and Agriculture, 80(12), 1744–1756.

[53] Das, L., Bhaumik, E., Raychaudhuri, U. and Chakraborty, R. 2012. Role of nutraceuticals in human health. Journal of Food Science and Technology, 49, 173–183.

[54] Ronis, M.J., Pedersen, K.B. and Watt, J. 2018. Adverse effects of nutraceuticals and dietary supplements. Annual Review of Pharmacology and Toxicology, 58, 583–601.

[55] Mazzanti, G., Menniti-Ippolito, F., Moro, P.A., Cassetti, F., Raschetti, R., Santuccio, C. and Mastrangelo, S. 2009. Hepatotoxicity from green tea: A review of the literature and two unpublished cases. European Journal of Clinical Pharmacology, 65, 331–341.

[56] Reid, C., Smith, K. and Kocha BSc, W. 2004. Quercetin/bromelain use associated with increased liver enzymes. Journal of Oncology Pharmacy Practice, 10(4), 225–227.

[57] Badger, T.M., Gilchrist, J.M., Pivik, R.T., Andres, A., Shankar, K., Chen, J.R. and Ronis, M.J. 2009. The health implications of soy infant formula. The American Journal of Clinical Nutrition, 89(5), 1668S–1672S.

[58] Chen, A. and Rogan, W.J. 2004. Isoflavones in soy infant formula: A review of evidence for endocrine and other activity in infants. Annual Review of Nutrition, 24, 33–54.

[59] Akingbemi, B.T., Braden, T.D., Kemppainen, B.W., Hancock, K.D., Sherrill, J.D., Cook, S.J., He, X. and Supko, J.G. 2007. Exposure to phytoestrogens in the perinatal period affects androgen secretion by testicular Leydig cells in the adult rat. Endocrinology, 148(9), 4475–4488.

[60] Allred, C.D., Allred, K.F., Ju, Y.H., Virant, S.M. and Helferich, W.G. 2001. Soy diets containing varying amounts of genistein stimulate growth of estrogen-dependent (MCF-7) tumors in a dose-dependent manner. Cancer Research, 61(13), 5045–5050.

[61] McCarver, G., Bhatia, J., Chambers, C., Clarke, R., Etzel, R., Foster, W., Hoyer, P., Leeder, J.S., Peters, J.M., Rissman, E. and Rybak, M. 2011. NTP-CERHR expert panel report on the developmental toxicity of soy infant formula. Birth Defects Research Part B: Developmental and Reproductive Toxicology, 92(5), 421–468.

[62] Ronis, M.J., Gomez-Acevedo, H., Blackburn, M.L., Cleves, M.A., Singhal, R. and Badger, T.M. 2016. Uterine responses to feeding soy protein isolate and treatment with 17β-estradiol differ in ovariectomized female rats. Toxicology and Applied Pharmacology, 297, 68–80.

[63] Miousse, I.R., Sharma, N., Blackburn, M., Vantrease, J., Gomez-Acevedo, H., Hennings, L., Shankar, K., Cleves, M.A., Badger, T.M. and Ronis, M.J. 2013. Feeding soy protein isolate and treatment with estradiol have different effects on mammary gland morphology and gene expression in weanling male and female rats. Physiological Genomics, 45(22), 1072–1083.

[64] Ronis, M., Hennings, L., Gomez-Acevedo, H. and Badger, T. 2014. Different responses to soy and estradiol in the reproductive system of prepubertal male rats and neonatal male pigs (373.5). The FASEB Journal, 28, 373–375.

[65] Binns, C.W., Lee, M.K. and Lee, A.H. 2018. Problems and prospects: Public health regulation of dietary supplements. Annual Review of Public Health, 39, 403–420.

[66] Europe Nutrition And Supplements Market Size, Share & Trends Analysis Report, By Product (Sports Nutrition, Fat Burners, Dietary Supplements), By Consumer Group, By Formulation, By Sales Channel, And Segment Forecasts, 2023 – 2030. Available online: https://www.grandviewresearch. com/industry-analysis/europe-nutrition-supplements-market (accessed on 9 October 2022).

[67] Komala, M.G., Ong, S.G., Qadri, M.U., Elshafie, L.M., Pollock, C.A. and Saad, S. 2023. Investigating the regulatory process, safety, efficacy and product transparency for nutraceuticals in the USA, Europe and Australia. Foods, 12(2), 427.

[68] Santini, A. and Novellino, E. 2017. To nutraceuticals and back: Rethinking a concept. Foods, 6(9), 74.

[69] Brownie, S. 2005. The development of the US and Australian dietary supplement regulations: What are the implications for product quality? Complementary Therapies in Medicine, 13(3), 191–198.

[70] Low, T.Y., Wong, K.O., Yap, A.L., De Haan, L.H. and Rietjens, I.M. 2017. The regulatory framework across international jurisdictions for risks associated with consumption of botanical food supplements. Comprehensive Reviews in Food Science and Food Safety, 16(5), 821–834.

[71] Eussen, S.R., Verhagen, H., Klungel, O.H., Garssen, J., Van Loveren, H., Van Kranen, H.J. and Rompelberg, C.J. 2011. Functional foods and dietary supplements: Products at the interface between pharma and nutrition. European Journal of Pharmacology, 668, S2–S9.

[72] Siro, I., Kápolna, E., Kápolna, B. and Lugasi, A. 2008. Functional food. Product development, marketing and consumer acceptance – A review. Appetite, 51(3). 456–467.

[73] Ramalingum, N. and Mahomoodally, M.F. 2014. The therapeutic potential of medicinal foods. Advances in Pharmacological and Pharmaceutical Sciences, 2014, 1–18.

[74] Gulati, O.P. and Ottaway, P.B. 2006. Legislation relating to nutraceuticals in the European Union with a particular focus on botanical-sourced products. Toxicology, 221(1), 75–87.

Imad Ahmad*

# Chapter 2
# Nutraceutical properties of bioactive peptides

**Abstract:** Bioactive proteins and peptides are specific amino acid fragments that modify the function of human body and may ultimately influence human health. Food-derived bioactive peptides are associated with numerous health benefits. Naturally occurring peptides have an identified role in angiogenesis, cell proliferation, cell migration, inflammation, melanogenesis, and protein regulation. Such molecular mechanisms underlie the physiological processes of defense, growth, homeostasis, immunity, reproduction, and stress. This chapter provides an insight into the basics of peptides, their natural sources, their therapeutics, and their mechanisms with examples from preclinical studies. However, the discussion is focused on those pathologies that have affected human lifestyle in a very adverse manner, including cancer, hypertension, diabetes, and some other associated ailments. Despite a proven efficacy, peptides also present a challenge to translational medicine and pharmacy regarding formulations and bioavailability. Nutritional intervention can significantly contribute to the challenges related to modern medicines. Therefore, a nutraceutical-based approach can provide a solution to confirm peptide therapeutics.

# 1 Introduction

The polymeric combination of amino acids via peptide bonds give birth to a molecule with a short amino acid chain and distinct properties, known as peptides. The word "peptide" is derived from *Peptos* (Greek), which refers to "digested" [1], as majority of bioactive peptides are obtained after protein digestion. Bioactive peptides are specific fragments of proteins that modify the function of human body and ultimately influence human health. In the presence of proteolytic enzymes, proteins are fragmented at specific sites to obtain bioactive peptides. Due to their small size, they lack protein–protein interactions, possess tissue affinity, and target specificity [2]. Naturally occurring peptides present inside the human body are known for their role in angiogenesis, cell communication, cell proliferation, cell migration, inflammation, melanogenesis, and protein regulation. These physiological processes result in alleviating stress, body defense and immunity, promoting growth, maintaining homeostasis, and normal reproduction [3, 4].

---

*Corresponding author: Imad Ahmad**, Department of Pharmacy, The Professional Institute of Health Sciences Mardan, Khyber Pakhtunkhwa, Pakistan; Department of Pharmacy, Abdul Wali Khan University Mardan, Khyber Pakhtunkhwa, Pakistan, e-mail: imadahmad4574@gmail.com

https://doi.org/10.1515/9783111317601-002

Bioactive peptides are generally 3–20 amino acid residues in length. Animal and plant kingdoms are known to contain potential bioactive protein sequences including microbes and marine sources [5].

## 1.1 Historical background

Whenever it came to discussion, Emil Fischer and Hofmeister were the first to describe peptides in the early nineteenth century. The first peptide synthesis was reported by *Fischer* and *Fourneau* in 1901 [6]. In his lectures, Fischer described glycyl–glycine as the first peptide and also peptide moieties of variable compositions like polypeptides, tripeptides, and dipeptides [4]. Later, researchers identified more from natural origin, synthesized new peptides, and revealed about their physiological functions. Besides the growing library of natural peptides, different synthetic peptides were also developed and studied for a possible biological activity.

## 1.2 Recent advances in peptide therapeutics

Recent scientific evidence has identified numerous peptides that modulate the physiological functions. This depends upon their native protein composition, primary structure, and sequence of amino acids. Nowadays, peptides of routine or special food items are being considered alternative to pharmacological treatments. Natural peptides are known for their effective pharmacological effects with diverse mechanisms [7]. In this regard, hypertension, diabetes mellitus type 2 (T2DM), and oxidative stress present challenging interest to researchers. Predominantly, diabetes is a serious concern, and it is estimated to affect approximately 366 million people by 2030, being a very attractive target for researchers [8].

Antidiabetic peptides usually have insulin mimetic action, or they inhibit alpha-glucosidase, alpha-amylase, glucose transport system, and dipeptidyl peptidase-IV (DPP-IV). Among these, DPP-IV is an aminodipeptidase, whose function is to cleave glucagon-like peptide 1 (GLP1) and glucose-dependent insulinotropic peptide, collectively known as incretins [9, 10]. Antioxidant peptides inactivate reactive oxygen species, scavenge free radicals, chelate pro-oxidative transition metals, and promote the activities of intracellular antioxidant enzymes [11]. Antihypertensive peptides derived from food act by inhibiting the angiotensin-converting enzyme (ACE), promoting nitric oxide-mediated vasodilation, or modifying cyclo-oxygenase or endothelin-1 [12].

## 1.3 Sources of peptides

Peptides have numerous natural sources, including plants (beans, cereals, seeds, and seed meal), animals (dairy, meat, and egg), microbes, and marine sources. Exendin-4 is an antidiabetic peptide which was first isolated from *Heloderma suspectum* (*Gila monster*). However, its synthetic version "Exenatide" is now available in the market. Several antimicrobial peptides, including defensins, lectins, protease inhibitors, snaking, thionin-like peptides, and vicilin-like peptides, are isolated from plants of the Solanaceae family. Some of the most common genera having antimicrobial peptides are *Brugmansia, Capsicum, Datura, Nicotiana, Petunia, Salpichora, Solanum*, and *Withania* [13]. *Oryza sativa* L., *Glycine max* L., *Lupinus albus* L., *Amaranthus hypochondriacus* L., *Linum usitatissimum* L., *Momordica charantia* L., *Telfairia occidentalis* Hook.f., *Citrullus lanatus* L., *Lagenaria siceraria* Molina., and *Cucurbita moschata* Duchesne. are some of the widely studied species for bioactive peptide isolation [10, 14].

Among marine sources, peptides have been isolated from marine fungi. Mostly studied genera include *Ascotricha, Acremonium, Clonostachys, Emericella, Microsporum, Exserohilum, Penicillium, Ceratodictyon, Scytalidium, Talaromyces, Stachylidium*, and *Zygosporium* [15, 16].

Peptides have also been isolated from dairy products. Milk may have bioactive peptides either in natural form or enzymatically released from their parental proteins. Bioactive peptides are obtained from various dairy species like camel, goat, mithun, and sheep [17]. Peptides from cow milk possess antidiabetic, antioxidant, and antihypertensive properties [18]. Fermented camel milk (*Camelus dromedarius*) is known for antidiabetic, anti-inflammatory, and ACE inhibitory activities [19]. Microbially fermented dairy products are also a good source of bioactive peptides. Several studies are carried out with *Lactobacillus bulgaricus, Lactobacillus delbrueckii, Lactobacillus helveticus, Streptococcus salivarius*, and *Streptococcus thermophilus* [20].

## 1.4 Extraction and isolation

Once the source of a peptide is identified, a challenge is how to isolate it. The extraction of a protein is carried out through conventional and advanced techniques, like microwave-assisted extraction, enzyme-assisted extraction, pulsed electric field-assisted extraction, pressurized solvent extraction, ultrasonic-assisted extraction, and supercritical fluid extraction [21, 22].

One of the most commonly used methods for the preparation of biologically active peptides is enzymatic hydrolysis. Different protease enzymes used to hydrolyze proteins of animal and plant origins include Alcalase, Flavourzyme, papain, pepsin, thermoase, and trypsin [23, 24].

Generally, the raw material is pretreated or processed as crushing and screening to obtain the protein powder. Then the protein is hydrolyzed enzymatically, after which

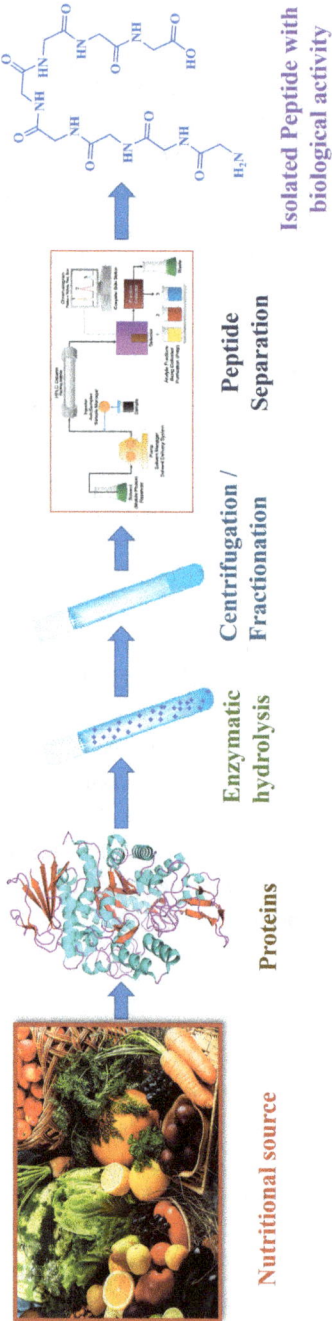

**Figure 1:** Isolation of a bioactive peptide from a nutritional source.

the enzyme is inactivated and removed by centrifugal separation. The separated poly-peptide solution is dried for the purpose to obtain the crude peptide in powder form. A general scheme of protein hydrolysis, fractionation, separation, and isolation of a bioactive peptide is shown in Figure 1.

# 2 Biologically active proteins and peptides

## 2.1 Antiaging proteins and peptides

### 2.1.1 Soybean

Soy oligopeptides (*Glycine max* L.) are composed of three to six amino acid residues. Their size may range from 300 to 700 kDa. These are obtained from soybean (*Glycine max* L.) proteins. They have several bioeffects like antioxidant, antihypertensive, and anti-hyperlipidemic. Topical application of soy oligopeptides increased Bcl-2 expression; while decreased apoptotic cells, Bax and p53 protein expressions, cyclobutene–pyrimidine dimer-positive cells, and sunburn cells in the epidermal layer of UVB-irradiated foreskin. Furthermore, soy oligopeptides offered protection of the human skin against UVB-induced photo damage in nine healthy male volunteers. A pseudo-randomized study conducted on 10 women with soybean peptide increased the synthesis of glycosamino-glycan and collagen synthesis. These regenerative effects favor the antiaging potential of soy peptide [25].

### 2.1.2 Rice

*Oryza sativa* L. (black rice oligopeptides) inhibit matrix metalloproteinases and stimu-late gene expression of hyaluronan synthase-2 in human keratinocytes. Three new peptides from rice bran protein possess C-terminal tyrosine residue, capable of inhib-iting tyrosinase-mediated monophenolase reactions. Another peptide from the same source LQPSHY inhibited melanogenesis in mouse melanoma cells, improved mela-nin-related skin conditions, and without any cytotoxicity [26]. Proteins from rice bran possess potent tyrosinase inhibitory property. Niosome entrapment of these bioactive peptides is a clinically proven anti-aging formulation [27, 28].

### 2.1.3 Soybean, oyster, and sea cucumber

Degenerative diseases are associated with aging due to detrimental effects of free rad-icals on cells and their biochemical environment. A bioassay-guided isolation and characterization identified antiaging effects of oyster (OPH), soybean (SPH), and sea cucumber (SCPH) protein hydrolysates using Alcalase. All the three reversed D-galactose-induced aging, oxidative stress, and associated learning and memory impairments (SPH > OPH > SCPH). Chromatographic purification of SPH identified two peptides, WPK and AYLH, as the active components responsible for *in vivo* and *in vitro* antioxidant and antiaging activities. These peptides eased hydrogen peroxide-induced oxidative damage in PC12 cells *in vitro*. Therefore, WPK and AYLH from PHS are effective antioxidant agents to be developed as nutraceuticals for aging-related learning and memory impairment, and oxidative stress [29].

## 2.2 Anticancer peptides

Cancer is one of the most devastating noncommunicable diseases accounting for millions of deaths. Currently, conventional chemotherapy remains the mainstay of cancer treatment. However, the adverse effects on normal cells and multidrug-resistant cancer are now more challenging. This alarming situation calls for urgent need to develop novel therapeutics capable of overcoming the unwanted effects as well as drug resistance. Therefore, small peptides that exhibit cancer-selective toxicity offer a new era of cancer therapeutics. Peptides with high specificity and low toxicity make them the preferred choice now as compared to small molecules and antibody-mediated therapies [30, 31]. Here are few of these examples from peptides that can be used in cancer therapeutics. Table 1 summarizes the sequences, pharmacological mechanism, and experimental model of anticancer peptides.

### 2.2.1 Bean (Azufrado Higuera)

Peptides from nondigestible fractions of common bean (*Phaseolus vulgaris* L.) GLTSK, LSGNK, GEGSGA, MPACGSS, and MTEEY possess antiproliferative action demonstrated in human colorectal cancer cells via cell cycle arrest or apoptosis. HCT116 cell line was the most sensitive to bean Azufrado Higuera with an $IC_{50}$ value of 0.53 mg/mL of peptide extracts. The expression of p53 in HCT116 cells was increased along with the expression of cell cycle regulation proteins cyclin B1 and p21. There was modification in the expression of mitochondrial-activated apoptotic proteins BAD, BIRC7, c-casp3, cytC, and Survivin [32]. An increase in mitochondrial-activated apoptotic proteins leads to cell death [33].

## 2.2.2 Bovine β-casein

PGPIPN is a hexapeptide derived from bovine β-casein with the ability to inhibit the proliferation of human ovarian cancer cells (SKOV3) and primary ovarian cancer cells *in vitro*. The effect was consistent in xenograft mice ovarian cancer model where it decreased the tumor growth rate in a dose-dependent manner. Further study revealed that the partial contribution to antitumor effect of PGPIPN was due to cell apoptosis and inhibition of BCL2 pathway. Such effects promulgate PGPIPN as a potential anticancer agent against ovarian cancer or other types of cancer [34].

## 2.2.3 Spirulina

Spirulina (*Arthrospira platensis*) is a microbial source of proteins with hydrolyzable bioactive peptides. Gastrointestinal endopeptidase hydrolysis with pepsin, trypsin, and chymotrypsin of its whole protein's fraction showed highest antiproliferative activity against three cancer cell lines MCF-7 (31.25 µg/mL), HepG-2 (36.42 µg/mL), and SGC-7901 (48.25 µg/mL). A new peptide from spirulina HVLSRAPR possesses strong inhibitory activity against HT-29 cancer cells with an $IC_{50}$ value of 99.88 µg/mL [35].

## 2.2.4 Olive seed proteins

Thermolysin-hydrolyzed peptides (MW < 3 kDa) from olive (*Olea europaea* L.) seed proteins possess antitumor activity. They increased the adhesion capacity of tumor cells, decreased metastasis, and seized cell cycle in the S-phase. The identified peptide LLPSY exhibited this activity against PC-3 and MDA-MB-468 [36].

## 2.2.5 Rapeseed

Rapeseed (*Brassica campestris* L.) contains a peptide component of possible amino acid sequence WTP (408.2 Da) that possesses the apoptotic property. Addition of this peptide led to morphological changes in HepG2 cells, induced apoptosis, and ultimately inhibited their proliferation [37].

## 2.2.6 Chickpea

Chickpea (*Cicer arietinum* L.) is one of the common dietary sources of proteinaceous food. A peptide RQSHFANAQP derived from its albumin hydrolysate exhibited efficient bioactivity against cell viability. $EC_{50}$ values for inhibitory activity against MCF-7 and

MDA-MB-231 cells were 2.38 and 1.50 µmol/mL, respectively. There was a dose-dependent increase in the concentration of p53 protein. This leads to identification of a peptide with antiproliferative activity against breast cancer cells by increasing p53 [38].

**Table 1:** Sequence, mechanism, and experimental model of anticancer proteins and peptides.

| Protein/peptide source | Sequence | Mechanism | Model/cell line | References |
|---|---|---|---|---|
| Common bean (Azufrado Higuera) | GLTSK, LSGNK, GEGSGA, MPACGSS, and MTEEY | Increased p53, cyclin B1 and p21, cell cycle arrest, or apoptosis | Human colorectal cancer cell HCT116 | [32] |
| Bovine β-casein | PGPIPN | Cell apoptosis and inhibition of BCL2 pathway | Human ovarian cancer cell line (SKOV3), primary ovarian cancer cells, *in vivo* | [34] |
| *Arthrospira platensis* | HVLSRAPR | | MCF-7, HepG-2, SGC-7901, and HT-29 | [35] |
| Olive seed (*Olea europaea*) | LLPSY | S-phase cell cycle arrest, adhesion of tumor cells, and decrease metastasis | PC-3 and MDA-MB-468 | [36] |
| Rapeseed (*Brassica campestris* L.) | WTP | Apoptosis | HepG2 | [37] |
| Chickpea (*Cicer arietinum* L.) | RQSHFANAQP | Merge DNA with domain of p53 protein, expression of p53 protein, inhibit cell proliferation | MCF-7 and MDA-MB -231 | [38] |

## 2.3 Anticholesterolemic peptides

Anticholesterolemic peptides, their sequence, mechanism, and experimental models are summarized in Table 2.

### 2.3.1 Soybean

Three peptides such as IAVPGEVA, IAVPTGVA, and LPYP obtained from (soy glycinin hydrolysis) possess hypocholesterolemic activity. They are competitive inhibitors of hydroxymethylglutaryl CoA (HMG CoA) reductase, which is the rate-limiting enzyme of cholesterol biosynthesis. In HepG2 cells, peptides IAVPGEVA, IAVPTGVA, and LPYP restricted HMG CoA reductase catalysis and modulated cholesterol production. This was due to activation of the LDLR-SREBP2 (low-density lipoprotein receptor/regulatory element-binding protein 2) pathway, which leads to increased ability of HepG2 cells to uptake the low-density lipoprotein. Furthermore, AMPK and ERK 1/2 pathways were also activated [39]. Another novel peptide Soystatin (VAWWMY), derived from soybean protein, has the ability to prevent cholesterol absorption. It binds the bile acid better than casein tryptic hydrolysate (CTH) or soybean protein peptic hydrolysate (SPPH). Its bile acid binding is as potent as hypocholesterolemic drug, cholestyramine. In the same instance, VAWWMY reduced the micellar solubility of cholesterol better than SPPH or CTH [40].

### 2.3.2 11S-soy globulin

A peptide isolated from pepsin hydrolysate of 11S-globulin with amino acid sequence of IAVPGEVA (755.2 Da) proved to be hypocholesterolemic via bile-acid-binding analysis and *in vitro* inhibition of HMG CoA reductase. The effect of lowering cholesterol content is described by bile acids binding to hydrolysate peptides. This prevents them from re-absorption and provokes the transformation of cholesterol [41].

### 2.3.3 Lupin protein

Two peptides LILPKHSDAD and LTFPGSAED are obtained from lupin (*Lupinus albus* L.) protein with absorptive ability into Caco-2 cells. These peptides inhibit HMG CoA reductase functionality *in vitro* and halt cholesterol biosynthesis in HepG2 cells. This leads to improvement in LDLR levels, which further promoted the uptake of extracellular LDL creating hypocholesterolemia. However, only LILPKHSDAD peptide decreased the pro-protein convertase subtilisin/kexin type 9 (PCSK9) and HNF1-alpha protein levels which reduce mature PCSK9 secretion by HepG2 cell lines [42].

### 2.3.4 Hempseed

Hempseed's (*Cannabis sativa* L.) total protein hydrolysate prepared with pepsin possesses a total of 90 peptides from 33 proteins. The hydrolysate inhibited HMG CoA reductase catalytic activity in a dose-dependent manner at 0.1–1.0 mg/mL. Protein levels of HMG CoA reductase, LDLR, and SREBP2 were raised in HepG2 cells. However, switching the AMPK pathway (phosphoadenosine monophosphate-activated protein kinase) inactivated HMG CoA reductase via phosphorylation. The uptake of extracellular LDL was increased and PCSK9 level also appeared raised. These mechanisms are similar with that of statins [43].

**Table 2:** Anticholesterolemic peptides, their sequences, mechanisms, and experimental models.

| Peptides | Sequences | Mechanisms | Experimental models | References |
|---|---|---|---|---|
| Soy glycinin (*Glycine max* L.) | LPYP, IAVPGEVA, and IAVPTGVA | Inhibit HMG CoA reductase,[1] activation of LDLR-SREBP2 and AMPK and ERK 1/2 pathways | HepG2 cells | [39] |
| Soy 11S-globulin (*Glycine max* L.) | IAVPGEVA | Inhibit HMG CoA reductase,[1] bile acids binding to prevent reabsorption | *In vitro* | [41] |
| Lupin protein (*Lupinus albus*) | LILPKHSDAD and LTFPGSAED | Inhibit HMG CoA reductase,[1] improvement in LDLR protein levels, decreased PCSK9 and HNF1-alpha protein levels | *In vitro*, HepG2 cells | [42] |
| Soystatin (*Glycine max* L.) | VAWWMY | Inhibit cholesterol absorption, bile acid binding, reduced micellar solubility of cholesterol | *In vivo* | [40] |

[1]3-Hydroxy-3-methylglutaryl CoA reductase.

## 2.4 Antioxidant peptides

The free radical theory states that the overproduction of free radicals arising from mitochondrial metabolic processes leads to serious complications, including but not limited to cardiac diseases, diabetes, aging, and cancer [44]. The presence of amino acid residues like alanine, cysteine, leucine, histidine, lysine, methionine, proline, tryptophan, tyrosine, and valine confirms the antioxidant features of peptides [45]. Amino acids with hydrophobic side chains favor the miscibility of peptides with fat solvents. This property supports its access to hydrophobic radical species and polyunsaturated fatty acids elaborating its role in oxidative stress. The antioxidant peptides, their sequences, mechanisms, and experimental models are summarized in Table 3.

### 2.4.1 Quinoa

Quinoa (*Chenopodium quinoa* Willd.) is a traditional seed crop cultivated in the Andean region since decades. Quinoa grains and flour are commonly consumed by humans and animals. *Lactobacillus plantarum* T0A10-fermented quinoas possess low-molecular-weight peptides FTLIIN, IVLVQEG, LENSGDKKY, TLFRPEN, and VGFGI, having antioxidant activity [46].

### 2.4.2 Kluyveromyces marxianus

*Kluyveromyces marxianus* is aerobic lactose-fermenting yeast found in milk/dairy products. A peptide fragment obtained through sonicated-enzymatic (trypsin and chymotrypsin) hydrolysate of *K. marxianus* protein exhibited antioxidant activity. Peptide
with amino acid sequence of VLSTSFPPK and MW of 1,118 Da possesses DPPH and ABTS
free radical scavenging activity. However, the highest antioxidant activity was against
DPPH with $IC_{50}$ value of 15.20 μM [47].

### 2.4.3 Sardinella aurita

A bioassay-guided fractionation and isolation leads to identification of bioactive peptides
LHY, LARL, GGE, GAH, GAWA, PHYL, GLAAW from *Sardinella aurita* (round sardinella)
which possess antioxidant activity. Highest activity was demonstrated by LARL with
DPPH radical scavenging activity [24].

### 2.4.4 Pecan

Hydrolysis of a pecan (*Carya illinoinensis* (Wangenh.) K. Koch) protein isolate with Alcalase leads to generation of antioxidant peptides. Ultrafiltration identified peptide
LAYLQYTDFETR with an intrinsic ability of ABTS radical and DPPH radical scavenging
activity [48].

### 2.4.5 Walnut

Walnut (*Juglans regia* L.) is one of the most widespread tree nuts in the world. Several
biological activities such as antiatherogenic, anti-inflammatory, antimutagenic, and antioxidant properties have been attributed to it. The purified peptide, ADAF (423.23 Da),
effectively quenched DPPH and hydroxyl radicals comparable to those of reduced glutathione [49].

### 2.4.6 Ark shell

Peptides MCLDSCLL and HPLDSLCL from ark shell (*Scapharca subcrenata*) possess antioxidant activity against DPPH, ABTS, ORAC, and reducing power assay. MCLDSCLL
possesses the highest scavenging activity against DPPH, while HPLDSLCL strongly inhibited copper-catalyzed human LDL oxidation [50].

### 2.4.7 Rice bran

Rice (*Oryza sativa* L.) by-products are cheap and abundant source of bioactive peptides. It is the fundamental staple food for major part of world's population. Two co-products obtained from rice include bran and broken rice. Rice bran peptides YVAQGEGVVA and YLAGMN possess strong antioxidative activity against ABTS [47].

### 2.4.8 Yellow croaker

Yellow croaker (*Pseudosciaena polyactis*) is a source of antioxidant peptides like GPEGPMGLE, EGPFGPEG, and GFIGPTE. $EC_{50}$ values of DPPH radical scavenging activities were 0.59, 0.37, and 0.45 mg/mL, while $EC_{50}$ values against superoxide anion radical were 0.62, 0.47, and 0.74 mg/mL. Against hydroxyl radical, the $EC_{50}$ values were lowest of all, as 0.45, 0.33, and 0.32 mg/mL [51].

### 2.4.9 Monkfish

Three peptides were isolated from defatted muscle proteins of monkfish (*Lophius litulon*) after peptic-tryptic hydrolysis. Two peptides EDIVCW and YWDAW block lipid peroxidation with glutathione equivalent potency in the linoleic acid model system. DPPH radical scavenging was exhibited by EDIVCW, MEPVW, and YWDAW with $EC_{50}$ values of 0.39, 0.62, and 0.51 mg/mL, respectively. Against hydroxyl radical, the $EC_{50}$ values were 0.61, 0.38, and 0.32 mg/mL, while $EC_{50}$ values for scavenging superoxide anion radicals were 0.760, 0.940, and 0.480 mg/mL, respectively [52].

### 2.4.10 Cheddar cheese

Cheddar cheese prepared with *Lactobacillus helveticus* 1.0612 is a source to obtain antioxidant peptides VLPVPQK, EDVPSER, EMPFPK, KEMPFPK, and SDIPNPIGSENSEK [54].

**Table 3:** Antioxidant peptides, their sequences, mechanisms, and experimental models.

| Sources | Sequences | Targets | Models | References |
|---------|-----------|---------|--------|------------|
| *Sardinella aurita* | HY, LARL, GGE, GAH, GAWA, PHYL, GLAAW | DPPH | *In vitro* | [24] |
| Pecan meal (*Carya illinoinensis*) | LAYLQYTDFETR | DPPH and ABTS | *In vitro* | [48] |

**Table 3** (continued)

| Sources | Sequences | Targets | Models | References |
|---|---|---|---|---|
| *Juglans regia* L. | ADAF | DPPH, lipid peroxidation, hydroxyl radical, ferric reducing power, $Fe^{2+}$ chelating | *In vitro* | [49] |
| Ark shell (*Scapharca subcrenata*) | MCLDSCLL, HPLDSLCL | DPPH, ABTS, ORAC, and reducing power | *In vitro* | [50] |
| *Kluyveromyces marxianus* | VLSTSFPPK | DPPH and ABTS | *In vitro* | [47] |
| Rice bran (*Oryza sativa*) | YLAGMN | ABTS | *In vitro* | [47] |
| *Pseudosciaena polyactis* | GPEGPMGLE, EGPFGPEG, and GFIGPTE | Superoxide anion radical, DPPH, and hydroxyl radical | *In vitro* | [51] |
| *Lophius litulon* | EDIVCW, MEPVW, and YWDAW | Superoxide anion radical, DPPH, hydroxy radical, and lipid peroxidation | *In vitro* | [52] |
| *Corylus heterophylla* Fisch. | AVKVL, YLVR, and TLVGR | – | – | [53] |
| Cheddar cheese | EMPFPK, KEMPFPK, and SDIPNPIGSENSEK | – | – | [54] |

## 2.5 Hypotensive peptides

Health benefits of functional foods are early known in blood pressure homeostasis through dietary approaches. Hypertension is one of the fatal risk factors for cardiovascular complications. The underlying mechanisms identified include ACE inhibition, modification of endothelin-1 and cyclooxygenase, and expression of nitric oxide. Peptides with blood pressure lowering mechanisms are summarized in Table 4 and discussed in detail below. The antihypertensive mechanisms are shown in Figure 2.

### 2.5.1 Sunflower

Sunflower (*Helianthus annuus* L.) seed proteins hydrolyzed with pepsin and pancreatin are a potential source of ACE inhibitory peptides. Sunflower protein isolates and their peptic-pancreatic protein hydrolysates are ACE inhibitors and can reduce high blood pressure. An ACE inhibitory peptide with a sequence of FVNPQAGS responds to helianthinin, a main storage 11S-globulin protein obtained from sunflower seeds [55].

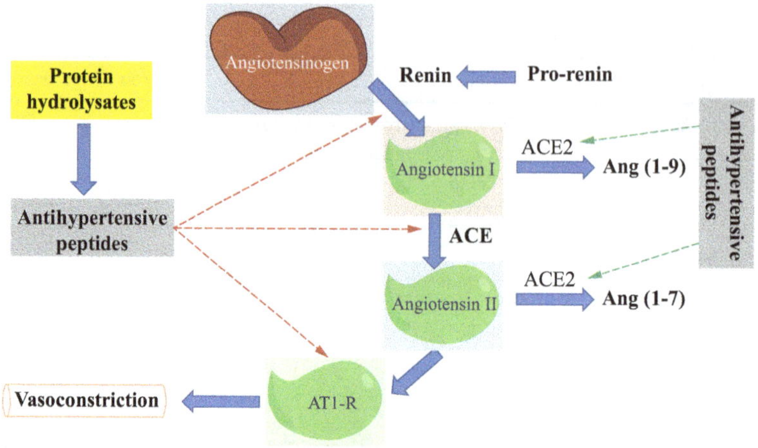

**Figure 2:** Hypotensive mechanism of peptides by targeting angiotensin converting enzyme-2 (ACE2) and angiotensin-1 receptor (AT1-R) to inhibit vasoconstriction and relieve hypertension. Green dotted lines indicate positive stimulation, while red dotted lines imply inhibition of the respective pathways.

## 2.5.2 Ginkgo

Alcalase, dispase, trypsin, and Flavourzyme hydrolysate of the extracted *Ginkgo biloba* seeds protein isolate (GPI) demonstrated ACE inhibitory ability. In the hydrolysate, components having MW of <1 kDa inhibited ACE with an $IC_{50}$ value of 0.2227 mg/mL. Further purification and identification conferred three new ACE inhibitory peptides. The first peptide TNLDWY was a noncompetitive inhibitor with $IC_{50}$ of 1.932 mM. RADFY and RVFDGAV demonstrated competitive inhibition with $IC_{50}$ of 1.35 and 1.006 mM, respectively [56].

## 2.5.3 Kluyveromyces marxianus

A sonicated-enzymatic (trypsin and chymotrypsin) hydrolysates of *Kluyveromyces marxianus* protein exhibited ACE inhibitory peptides. Further fractionation with ultrafiltration and RP-HPLC techniques yielded two new peptides, LPESVHLDK (MW = 1,180 Da) and VLSTSFPPK (MW = 1,118 Da) [47].

## 2.5.4 Wheat

Similarly, wheat gluten peptides obtained from *Triticum aestivum* L., including APSY, LY, LVS, RGGY, and YQ, possess antihypertensive effect. They exhibited ACE inhibitory activity with $IC_{50}$ values of 1.47, 0.31, 0.60, 1.48, and 2.00 mmol/L, respectively [57].

### 2.5.5 Olive flounder

IVDR, VASVI, and WYK obtained from surimi, which is a refined fish myofibrillar protein and a functional food ingredient of olive flounder (*Paralichthys olivaceus*), inhibit ACE with $IC_{50}$ values of 46.90, 32.66, and 32.97 µM, respectively [58]. These peptides increase the phosphorylation of protein kinase B and endothelial nitric oxide synthase which increased nitric oxide production in human umbilical vein endothelial cells. In spontaneously hypertensive rat model, per oral administration of the aforementioned peptides decreased blood pressure, confirming Akt/eNOS pathway. Surimi made from *P. olivaceus* is a source of novel antihypertensive peptides that possess health restoring properties [59].

### 2.5.6 Antarctic krill

Antarctic krill (*Euphausia superba*) is a source of antihypertensive peptides. A protein hydrolysate of krill possesses eight peptides with sequences FAS, FQK, FRKE, VD, WF, YKD, YRK, and YRKER. Peptides WF and FAS possess lowest $IC_{50}$ values of $0.32 \pm 0.05$ mg/mL and $0.15 \pm 0.02$ mg/mL against ACE activity than the other six peptides. Furthermore, peptides WF, YRK, FQK, and FAS dose-dependently increased nitric oxide concentration and reduced endothelin-1 in human umbilical vein endothelial cells. These two peptides also reversed the norepinephrine-induced decreased production of nitric oxide and norepinephrine-induced effect on endothelin-1 production. These mechanisms support the therapeutics of isolated antihypertensive peptides in endothelial cell dysfunction, even if induced by norepinephrine [59].

### 2.5.7 Hen

Muscle protein hydrolysate of hen (*Gallus gallus domesticus*), prepared by thermoase, provides the starting material for synthesizing hypotensive peptides. Peptides with ACE2 upregulatory (ACE2u) activity are recent targets for treating hypertension. Peptides from the hen muscle protein hydrolysate possess both ACE inhibitory and ACE2u activities. Four peptides VAQWRTKYETDAIQRTEELEEAKKK, VHPKESF, VKW, and VVHPKESF increased ACE2 expression by 0.52–0.84 folds. Fractionation and ultrafiltration led to the identification of five potent ACE-inhibiting peptides VRP (0.64), LKY (0.81), VRY (5.77), KYKA (2.87), and LKYKA (0.034) with $IC_{50}$ values in µg/mL. Among all these peptides, VKW inhibited ACE activity and upregulated ACE2. However, all these peptides were found to be susceptible to simulated gastrointestinal degradation except VRP [60].

### 2.5.8 Spent grain

One of the major proteinaceous by-products of Chinese liquor industry is the distilled spent grains. Ultrafiltration and reversed-phase HPLC led to the identification of new peptides from its prolamin hydrolysates. Those two peptides, AVQ and YPQ, inhibited ACE with $IC_{50}$ values of 181 μM and 220 μM, respectively [61].

### 2.5.9 Whey/milk protein

Similarly, EKVNELSK, LLYQEPVLGPVR, and MKP were obtained from casein hydrolysate of whey/milk protein. They were found to be ACE inhibitors with $IC_{50}$ value of 6.0, 0.43, and 5.0 μM, respectively [62]. Peptides IPP (1.23), IIAE (128), LVYPFP (97), and LIVTQ (113) from whey/milk protein inhibited ACE with $IC_{50}$ values mentioned in μg/mL, respectively [63].

### 2.5.10 Naked oat

*Avena nuda* var. *nuda* (L.), commonly known as naked oat, possesses a globulin whose peptide SSYYPFK inhibits ACE with $IC_{50}$ values of 91.82 μM [64].

### 2.5.11 Amaranth grains

Amaranth grains (*Amaranthus caudatus* L.) possess two peptides SFNLPILR and AFEDG-FEWVSKF that inhibit ACE and can relieve vasoconstriction [65].

**Table 4:** Hypotensive peptides, their sequences, mechanisms, and experimental models.

| Sources | Sequences | Targets | Models | References |
|---|---|---|---|---|
| Sunflower (*Helianthus annuus* L.) | FVNPEAGS | ACE-I inhibition | *In vitro* | [55] |
| *Ginkgo biloba* seeds | TNLDWY, RADFY, and RVFDGAV | ACE-I inhibition | *In vitro* | [56] |
| *Kluyveromyces marxianus* | VLSTSFPPK and LPGSVHLAL | ACE-I inhibition | *In vitro* | [47] |
| Wheat gluten (*Triticum aestivum*) | APSY, LY, LVS, RGGY, and YQ | ACE-I inhibition | *In vitro* | [57] |

**Table 4** (continued)

| Sources | Sequences | Targets | Models | References |
|---|---|---|---|---|
| Amaranth grains | SFNLPILR and AFEDGFEWVSKF | ACE-I inhibition | *In vitro* | [65] |
| Olive flounder (*Paralichthys olivaceus*) | IVDR, VASVI, and WYK | Promoted nitric oxide production, and enhanced phosphorylation of endothelial nitric oxide synthase and protein kinase B | Human umbilical vein endothelial cells, *in vivo* model | [59] |
| *Euphausia superba* (Antarctic krill) | FAS, FQK, FRKE, VD, WF, YKD, YRK, and YRKER | ACE inhibition, increased nitric oxide concentration and decreased endothelin-1, reversed norepinephrine-induced vasoconstriction | *In vitro*, human umbilical vein endothelial cells | [59] |
| Hen (*Gallus gallus domesticus*) | LKY, KYKA, LKYKA, VRY, and VRP | ACE inhibition | *In vitro* | [60] |
| Hen (*Gallus gallus domesticus*) | VKW, VHPKESF, VVHPKESF, and VAQWRTKYETDAIQRTEELEEAKKK | Increased ACE2 expression | – | [60] |
| Whey/milk protein | EKVNELSK, MKP, and LLYQEPVLGPVR | ACE-I inhibition | – | [62] |
| Whey/milk protein | IPP, IIAE, LVYPFP, and LIVTQ | ACE-I inhibition | – | [63] |
| *Avena nuda* (naked oat) globulin | SSYYPFK | ACE-I inhibition | – | [64] |

## 2.6 Antidiabetic peptides

T2DM is a devastating complex disorder of insulin insufficiency or insulin resistance. Due to its increasing incidence, it presents a considerable social and healthcare as well as economic burden. Due to its limitations of the existing therapeutics, bioactive peptides are becoming an alternative to small organic molecules. Numerous functional foods have demonstrated diabetes treatment in preclinical studies, where the underlying agent was an antidiabetic bioactive peptide. Table 5 describes the original sources, sequences, and mechanisms of antidiabetic peptides.

### 2.6.1 Egg yolk

Four peptides were obtained from an egg yolk protein by-product after ethanolic extraction of phospholipids lysed with pepsin. These novel peptides were found with antidiabetic mechanisms like α-glucosidase inhibition and DPP-IV inhibition. Their MW range from 1,210.62 to 1,677.88 Da and include apolipoprotein B (YINQMPQKSRE and YINQMPQKSREA), apovitellenin-1 (YIEAVNKVSPRAGQF), and vitellogenin-2 (VTGRFAGH-PAAQ). The peptide VTGRFAGHPAAQ was the most potent α-glucosidase inhibitory agent with $IC_{50}$ value of 365.4 µg/mL [66].

### 2.6.2 Egg white protein

Eight antidiabetic peptides were obtained from egg white protein. Alcalase hydrolysates possess α-glucosidase and α-amylase inhibitory activities. The peptide RVPSLM exhibited an $IC_{50}$ value of 23.07 µmol/L against α-glucosidase and did not modify the catalytic efficiency of α-amylase [67]. Yu et al. found another α-glucosidase and α-amylase inhibitory peptide KLPGF in egg proteins [67]. This inhibition leads to reduction in postprandial hyperglycemia. Protein wastes of defatted egg yolk particles also possess bioactive peptides. The action of Asian pumpkin enzyme on defatted egg yolk particles leads to the formation of peptide extracts. A biopeptide LAPSLPGKPKPD was identified, which possesses α-glucosidase inhibitory activity. This peptide can be used to control postprandial hyperglycemia [66].

### 2.6.3 Salmon

Harnedy et al. found the antidiabetic activity of agelatinous protein from salmon (*Salmo salar*) skin and protein hydrolysates of cutaneous cuttings. Alcalase and Flavourzyme-induced *Salmo* skin protein hydrolysates possess DPP-IV inhibition property and stimulate digestion. Furthermore, it exerts insulin-stimulating action through Katp channel-dependent pathway along with inhibition of GLP1 secretion and DPP-IV catalysis [68]. Li-Chan et al. found DPP-IV inhibitory peptides, GPAE (372.4 Da) and GPGA (300.4 Da), in *Salmo* gelatin. The $IC_{50}$ values for dose-dependent inhibition of DPP-IV were 49.6 and 41.9 µM, respectively. Hence, Atlantic salmon skin gelatin serves as a functional food and a nutraceutical against noninsulin-dependent diabetes [69].

### 2.6.4 Soy protein

Another promising natural ingredient for nutraceutical applications is soy protein hydrolysate prepared with alkaline proteinase which possesses potent α-glucosidase and DPP-IV suppression potential as compared with those from papain-tryptic digestion. $IC_{50}$ values for inhibition of α-glucosidase enzymatic action by LLPLPVLK, SWLRL, and WLRL were 237.43, 182.05, and 162.29 μmol/L, respectively [70].

### 2.6.5 Common bean

Common bean is one of the important sources of proteinaceous diet. Scientifically known as *Phaseolus vulgaris* L. is grown worldwide, due to its abundance in nutritional proteins. Several studies reported biological activities of its hydrolysates and peptides. "Eat three beans a day, don't take medicine for years" is an old Chinese rationalization of the traditional uses of beans to maintain well-being of the humankind. One of its peptides KTYGL possesses DPP-IV inhibitory potential which was demonstrated in an *in vitro* experiment [71]. A type of beans known as Pinto bean possesses α-amylase inhibitory peptides PPHMLP, PPMHLP, PPHMGGP, PLPPHALL, and PAPFPSPHTP [72].

### 2.6.6 Pea

Pea oligopeptides (*Pisum sativum* L.) exhibited hypoglycemic effect in two different diabetic mice models: high-fat diet and streptozotocin-induced. Four peptides such as ALP, LLP, VLP, and SP were identified as having a proline amino acid at the carboxyl terminal (C-terminus) and DPP-IV inhibitory action. Administration of pea oligopeptide concomitantly with metformin for 4 weeks improved glucose tolerance, promoted glycogenesis, as well as preserved liver and kidney microstructures [73].

### 2.6.7 Other sources

A peptide LAPSLPGKPKPD from the pumpkin seed (*Cucurbita ficifolia* Bouché) has antidiabetic activity *in vitro*. The mechanism identified was inhibition of α-glucosidase and DPP-IV [74]. Rice bran peptides LP, IP, MP, and VP also have inhibitory action against DPP-IV [75]. Similarly, mulberry (*Morus alba* L.) also contains peptides with α-glucosidase and α-amylase inhibitory potentials [76].

**Table 5:** Antidiabetic peptides, their sequences, mechanisms, and experimental models.

| Sources | Sequences | Models | Target receptors | References |
|---|---|---|---|---|
| Egg yolk protein by-product | YINQMPQKSRE, YINQMPQKSREA, YIEAVNKVSPRAGQF, VTGRFAGHPAAQ | AG, DPP-IV | *In vitro* | [66] |
| Egg white protein | RVPSLM | AG | *In vitro* | [67] |
| Egg protein | KLPGF | AG, α-amylase | *In vitro* | [67] |
| Defatted egg yolk particles | LAPSLPGKPKPD | AG | *In vitro* | [66] |
| Atlantic salmon skin (*Salmo salar*) | GPAE and GPGA | DPP-IV | *In vitro* | [68, 69] |
| Soy protein (*Glycine max*) | WLRL, SWLRL, and LLPLPVLK | AG | *In vitro* | [70] |
| Common bean (*Phaseolus vulgaris* L.) | KTYGL | DPP-IV | *In vitro* | [71] |
| Mulberry (*Morus alba*) | WGYENAATYFWQTV | AG | *In vitro* | [76] |
| Pumpkin seed (*Cucurbita ficifolia*) | LAPSLPGKPKPD | AG and DPP-IV | *In vitro* | [74] |
| Rice bran (*Oryza sativa*) | LP, IP, MP, and VP | DPP-IV | *In vitro* | [75]. |
| Pinto bean (*Phaseolus vulgaris* cv. *Pinto*) | PPHMLP, PPMHLP, PPHMGGP, PLPPHALL, and PAPFPSPHTP | α-Amylase | *In vitro* | [72]. |
| Pea (*Pisum sativum*) | ALP, LLP, VLP, and SP | DPP-IV | *In vitro, in vivo* | [73] |

AG, α-glucosidase enzyme.

# 3 Challenges and prospects

The future of bioactive peptides is trending upward, but still several challenges need to be addressed. More than 80 peptide-based drugs are available on the market for treatment of cancer, and cardiovascular, endocrine, and metabolic disorders. Approximately, 150 peptides are in clinical trials and nearly 600 are in the ongoing preclinical studies, which present strong evidence in favor of peptide therapeutics. A high cost of processing, low stability, and pharmacokinetic limitations like low absorption fraction, low bioavailability, and short half-life necessitate more systematic and novel strategies to overcome these factors. Modern-day processing techniques are now available, each one with its own limitations. However, they are applied successfully to accommodate bioactive peptides in the therapeutic arena. Oral route of drug administration is the most common and preferable noninvasive method. A drug with small molecular size that survives the harsh enzymatic environment of gastrointestinal tract and reaches its site of absorption is only a successful drug candidate.

Summarizing the discussion, it can be said that protein-derived bioactive peptides modulate the physiology by maintaining human body homeostasis. Peptide structure and its amino acid sequence influence its function. Therefore, during the production process, the stability of peptides must be considered. This further determines the bioavailability of a peptide, its target site in the body. The future research focus should also consider the industrial production of stable and chemically suitable bioactive peptides to be used in therapeutics.

# 4 Summary

Increasing evidence supports the importance of food-derived bioactive peptides and functional foods in several diseases. Bioactive peptides could be utilized as antiaging, antioxidant, anticancer, antidiabetic, hypocholesterolemic, hypotensive, and other diseases. They can be administered as food supplements, or developed into lead compounds through the drug discovery and development process. Even though studies are described with identified mechanisms, their effect on human body is still to be studied actively. The translation of scientific studies to marketing of these bioactive peptides needs focus. However, in some cases, more targeted cellular assays and clinical trials need to be carried out at the account of efficacy and safety. These studies will eventually bring peptides as nutraceuticals to the mankind for improving and restoring human health.

# References

[1]    Ahsan, H. 2019. Immunopharmacology and immunopathology of peptides and proteins in personal products. Journal of Immunoassay and Immunochemistry, 40(4), 439–447.
[2]    Möller, N.P., et al. 2008. Bioactive peptides and proteins from foods: Indication for health effects. European Journal of Nutrition, 47(4), 171–182.
[3]    Schagen, S.K. 2017. Topical peptide treatments with effective anti-aging results. Cosmetics, 4(2), 16.
[4]    Zhang, L. and Falla, T.J. 2009. Cosmeceuticals and peptides. Clinics in Dermatology, 27(5), 485–494.
[5]    Moughan, P. 2009. Digestion and absorption of proteins and peptides. In *Designing Functional Foods*. Elsevier. pp. 148–170.
[6]    Fischer, E., Fourneau, E. 1901. *Über Einige Derivate Des Glykocolls*. Vol. 34, Ber, pp. 2868–2877.
[7]    Elam, E., et al. 2021. Recent advances on bioactive food derived anti-diabetic hydrolysates and peptides from natural resources. Journal of Functional Foods, 86, 104674.
[8]    Saeedi, P., et al. 2019. Global and regional diabetes prevalence estimates for 2019 and projections for 2030 and 2045: Results from the international diabetes federation diabetes Atlas. Diabetes Research and Clinical Practice, 157, 107843.

[9]    You, H., et al. 2022. *Identification of Dipeptidyl Peptidase IV Inhibitory Peptides from Rapeseed Proteins.* Vol. 160, LWT, p. 113255.

[10]   Patil, S.P., et al. 2020. Plant-derived bioactive peptides: A treatment to cure diabetes. International Journal of Peptide Research and Therapeutics, 26, 955–968.

[11]   Guidea, A., et al. 2020. Comprehensive evaluation of radical scavenging, reducing power and chelating capacity of free proteinogenic amino acids using spectroscopic assays and multivariate exploratory techniques. Spectrochimica Acta Part A: Molecular and Biomolecular Spectroscopy, 233, 118158.

[12]   Samaei, S.P., et al. 2021. Antioxidant and angiotensin I-converting enzyme (ACE) inhibitory peptides obtained from alcalase protein hydrolysate fractions of hemp (*Cannabis sativa* L.) bran. Journal of Agricultural and Food Chemistry, 69(32), 9220–9228.

[13]   Afroz, M., et al. 2020. Ethnobotany and antimicrobial peptides from plants of the Solanaceae family: An update and future prospects. Frontiers in Pharmacology, 11, 565.

[14]   Wen, C., et al. 2020. Plant protein-derived antioxidant peptides: Isolation, identification, mechanism of action and application in food systems: A review. Trends in Food Science & Technology, 105, 308–322.

[15]   Youssef, F.S., et al. 2019. A comprehensive review of bioactive peptides from marine fungi and their biological significance. Marine Drugs 17(10), 559.

[16]   Cunha, S.A. and Pintado, M.E. 2022. Bioactive peptides derived from marine sources: Biological and functional properties. Trends in Food Science & Technology, 119, 348–370.

[17]   Guha, S., et al. 2021. A comprehensive review on bioactive peptides derived from milk and milk products of minor dairy species. Food Production, Processing and Nutrition, 3(1), 1–21.

[18]   Alu'datt, M.H., et al. 2021. Characterization and biological properties of peptides isolated from dried fermented cow milk products by RP-HPLC: Amino acid composition, antioxidant, antihypertensive, and antidiabetic properties. Journal of Food Science, 86(7), 3046–3060.

[19]   Shukla, P., et al. 2022. Exploring the potential of Lacticaseibacillus paracasei M11 on antidiabetic, anti-inflammatory, and ACE inhibitory effects of fermented dromedary camel milk (*Camelus dromedaries*) and the release of antidiabetic and anti-hypertensive peptides. Journal of Food Biochemistry, e14449.

[20]   Chai, K.F., Voo, A.Y. and Chen, W.N. 2020. Bioactive peptides from food fermentation: A comprehensive review of their sources, bioactivities, applications, and future development. Comprehensive Reviews in Food Science and Food Safety, 19(6), 3825–3885.

[21]   Vásquez, V., Martínez, R. and Bernal, C. 2019. Enzyme-assisted extraction of proteins from the seaweeds Macrocystis pyrifera and Chondracanthus chamissoi: Characterization of the extracts and their bioactive potential. Journal of Applied Phycology, 31, 1999–2010.

[22]   Umego, E.C., et al. 2021. Ultrasonic-assisted enzymolysis: Principle and applications. Process Biochemistry, 100, 59–68.

[23]   Mora, L. and Toldrá, F. 2022. Advanced enzymatic hydrolysis of food proteins for the production of bioactive peptides. Current Opinion in Food Science, 100973.

[24]   Bougatef, A., et al. 2010. Purification and identification of novel antioxidant peptides from enzymatic hydrolysates of sardinelle (Sardinella aurita) by-products proteins. Food Chemistry, 118(3), 559–565.

[25]   Andre-Frei, V., et al. 1999. A comparison of biological activities of a new soya biopeptide studied in an in vitro skin equivalent model and human volunteers. International Journal of Cosmetic Science, 21(5), 299–311.

[26]   Sim, G.-S., et al. 2007. Black rice (*Oryza sativa* L. var. japonica) hydrolyzed peptides induce expression of hyaluronan synthase 2 gene in HaCaT keratinocytes. Journal of Microbiology and Biotechnology, 17(2), 271–279.

[27]   Manosroi, A., et al. 2012. Anti-aging efficacy of topical formulations containing niosomes entrapped with rice bran bioactive compounds. Pharmaceutical Biology, 50(2), 208–224.

[28] Ochiai, A., et al. 2016. Rice bran protein as a potent source of antimelanogenic peptides with tyrosinase inhibitory activity. Journal of Natural Products, 79(10), 2545–2551.

[29] Amakye, W.K., et al. 2021. Bioactive anti-aging agents and the identification of new anti-oxidant soybean peptides. Food Bioscience, 42, 101194.

[30] Tyagi, A., et al. 2015. CancerPPD: A database of anticancer peptides and proteins. Nucleic Acids Research, 43(D1), D837–D843.

[31] Gabernet, G., et al. 2016. Membranolytic anticancer peptides. MedChemComm, 7(12), 2232–2245.

[32] Vital, D.A.L., et al. 2014. Peptides in common bean fractions inhibit human colorectal cancer cells. Food Chemistry, 157, 347–355.

[33] Cook, S.J., et al. 2017. Control of cell death and mitochondrial fission by ERK 1/2 MAP kinase signalling. The FEBS Journal, 284(24), 4177–4195.

[34] Wang, W., et al. 2013. PGPIPN, a therapeutic hexapeptide, suppressed human ovarian cancer growth by targeting BCL2. PLoS One, 8(4), e60701.

[35] Wang, Z. and Zhang, X. 2017. Isolation and identification of anti-proliferative peptides from Spirulina platensis using three-step hydrolysis. Journal of the Science of Food and Agriculture, 97(3), 918–922.

[36] Vásquez-Villanueva, R., et al. 2018. In vitro antitumor and hypotensive activity of peptides from olive seeds. Journal of Functional Foods, 42, 177–184.

[37] Wang, L., et al. 2016. Separation and purification of an anti-tumor peptide from rapeseed (*Brassica campestris* L.) and the effect on cell apoptosis. Food & Function, 7(5), 2239–2248.

[38] Xue, Z., et al. 2015. Antioxidant activity and anti-proliferative effect of a bioactive peptide from chickpea (*Cicer arietinum* L.). Food Research International, 77, 75–81.

[39] Lammi, C., Zanoni, C. and Arnoldi, A. 2015. IAVPGEVA, IAVPTGVA, and LPYP, three peptides from soy glycinin, modulate cholesterol metabolism in HepG2 cells through the activation of the LDLR-SREBP2 pathway. Journal of Functional Foods, 14, 469–478.

[40] Nagaoka, S., et al. 2010. Soystatin (VAWWMY), a novel bile acid-binding peptide, decreased micellar solubility and inhibited cholesterol absorption in rats. Bioscience, Biotechnology, and Biochemistry 74(8), 1738–1741.

[41] Pak, V., et al. 2005. Isolation and identification of peptides from soy 11S-globulin with hypocholesterolemic activity. Chemistry of Natural Compounds, 41, 710–714.

[42] Zanoni, C., et al. 2017. Investigations on the hypocholesterolaemic activity of LILPKHSDAD and LTFPGSAED, two peptides from lupin β-conglutin: Focus on LDLR and PCSK9 pathways. Journal of Functional Foods, 32, 1–8.

[43] Zanoni, C., et al. 2017. Hempseed peptides exert hypocholesterolemic effects with a statin-like mechanism. Journal of Agricultural and Food Chemistry, 65(40), 8829–8838.

[44] Valko, M., Leibfritz, D., Moncol, J., Cronin, M.T., Mazur, M. and Telser, J. 2007. Free radicals and antioxidants in normal physiological functions and human disease. The International Journal of Biochemistry & Cell Biology, 39(1), 44–84.

[45] Salmenkallio-Marttila, M., Katina, K. and Autio, K. 2001. Effects of bran fermentation on quality and microstructure of high-fiber wheat bread. Cereal Chemistry, 78(4), 429–435.

[46] Rizzello, C.G., et al. 2017. Improving the antioxidant properties of quinoa flour through fermentation with selected autochthonous lactic acid bacteria. International Journal of Food Microbiology, 241, 252–261.

[47] Mirzaei, M., et al. 2018. Production of antioxidant and ACE-inhibitory peptides from Kluyveromyces marxianus protein hydrolysates: Purification and molecular docking. Journal of Food and Drug Analysis, 26(2), 696–705.

[48] Hu, F., et al. 2018. Identification and hydrolysis kinetic of a novel antioxidant peptide from pecan meal using Alcalase. Food Chemistry, 261, 301–310.

[49] Chen, N., et al. 2012. Purification and identification of antioxidant peptides from walnut (*Juglans regia* L.) protein hydrolysates. Peptides, 38(2), 344–349.

[50] Jin, J.-E., Ahn, C.-B. and Je, J.-Y. 2018. Purification and characterization of antioxidant peptides from enzymatically hydrolyzed ark shell (*Scapharca subcrenata*). Process Biochemistry, 72, 170–176.

[51] Wang, W.-Y., et al. 2020. Antioxidant peptides from collagen hydrolysate of redlip croaker (*Pseudosciaena polyactis*) scales: Preparation, characterization, and cytoprotective effects on H2O2-damaged HepG2 cells. Marine Drugs 18(3), 156.

[52] Hu, X.-M., et al. 2020. Antioxidant peptides from the protein hydrolysate of monkfish (Lophius litulon) muscle: Purification, identification, and cytoprotective function on HepG2 cells damage by H2O2. Marine Drugs, 18(3), 153.

[53] Liu, C., et al. 2018. Exploration of the molecular interactions between angiotensin-I-converting enzyme (ACE) and the inhibitory peptides derived from hazelnut (*Corylus heterophylla* Fisch.). Food Chemistry, 245, 471–480.

[54] Yang, W., et al. 2021. *Identification of Antioxidant Peptides from Cheddar Cheese Made with Lactobacillus Helveticus.* Vol. 141, LWT, p. 110866.

[55] Megías, C., et al. 2004. Purification of an ACE inhibitory peptide after hydrolysis of sunflower (*Helianthus annuus* L.) protein isolates. Journal of Agricultural and Food Chemistry, 52(7), 1928–1932.

[56] Ma, -F.-F., et al. 2019. Three novel ACE inhibitory peptides isolated from Ginkgo biloba seeds: Purification, inhibitory kinetic and mechanism. Frontiers in Pharmacology, 9, 1579.

[57] Liu, X., et al. 2021. Identification, characterization and antihypertensive effect in vivo of a novel ACE-inhibitory heptapeptide from defatted areca nut kernel globulin hydrolysates. Molecules, 26(11), 3308.

[58] Oh, J.-Y., et al. 2020. Anti-hypertensive activity of novel peptides identified from olive flounder (*Paralichthys olivaceus*) surimi. Foods, 9(5), 647.

[59] Zhao, Y.-Q., et al. 2019. Eight antihypertensive peptides from the protein hydrolysate of Antarctic krill (*Euphausia superba*): Isolation, identification, and activity evaluation on human umbilical vein endothelial cells (HUVECs). Food Research International, 121, 197–204.

[60] Fan, H. and Wu, J. 2021. Purification and identification of novel ACE inhibitory and ACE2 upregulating peptides from spent hen muscle proteins. Food Chemistry, 345, 128867.

[61] Wei, D., Fan, W. and Xu, Y. 2019. In vitro production and identification of angiotensin converting enzyme (ACE) inhibitory peptides derived from distilled spent grain prolamin isolate. Foods, 8(9), 390.

[62] Okagu, I.U., Ezeorba, T.P., Aham, E.C., Aguchem, R.N. and Nechi, R.N. 2022. Recent findings on the cellular and molecular mechanisms of action of novel food-derived antihypertensive peptides. Food Chemistry: Molecular Sciences, 4, 100078.

[63] Chamata, Y., Watson, K.A. and Jauregi, P. 2020. Whey-derived peptides interactions with ACE by molecular docking as a potential predictive tool of natural ACE inhibitors. International Journal of Molecular Sciences, 21(3), 864.

[64] Zheng, Y., et al. 2020. Isolation of novel ACE-inhibitory peptide from naked oat globulin hydrolysates in silico approach: Molecular docking, in vivo antihypertension and effects on renin and intracellular endothelin-1. Journal of Food Science, 85(4), 1328–1337.

[65] Nardo, A.E., et al. 2020. Amaranth as a source of antihypertensive peptides. Frontiers in Plant Science, 11, 578631.

[66] Zambrowicz, A., et al. 2015. Multifunctional peptides derived from an egg yolk protein hydrolysate: Isolation and characterization. Amino Acids, 47, 369–380.

[67] Yu, Z., et al. 2011. Novel peptides derived from egg white protein inhibiting alpha-glucosidase. Food Chemistry, 129(4), 1376–1382.

[68] Harnedy, P.A., et al. 2018. Atlantic salmon (*Salmo salar*) co-product-derived protein hydrolysates: A source of antidiabetic peptides. Food Research International, 106, 598–606.

[69] Li-Chan, E.C., et al. 2012. Peptides derived from Atlantic salmon skin gelatin as dipeptidyl-peptidase IV inhibitors. Journal of Agricultural and Food Chemistry, 60(4), 973–978.

[70]   Wang, R., et al. 2019. Preparation of bioactive peptides with antidiabetic, antihypertensive, and antioxidant activities and identification of α-glucosidase inhibitory peptides from soy protein. Food Science & Nutrition, 7(5), 1848–1856.

[71]   Mojica, L., Luna-Vital, D.A. and González de Mejía, E. 2017. Characterization of peptides from common bean protein isolates and their potential to inhibit markers of type-2 diabetes, hypertension and oxidative stress. Journal of the Science of Food and Agriculture, 97(8), 2401–2410.

[72]   Ngoh, -Y.-Y. and Gan, C.-Y. 2016. Enzyme-assisted extraction and identification of antioxidative and α-amylase inhibitory peptides from Pinto beans (*Phaseolus vulgaris* cv. Pinto). Food Chemistry, 190, 331–337.

[73]   Wei, Y., et al. 2019. Hypoglycemic effects and biochemical mechanisms of Pea oligopeptide on high-fat diet and streptozotocin-induced diabetic mice. Journal of Food Biochemistry, 43(12), e13055.

[74]   Fan, S., et al. 2014. Optimization of preparation of antioxidative peptides from pumpkin seeds using response surface method. PLoS One, 9(3), e92335.

[75]   Hatanaka, T., et al. 2012. Production of dipeptidyl peptidase IV inhibitory peptides from defatted rice bran. Food Chemistry, 134(2), 797–802.

[76]   Jha, S., et al. 2018. In vitro antioxidant and antidiabetic activity of oligopeptides derived from different mulberry (*Morus alba* L.) cultivars. Pharmacognosy Research, 10(4).

Munazza Kiran*, Almas Jahan, Sammina Mahmood, Hammad Ullah

# Chapter 3
# Nutraceutical properties of dietary lipids

**Abstract:** Lipids have been described as harmful components of food due to their high caloric value and some unhealthy effects, in particular their association with cardiometabolic disorders. In recent decades, the research mainly focused on exploring the beneficial effects of plant's bioactive components, where lipids have been reported with associated health benefits. Thus, they emerged as one of the functional foods. They help the host to decrease the risk of chronic pathologies like obesity, cardiovascular disorders, cancer, and neurological ailments. Initial observations shed light on omega-3-fatty acids and other dietary lipids for their possible health benefits but thereafter much attention has been paid to unsaturated fatty acids and phytosterols for their noncalorific roles in prevention and treatment of certain diseased states. Regular intake of these lipids in adequate amount may not only suppress hypercholesterolemia, systemic inflammation, and neurodegeneration but also play a critical role in other signaling pathways in the body like cell survival, cell proliferation, and apoptosis. This chapter focuses on dietary lipids from different sources and their potential nutraceutical benefits providing a roadmap for the possibility of their incorporation in human diet as functional foods.

## 1 Introduction

Lipids are an important class of biological compounds that perform several roles in various cellular activities including energy storage and involvement in signaling cascades. Besides these, they are also a vital structural component of biological membranes. In a cell, their diversity is comparable to proteins, where a cell expresses both classes to regulate transport and metabolic processes [1]. Their diversity rather than quantity also makes them significant from health and disease prevention perspectives.

Over several decades, lipids have been identified not only as an energy source but also have multiple other beneficiary roles in several physiological processes [2]. First, intake of dietary fats was considered to be associated with weight gain and

*Corresponding author: Munazza Kiran**, Department of Botany, Division of Science and Technology, University of Education, Lahore, Pakistan, e-mail: munazza.kiran@ue.edu.pk
**Almas Jahan, Sammina Mahmood**, Department of Botany, Division of Science and Technology, University of Education, Lahore, Pakistan
**Hammad Ullah,** Department of Pharmacy, University of Naples Federico II, Via Domenico Montesano 49, 80131 Naples, Italy

https://doi.org/10.1515/9783111317601-003

many health problems; however, in the year 1929, the notion was challenged by providing the evidence that rats that were given a fat-free diet showed retarded growth, scaly skin, and a high mortality rate, whereas these symptoms reversed when they were fed with a diet containing linoleic acid (LA). Therefore, LA was identified not only as an "essential nutrient" but the term "essential fatty acids" was also ascribed to them [3]. A few years later, Nunn and Smedley-Maclean [4] showed that rats feeding on a fat-free diet exhibited very low levels of arachidonic acid (ARA). However, supplementing their diet with methyl linoleate normalized their ARA levels. Further studies demonstrated that LA acts as a precursor for the synthesis of ARA, which is an essential primary unsaturated fatty acid in the diet [5]. Similarly, omega-3 polyunsaturated fatty acids ($n$-3 PUFAs) including eicosapentaenoic acid (EPA) and docosahexaenoic acid (DHA) were considered another potential class of essential fatty acids with health benefits [2].

Functional lipids can be described as functional foods or nutraceuticals, resembling traditional foods that are included in normal diet, and besides providing fundamental nutritional benefits, they also decrease the risks of various chronic illnesses along with many physiological benefits [6]. Their roles in prevention and treatment of various health conditions like blood pressure, bone health, obesity, diabetes, cardiovascular health, and psychological depression have been recorded in recent studies [1]. Thus, current studies have demonstrated lipids as functional foods or nutraceuticals that are not only essential from nutrition point of view but also possess various health benefits including reduced risk of many diseases.

# 2 Classification of lipids

Lipids comprise structurally and functionally heterogeneous molecules; thus, there are differences with reference to the scope and organization of recent classification schemes. The most generally accepted lipid classification scheme is based on their chemical structure (Figure 1).

## 2.1 Simple lipids

Those yielding maximum two distinct entities upon hydrolysis are known as simple lipids [7]. These are esters of alcohols and fatty acids. Examples include triacylglycerols (TAGs) and waxes.

## LIPIDS

| SIMPLE | COMPLEX | DERIVED |
|---|---|---|
| *Esters of alcohols + Fatty acids* | *Esters of alcohols + Fatty acids + Additional groups* | *Hydrolysis of simple/complex lipids* |
| ➡ TAGs | ➡ Phospholipids | ➡ Sterols |
| ➡ Waxes | ➡ Glycolipids | ➡ Fatty acid |
|  |  | ➡ Eicosanoids |
|  |  | ➡ Terpenes |

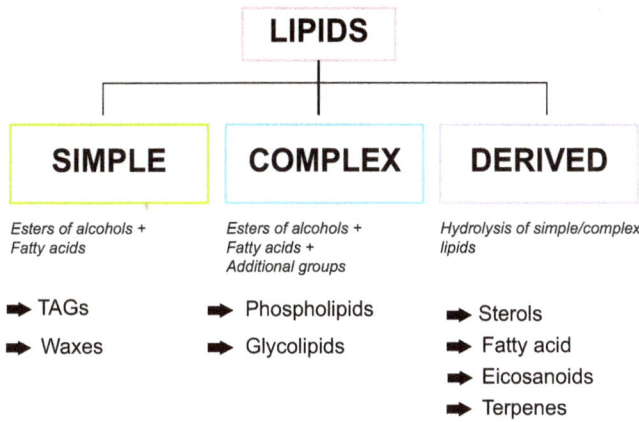

**Figure 1:** General classification of lipids.

### 2.1.1 Triacylglycerols (TAGs)

These are hydrophobic, nonpolar, energy-storing lipids that are classified as "neutral lipids." TAGs are made up of fatty acid chains esterified by a glycerol molecule (Figure 2). Simple TAGs comprised three identical fatty acids linked to a glycerol molecule, whereas complex TAGs comprised two or three different fatty acids. TAGs exist as fats or oils. TAGs consisting of longer saturated fatty acid chains exhibit higher melting points and are solids at room temperature, while those with a higher degree of unsaturation remain as oils. Animal fats are rich in saturated fatty acids except fish fat, whereas plant fats have a higher content of unsaturation in their fatty acid chains; therefore, they exist as oils and are one of the major sources of dietary TAGs [2].

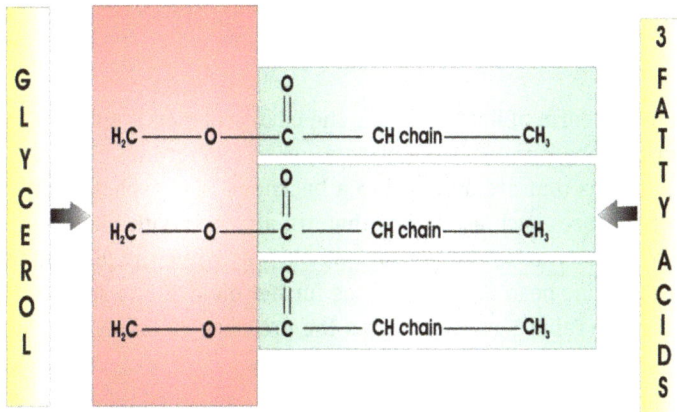

**Figure 2:** Chemical structure of triacylglycerols.

### 2.1.2 Waxes

These are esters of long-chain alcohols and long-chain fatty acids that make them extremely hydrophobic (Figure 3). They also have high melting temperatures (>60 °C). Owing to these properties, waxes are the lipid components in the outer coatings of animals and insects. They can be natural as well as synthetic. Natural waxes include animal, vegetable, and mineral waxes. Some examples include lanolin, beeswax, candelilla wax, and carnauba wax. They found their applications in several industries, including food and cosmetics [2].

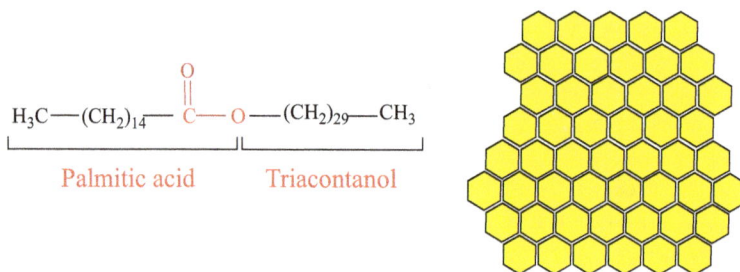

$$H_3C\text{---}(CH_2)_{14}\text{---}\overset{\overset{\displaystyle O}{\|}}{C}\text{---}O\text{---}(CH_2)_{29}\text{---}CH_3$$

Palmitic acid          Triacontanol

**Figure 3:** Chemical skeleton of beeswax.

## 2.2 Complex lipids

These yield three or more entities upon hydrolysis [8]. They are also esters of alcohols and fatty acids but with additional groups. They are further classified as phospholipids and glycolipids.

### 2.2.1 Phospholipids

These are the most abundant form of lipids found in the biological membranes. Their amphipathic nature brings about the specific orientation of membranes. They are composed of fatty acid chains that are attached to a backbone, a phosphate, and a hydroxyl group or organic bases such as choline that are attached with phosphate. The hydrocarbon chains of fatty acids act as hydrophobic tails while the phosphate group makes up the hydrophilic head [2]. They can be further divided into sphingolipids and phosphoglycerides (Table 1), depending on the presence of sphingosine or glycerol (Figure 4).

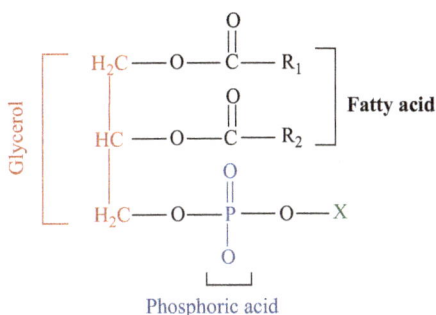

**Figure 4:** Chemical structure of phosphoglycerides.

**Table 1:** Phosphoglycerides and their biological functions.

| Name | Biological functions |
| --- | --- |
| Phosphatidic acid | – It acts as a precursor for other phospholipids. |
| | – It acts as a lipid second messenger. |
| | – It acts as a modulator of membrane shape due to its structural properties. |
| | – In the nucleus, it behaves as mitogen, influencing gene expression regulation, and in the Golgi membrane, it plays a role in membrane trafficking. |
| | – It is essential in numerous cellular functions such as signal transduction, cytoskeleton organization, and cell proliferation. |
| Phosphatidylinositol | – It acts as a precursor of phosphoinositides (PPIns), phosphorylated derivatives of phosphatidylinositol. |
| | – It is known to play an important role in the life cycle of a cell. |
| Phosphatidylethanolamine | – It is present in mammalian cells. |
| | – It is involved in apoptosis and cell signaling. |
| | – It might also play a role in hepatic lipoprotein secretion. |
| | – It is required for contractile ring disassembly at the cleavage furrow of mammalian cells during cytokinesis. |
| | – It helps in membrane fission and fusion. |
| | – It is the donor of the ethanolamine moiety of the glycosylphosphatidylinositol anchors of many cell surface signaling proteins. |
| | – It is a precursor of anandamide (*N*-arachidonoylethanolamine), which is a ligand for cannabinoid receptors in the brain. |
| | – In the yeast *Saccharomyces cerevisiae*, it is involved in the delivery of cytoplasmic proteins to the vacuole. |

**Table 1** (continued)

| Name | Biological functions |
|---|---|
| Phosphatidylglycerol | – In plants, photosynthetic transport of electrons, the development of chloroplasts, and tolerance to chilling.<br>– In terrestrial organisms, it has a fluidizing effect on the lung surfactant.<br>– It prevents alveolar epithelial apoptosis and profibrotic stimulation.<br>– It suppresses proliferation of respiratory viral infection.<br>– In healthy mammalian systems, it has been shown to activate RNA synthesis and a nucleus PKC but inhibit platelet-activating factor.<br>– It is also produced in response to viral infection and can be used by the virus to prepare its own membrane.<br>– In plants, it is required for embryo development.<br>– In bacteria, it helps in protein folding and protein binding. It also activates a glycerol phosphate acyl transferase. |
| Phosphatidylcholine (lecithin) | – It increases the solubility of biliary cholesterol and lowers the phospholipid concentrations in patients with gallstones than in those without gallstones.<br>– It facilitates the transport of triacylglycerol out of the liver.<br>– It is an important component of the mucosal layer of the colon and acts as a surfactant within the mucus to create a hydrophobic surface to prevent bacterial penetrance. |
| Cardiolipin | – It plays a central role in mitochondrial metabolism, by maintaining the proper architecture and morphology of the mitochondrial membranes and by regulating the activity of a variety of proteins and enzymes involved in mitochondrial function.<br>– It is required for full activity of respiratory chain complexes, as well as for their assembly and stabilization into supercomplexes, essential for a more efficient electron/proton flux, and hence for ATP synthesis.<br>– It is an important player in mitochondrial fusion–division dynamics, in mitochondrial biogenesis and protein import, and in regulation of mitophagy and cell death processes. |
| Phosphatidylserine | – It involves in apoptosis and cell signaling.<br>– It acts as a cofactor that activates several key signaling proteins, including protein kinase C, neutral sphingomyelinase, and cRaf1 protein kinase, as well as $Na^+/K^+$ ATPase and dynamin-1. |

## 2.2.2 Glycolipids

Lipids attached to carbohydrates are known as glycolipids. They also constitute a vital component of cell membranes and are involved in several important functions, such as immunity and cell-to-cell recognition. Glycolipids contain sphingosine as compared to phospholipids that have glycerol lacking a phosphate group [2].

**Table 2:** Glycolipids and their biological functions.

| Name | Biological functions |
| --- | --- |
| Glucocerebroside | – It plays an important role in the development of multidrug resistance by cancer cells.<br>– It is able to provoke immune reaction and acts as a self-antigen in Gaucher disease. |
| Galactosylcerebroside | – It plays a key role in the maintenance of proper structure and stability of myelin and differentiation of oligodendrocytes.<br>– It plays important roles in the development of multidrug resistance by cancer cells.<br>– It is an important cellular receptor for HIV-1. |
| Gangliosides | – They are involved in neuronal differentiation and development.<br>– They act as a modulator of neuronal calcium.<br>– They are involved in cell-to-cell recognition, adhesion, and signal transduction within specific cell microdomains termed caveolae, lipid rafts, and with other membrane components such as sphingomyelin and cholesterol.<br>– They are involved in the modulation of intracellular and intranuclear calcium homeostasis and the ensuing cellular functions.<br>– They are involved in the functional maturation of cochlea during early development.<br>– They can be utilized as biomarkers for cancer stem cells and as target for the treatment of brain cancer. |

## 2.3 Derived lipids

These are derived from simple and complex lipids as a result of hydrolysis.

### 2.3.1 Sterols

These are amphipathic molecules and are ubiquitously produced in eukaryotic cells. Cholesterol ($C_{27}H_{45}OH$, Figure 5) is the most common sterol present in animal tissues. As a part of cell membrane, cholesterol plays a significant role in membrane fluidity and acts as a precursor of steroid hormones such as testosterone, progesterone cortisol, vitamin D, and estradiol [8]. The recommended cholesterol level in serum is <200 mg/dL. Plant-based sterols are called phytosterols which includes brassicasterols, sitosterols, and stigmasterols. They share a structural similarity with cholesterol but the nature of their side chain is different. The addition of phytosterols in food products poses cholesterol-lowering properties.

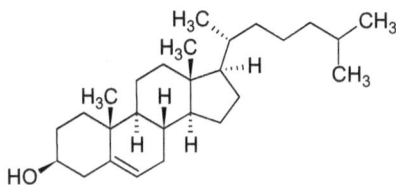

**Figure 5:** Chemical structure of cholesterol.

## 2.3.2 Fatty acids

Fatty acids are composed of long hydrocarbon chains (ranges from 4 to 24 C atoms) with carboxyl group at one end and methyl group at the other; thus, they have polar as well as nonpolar ends rendering them an amphipathic nature (Figure 6). Palmitic acid (16:0) is the most widely distributed fatty acid in microorganisms, animals, and plants. Fatty acids can be broadly categorized into saturated and unsaturated fatty acids (based on the presence and absence of carbon–carbon double bond). Unsaturated fatty acids can be further classified as (i) monounsaturated fatty acids and (ii) polyunsaturated fatty acids. A few PUFAs are synthesized only in phytoplanktons and plants but not in mammals; thus, a plant-based diet is an essential source for such fatty acids. These include α-linolenic acid (ALA) and LA, which are found in all dietary fats and vegetable oils.

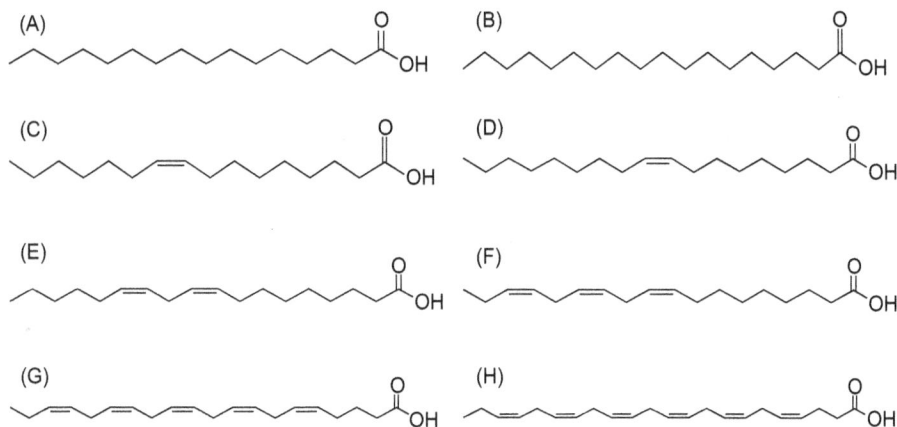

**Figure 6:** Chemical structure of saturated, monounsaturated, and polyunsaturated fatty acids: **(A)** palmitic acid (16:0), **(B)** stearic acid (18:0), **(C)** palmitoleic acid (16:1$^{\Delta9}$), **(D)** oleic acid (18:1$^{\Delta9}$), **(E)** linoleic acid (18:2$^{\Delta9,12}$), **(F)** α-linolenic acid (18:3$^{\Delta9,12,15}$), **(G)** eicosapentaenoic acid (20:5$^{\Delta5,8,11,14,17}$), and **(H)** docosohexaenoic acid (22:6$^{\Delta4,7,10,13,16,19}$).

### 2.3.3 Eicosanoids

These are oxygenated fatty acids ($C_{20}$) with methylene-interrupted cis-bonds including leukotrienes, thromboxanes, and prostaglandins. Besides acting as local hormones, they regulate several biological processes such as bronchodilation, bronchoconstriction, smooth muscle contraction, and inflammation.

### 2.3.4 Terpenes

The term "terpene" is used for $C_{10}H_{18}$ hydrocarbons, found in turpentine oil. "Isoprene rule" was first proposed by Otto Wallach, stating that terpenes are made up of isoprene subunits (Figure 7). It was later suggested by Ingold that these units are linked from head to tail. Based on isoprene units, terpenes are further classified as monoterpenes, sesquiterpenes, diterpenes, sesterterpenes, triterpenes, and carotenoids. Further details of these classes are given in Table 3.

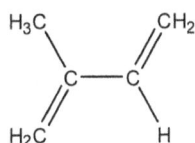

**Figure 7:** One isoprene unit.

**Table 3:** Classification of terpenes.

| Classes | C atoms | Isoprene units |
|---|---|---|
| Monoterpenes | 10 | 2 |
| Sesquiterpenes | 15 | 3 |
| Diterpenes | 20 | 4 |
| Sesterterpenes | 25 | 5 |
| Triterpenes | 30 | 6 |
| Tetraterpenes (carotenoids) | 40 | 8 |

Among these, monoterpenes and sesquiterpenes are major parts of essential oils. Triterpenoids are the largest class and these include lanosterol and squalene acting as steroid precursors in both animals and plants. Carotenoids include "carotenes" and "xanthophylls."

## 2.4 Functional lipids

Though there is no proper definition of functional lipids, they can be described as a subset of functional foods that are consumed as a part of regular diet but besides providing fundamental nutritional benefits to the body, they also offer various physiological advantages and may also lower the risk of chronic diseases [6]. Examples include n-3 PUFAs (i.e., ALA, DHA, and EPA), n-6 PUFAs (i.e., LA, γ-LA, and conjugated LA (CLA)), triglycerides oils, and phytosterols.

### 2.4.1  n-3 PUFAs

n-3 PUFAs can be found in both marine and terrestrial environments. Adequate intake of n-3 PUFAs helps in reducing inflammation as well as decreasing the risk of chronic illnesses such as arthritis, cancer, and heart diseases. They can also regulate glucose tolerance, blood pressure, blood coagulation, and nervous system growth/repair and its functions.

They are also known as "vitamin F" [9]. Recently, marine fatty fish like mullet, mackerel, and salmon are considered primary dietary sources of DHA and EPA for human diet. In addition, their alternative sources such as microalgae, plants, fungus, and bacteria are being researched for commercial production [10]. On the other hand, sources of ALA include vegetables such as tomato, broccoli, spinach, and meat [1].

The nutraceutical properties attributed to n-3 PUFAs intake are as follows.

#### 2.4.1.1 Effect on cardiovascular disease

n-3 PUFAs are found to have a dramatic effect on lowering the risk of acute fatal cardiac arrhythmias in individuals having coronary heart disease (CHD) [11]. Epidemiologic studies have indicated that those at the risk of CHD benefit greatly by the intake of n-3 PUFAs sourced from seaweed and plants. Though the optimum dosage is still undefined, research has suggested that the intake of DHA + EPA of 0.5–1.8 g/day is extremely useful [12]. Consumption of DHA + EPA has been found to reduce the atherogenic and inflammatory pathways by regulating their genes [13]. Besides their role in lowering the risk of sudden cardiac death after an acute heart attack or heart failure, some contradictions have also been reported regarding the usage of DHA + EPA [14–16].

A study in patients with diabetes mellitus type 2 (T2DM) showed that EPA supplementation of 1,800 mg/day dose reduced carotid intimal-medical thickness and improved brachial ankle pulse wave velocity, which leads to reduced atherosclerosis and increased endothelial function [17]. Moreover, n-3 PUFAs with long chains are found to have an essential function in stabilizing atherosclerosis and reducing the risk of cardiovascular events. Study has also confirmed that patients with acute coronary syndrome were found to have low EPA as well as their elongated metabolite blood levels [18]. So far, the most compelling results have been obtained from marine-derived *n*-3 fatty acids sourced from fish or supplements. On the other hand, the cardioprotective effects of ALA still need to be established. Conclusively, supplementation is a viable option; however, a diet-based approach is more preferable for *n*-3 fatty acids.

### 2.4.1.2 Effect on immune cells

Dietary *n*-3 PUFAs have been found to improve specific immunological activities in a few immune cell types [19]. They change three major biological aspects of macrophage, namely capability of phagocytosis, chemokine and cytokine synthesis, and polarization into activated macrophages [1]. In addition, studies have shown that its supplementation is useful in T-mediated illnesses such as asthma [20] and autoimmune hepatitis [21].

### 2.4.1.3 Effect on fetal/neonatal development

Dietary or supplementary intake of *n*-3 PUFAs throughout pregnancy and postpartum period has been linked with several health advantages. It has been observed that consumption of *n*-3 PUFAs improves neurological development in infants [22]. A study conducted by Hibbeln et al. [23] indicated that children of those mothers who did not include seafood in their diet had higher risks of suboptimal or adverse outcomes resulting in behavioral issues at age 7, lowest quartile for verbal and performance IQ at age 8, and poor scoring in early developmental tests for evaluating social, communication, and fine motor skills in contrast to the children of the mothers who included seafood in their diet during pregnancy. Moreover, intake of *n*-3 PUFAs was also found to be essential during the nursing period as well. During pregnancy and breastfeeding period, supplementary fish oil lowers the risk of newborn allergies. Thus, all these evidences show that consumption of *n*-3 PUFAs during pregnancy is critical not only for fetal neurodevelopment and growth, but for a child's future brain growth as well whereby its deficiency is strongly linked to poor behavioral and developmental scores in children.

### 2.4.1.4 Effect on Alzheimer's disease

Alzheimer's disease is a neurodegenerative disease in elders and is considered as the root cause of dementia. Consumption of DHA has been linked with mental health and neurodevelopment, especially during prenatal brain developmental stages [24]. Studies on animals have shown that DHA deficiency in brain tissues results in behavioral impairments, ultimately leading to cognitive dysfunction and neurodegeneration similar to Alzheimer's patients [25–27]. Reduced levels of DHA have a negative effect on glutamate, which is a key excitatory neurotransmitter contributing toward integrity of brain functions regarding learning and memorizing things [28]. Various studies over the last decade have shown the effectiveness of $n$-3 PUFA supplementation in earlier stages of cognitive impairment. Studies have also shown that diets rich in plant-derived ALA and fish/shellfish-derived long-chain PUFAs might prevent Parkinson's and Alzheimer's diseases [29].

### 2.4.2 *n*-6 PUFAs

$n$-6 PUFAs possess two primary physiological roles [30]:
1. They act as an important component of a structural framework in membranes that can also modulate membrane functions.
2. They act as a precursor of eicosanoids that are involved in regulating pulmonary and renal functions, inflammatory responses, and vascular tone.

The dietary sources of $n$-6 PUFAs include seeds, nuts, and vegetable oils. LA is one of the major dietary $n$-6 PUFAs. After its ingestion, LA can be transformed further into longer chains like ARA, by elongation and desaturation processes. Food sources of ARA include fish, meat, eggs, seafood, and poultry while CLA can be obtained from dairy products and ruminant meat [1].

The nutraceutical properties attributed to $n$-6 PUFA supplementation are as follows:

#### 2.4.2.1 Linolenic acid
It is a very frequently occurring PUFA in human diet. After absorption, LA may be used for various purposes. It can be used as a source of energy like every fatty acid. It can also be esterified to form polar and neutral lipids, including cholesterol esters, TAGs, and phospholipids. Besides these, being a component of phospholipids, LA can function in the maintenance of membrane fluidity. Moreover, when it is liberated from phospholipids in membranes, it may be oxidized to various cellular signaling derivatives [31]. Various studies of clinical trials have demonstrated a negative correlation between tissue levels of $n$-6 PUFAs, particularly LA and cardiovascular risk. Additionally, improvement in insulin resistance and long-term glycemic management has also been observed under high dietary intake of $n$-6 PUFAs, including LA [32].

#### 2.4.2.2 Arachidonic acid
ARA is incorporated in cytosolic phospholipids adjacent to endoplasmic reticulum membranes having proteins required for the biosynthesis of phospholipids and then allocating to various biological membranes [33]. Being a basic component of cellular structure, ARA plays a vital role during growth and development. Furthermore, they also play their part in the events occurring due to injury or cell damage. These PUFAs influence various ion channels and enzymes. They are also involved in apoptotic, ne-

crotic, and death responses of cells occurring during embryogenesis. Thus, they impart a significant pharmacological and physiological impact in newborn [34].

It has been observed through various studies that autistic children have low levels of blood PUFAs, specifically of ARA. However, a significant improvement was also observed after consumption of dietary PUFAs. In previous studies, PUFAs particularly ARA, have been shown to exhibit tumoricidal action in both *in vitro* and *in vivo* studies. Free ARA showed inhibiting effects on human cervical cancer and methyl cholanthrene-induced sarcoma cells. Lipid peroxidation and superoxide anion synthesis were enhanced by free ARA in tumor cells, indicating a probable linkage between these PUFAs and tumoricidal activity of cells by increased free radicals [35, 36]. Furthermore, during resistance exercise, its supplementation improves anaerobic capacity and reduces inflammatory responses; however, it was not effective in increasing muscle mass and strength [37].

### 2.4.2.3 Conjugated linoleic acid

As minor lipids, CLA is gaining attention in treating major human illnesses, including atherosclerosis, cancer, and diabetes. In recent years, CLA supplementation is also under practice in sports to reduce body fats and improve performance [38]. Initially, this fatty acid was known to have anticarcinogenic properties but now studies have confirmed its antiatherosclerotic and antiobesogenic characteristics as well [39]. These fatty acids can affect gene transcription processes by interacting with the transcription factors and nuclear receptors [40]. CLA aids in reduced food intake, preadipocyte differentiation and proliferation, lipogenesis and enhanced energy expenditure as well as fat burning, and lipolysis. As per reports on human and animal studies, CLA administration results in loss of body fat and total weight and also lowers the plasma levels of total and low-density lipoprotein (LDL) cholesterol. Moreover, it also exhibits anti-inflammatory effects on body.

The anticarcinogenic property of CLA is due to its ability to inhibit tumor development. Its immediate action suppresses the growth of both benign and malignant tumors [41]. Its various isomers act by regulating cell cycle, apoptosis, and ARA metabolism. It has been demonstrated through studies that 10-CLA is one of the bioactive isomers of CLA, which is shown to influence alternation in bodyweight linked to T2DM [42]. As a powerful antiatherogenic fatty acid, CLA activates peroxisome proliferator-activated receptor in model animals. LDL/HDL (high-density lipoprotein) cholesterol and total cholesterol/HDL cholesterol levels were remarkably lower in rabbits fed with CLA [43, 44]. Studies have also demonstrated that 9-CLA isomer favors the development of mineralized bone nodules by using human bone cells, thus showing positive effects on bone health [45].

### 2.4.3 *n*-6/*n*-3 PUFA ratio

A healthy ratio of *n*-6/*n*-3 PUFAs prevents excessive inflammatory responses, which may cause autoimmune disease and tissue damage. A high *n*-6/*n*-3 PUFAs ratio has been found to be crucial to cause low-grade inflammation. It has been hypothesized that overconsumption of LA on the one hand and deficiency of DHA and EPA on the other hand make the population to become prothrombotic and proinflammatory [46, 47]. Research has indicated that high consumption of *n*-6 PUFAs and high *n*-6/*n*-3 PUFAs ratio leads to weight growth, while high *n*-3 PUFAs lower the risk of weight gain. Therefore, reducing the LA/ALA in animals resulted in reducing the weight gain and subsequent obesity [1].

A diet rich in plant oils and reduced consumption of sea food results in raising the *n*-6/*n*-3 PUFA ratio, as plant oils are high in *n*-6 PUFAs while seafood has high contents of *n*-3 PUFAs. The optimum ratio of *n*-6/*n*-3 PUFAs to preserve human health is 1:5–10 [48]. Therefore, it is very critical to consume a balanced diet having more *n*-3 PUFAs and less *n*-6 PUFAs, and for this purpose, switching from dietary vegetable oils containing *n*-6 PUFAs such as sunflower, corn, safflower, soybean oils and cotton seed oils to the oils containing *n*-3 PUFAs such as rapeseed, chia, perilla, and flax seed oil. Moreover, consumption of more monounsaturated oils like hazelnut oil, olive oil, or monounsaturated sunflower oil also ensures a healthy diet. Besides this, increased fish consumption and reduced red meat consumption will also maintain high *n*-3 PUFAs as compared to *n*-6 PUFAs [49].

### 2.4.4 Medium-chain triglycerides (MCTs)

As the name depicts, medium-chain triglycerides (MCTs) are lipids of medium-chain length with aliphatic tails of 6–12 carbons. Due to their short length, they are quickly hydrolyzed and easily absorbed during digestion as compared to other long-chain fatty acids. Nonhydrolyzed MCTs are absorbed by the intestinal cells [1]. MCTs account for more than half of the fats in coconut oil; hence, it serves as a rich source of its own derivation. In addition, dairy products and palm oils are also dietary sources rich in MCTs [50].

For decades, they have been used to treat many digestive and metabolic disorders such as fat malabsorption, pancreatic insufficiency, severe hyperchylomicronemia, complete parenteral feeding, and poor lymphatic chylomicron transfer. Apart from this, they are also included in infant milk formulae, which signify their importance in neonates' diet [51]. Studies have also confirmed their potential in fighting against steatogenic and adipogenic ailments. An experimental study was conducted where healthy rats fed with MCT-rich diets showed less fat deposition when compared with those rats that fed on long-chain triglycerides (LCTs). It was also worth noting that less fat deposition was not associated with altered whole body protein content or absorption [52]. In another study, women who were on a 1-month diet with MCT supplementation exhibited better energy expenditure and fat oxidation as compared to the women on LCT-rich diet. In addition, their tendency to lose weight was also better than those having an MCT-rich diet [53].

High lactate levels during exercise cause a negative effect on muscle performance. However, MCT aids in reducing lactate accumulation during exercise. Previous studies have also confirmed that prior to cycling; athletes who consumed 6 g of MCTs with diet had reduced lactate contents in their muscles and thus showed improved performance than those consuming LCT [54]. MCT capric acid has been shown to have comparatively better seizure control than some commonly used antiepileptic medicines [55]. MCT ketogenic diet has been considered as an additional energy source and it also improves the survival of brain cells. It also functions in inhibiting a brain receptor which is involved in memory loss. Studies have shown that 20–70 g of MCT supplementation, including capric acid, might help in relieving mild-to-severe Alzheimer's symptoms [56, 57].

Another study has confirmed that coconut oil, which is rich in MCT, reduced the growth of *Candida albicans* by 25%, which is a causative agent of various skin diseases. Coconut oil has also been shown to have potential of inhibiting the growth of *Clostridium difficile*, which is a disease-causing bacterium [58, 59].

### 2.4.5 Phytosterols

The term "sterol" is typically referred to the fused ring structure cyclopentanophenanthrine plus an alcohol moiety (Figure 8), which are further classified into sterols (containing double bond at C-5) and stanols (saturated sterols). As compared to sterols, stanols are less abundant in nature [60, 61]. More than 200 types of phytosterols have been identified in plants, where β-sitosterol, campesterol, and stigmasterol are the most abundant ones. The average daily intake of phytosterols in Western countries is about 250 mg/day, mainly derived from nuts, seeds, cereals, fruits, vegetables, and edible oils [62].

Plant sterols are more hydrophobic as compared to cholesterol and display a high affinity toward fat-digesting micelles. This property makes them to displace cholesterol from micelles, thus reducing absorption of intestinal cholesterol [63]. According to previous studies, supplementary phytosterols decrease LDL cholesterol and total cholesterol. Various clinical trials have also confirmed that supplementary phytosterols in the form of tablets and capsules have similar effects as those that are obtained through fortified meals [64–68].

It has been assumed through studies that decreased cholesterol micellization by phytosterols is caused due to its hydrophobic properties which displace cholesterol and result in much fecal loss. However, their main drawback seems to be their interference with absorption of carotenoids, which can be compensated through diet.

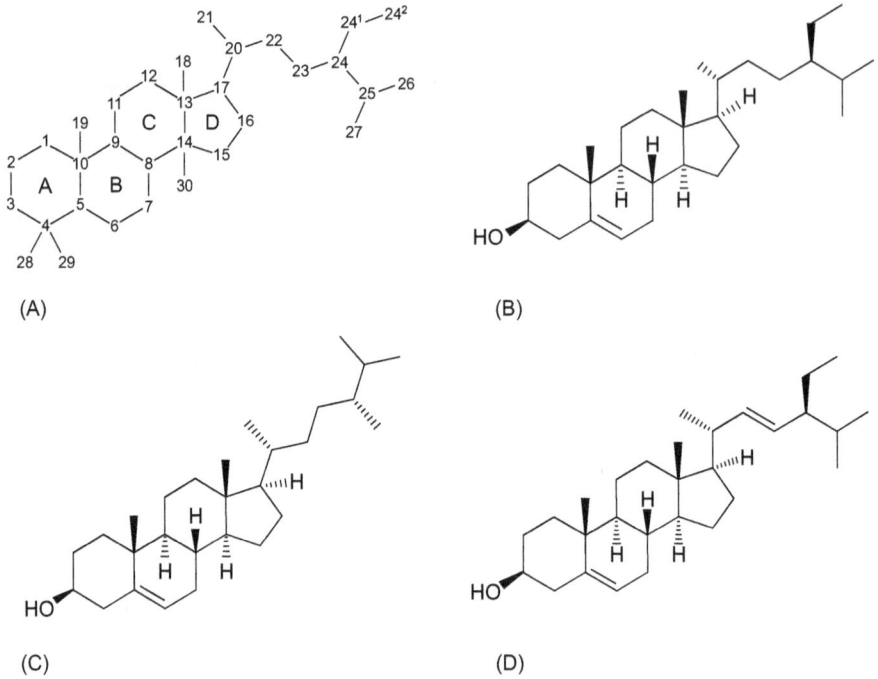

**Figure 8:** Chemical structures of phytosterols. **(A)** Basic skeleton of sterols, **(B)** β-sitosterol, **(C)** campesterol, and **(D)** stigmasterol.

Moreover, phytosterols are also known to exhibit anticancer properties and modulation of the immune system [69].

# 3 Summary

Dietary components consisting of functional lipids have nutraceutical potential affecting human health by decreasing the risk of many diseases and improving the quality of life. They are not only readily available in nature but are also regarded as very cost-effective dietary nutraceuticals present in the food chain. Due to the vast variety of functional lipids, further research is needed to determine a healthy amount for the dietary intake, dosages, and long-term effects of each lipid molecule.

**Acknowledgments:** Authors are thankful to Syed Muhammad Yasir Naseem, Averon Trading (pvt.) Limited, for reviewing and editing this chapter.

# References

[1]  Bhat, S.S. 2021. Functional lipids as nutraceuticals: A review. International Journal of Science and Healthcare Research, 6(4), 111–123.

[2]  Kumar, S., Sharma, B., Bhadwal, P., Sharma, P. and Agnihotri, N. 2018. Lipids as nutraceuticals: A shift in paradigm. In *Therapeutic Foods*. Academic Press, pp. 51–98.

[3]  Mukhopadhyay, R. 2012. Essential fatty acids: The work of George and Mildred Burr. Journal of Biological Chemistry, 287(42), 35439–35441.

[4]  Nunn, L.C.A. and Smedley-Maclean, I. 1938. The nature of the fatty acids stored by the liver in the fat-deficiency disease of rats. Biochemical Journal, 32(12), 2178.

[5]  Mead, J.F., Steinberg, G. and Howtron, D.R. 1953. Metabolism of essential fatty acids. Incorporation of acetate into arachidonic acid. Journal of Biological Chemistry, 205, 683–689.

[6]  Moreau, R.A. 2011. An overview of functional lipids. In *102nd AOCS Annual Meeting and Expo*. Cincinnati, Ohio, USA: Duke Energy Center.

[7]  Fahy, E., Cotter, D., Sud, M. and Subramaniam, S. 2011. Lipid classification, structures and tools. Biochimica et Biophysica Acta (BBA) – Molecular and Cell Biology of Lipids, 1811(11), 637–647.

[8]  Cruz, P.M., Mo, H., McConathy, W.J., Sabnis, N. and Lacko, A.G. 2013. The role of cholesterol metabolism and cholesterol transport in carcinogenesis: A review of scientific findings, relevant to future cancer therapeutics. Frontiers in Pharmacology, 4, 119.

[9]  De Filippis, A.P. and Sperling, L.S. 2006. Understanding omega-3's. American Heart Journal, 151(3), 564–570.

[10]  Adarme-Vega, T.C., Lim, D.K., Timmins, M., Vernen, F., Li, Y. and Schenk, P.M. 2012. Microalgal biofactories: A promising approach towards sustainable omega-3 fatty acid production. Microbial Cell Factories, 11(1), 1–10.

[11]  Jain, A.P., Aggarwal, K.K. and Zhang, P.Y. 2015. Omega-3 fatty acids and cardiovascular disease. European Review for Medical and Pharmacological Sciences, 19(3), 441–445.

[12]  Kris-Etherton, P.M., Harris, W.S. and Appel, L.J. 2003. Omega-3 fatty acids and cardiovascular disease: New recommendations from the American Heart Association. Arteriosclerosis, Thrombosis, and Vascular Biology, 23(2), 151–152.

[13]  Bouwens, M., Van De Rest, O., Dellschaft, N., Bromhaar, M.G., De Groot, L.C., Geleijnse, J.M., Müller, M. and Afman, L.A. 2009. Fish-oil supplementation induces antiinflammatory gene expression profiles in human blood mononuclear cells. The American Journal of Clinical Nutrition, 90(2), 415–424.

[14]  Marchioli, R., Barzi, F., Bomba, E., Chieffo, C., Di Gregorio, D., Di Mascio, R., et al. 2002. Early protection against sudden death by n-3 polyunsaturated fatty acids after myocardial infarction: Time-course analysis of the results of the Gruppo Italiano per lo Studio della Sopravvivenza nell'Infarto Miocardico (GISSI)-Prevenzione. Circulation, 105(16), 1897–1903.

[15]  Kris-Etherton, P.M., Harris, W.S. and Appel, L.J. 2002. Fish consumption, fish oil, omega-3 fatty acids, and cardiovascular disease. Circulation, 106(21), 2747–2757.

[16]  Tavazzi, L., Maggioni, A.P., Marchioli, R., Barlera, S., Franzosi, M.G., Latini, R., Lucci, D., Nicolosi, G.L., Porcu, M. and Tognoni, G. 2008. Effect of n-3 polyunsaturated fatty acids in patients with chronic heart failure (the GISSI-HF trial): A randomised, double-blind, placebo-controlled trial. Lancet (London, England), 372(9645), 1223–1230.

[17]  Mita, T., Watada, H., Ogihara, T., Nomiyama, T., Ogawa, O., Kinoshita, J., Shimizu, T., Hirose, T., Tanaka, Y. and Kawamori, R. 2007. Eicosapentaenoic acid reduces the progression of carotid intima-media thickness in patients with type 2 diabetes. Atherosclerosis, 191(1), 162–167.

[18]  Amano, T., Matsubara, T., Uetani, T., Kato, M., Kato, B., Yoshida, T., Harada, K., Kumagai, S., Kunimura, A., Shinbo, Y., Kitagawa, K., Ishii, H. and Murohara, T. 2011. Impact of omega-3 polyunsaturated fatty acids on coronary plaque instability: An integrated backscatter intravascular ultrasound study. Atherosclerosis, 218(1), 110–116.

[19]  Gutiérrez, S., Svahn, S.L. and Johansson, M.E. 2019. Effects of omega-3 fatty acids on immune cells. International Journal of Molecular Sciences, 20(20), 5028.

[20]  Farjadian, S., Moghtaderi, M., Kalani, M., Gholami, T. and Teshnizi, S.H. 2016. Effects of omega-3 fatty acids on serum levels of T-helper cytokines in children with asthma. Cytokine, 85, 61–66.

[21]  Li, Y., Tang, Y., Wang, S., Zhou, J., Zhou, J., Lu, X., Bai, X., Wang, X.Y., Chen, Z. and Zuo, D. 2016. Endogenous n-3 polyunsaturated fatty acids attenuate T cell-mediated hepatitis via autophagy activation. Frontiers in Immunology, 7, 350.

[22]  Makrides, M., Gibson, R.A., McPhee, A.J., Yelland, L., Quinlivan, J., Ryan, P. and DOMInO Investigative Team, A.T. 2010. Effect of DHA supplementation during pregnancy on maternal depression and neurodevelopment of young children: A randomized controlled trial. JAMA, 304(15), 1675–1683.

[23]  Hibbeln, J.R., Davis, J.M., Steer, C., Emmett, P., Rogers, I., Williams, C. and Golding, J. 2007. Maternal seafood consumption in pregnancy and neurodevelopmental outcomes in childhood (ALSPAC study): An observational cohort study. The Lancet, 369(9561), 578–585.

[24]  Moreira, J.D., Knorr, L., Ganzella, M., Thomazi, A.P., de Souza, C.G., de Souza, D.G., Pitta, C.F., Mello E Souza, T., Wofchuk, S., Elisabetsky, E., Vinade, L., Perry, M.L.S. and Souza, D.O. 2010. Omega-3 fatty acids deprivation affects ontogeny of glutamatergic synapses in rats: Relevance for behavior alterations. Neurochemistry International, 56(6–7), 753–759.

[25]  Lim, G.P., Calon, F., Morihara, T., Yang, F., Teter, B., Ubeda, O., Salem Jr, N., Frautschy, S.A. and Cole, G.M. 2005. A diet enriched with the omega-3 fatty acid docosahexaenoic acid reduces amyloid burden in an aged Alzheimer mouse model. Journal of Neuroscience, 25(12), 3032–3040.

[26]  Fernandes, J.S., Mori, M.A., Ekuni, R., Oliveira, R.M.W. and Milani, H. 2008. Long-term treatment with fish oil prevents memory impairments but not hippocampal damage in rats subjected to transient, global cerebral ischemia. Nutrition Research, 28(11), 798–808.

[27]  Pomponi, M. and Pomponi, M. 2008. DHA deficiency and Alzheimer's disease. Clinical Nutrition, 27(1), 170.

[28]  Su, H.M. 2010. Mechanisms of n-3 fatty acid-mediated development and maintenance of learning memory performance. The Journal of Nutritional Biochemistry, 21(5), 364–373.

[29]  Sofi, F., Abbate, R., Gensini, G.F. and Casini, A. 2010. Accruing evidence on benefits of adherence to the Mediterranean diet on health: An updated systematic review and meta-analysis. The American Journal of Clinical Nutrition, 92(5), 1189–1196.

[30]  Mori, T.A. and Hodgson, J.M. 2013. Fatty acids: Health effects of omega-6 polyunsaturated fatty acids. In *Encyclopedia of Human Nutrition*. Elsevier, pp. 209–214.

[31]  Whelan, J. and Fritsche, K. 2013. Linoleic acid. Advances in Nutrition, 4(3), 311–312.

[32]  Marangoni, F., Agostoni, C., Borghi, C., Catapano, A.L., Cena, H., Ghiselli, A., Vecchia, C.L., Lercker, G., Manzato, E., Pirillo, A., Riccardi, G., Rise, P., Visioli, F. and Poli, A. 2020. Dietary linoleic acid and human health: Focus on cardiovascular and cardiometabolic effects. Atherosclerosis, 292, 90–98.

[33]  Vance, J.E. 1998. Eukaryotic lipid-biosynthetic enzymes: The same but not the same. Trends in Biochemical Sciences, 23(11), 423–428.

[34]  Ordway, R.W., Singer, J.J. and Walsh Jr, J.V. 1991. Direct regulation of ion channels by fatty acids. Trends in Neurosciences, 14(3), 96–100.

[35]  Sagar, P.S. and Das, U.N. 1995. Cytotoxic action of cis-unsaturated fatty acids on human cervical carcinoma (HeLa) cells in vitro. Prostaglandins, Leukotrienes and Essential Fatty Acids, 53(4), 287–299.

[36]  Ramesh, G. and Das, U.N. 1998. Effect of cis-unsaturated fatty acids on Meth-A ascitic tumour cells in vitro and in vivo. Cancer Letters, 123(2), 207–214.

[37]  Roberts, M.D., Iosia, M., Kerksick, C.M., Taylor, L.W., Campbell, B., Wilborn, C.D., Harvey, T., Cooke, M., Rasmussen, C., Greenwood, M., Wilson, R., Jitomir, J., Willoughby, D. and Kreider, R.B. 2007. Effects of arachidonic acid supplementation on training adaptations in resistance-trained males. Journal of the. International Society of Sports Nutrition, 4(1), 21.

[38] Barone, R., Macaluso, F., Catanese, P., Marino Gammazza, A., Rizzuto, L., Marozzi, P., et al. 2013. Endurance exercise and conjugated linoleic acid (CLA) supplementation up-regulate CYP17A1 and stimulate testosterone biosynthesis. PloS One, 8(11), e79686.

[39] Hartigh, L.J.D. 2019. Conjugated linoleic acid effects on cancer, obesity, and atherosclerosis: A review of pre-clinical and human trials with current perspectives. Nutrients, 11(2), 370.

[40] Sampath, H. and Ntambi, J.M. 2005. Polyunsaturated fatty acid regulation of genes of lipid metabolism. Annual Review of Nutrition, 25, 317–340.

[41] Belury, M.A., Moya-Camarena, S.Y., Lu, M., Shi, L., Leesnitzer, L.M. and Blanchard, S.G. 2002. Conjugated linoleic acid is an activator and ligand for peroxisome proliferator-activated receptor-gamma (PPARγ). Nutrition Research, 22(7), 817–824.

[42] Belury, M.A., Mahon, A. and Banni, S. 2003. The conjugated linoleic acid (CLA) isomer, t10c12-CLA, is inversely associated with changes in body weight and serum leptin in subjects with type 2 diabetes mellitus. The Journal of Nutrition, 133(1), 257S–260S.

[43] Brown, J.M. and McIntosh, M.K. 2003. Conjugated linoleic acid in humans: Regulation of adiposity and insulin sensitivity. The Journal of Nutrition, 133(10), 3041–3046.

[44] Toomey, S., McMonagle, J. and Roche, H.M. 2006. Conjugated linoleic acid: A functional nutrient in the different pathophysiological components of the metabolic syndrome? Current Opinion in Clinical Nutrition & Metabolic Care, 9(6), 740–747.

[45] Platt, I., Rao, L.G. and El-Sohemy, A. 2007. Isomer-specific effects of conjugated linoleic acid on mineralized bone nodule formation from human osteoblast-like cells. Experimental Biology and Medicine, 232(2), 246–252.

[46] DiNicolantonio, J.J. and O'Keefe, J.H. 2018. Importance of maintaining a low omega-6/omega-3 ratio for reducing inflammation. Open Heart, 5(2), e000946.

[47] DiNicolantonio, J.J. and O'Keefe, J. 2019. Importance of maintaining a low omega-6/omega-3 ratio for reducing platelet aggregation, coagulation and thrombosis. Open Heart, 6(1), e001011.

[48] Abedi, E. and Sahari, M.A. 2014. Long-chain polyunsaturated fatty acid sources and evaluation of their nutritional and functional properties. Food Science & Nutrition, 2(5), 443–463.

[49] Simopoulos, A.P. 2016. An increase in the omega-6/omega-3 fatty acid ratio increases the risk for obesity. Nutrients, 8(3), 128.

[50] Bach, A. and Babayan, V.K. 1982. Medium-chain triglycerides: An update. The American Journal of Clinical Nutrition, 36(5), 950–962.

[51] Marten, B., Pfeuffer, M. and Schrezenmeir, J. 2006. Medium-chain triglycerides. International Dairy Journal, 16(11), 1374–1382.

[52] Ling, P.R., Hamawy, K.J., Moldawer, L.L., Istfan, N., Bistrian, B.R. and Blackburn, G.L. 1986. Evaluation of the protein quality of diets containing medium-and long-chain triglyceride in healthy rats. The Journal of Nutrition, 116(3), 343–349.

[53] St-Onge, M.P., Bourque, C., Jones, P.J.H., Ross, R. and Parsons, W.E. 2003. Medium-versus long-chain triglycerides for 27 days increases fat oxidation and energy expenditure without resulting in changes in body composition in overweight women. International Journal of Obesity, 27(1), 95–102.

[54] Nosaka, N., Suzuki, Y., Nagatoishi, A., Kasai, M., Wu, J. and Taguchi, M. 2009. Effect of ingestion of medium-chain triacylglycerols on moderate-and high-intensity exercise in recreational athletes. Journal of Nutritional Science and Vitaminology, 55(2), 120–125.

[55] Chang, P., Terbach, N., Plant, N., Chen, P.E., Walker, M.C. and Williams, R.S. 2013. Seizure control by ketogenic diet-associated medium chain fatty acids. Neuropharmacology, 69, 105–114.

[56] Cunnane, S.C., Courchesne-Loyer, A., St-Pierre, V., Vandenberghe, C., Pierotti, T., Fortier, M., Croteau, E. and Castellano, C.A. 2016. Can ketones compensate for deteriorating brain glucose uptake during aging? Implications for the risk and treatment of Alzheimer's disease. Annals of the New York Academy of Sciences, 1367(1), 12–20.

[57] Augustin, K., Khabbush, A., Williams, S., Eaton, S., Orford, M., Cross, J.H., Heales, S.J.R., Walker, M.C. and Williams, R.S. 2018. Mechanisms of action for the medium-chain triglyceride ketogenic diet in neurological and metabolic disorders. The Lancet Neurology, 17(1), 84–93.

[58] Ogbolu, D.O., Oni, A.A., Daini, O.A. and Oloko, A.P. 2007. In vitro antimicrobial properties of coconut oil on Candida species in Ibadan, Nigeria. Journal of Medicinal Food, 10(2), 384–387.

[59] Shilling, M., Matt, L., Rubin, E., Visitacion, M.P., Haller, N.A., Grey, S.F. and Woolverton, C.J. 2013. Antimicrobial effects of virgin coconut oil and its medium-chain fatty acids on Clostridium difficile. Journal of Medicinal Food, 16(12), 1079–1085.

[60] Phillips, K.M., Ruggio, D.M. and Ashraf-Khorassani, M. 2005. Phytosterol composition of nuts and seeds commonly consumed in the United States. Journal of Agricultural and Food Chemistry, 53(24), 9436–9445.

[61] Shepherd, J., Cobbe, S.M., Ford, I., Isles, C.G., Lorimer, A.R., Macfarlane, P.W., et al. 1995. Prevention of coronary heart disease with pravastatin in men with hypercholesterolemia. New England Journal of Medicine, 333, 1301–1308.

[62] Hovenkamp, E., Demonty, I., Plat, J., Lütjohann, D., Mensink, R.P. and Trautwein, E.A. 2008. Biological effects of oxidized phytosterols: A review of the current knowledge. Progress in Lipid Research, 47, 37–49.

[63] Plat, J. and Mensink, R.P. 2001. Effects of plant sterols and stanols on lipid metabolism and cardiovascular risk. Nutrition, Metabolism, and Cardiovascular Diseases: NMCD, 11(1), 31–40.

[64] Rideout, T.C., Chan, Y.M., Harding, S.V. and Jones, P.J. 2009. Low and moderate-fat plant sterol fortified soymilk in modulation of plasma lipids and cholesterol kinetics in subjects with normal to high cholesterol concentrations: Report on two randomized crossover studies. Lipids in Health and Disease, 8, 1–7.

[65] Maki, K.C., Lawless, A.L., Reeves, M.S., Dicklin, M.R., Jenks, B.H., Shneyvas, E.D. and Brooks, J.R. 2012. Lipid-altering effects of a dietary supplement tablet containing free plant sterols and stanols in men and women with primary hypercholesterolaemia: A randomized, placebo-controlled crossover trial. International Journal of Food Sciences and Nutrition, 63(4), 476–482.

[66] Maki, K.C., Lawless, A.L., Reeves, M.S., Kelley, K.M., Dicklin, M.R., Jenks, B.H., Shneyvas, E. and Brooks, J.R. 2013. Lipid effects of a dietary supplement softgel capsule containing plant sterols/stanols in primary hypercholesterolemia. Nutrition, 29(1), 96–100.

[67] Padro, T., Vilahur, G., Sánchez-Hernández, J., Hernández, M., Antonijoan, R.M., Perez, A. and Badimon, L. 2015. Lipidomic changes of LDL in overweight and moderately hypercholesterolemic subjects taking phytosterol- and omega-3-supplemented milk. Journal of Lipid Research, 56(5), 1043–1056.

[68] Ras, R.T., Fuchs, D., Koppenol, W.P., Garczarek, U., Greyling, A., Keicher, C., et al. 2015. The effect of a low-fat spread with added plant sterols on vascular function markers: Results of the investigating vascular function effects of plant sterols (INVEST) study. The American Journal of Clinical Nutrition, 101(4), 733–741.

[69] Vilahur, G., Ben-Aicha, S., Diaz-Riera, E., Badimon, L. and Padró, T. 2019. Phytosterols and inflammation. Current Medicinal Chemistry, 26(37), 6724–6734.

Arif Ali*, Mac Dionys Rodrigues da Costa, Emanuel Paula Magalhães,
Alice Maria Costa Martins

# Chapter 4
# Biological importance of vitamins and minerals

**Abstract:** Vitamins and minerals play an important role in regulation and maintenance of important biological functions. They contribute to numerous cellular functions, including energy metabolism pathways, DNA synthesis, oxygen transport, and central nervous system functions. Their deficiency leads to adverse health effects; therefore, diet that is rich in essential vitamins and minerals is important to be included in daily routine in order to promote health and protect from various diseases. Vitamins are categorized into water-soluble and fat-soluble vitamins, while minerals are categorized into macrominerals, microminerals, and ultratrace minerals. This chapter focuses on the chemistry, sources, and biological importance of vitamins and minerals.

## 1 Introduction

Vitamins and minerals are vital micronutrients for the maintenance of numerous physiological functions in the human body [1]. The human body is unable to synthesize these micronutrients; thus, they are dependent on other sources such as diet and supplements to fulfill their needs [1, 2]. The significance of vitamins and minerals on human health was identified almost 100 years ago, and their role is every bit still relevant today. The exact amount of micronutrient required for human nutrition is still debatable, and for these purposes, various research and recommendation guidelines are used [3].

Vitamins are organic in nature; each vitamin has a unique function and role in the human body. Usually, it participates in the enzymatic system pathway (cofactor/coenzyme) and catalyzes the biochemical reactions. Vitamins play a vital role in the metabolism of carbohydrates, proteins, and lipids. The human body needs vitamins for growth, development, and reproduction in trace amount [4]. Based on their chemical nature and solubility, they are classified into water-soluble vitamins (vitamin B complex and vitamin C) and fat-soluble vitamins (vitamins A, D, E, and K) [5]. The link

*Corresponding author: Arif Ali, Postgraduate Program in Pharmacology, Federal University of Ceará, Fortaleza, Ceará, Brazil, e-mail: arifpak@alu.ufc.br
**Mac Dionys Rodrigues da Costa, Emanuel Paula Magalhães**, Postgraduate Program in Pharmaceutical Sciences, Federal University of Ceará, Fortaleza, Ceará, Brazil
**Alice Maria Costa Martins**, Department of Clinical and Toxicological Analysis, Federal University of Ceará, Fortaleza, Ceará, Brazil

https://doi.org/10.1515/9783111317601-004

of vitamins and health regulation falls into three categories of evidence: (a) metabolic dysfunctions due to nutritional deficiency, which can be corrected with inclusion of required vitamin as each vitamin has a defined role; (b) the second link is based on epidemiological evidence (analyzing dietary style and identifying disease); (c) the other evidence is provided by randomized controlled clinical trials, which show direct observation/impact of a vitamin/nutrient on specific set of population [6]. The deficiency of vitamins occurs either due to primary causes (poor diet/vitamin-deficient nutrition) or secondary causes (disease, alcohol use, and smoking) [7].

Minerals, on the other hand, are essential inorganic elements, which also play a significant role in human development and growth. They are broadly classified into macrominerals (calcium, phosphorus, sodium, and chloride) and microminerals (iron, copper, cobalt, potassium, magnesium, iodine, zinc, manganese, molybdenum, fluoride, chromium, selenium, and sulfur) [8]. These minerals are involved in numerous biological processes, including energy storage, inflammatory pathways, metabolic processes, oxygen transport, regulation of cardiovascular functions, bone health, and immune system modulations [9]. The potential role of minerals in health regulation carries immense importance. They are present in the form of salts inside the human body and comprise around 4–6% of human body weight. Moreover, they provide strength to bones and structure to body, and contribute majorly to metabolic processes (electrolytes, enzyme cofactors, and as constituents of other organic molecules) [10].

Many vitamins and minerals coordinate with each other and synergize their function. The synergizing effects need to be analyzed, which may assist in combating diseases such as cardiometabolic diseases and osteoporosis [11]. For example, a study evaluated the synergistic effects of magnesium and vitamin $B_6$ for the relief of anxiety-related premenstrual symptoms, which demonstrates modest synergistic effect [12]. Both these micronutrients have immense importance in regulating human health, as extensively evidenced from the available literature [13].

# 2 Water-soluble vitamins

The water-soluble vitamins are organic molecules having immense importance in growth, development and physiological functions [14]. They include a broad range of compounds with individualized characteristics. Most of these vitamins exist in nature in the form of two or more vitamins having same biological properties but different physiochemical characteristics [15]. The water-soluble vitamins comprise vitamin C and vitamin B complex. Vitamin B complex includes thiamin, riboflavin, niacin, pantothenic acid, pyridoxine, folate, and cobalamin. The major sources of vitamin B complex and vitamin C are vegetables, fruits, meat, legumes, cereals, and eggs [16].

## 2.1 Vitamin C

Vitamin C, also known as ascorbic acid, acts as cofactors for 15 million enzymes because of its electron-donating characteristic [17]. Vitamin C is a highly polar and water-soluble molecule. It exists in "ascorbate anion" form, due to its acidic nature of two attached –OH groups, causing ionization at physiologic pH [18]. It also contains a lactone ring along with electron-rich 2-en-2,3-one moiety (Figure 1), which contributes both to its antioxidant and physiochemical properties [19]. Humans are unable to synthesize vitamin C because of genetic mutation in gene for gluconolactone oxidase. Thus, it must be obtained from the diet to fulfill the normal physiological processes in human body [20, 21].

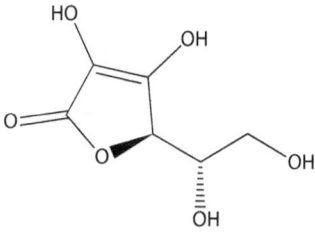

**Figure 1:** Chemical structure of ascorbic acid.

Fruits and vegetables are plentiful dietary sources of vitamin C. The amount of vitamin C in each food varies; however, in some species, the content reaches to unit of grams per 100 g of weight. The high amount of vitamin C in food may be related to the fact that its main precursor is sugar, which is a common compound available in most species. Almost all plant species have been reported to synthesize vitamin C [22]. Edible plants that are rich sources of vitamin C include citrus fruits, peppers, strawberries, brussels sprouts, guava, potatoes, tomatoes, kiwifruits, and green leafy vegetables (broccoli, cauliflower, and cabbage) [23, 24]. It is estimated that 90% of the population fulfill their daily intake of vitamin C requirement from vegetables and fruit sources. The type of processing, storage conditions, and preparation of food affect the amount of vitamin C. For example, five to nine servings of fresh or frozen fruits and vegetables (minimally processed) contain around 200 mg of vitamin C [25]. The amount of vitamin C from the animal source is comparatively low and it is insufficient for daily requirement. However, animal livestock liver and fresh eggs are the rich sources of vitamin C [26].

### 2.1.1 Biological importance

Vitamin C is one of the important nutrient components linked with significant physiological/biological functions in the human body. It has known biological activities, the most important of which is the prevention of the pathogenesis of several diseases due to its strong antioxidant role [27, 28]. Vitamin C acts both as a reducing agent and a donor antioxidant by forming ascorbate radical and dehydroascorbic acid (DHA). The ascorbate form produces hydrogen peroxide ($H_2O_2$) as a result of pH-dependent autoxidation. This role of ascorbate is being exploited in treatment of tumors acting as a prodrug for generation of $H_2O_2$ to kill tumor cells [29]. The ascorbate form also acts as a cofactor for various types of enzymes, for instance, hydroxylases [30].

Vitamin C induces its anticancer effects by modulating tumor microenvironment. It regulates the immune cells, pro-oxidation progression, and epigenetic activity in tumor microenvironment [31]. At normal concentration, vitamin C acts as an antioxidant; however, at higher concentration, it induces oxidative stress and apoptosis in cells. Vitamin C binds with labile iron in tumor cell microenvironment generating $H_2O_2$, which further results in production of hydroxyl radical and reactive oxygen species (ROS) causing cell death of tumor cells [32, 33]. Targeting the iron in tumor microenvironment via vitamin C is a novel strategy to treat cancerous diseases [31]. In other studies, high dose of vitamin C intake in cancer patients has shown an immunomodulatory effect by enhancing the action of natural killer (NK) cells. Moreover, it was also found that the intake of high dose of vitamin C was associated with increased activity of B- and T-cells in chemically exposed cancer patients [34, 35].

In addition to the above biological activities, vitamin C also possesses the antimicrobial activity. Various studies have shown vitamin C combat or prevent different bacterial infections. It prevents gastric diseases (such as gastritis, duodenal ulcer, and carcinoma) by effectively inhibiting *Helicobacter pylori* growth in microaerophilic setting [36, 37]. It has also shown inhibitory effects against other bacterial strains such as *Salmonella enterica*, *Salmonella enterica* [38, 39], and *Campylobacter jejuni* [40]. Moreover, vitamin C, specifically in the form of DHA, also possesses antiviral activity against herpes simplex type 1, polio virus, and rabies virus [41]. It also has antiparasitic activity, for example, against *Trypanosoma cruzi*, *Plasmodium yoelii* 17XL, and *Plasmodium berghei* [42, 43]. The antifungal activity of vitamin C was reported against *Candida albicans* [44].

The role of vitamin C has been evaluated in prevention and treatment of various cardiovascular diseases (CVDs) such as coronary heart disease, heart failure, hypertension, and cerebrovascular diseases. Vitamin C prevents CVDs by potentiating the synthesis of nitric oxide (NO), which helps in maintaining blood vessel tone, reduce vasodilation, and reduce blood pressure [45–47]. Moreover, vitamin C induces antioxidant effects, thereby preventing low-density lipoprotein (LDL) oxidation resulting in decreased risk of CVDs [48]. Vitamin C also minimizes the monocyte adhesion with endothelium, which results in lower risk of atheroma's ultimately preventing atherosclerosis [49]. Vitamin C reduces the incidence of CVDs via multiple mechanisms, each one contributing to the beneficial effect on health in general.

Vitamin C is present in the brain in highly concentrated amounts regulating important neuronal functions and central nervous system (CNS) homeostasis. It possesses the neuroprotective effects against various neurodegenerative diseases such as Alzheimer's disease (AD), multiple sclerosis, and amyotrophic lateral sclerosis. It induces its neuroprotective effects via improving neurogenesis, synaptic plasticity, and biochemical functioning [50]. Similarly, vitamin C also showed a nephroprotective effect against various nephrotoxic agents, including cisplatin [51, 52], colistin [53], and amikacin [54]. It also has nephroprotective effects in renal ischemia/reperfusion-induced acute kidney injury [55].

### 2.1.2 Impact of ascorbic acid deficiency

The severe deficiency of vitamin C may lead to scurvy disease, resulting in reduced osteoblastic activity and dysfunction in collagen synthesis. These manifestations result in increased damage to blood vessels accompanied by hemorrhagic phenomena [56]. The recent data shows high prevalence of vitamin C deficiency, specifically in under-developed countries along with particular groups of population in developed countries. The main factor responsible for vitamin C deficiency is dietary style, geographic region, climate, staple food (grains), and minimal use of supplements. Similarly, the health of an individual also plays an important role; for example, body weight, pregnancy, smoking, and chronic disease affect the status of vitamin C in the body [57].

## 2.2 Vitamin B$_1$

Vitamin B$_1$ (thiamine) consists of two heterocyclic rings bridged by methylene. The rings attached are substituted pyrimidine and substituted thiazole (Figure 2). Vitamin B$_1$ mostly occurs in free form in living organisms. It has four phosphorylated derivatives, including thiamine monophosphate, thiamine diphosphate (TDP), thiamine triphosphate, and adenosine thiamine triphosphate. Among these TDP is the main derivative, which functions as a cofactor for various enzymes such as pyruvate dehydrogenase, transketolase, and pyruvate decarboxylase [58].

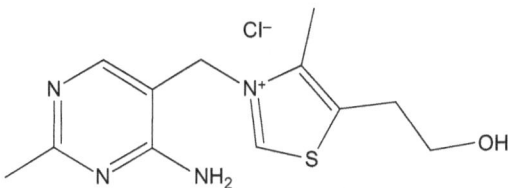

**Figure 2:** Chemical structure of thiamine.

All animals, including humans, are unable to synthesize vitamin $B_1$; only plants, fungi, and some bacteria have this ability [59, 60]. Therefore, humans are dependent on diet and supplements to continuously obtain this essential vitamin. Pulses, whole grains, bread, fish, and meat are some of the common sources of vitamin $B_1$ [61]. It is a highly water-soluble vitamin; thus, significant amount of vitamin $B_1$ may be lost during cooking processes. Similarly, pasteurization of milk also reduces vitamin $B_1$ content by up to 20%. Vitamin $B_1$ is commonly available in the form of supplements to promote health and wellness in special population, such as the baby milk formulas (added with vitamin $B_1$) and multivitamins [62].

### 2.2.1 Biological importance

Vitamin $B_1$ plays a significant role in energy generation from glucose. Because of this role, it is particularly important for proper regulation of brain function and metabolism [63]. The continuous supply of energy to the brain prevents premature aging. It helps in maintaining biochemical energy processes such as pentose phosphate pathway, glycolysis, and Krebs cycle. These energy processes ultimately generate adenosine triphosphate (ATP) or nicotinamide adenine dinucleotide phosphate, which are essential for the constant function of nerves and other cellular developments [64, 65]. It is also indirectly involved in nucleic acid, neurotransmitter, and myelin synthesis, where myelin is essential for nerve conduction velocity [64]. Vitamin $B_1$ is also involved in membrane (axoplasmic, mitochondrial, and synapse membranes) structure and function regulation, where it prevents agent-induced cytotoxicity and adjusts membrane sites. Moreover, it plays a significant role in synaptic transmission and neuron growth [66].

### 2.2.2 Impact of thiamine deficiency

The two major clinical manifestations of vitamin $B_1$ deficiency include CVDs and nervous diseases, referred to as "wet beriberi" and "dry beriberi or Wernicke–Korsakoff syndrome," respectively. Wet beriberi refers to prominent myocardial disease, causing an increase in cardiac output leading to peripheral vasodilation accompanied by warm extremities. To recompensate these complications, tachycardia, increased pulse pressure, sweating, and retention of salt and water occur in the kidneys; however, when the body does not recompensate, it ultimately advances to heat failure. The dry beriberi symptoms appear bilaterally and symmetrically, principally in lower extremities together with burning in toes, muscle cramps in calves, and plantar dysesthesia. Moreover, if the deficiency is not fulfilled, it further leads to loss of knee jerk reflex and atrophy of calf and thigh muscles [66]. The main physiological function of vitamin $B_1$ is the regulation of metabolism, where its deficiency may result in failure of the host body to produce ATP and neurotransmitters. The CNS and heart are affected the most as both are highly dependent on ATP generation via oxidative decarboxylation [67].

There are multiple causes of vitamin $B_1$ deficiency disorders, including famine, dependency on staple food, diet containing low vitamin $B_1$, and food preparation processes (milling grains, heating, and washing milled rice). To avoid vitamin $B_1$ deficiency, there is a crucial need to identify exact cause that puts the body at increased risk of deficiency. Similarly, scheduling awareness programs on a regular basis among the masses to educate them about the importance of vitamin $B_1$ can also reduce the risk of vitamin $B_1$ deficiency [68].

## 2.3 Vitamin $B_2$

Riboflavin is known as vitamin $B_2$, lactoflavin, ovoflavin, and hepatoflavin based on its source. The flavin refers to a chemical entity comprising isoalloxazine ring with two methyl groups attached at position numbers 7 and 8 and another substituent at position 10. The isoalloxazine ring gives redox characteristics to the whole molecule (Figure 3). Vitamin $B_2$ is found in free form in nature as 5'-phosphate or "flavin mononucleotide (FMN) and flavin adenine dinucleotide (FAD)." Both FMN and FAD have no roles as a precursor for nucleotide synthesis; the term just exists in their name and is often misleading [69].

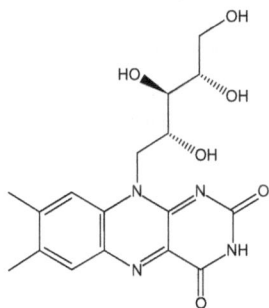

**Figure 3:** Chemical structure of riboflavin.

Vitamin $B_2$ is found in a number of natural food sources, including milk, calf liver, eggs, fish, dry fruits, legumes, integral rice, dark green leafy vegetables, mushrooms, and dietary products (cheese and yogurt) [70]. The contents of vitamin $B_2$ in food may not be affected by heating and cooking as it is heat stable, but exposure to light may impact its content due to its photosensitive nature [71]. Vertebrates including humans poorly stored vitamin $B_2$; therefore, they are mainly dependent on healthy diet and supplements to regularly acquire the essential amount of the vitamin.

### 2.3.1 Biological importance

Vitamin $B_2$ has protective and therapeutic effects in numerous diseases, including ischemia/reperfusion-induced injuries, that is, cerebral or renal ischemia [72–75], malarial disease [76], cancer [77], diabetes [78], migraine [79], cataract [80], premenstrual syndrome [81], osteoporosis [82], neuropathy [83], CVDs [84], and anemia [85]. Vitamin $B_2$ has antioxidant properties, which are thought to be mainly responsible for its protective and therapeutic characteristics [86]. The accumulative evidence shows that oxidative stress is the principal pathological initiator of numerous diseases such as renal and cerebral ischemia, CVDs, and diabetes [87, 88]. Vitamin $B_2$ induces its antioxidant effect by increasing the synthesis of extracellular matrix, which results in decreased ROS generation. Similarly, it enhances the level of antioxidant enzymes such as superoxide dismutase (SOD), catalase, and glutathione peroxidase during pathological states [86]. Moreover, it also regulates the conversion of oxidized glutathione to the reduced form of glutathione, which then serves as the main antioxidant in various pathways [89].

### 2.3.2 Impact of riboflavin deficiency

Vitamin $B_2$ deficiency can cause impairment of normal body functioning and may lead to multisystem disorders in humans, for example,, neuromuscular disorders, anemia, abnormal fetal development, migraines, growth retardation, and various CVDs. Deficiency during pregnancy is linked with increased risk of congenital heart diseases in the newborn. Lactating mothers, infants who are on phototherapy, and individuals who have diseases like celiac disease or alcoholics or taking antipsychotics, antimalarials, and cancer drugs are at increased risk of vitamin $B_2$ deficiency [90].

## 2.4 Vitamin $B_3$

Two terms are used for the niacin, that is, nicotinic acid (NA) and nicotinamide (Figure 4) [91]. Both nicotinic and nicotinamide are colorless, stable in dry form, and soluble in water. The NA is zwitterionic in nature and is negatively charged at the carboxylic function at high pH and positively charged at pyridinyl nitrogen at low pH. These properties give amphoteric nature to the NA molecule because it results in salt formation on reaction with acids and bases [92].

Niacin is commonly found both in plants and animals . The plant-based sources of niacin include alfalfa hay, beans, corn, citrus pulp, and cotton seed meal. Its absorption is increased in the presence of alkaline solutions. Niacin in free form can also be found in fish, cattle liver, and meat; however, these sources contain high con-

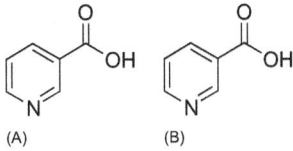

**Figure 4:** Chemical structure of vitamin B$_3$: **(A)** nicotinic acid and **(B)** nicotinamide.

centrations of NAD$^+$ and NADP that act as precursors for niacin following digestion processes [93, 94].

### 2.4.1 Biological importance

Both NA and nicotinamide act as precursors for pyridine coenzymes NAD$^+$ and NADP synthesis. The NAD$^+$ and NADP are involved in important metabolic processes, including metabolism of carbohydrates, lipids, and proteins. Vitamin B$_3$ plays a significant role in energy generation and metabolism, maintaining healthy skin and proper circulation, regulating secretion (stomach fluids and bile), and maintaining neuronal health [95]. NA in large doses has beneficial effects on serum lipid levels, that is, decreasing level of serum cholesterol and triglycerides. The high dose of niacin supplementation may cause a flushy and itchy skin, especially on the face, chest, and arms, which can be avoided with delayed release formulation [96].

### 2.4.2 Impact of niacin deficiency

Vitamin B$_3$ deficiency may lead to pellagra, presenting with 3D symptoms (dermatitis, dementia, and diarrhea) and pretty high rates of mortality. The initial symptoms of its deficiency appear as anxiety, poor concentration, fatigue, and mental problems including depression, dementia, and delirium. Similarly, gastrointestinal symptoms include nausea, vomiting, poor appetite, constipation, frequent diarrhea, and other complications such as cheilosis, and glossitis may appear. The mortality occurs due to unavailability of energy molecules, which are generated due to coenzymatic reactions in the presence of vitamin B$_3$ [97]. Its deficiency may also lead to unavailability of NAD$^+$ during gestational period, which may link with congenital malformation [98].

## 2.5 Vitamin B$_5$

Vitamin B$_5$ (pantothenic acid) exists in two isomeric forms, that is, dextrorotatory (D) and levorotatory (L), among which dextrorotatory is most active biologically. Chemically, vitamin B$_5$ consists of five carbon atoms having carboxylic acid as functional groups (Figure 5). Vitamin B$_5$ in free form has crystalline morphology with white color, and is completely soluble in water and partially soluble or insoluble in organic solvents. Its complex form makes an active part of coenzyme A, playing an important role in several biological activities in the body [99, 100].

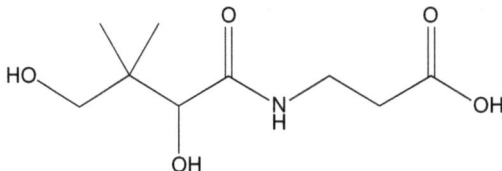

**Figure 5:** Chemical structure of pantothenic acid.

Vitamin $B_5$ is found in large number of foods including meats (abundantly in liver, kidney, brain, and heart), vegetables (potato and broccoli), fruits (avocado), mushrooms, milk, eggs, nuts (almonds and peanuts), whole grains (oatmeal, brown rice, and breads), and legumes (lentils, beans, and chickpeas). Food processing may decrease the content of vitamin $B_5$. It is also produced in the gut by the action of microorganisms, including *Escherichia coli* and *Salmonella typhimurium* [101].

### 2.5.1 Biological importance

Vitamin $B_5$ acts as essential precursors for coenzyme-A and acyl carrier protein (ACP), which further regulates the metabolism of carbohydrates, lipids, proteins, and nucleic acids [102]. The dietary supplementation of vitamin $B_5$ has been found to be effective in healthy adults having acne lesions. It reduced the inflammatory blemishes attached with acne along with good tolerability and safety profile. Vitamin $B_5$ ameliorates acne due to its antibacterial and skin softening activities, as it is converted to coenzyme A, which ultimately regulates the epidermal barrier function via proliferation and differentiation of keratinocytes [100, 103]. The *in vivo* effects of vitamin $B_5$ are linked to its coenzymatic role in forming coenzyme-A and ACP. These enzymes are involved in the synthesis of numerous secondary metabolites, including polyisoprenoid and acetylated compounds, steroid molecules, acetylated neurotransmitters, and prostaglandins. The main role of vitamin $B_5$ is associated with biochemical functions that require coenzyme-A as substrate and co-substrate [104].

### 2.5.2 Impact of pantothenic acid deficiency

The deficiency states of vitamin $B_5$ may result in clinical symptoms such as insomnia, numbness, burning of feet and hands, impaired immune response, fatigue, decreased eosinopenia response, and increased sensitivity of insulin resulting in hypoglycemia [105]. The deficiency of vitamin $B_5$ in humans is rare, except in people with malnutrition, as normally it is 6 weeks in severe depletion of the vitamin. The main reason of rare deficiency is its high abundance in the food sources, both in meats and vegetables. The deficiency usually arises in cases of hunger strikes or other critical salutations [104].

## 2.6 Vitamin $B_6$

Vitamin $B_6$ consists of a family of vitamin species having a scaffold of 2-methylpyridine, 3-hydroxypyridine, and 5-hydroxymethylpyridine. Each form of vitamin $B_6$ differs from each other based on its C-4 substituent, which is either $-CHO$, $-CH_2OH$, or $-CH_2NH_2$ termed as pyridoxal, pyridoxine, and pyridoxamine, respectively (Figure 6). Moreover, these forms also have the ability to exist as phosphate esters [106, 107]. Pyridoxal and pyridoxamine can be transformed into aldehyde forms, playing a significant role as coenzymes in numerous biochemical functions such as amino acid metabolism, glycogen breakdown, and synthesis of neurotransmitters. The phosphate forms are photosensitive and often oxidized when exposed to light [107].

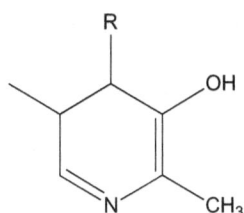

**Figure 6:** Chemical structure of vitamin $B_6$. R = $CH_2OH$ (pyridoxine); R = CHO (pyridoxal); R = $CH_2NH_2$ (pyridoxamine).

The common dietary sources of vitamin $B_6$ include fortified cereals, meats (liver, beef, poultry, and fish), potatoes, and fruits (banana and watermelons). The processing of food decreases the content of vitamin $B_6$ on average 30–40%; however, pyridoxine is more heat resistant compared to its other forms. The animal-derived sources contain vitamin $B_6$ in the form of pyridoxal 5'-phosphate (PLP) and pyridoxamine phosphate, while the plant-derived sources contain vitamin $B_6$ in the form of pyridoxine 5'-phosphate and 4'-$O$-($\beta$-D-glycopyranosyl) pyridoxine [108].

### 2.6.1 Biological importance

The active form of vitamin $B_6$ (PLP) acts as a cofactor in more than 150 biochemical enzymatic reactions such as transamination, $\alpha$-decarboxylation, oxidoreductions, and replacement reaction [109]. It plays an important role in fatty acid metabolism by catalyzing the synthesis of polyunsaturated fatty acids. Similarly, it acts as a cofactor of enzyme PLP-dependent glycogen phosphorylase during glycogen degradation [110]. Vitamin $B_6$ regulates most of the metabolic processes either directly or indirectly, including amino acid metabolism. It also plays a role in the metabolism of one-carbon compound (i.e., 5-methyltetrahydrofolate and $S$-adenosylmethionine), which may help in methionine recycle process and balance homocysteine concentration, thereby contributing the synthesis of purine and thymidylate. The purine nucleotide and thymidylate turn as building blocks for the synthesis and repair of DNA [111]. Vitamin $B_6$ antagonizes the harmful endogenous reactive intermediates (ROS, reactive carbonyl species, and transition metal ions), which act as the main pathological mediators in various diseases such as diabetes, CVDs, and neurodegenerative disorders. Chemical features of vitamin $B_6$ allow it to act as a potent antioxidant, metal chelator, and free radical scavenger to counteract the pathological response in these kinds of diseases [112].

### 2.6.2 Impact of pyridoxine deficiency

Vitamin $B_6$ acts as a cofactor in numerous enzymatic reactions, and its deficiency affects these reactions, resulting in drastic clinical crisis such as dermatitis, anemia, nervous system dysfunction, stomatitis, and glossitis [111, 112]. The malabsorption, irritable bowel disease, gastric intestinal disturbances, kidney failure, pregnancy, and drugs (isoniazid, cycloserine, and penicillamine) are the main factors responsible for the deficiency of vitamin $B_6$. The elderly patients are at greater risk toward vitamin $B_6$ deficiency because of reduced food intake, impaired absorption, and phosphorylation [113].

## 2.7 Vitamin $B_9$

Vitamin $B_9$ (folic acid) is found as an orange-yellow color compound. It is odorless, completely soluble in water, and partially soluble in organic solvents like methanol and ethanol [114]. Vitamin $B_9$ consists of three moieties: pteridine ring, *para*-aminobenzoic acid (PABA), and glutamic acid. The pteridine ring and PABA are linked to each other via a methylene bridge, while PABA is bridged by a peptide bond with glutamic acid to complete the structure of folic acid [115] (Figure 7).

Vitamin $B_9$ is most commonly found in large green leafy plants, for example, spinach, turnip, mustard, lettuce, and even grass. Similarly, asparagus brussels sprouts, beans, and peas are the other common and abundant sources of vitamin $B_9$ [116]. In some countries, various foods like flour, energy bars, drinks, and cereals are fortified with vitamin $B_9$ to fulfill the daily requirement as it plays an essential role in life-dependent processes [117].

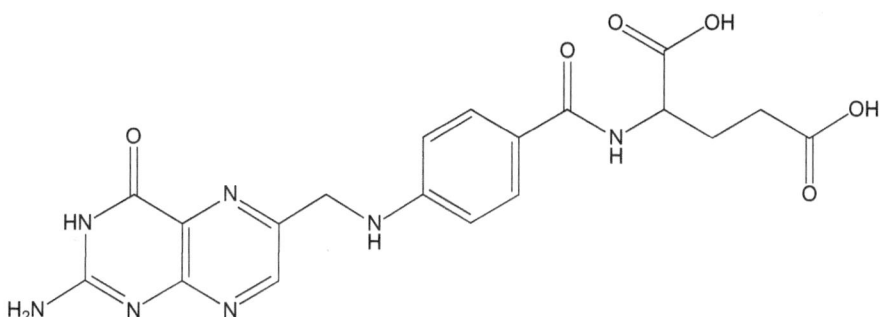

**Figure 7:** Chemical structure of folic acid.

### 2.7.1 Biological importance

Vitamin $B_9$ is essential for numerous physiological processes in the human body such as during synthesis of nucleotides, metabolism of homocysteine, and in regulation of DNA. Amino acids such as glycine, serine, and histidine also required vitamin $B_9$ for their metabolism [118]. Vitamin $B_9$ plays an important role in the synthesis of deoxythymidine monophosphate from its precursor deoxyuridine monophosphate (dUMP). In case of vitamin $B_9$ deficiency, the dUMP remains accumulated and leads to insertion of uracil in DNA in place of thymine. The disproportionate uracil insertion in DNA causes point mutation, chromosome splintering, micronucleus formation, and single- or double-strand breakage [119, 120]. It also plays a significant role in the synthesis of red blood cells (RBCs), by promoting the synthesis of heme, regulation of gene expression, and DNA synthesis [121]. Vitamin $B_9$ has beneficial effects on endothelial function (antiatherosclerotic effect, inhibiting lipid peroxidation and stimulating NO production) by decreasing the risk of CVDs [122, 123].

### 2.7.2 Impact of folate deficiency

Deficiency of vitamin $B_9$ occurs mostly due to poor eating habits, increased turnover during pregnancy and lactation, gastrointestinal malabsorption, and in some diseases like psoriasis and other blood disorders. Vitamin $B_9$ deficiency is also caused by drugs such as anticonvulsants, folate antagonists, and oral contraceptives. Its deficiency may lead to the increased concentration of homocysteine in plasma, which possesses a major risk for various CVDs [124]. Homocysteinemia, neural tube defects, anemia, mental disorders, and cancer are health issues associated with folate deficiency [125].

## 2.8 Vitamin B$_{12}$

Vitamin B$_{12}$ has a complex structure consisting of corrin ring with a central cobalt atom. At the right angle of corrin ring is 5,6-dimethylbenzimidazole having two nitrogen atoms, one bridged with cobalt atom while the other with ribose (Figure 8). Vitamin B$_{12}$ has various forms based on the chemical attachment of different groups, including cyanide, hydroxy, deoxyadenosyl, and methyl attached with the central cobalt atom, which are termed as cyanocobalamin, hydroxocobalamin, adenosyl-cobalamin, and methylcobalamin, respectively [126–128]. It is a light-sensitive vitamin and may convert from one form to another on exposure to light. It is freely soluble in water that makes a stable solution at normal temperature [126, 128].

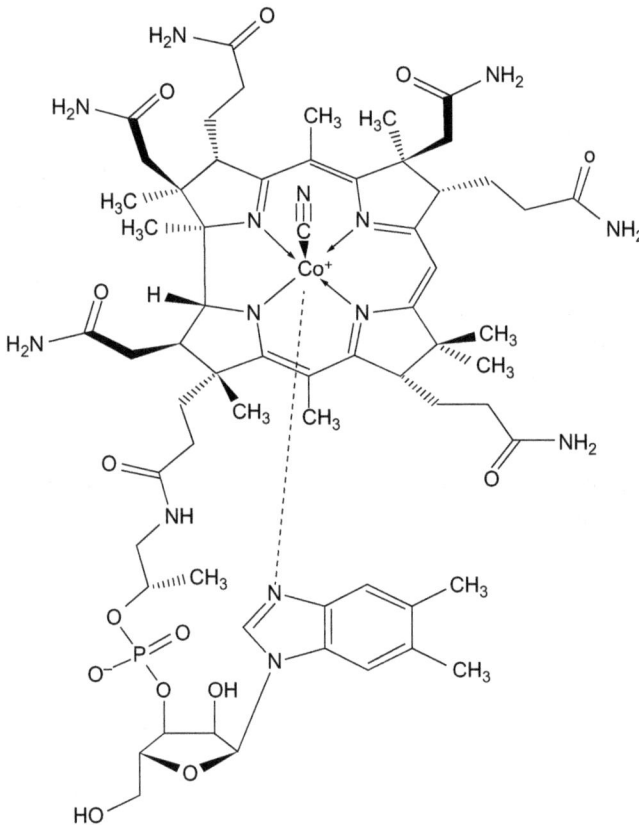

**Figure 8:** Chemical structure of cobalamin.

Vitamin B$_{12}$ is commonly obtained mainly from animal products including meats, eggs, fish, and dairy products, as animals depend on its synthesis via action of intestinal flora [129, 130]. Among herbal sources, only leguminous plants may contain some amounts of vitamin $B_{12}$ because of their symbiotic association with microorganisms. In some countries, the food is often fortified with vitamin $B_{12}$, for example, in the USA, ready-to-eat cereals are fortified with vitamin $B_{12}$ in addition to other vitamins. Some species of wild mushrooms were also found to contain vitamin $B_{12}$, for example, *Boletus* sp., *Macrolepiota procera*, *Pleurotus ostreatus*, and *Morchella conica*. The edible algae (dried green laver and purple laver) also serve as dietary sources of vitamin $B_{12}$ [131].

### 2.8.1 Biological importance

Vitamin $B_{12}$ has a diverse role in the human body. It plays an important role as a cofactor in numerous biochemical reactions. The enzymes in which it is involved as a cofactor are grouped into three subfamilies, including adenosylcobalamin-dependent isomerases, methylcobalamin-dependent methyltransferases, and $B_{12}$-dependent reductive dehalogenases. These enzymes regulate isomerization reactions via radical catalysis [132]. Vitamin $B_{12}$ as a cofactor regulates methionine synthase, which is essential for the synthesis of nucleotides (purines and pyrimidines) [133]. Vitamin $B_{12}$ is essential for erythropoiesis, nerve myelination, and regulation of physiological functions of the CNS [134].

### 2.8.2 Impact of cobalamin deficiency

Vitamin $B_{12}$ deficiency has major adverse outcomes on health, among which megaloblastic anemia is the most prevalent. Its deficiency leads to demyelination of numerous neurons, specifically in thoracic, cervical lateral columns of spinal cord, and in white matter of brain regions resulting in neurological dysfunctions. Other health-related conditions that are linked with vitamin $B_{12}$ deficiency include glossitis, gastrointestinal disturbance, infertility, thrombosis, and hyperpigmentation. These conditions can be reversed by supplementing the body with vitamin $B_{12}$ [135]. The special population like elder individuals, children, infants, and women (especially during pregnancy) are at increased risk of its deficiency. The main factors responsible for deficiency of vitamin may include poor diet, malabsorption, metabolism disorders, and impaired cellular uptake of vitamin $B_{12}$ [136].

# 3 Fat-soluble vitamins

Liposoluble vitamins (A, D, E, and K) and their precursors (Figure 9) are obtained from dietary fats. During the digestion process, these vitamins are emulsified with bile salts and pancreatic enzymes, followed by the formation of micelles consisting of fat-soluble vitamins, cholesterol, phospholipids, and fatty acids, and absorbed through intestinal enterocytes via passive diffusion or specific transporters [137]. In the intestinal cells, micelles are hydrolyzed, re-esterified, and incorporated into lipoproteins, mainly chylomicrons and very LDLs (VLDLs); or linked to plasmatic proteins, as albumin and transthyretin; then released into blood circulation. It is then followed by lipoprotein lipase enzyme action, which releases fat-soluble vitamins in the liver, adipose tissue, or target tissues [138].

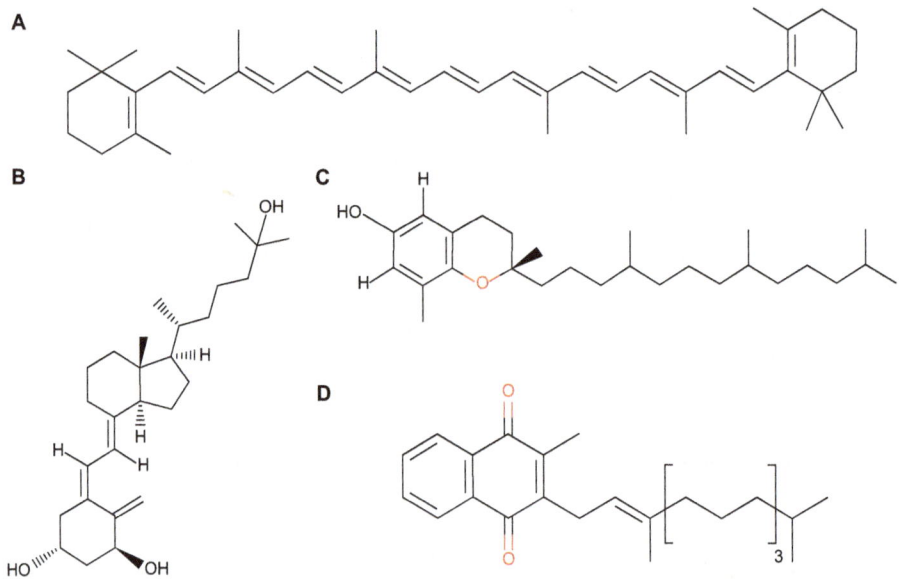

**Figure 9:** Chemical structures of main fat-soluble vitamins: **(A)** β-carotene, **(B)** 1,25-dihydroxycholecalciferol, **(C)** α-tocopherol, and **(D)** phylloquinone – vitamin $K_1$.

## 3.1 Vitamin A (retinol, retinal, retinoic acid, retinyl esters, and carotenoids)

Vitamin A, also termed β-carotenoid or retinoid, is a precursor of retinol. Its common sources include liver, vegetables, green and yellow fruits, dairy products, kidneys, eggs, carrots, egg yolk, fish liver oils, green leafy vegetables, kale, mangoes, red peppers, squash, spinach, and sweet potatoes. Their recommended daily intake is 700–900 μg or 2,333–3,000 IU [137, 139, 140]. Other precursor compounds, called as provitamin A (lycopene and lutein), are also obtained from plant sources [138, 141, 142]. Vitamin A and its precursors are mainly stored in the liver as retinyl esters, which are then released to the target tissues (epithelial tissue). It promotes differentiation of retinal cells, corneal wound healing, and maintains the production of substances that are related to visual function [137, 143]. Moreover, epithelial growth stimuli and collagen production promoted by vitamin A are useful for development of pharmaceutical preparations in treatment of skin illness (dermatology lesions, acne, and psoriasis), melasma, and anti-aging products [144–147].

Furthermore, vitamin A modulates the immune system through maintenance of epidermal and mucosal integrity (respiratory, digestive, and urinary systems), and stimulates mucosal secretion, thereby preventing pathogen adhesion and invasion [148]. Vitamin A influences the response of antibodies via immune cells (T-cells and NK cells) and regulates inflammatory cytokine (IL-1, IL-2, IL-6, IL-17, TNF-alpha, and IFN-gamma) levels [149–152]. Such effects of vitamin A on the immune system were employed in the development of a nasal spray for treatment of allergic rhinitis, reducing inflammatory infiltrate on nasal mucosa [153].

The antioxidant action of vitamin A and E supplementation is widely explored on evaluation of lipid disorders in children and adolescents, which showed that an increase in ingestion of these compounds may reduce cholesterol, VLDL, and triglyceride levels [154, 155]. Moreover, it is also associated with attenuation of hyperthyroidism symptoms [156].

## 3.2 Vitamin D (1,25-dihydroxycholecalciferol)

Vitamin D is also called as "sunshine vitamin." It represents mainly two forms of fat-soluble sterols: cholecalciferol (vitamin $D_3$), synthesized from a cholesterol precursor by ultraviolet activity, and ergocalciferol (vitamin $D_2$), obtained from yeast and plant sterols. However, these prohormones require two-hydroxylation reactions to form the active form of vitamin D [137]. Adequate levels of vitamin D (15–20 μg or 600–800 IU) can be obtained from multiple sources, including fish (salmon and trout), fish liver oil, eggs, cereals fortified with vitamin D, margarine, mushrooms, aquatic mammal liver, fortified dairy (milk and cheese), and grains [138–140].

The main effects of vitamin D are related to bone mineralization, calcium and phosphate balance, and immune response. Vitamin D influences the differentiation of osteoblast and osteoclasts, regulating bone mineralization. Moreover, the differentiation of intestinal cells and calcium-phosphate intestinal absorption is also regulated by its sterol isoforms. The intake of vitamin D may prevent osteoporosis, promote bone strength, and reduce the incidence of fractures in children and elderly subjects who may or may not perform physical exercises [138, 157–161].

The cellular development on thymus gland, cell proliferation and differentiation, and cytokine secretion requires vitamin D stimuli [148, 162, 163]. Some clinical trial findings showed that vitamin D supplementation is linked with reduced number of eosinophil levels while maintaining T-cell and leukocyte levels in cancer patients, acting as an immunomodulating agent, and showing its potential in prevention of auto-immune diseases [164–167].

## 3.3 Vitamin E (α-tocopherol)

Vitamin E (α-tocopherol) is present within the plasmatic membrane and it is commonly found in vegetable oils (olive oil and sunflower oil), avocado, nuts, almonds, fish, and breast milk [137–140]. Vitamin E possesses antioxidant effects by quenching free radicals that result from lipid peroxidation of unsaturated fatty acid. It also regulates platelet aggregation by promoting the release of prostacyclin and NO [168]. It coordinates with other vitamins such as vitamin C and vitamin $B_{12}$ to regulate multiple cellular processes, including metabolism and DNA synthesis, maintaining membrane function and cell homeostasis [138]. Vitamin E has the potential to prevent AD due to its antioxidant effect, as evidenced in a clinical trial. The 12- month vitamin E supplementation resulted in the improvement of memory and mood in treatment group [169, 170]. Moreover, other studies also showed an improvement of depression and anxiety with vitamin E supplementation [171].

Similarly, vitamin E and C co-supplementation in women (aged 15–45 years) with endometriosis reported a decreased level of pelvic pain, dysmenorrhea, and dyspareunia after 8 weeks of treatment, with a significant reduction of oxidative stress markers [172]. Moreover, vitamin E also improves the reproductive and immune functions in animals [148]. Previously, vaginal suppositories of vitamins D and E were used in women with breast cancer who presented tamoxifen-induced vaginal atrophy. After 8 weeks of treatment, women were presented with an increase in vaginal maturation index, and reduced effects on vaginal pH and genitourinary symptoms [173]. The co-supplementation of omega-3 and vitamin E improves antioxidant markers (catalase activity and glutathione levels) and visceral adiposity in obese/overweight women with polycystic ovary syndrome after 8 weeks when compared with placebo group [174, 175]. Vitamin E regulates RBC formation, prevention of platelet aggregation, and suppression of pro-inflammatory cytokines (IL-1, IL-6 and IL-8), which possibly contribute to atherosclerosis prevention [176, 177]. Moreover, oral vitamin C and E supplementation enhanced platelet-rich plasma-fibrin glue for wound healing in non-healing diabetic foot ulcer patients [178].

In another study, it was reported that after 12 months of supplementation with Tocovid 400 mg/day (a tocotrienol-rich vitamin E drug) decreased the neuropathy symptom in diabetes mellitus type 2 (T2DM) patients [179]. Similar effects were observed with the use of vitamin E (800 mg/day) in prevention of taxane-induced neuropathy [180]. It was evidenced that whey supplemented with vitamins E and D preserved muscle mass, increase strength, and improve quality on life in sarcopenic older adults [181]. During submaximal and resistance training, the use of vitamins A and E reduced muscular damage in elite athletes and mitigate gains in visceral adipose tissue [182–184].

## 3.4 Vitamin K (K$_1$, K$_2$, and K$_3$)

The adequate intake value of vitamin K is in the range of 90–120 µg/day, which can be obtained from a variety of sources, including green leafy vegetables (asparagus, beef, broccoli, and spinach) and vegetable oils (soybean, canola, and olive oils), breast milk, eggs, fish, and meat organs such as liver [137–140]. Vitamin K is essential for the activation of coagulation proteins (prothrombin and factors II, VII, IX, and X) and function of some proteins of bone mineralization. The three types of vitamin K include phylloquinone (K1), long-chain menaquinones (K2), and menadione (K3) [137, 138].

Osteocalcin and matrix Gla protein, two vitamin K-dependent proteins, are the main noncollagen proteins present in the bone, which exerts important function in bone mineralization, due to its influence on osteoblast and osteoclast activity [185]. These proteins are related as promising biomarkers for bone calcification in skeletal diseases [186]. Some studies support that supplementation of vitamins K and D enhances the bone mineral density and bone quality via enhancing osteoblast function and inhibiting bone absorption [187, 188]. Despite that, vitamin K-derived compounds are used in the development of new cancer chemosensitizers, due to their regulatory action on oxidative stress, inhibition of P-glycoprotein [189, 190]. In addition, a clinical trial study reported that topical vitamin K application in subjects who underwent electrocautery treatment showed reduction of wound healing time [191].

# 4 Minerals

Minerals are a class of chemical elements essential for the creation and maintenance of life on the Earth. Minerals are defined as "inorganic elements that are usually presented in the form of salts, either in their natural sources or in living beings." More than 20 elements are found in the human body, representing about 5–6% of body mass, which are responsible for various biological functions, such as cell formation and growth, enzymatic cofactor, neurotransmitters, osmotic agents, and pH buffering [192] (Figure 10).

Although minerals are included in the micronutrient category, the group of minerals can still be classified in different ways. One of the most used classifications divides the minerals into three groups, such as macrominerals, microminerals (trace elements), and ultra-trace minerals, according to the need for daily intake of each element. Macrominerals are needed by the human body in greater quantities, that is, '100 mg/day, while microminerals are required by the human body in amounts between 1 and 100 mg/day. Ultratrace elements are required in extremely low or unknown quantities [193].

The Reference Daily Intake (RDI) values established by the FDA for each element are summarized in Table 1. Macrominerals are represented by calcium (Ca), phosphorus (P), magnesium (Mg), sodium (Na), chlorine (Cl), and potassium (K). Microminerals or trace elements include iron (Fe), zinc (Zn), copper (Cu), and manganese (Mn). Meanwhile, the ultratrace minerals are iodide (I), selenium (Se), fluorine (F), molybdenum (Mo), chromium (Cr), arsenic (As), boron (B), nickel (Ni), silicon (Si), and vanadium (V) [194]. These elements can be obtained in different ways; however, they are mostly obtained through the ingestion of solid and liquid foods of animal and plant origin. It is important to highlight that the absorption and distribution of these minerals in the human body depend on some factors, such as the amount ingested, concentration, bioavailability, state of ionization, interactions with other foods, or medications [195].

**Table 1:** Minerals, Reference Daily Intake (RDI), and their food sources.

| Elements | | RDI | Food sources |
|---|---|---|---|
| Macrominerals | Ca | 1,300 mg | Dairy products (main), cabbage, broccoli, fruit, and enriched juices |
| | P | 1,250 mg | Dairy products, meat, fish, eggs, nuts, vegetables, and grains |
| | Mg | 420 mg | Chia and pumpkin seeds, spinach, vegetables, nuts, and whole grains |
| | Na Cl | 2,300 mg | They are usually added in the form of salt during food and beverage production |
| | K | 4,700 mg | Meat, poultry, fish, milk, fruits, potatoes, and nuts |
| Microminerals (trace elements) | Fe | 18 mg | Iron heme: meat and seafood; non-heme iron: nuts, beans, vegetables, and fortified grains |
| | Zn | 11 mg | Meat, fish, seafood, beans, nuts, and whole grains |
| | Cu | 0.9 mg | Liver, meat, shellfish, seeds, nuts, cereals, chocolate, and whole products |
| | Mn | 2.3 mg | Nuts, oysters, mussels, soy, rice, leafy vegetables, coffee, and tea |

**Table 1** (continued)

| Elements | | RDI | Food sources |
|---|---|---|---|
| Ultratrace minerals | I | 150 µg | Seaweed, fish, seafood, eggs, and products added from iodine salts |
| | Se | 55 µg | Brazil nuts, seafood, meat, cereals, grains, and dairy products |
| | F | Unknown | Tea, coffee, cereals, rice, milk, and dairy products |
| | Mo | 45 µg | Beans, cereal grains, leafy vegetables, beef liver, and milk |
| | Cr | 35 µg | Meat, grain, fruits, vegetables, nuts, spices, beer, wine, and grape juice |
| | As | Unknown | Fish and seafood, chicken, meat, rice, and seaweed |
| | B | Unknown | Fruits, tubers, vegetables, apples, wine, cider, and beer |
| | Ni | Unknown | Nuts, vegetables, breads, cereals, meat, and chocolate powder |
| | Si | Unknown | Beer, coffee, grains, and vegetables |
| | V | Unknown | Mushrooms, shellfish, black pepper, parsley, seeds, juices, and processed foods |

RDI values for adults and children ≥4 years according to the Food and Drug Administration (FDA). Ca, calcium; P, phosphorus; Mg, magnesium; Na, sodium; Cl, chlorine; K, potassium; Fe, iron; Zn, zinc; Cu, copper; Mn, manganese; I, iodide; Se, selenium; F, fluorine; Mo, molybdenum; Cr, chromium; As, arsenic; B, boron; Ni, nickel; Si, silicon; V, vanadium.

## 4.1 Macrominerals

### 4.1.1 Calcium

Calcium (Ca) is the major mineral found in the human body. In the form of hydroxyapatite, it is responsible for forming the bone matrix, representing about 99% of its mass. When absorbed in the intestine by active and passive processes, its absorption efficiency is inversely proportional to deficiency. In case of deficiency, active transport is favored by the actions of parathyroid hormone (PTH) and vitamin D, which act by increasing its absorption and decreasing its excretion by the kidneys, maintaining serum levels between 8.6 and 10.3 mg/dL [196, 197]. In addition to its structural function, calcium is involved in vasoconstriction and vasodilation, muscle contraction, neural transmission, and glandular secretion. Hypocalcemic conditions are common in postmenopausal women and individuals with lactose intolerance due to decreased estrogen and milk-derived food intake, predisposing to the occurrence of osteomalacia, osteoporosis, and fracture risk. Renal stones, renal failure, and decreased absorption of other essential minerals, such as Fe, Mg, P, and Zn, are common in cases of hypercalcemia [197]. Recent studies have shown that serum calcium levels have potential as a biomarker in several biological disorders, such as hypocalcaemia, which is known to be associated mainly with preeclampsia, and kidney and lung diseases [198]. Researchers also found an association between low calcium levels and multiple organ injury, septic shock, and higher mortality in COVID-19 [199].

### 4.1.2 Phosphorus

Similar to calcium, the greatest amount of phosphorus (P) in the human body is found in bones and teeth in the form of hydroxyapatite. Commonly this element is found in the form of phosphate ($PO_4^{3-}$), and inorganic phosphate is the best absorbed in the intestine passively and freely excreted in urine under the influence of PTH and vitamin D actions, which maintains its serum levels between 2.5 and 4.5 mg/dL. In addition to its structural bone function, phosphorus participates in the formation of genetic material, membrane phospholipids, kinase-mediated reactions, and physiological pH buffering [196, 200]. Deficiency of phosphorus is common in hyperthyroidism and diabetic ketoacidosis. However, risk groups, such as premature newborns and individuals in severe malnutrition, develop hypophosphatemia associated with muscle weakness, paresthesia, osteomalacia, and confusion. Recently, hyperphosphatemia has been associated with the development of CKDs (chronic kidney diseases) and CVDs [200].

### 4.1.3 Magnesium

Magnesium (Mg) is an extremely important mineral for the human body. In addition to its large participation in bone formation, its role as a cofactor in more than 300 types of enzymes is responsible for protein synthesis, muscle contraction, nerve transmission, glucose control, and blood pressure regulation. Its homeostatic balance is performed by the kidneys with vitamin D participation, maintaining its serum levels in the range of 1.7–2.2 mg/dL and, when in excess, is freely excreted by urine [196, 201]. Cases of hypomagnesemia are rare due to renal control. However, under special conditions, such as gastrointestinal diseases, T2DM, and chronic alcohol use, there may be excessive loss of this mineral, causing various systemic disorders ranging from muscle cramps and migraine to coronary arrhythmias and spasms. Hypermagnesemia, on the other hand, is common in cases of excessive supplementation, where high doses of magnesium cause toxic effects initially on the gastrointestinal system, such as diarrhea, nausea, vomiting, and paralytic ileus, before evolving to disorders of the cardiorespiratory system, which can lead to difficulty in breathing and cardiac arrest [201].

### 4.1.4 Sodium and chloride

Sodium (Na) and chloride (Cl) are the two main elements found in extracellular fluid. Together, they participate in muscle contraction, neural transmission, and maintenance of the hydroelectrolytic balance. While sodium is the main mineral responsible for maintaining blood pressure, chloride, in the form of hydrochloric acid, plays an important role in the digestion of gastric contents. About 98% of the salt ingested is absorbed by the small intestine and, when in excess, is freely excreted in the urine. Given the role of sodium on blood pressure, its serum levels are controlled by the renin–angiotensin–aldosterone system that promotes the resorption of this mineral in the kidneys, keeping them between 135 and 145 mmol/L [196]. Hyponatremia is a rare condition when it is not associated with hemodilution, since high amounts of sodium chloride (salt) are consumed daily in food. However, when present, it can pose serious risks to the nervous and cardiovascular system. Hypernatremia, on the other hand, is a risk condition for the development of hypertension, being associated with CVDs and CKDs [202, 203].

### 4.1.5 Potassium

Potassium (K) is the most abundant cation in the intracellular fluid, being responsible for maintaining intracellular water balance, muscle contraction, and neural transmission. It is passively absorbed, mainly in the small intestine, and excreted via urine, feces, and sweat. Serum levels of potassium are influenced by sodium metabolism, since they commonly perform antagonistic roles and are actively transported by the $Na^+/K^+$ ATPase pump against their concentration gradients [196, 204]. Body's normal functioning is most likely to disturb when potassium plasma levels are not in the healthy range, that is, 3.6–5.0 mmol/L. Hypokalemia in mild form can cause constipation, muscle weakness, and malaise, and in severe form can cause encephalopathy, glucose intolerance, and changes in cardiorespiratory function. Hyperkalemia, on the other hand, is rare and does not possess any major risk in individuals with adequate renal function; however, it can be life-threatening due to cardiac toxicity. Recent studies reinforce that higher potassium intake in healthy subjects could be a protective factor against hypertension, kidney stones, and glucose resistance, though it is alarming in individuals with heart diseases [204].

## 4.2 Microminerals

### 4.2.1 Iron

Iron (Fe) is essential for the formation of various proteins and cytochrome P450 enzymes (CYP450), mainly those responsible for transporting oxygen in the body, such as hemoglobin and myoglobin. Absorbed mainly in the form of heme iron ($Fe^{2+}$), the passage of this mineral from the upper part of the intestine to plasma occurs by the energy-dependent process. Furthermore, the maintenance of iron levels and iron storage protein, that is, ferritin, is essential for the maintenance of the human body, since its deficiency suppresses hematopoiesis, muscle growth, and neural and the production of hormones [196, 205]. The main clinical condition associated with iron deficiency is microcytic and hypochromic anemia, which is associated with fatigue, muscle weakness, impaired cognition, and decreased immune function. Generally, anemia is easily reversed with iron supplementation, except in cases of chronic diseases, cancer, or gastrointestinal disorders (celiac and Crohn's disease). Excess iron is rare and, when present, is usually associated with zinc deficiency and gastrointestinal disorders. However, extremely high doses (20 mg/kg) can cause corrosion in the intestinal wall and lead to great loss of fluids, developing shock [205].

### 4.2.2 Zinc

Zinc (Zn) is an essential mineral for the structure of the human body, since it acts as a catalyst in more than 100 types of enzymatic reactions, in the synthesis of genetic material, in immune function, in signaling, and in cell division. This mineral is absorbed mainly in the jejunum by the transcellular process, then transported in plasma by albumin, and excreted in feces, urine, semen, and sweat. Plasma zinc levels are clinically verified in several pathologies and should remain between 80 and 120 µg/dL in healthy subjects [196, 206]. Disorders associated with zinc deficiency are systemic and can cause skin sores, alopecia, diarrhea, loss of taste, recurrent infections, and cognitive changes. High intake is associated with nausea, dizziness, headache, gastric discomfort, vomiting, and loss of appetite [206].

### 4.2.3 Copper

Copper (Cu) is an essential mineral for various biological processes. It mainly acts as a cofactor of metalloenzymes (cuproenzymes), such as monoamine oxidases, diamine oxidases, ferroxidases, and copper SODs. Ceruloplasmin is the main ferroxidase responsible for transporting 95% of the copper present in the body and plays an important role in iron metabolism. Its absorption occurs mainly in the small intestine by saturable and nonsaturable mechanisms followed by transportation in the intracellu-lar medium by ATPase-type P proteins (ATP7A and ATP7B) that work to maintain plasma levels between 63.5 and 158.9 µg/dL and, when in excess, copper is excreted by the biliary and urinary tracts [196, 207].

Copper deficiency is rare and is more common in genetic diseases, such as Menkes' disease, which produces a mutant and inefficient ATP7A for the transport of copper. Copper deficiency is associated with anemia, hypopigmentation, dyslipide-mias, and bone changes. Excess copper, on the other hand, is associated with liver damage, and gastrointestinal and neurological disorders. The case of Wilson's disease, which consists of the ATP7B mutation, characterized by the accumulation of this min-eral in soft tissues, as the clinical sign known as Kayser–Fleischer rings observed in the eyes [207].

### 4.2.4 Manganese

Manganese (Mn) is known for its participation as a cofactor of enzymes, such as argi-nase, glutamine synthetase, phosphoenolpyruvate decarboxylase, and manganese SOD. For this reason, manganese is essential for the metabolism of carbohydrates, amino acids, cholesterol, and oxyreduction balance. Absorbed in the small intestine by active and passive mechanisms, manganese is transported into the bloodstream bound to plasma proteins and excreted mainly in feces. Its deficiency is rare, but when present, it can cause disorders of the skin, hair, bones, and metabolism. On the other hand, excess manganese is associated with high occupational exposures, where the main toxic effects are observed at the level of the CNS, manifesting in the form of tremors, insomnia, depression, delusions, and mood change [196, 208].

## 4.3 Ultratrace minerals

### 4.3.1 Iodine

Iodine (I) is an essential element for the metabolism and growth of the body, since it is a vital component for the synthesis and release of thyroid hormones, such as thy-roxine (T4) and triiodothyronine (T3), acting on the brain, muscles, heart, pituitary gland, and kidneys. The absorption of iodine occurs almost completely in the small intestine when it is in the form of iodide (I⁻) and, when in the bloodstream, it is stored mainly in the thyroid gland and excreted in urine and feces. The urinary concentra-tion of ‘100 µg/L is the indication of insufficient intake of this element. Its deficiency directly impacts the growth and development of the body, being associated with de-mentia, cretinism, atrophy, and delayed sexual maturation. Goiter is one of the most common clinical signs associated with disorders of this mineral, being observed to be associated with hypo- and hyperthyroidism [196, 209].

### 4.3.2 Selenium

Selenium (Se) is essential for the synthesis of selenoproteins, which play important roles in reproduction, metabolism of hormones T3 and T4, DNA synthesis, and protecting against ROS. Freely absorbed in the intestine in the inorganic (selenate and selenium) and organic (selenomethionine and selenocysteine) forms, these forms are quickly stored in the skeletal muscle and, when in excess, excreted in the urine. Plasma concentrations of ≥8 µg/dL are considered in the normal and healthy range. Its deficiency is associated with oxidative stress, heart problems, osteoarthritis, and iodo metabolism disorder. Excess selenium can cause fragility in nails and hair, as well as severe gastrointestinal and neurological symptoms [196, 210].

### 4.3.3 Fluorine

Fluorine (F), mainly in the form of fluoride ($F^-$), participates in the formation of protection of teeth and bones in the form of fluorohydroxyapatite. Due to its high affinity for calcium, its absorption in the small intestine can be greatly impaired, and generally 50% of what is absorbed accumulates in bone formation and the other 50% is excreted in the urine. Although the concentrations of this mineral are not investigated in clinical practice, the decrease in fluoride intake may be accompanied by the appearance of dental caries. Paradoxically, excess fluoride is associated with dental and skeletal fluorosis, which can cause dental spots and osteoporosis [196, 211].

### 4.3.4 Molybdenum

Molybdenum (Mo) is an essential element for the formation of molybdopterin, which acts as a cofactor of some enzymes such as sulfite oxidase, xanthine oxidase and aldehyde oxidase. In turn, these enzymes are responsible for metabolizing amino acids containing sulfur, sulfite, and heterocyclic compounds, including purines and pyrimidines. Although the mechanisms and sites of absorption are unknown, its transport through the bloodstream occurs in connection with erythrocytes and a macroglobulin, it is believed that its homeostatic regulation is performed by the kidneys. Disturbance associated with deficiency of molybdenum is not reported, except cases of genetic mutations that prevent the production of molybdopterin, which leads to inactivity of the enzyme sulfite oxidase and sulfite accumulation that may cause encephalopathy and death.

High molybdenum concentrations are rare due to rapid renal clearance of the mineral [196, 212].

### 4.3.5 Chromium

Chromium (Cr), in the form of chromium III ($Cr^{3+}$), is present in some foods, and only a small fraction ingested is absorbed by the intestine, stored in organs and soft tissues, and usually, most of what has been absorbed is excreted by the kidneys. Studies suggest that the chromium exerts activity on glucose metabolism by potentiating the actions of insulin. However, the mechanism of action of chromium is unknown. Physiological disorders associated with chromium levels are little reported in the literature, with only a few reports of excess supplementation, which was associated with weight loss, anemia, thrombocytopenia, liver dysfunction, renal failure, rhabdomyolysis, dermatitis, and hypoglycemia [196, 213].

### 4.3.6  Arsenic, boron, nickel, silicon, and vanadium

Arsenic (As), boron (B), nickel (Ni), silicon (Si), and vanadium (V) represent an extremely low portion of minerals in the human body and are little known about their biological functions.

Arsenic is associated with growth, reproduction, and gene expression, since its toxic effects affect virtually all biological systems and is associated with the development of cancers. Its inorganic form is widely absorbed by the intestine, transported to the liver, and biotransformed into arsenite. Much of the absorbed arsenic is rapidly excreted in the urine [196, 214].

Boron is involved in several cellular processes, such as reproduction, growth, immunity, and bone metabolism. Data on its absorption, distribution, metabolization, and excretion still need to be collected. Acute boron poisoning has been reported with gastrointestinal disorders, skin flushing, rash, arousal, seizures, depression, and vascular collapse [196, 215].

The role of nickel is possibly associated with metalloenzymes responsible for oxyreduction reactions and gene expression. In addition to being little absorbed by the intestine, its absorption can be affected by foods such as milk, coffee, tea, orange juice, and ascorbic acid. When in the bloodstream, nickel is transported by albumin and excreted by urine. Its toxic effects associated with adequate diet are unknown, but cases of acute poisoning with high amounts have been associated with nausea, abdominal pain, diarrhea, vomiting, and shortness of breath [196, 216].

Regarding silicon,, recent studies in animal model reinforce its role in bone formation and remodeling in association with vitamin D. Furthermore, silicon supplementation has become popular for strengthening nails, hair, and skin aesthetics. Well absorbed by the intestine, silicon tends to accumulate in some connective tissues such as aorta, bones, tendons, and skin. Its excretion occurs mainly in the urine. There is no evidence on the toxic effects of excessive silicon consumption [196, 217, 218].

Vanadium improves the action of insulin, stimulates cell proliferation and differentiation, and inhibits some ATPase enzymes. A small fraction of the ingested vanadium can be absorbed and accumulated in the liver, kidneys, and bones. Despite few studies on its toxic effects, there are reports of renal toxicity in animals and an association between high serum levels and CKD in children [196, 219].

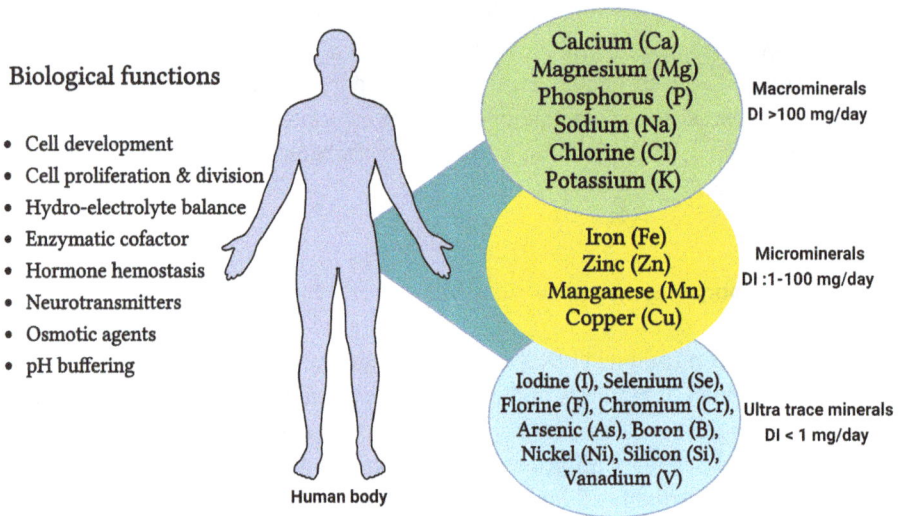

**Biological functions**

- Cell development
- Cell proliferation & division
- Hydro-electrolyte balance
- Enzymatic cofactor
- Hormone hemostasis
- Neurotransmitters
- Osmotic agents
- pH buffering

Human body

Calcium (Ca)
Magnesium (Mg)
Phosphorus (P)
Sodium (Na)
Chlorine (Cl)
Potassium (K)

Macrominerals
DI >100 mg/day

Iron (Fe)
Zinc (Zn)
Manganese (Mn)
Copper (Cu)

Microminerals
DI :1-100 mg/day

Iodine (I), Selenium (Se),
Florine (F), Chromium (Cr),
Arsenic (As), Boron (B),
Nickel (Ni), Silicon (Si),
Vanadium (V)

Ultra trace minerals
DI < 1 mg/day

**Figure 10:** Mineral classes and their biological importance.

# 5 Summary

Vitamins and minerals are essential for maintenance and regulation of normal physiological functions. They affect almost every cellular process in the human body, including the immune system, energy-generating cycles, growth and development, cell metabolism, DNA and RBC synthesis, and brain function. Each organism's vitamin and mineral requirement is different based on their age, sex, and medical conditions. Their deficiency is linked with life's critical conditions and disabilities. Multiple factors are responsible for vitamin and mineral deficiency in an organism such as poor diet, malabsorption, medications, and secondary diseases. All vitamins and minerals have their own role; however, there is a close interplay between these micronutrients signifying that all of them should be available in order to function properly.

# References

[1]    Gernand, A.D., Schulze, K.J., Stewart, C.P., West, K.P. and Christian, P. 2016. Micronutrient deficiencies in pregnancy worldwide: Health effects and prevention. Nature Reviews Endocrinology, 12(5), 274–289.

[2]    Godswill, A.G., Somtochukwu IV, I.A.O. and Kate, E.C. 2020. Health benefits of micronutrients (vitamins and minerals) and their associated deficiency diseases: A systematic review. International Journal of Food Sciences [Internet], 3(1), 1–32. [cited 2023 Feb 28] Available from: https://www.iprjb. org/journals/index.php/IJF/article/view/1024.

[3]    Agostoni, C., Bresson, J.L., Fairweather-Tait, S., Flynn, A., Golly, I., Korhonen, H., et al. 2010. Scientific opinion on principles for deriving and applying dietary reference values. EFSA Journal [Internet], 8(3), 1458. [cited 2023 Feb 28], Available from: https://onlinelibrary.wiley.com/doi/full/10.2903/j. efsa.2010.1458.

[4]    Akram, M., Munir, N., Daniyal, M., Egbuna, C., Găman, M.A., Onyekere, P.F., et al. 2020. Vitamins and minerals: Types, sources and their functions. Functional Foods and Nutraceuticals [Internet], [cited 2023 Feb 28] 149–172. Available from: https://link.springer.com/chapter/10.1007/978-3-030- 42319-3_9.

[5]    Baj, T. and Sieniawska, E. 2017. Vitamins. Pharmacognosy: Fundamentals, Applications and Strategy, 281–292.

[6]    Thomas, D.R. 2004. Vitamins in health and aging. Clinics in Geriatric Medicine [Internet], 20(2), 259–274. [cited 2023 Feb 28] Available from: http://www.geriatric.theclinics.com/article/ S0749069004000187/fulltext.

[7]    Bailey, R.L., Fulgoni, V.L., Keast, D.R. and Dwyer, J.T. 2012. Examination of vitamin intakes among US adults by dietary supplement use. Journal of the Academy of Nutrition and Dietetics, 112(5), 657–663.e4.

[8]    Soetan, K.O., Olaiya, C.O. and Oyewole, O.E. 2010. The importance of mineral elements for humans, domestic animals and plants: A review. African Journal of Food Science [Internet], 4(5), 200–222. [cited 2023 Feb 28] Available from: http://www.academicjournals.org/ajfs.

[9]    Heffernan, S., Horner, K., De Vito, G. and Conway, G. 2019. The role of mineral and trace element supplementation in exercise and athletic performance: A systematic review. Nutrients, 11(3), 696.

[10]   GUPTA, U.C. and Gupta, S.C. 2014. Sources and deficiency diseases of mineral nutrients in human health and nutrition: a review. Pedosphere, 24(1), 13–38.

[11]   Singh, V., Jain, S., Prakash, S. and Thakur, M. 2022. Studies on the synergistic interplay of vitamin D and K for improving bone and cardiovascular health. Current Research in Nutrition and Food Science Journal, 10(3), 840–857.

[12]   De Souza, M.C., Walker, A.F., Robinson, P.A. and Bolland, K. 2000. A synergistic effect of a daily supplement for 1 Month of 200 mg Magnesium plus 50 mg Vitamin B$_6$ for the relief of anxiety- related premenstrual symptoms: A randomized, double-blind, crossover study. Journal of Women's Health and Gender Based Medicine, 9(2), 131–139.

[13]   Zhang, F.F., Barr, S.I., McNulty, H., Li, D. and Blumberg, J.B. 2020. Health effects of vitamin and mineral supplements. British Medical Journal, 369, m2511.

[14]   Yaman, M., Çatak, J., Uğur, H., Gürbüz, M., Belli, İ., Tanyıldız, S.N., et al. 2021. The bioaccessibility of water-soluble vitamins: A review. Trends in Food Science and Technology, 109, 552–563.

[15]   Russell, L.F. 2012. *Food Analysis.* by HPLC. 3rd ed. Leo M.L. Nollet, Fidel Toldra, editors. CRC Press. pp. 325–442.

[16]   Lykstad, J. and Sharma, S. 2022. Biochemistry, Water Soluble Vitamins.

[17]   Padayatty, S. and Levine, M. 2016. Vitamin C: The known and the unknown and Goldilocks. Oral Diseases, 22(6), 463–493.

[18] Thiele, N., McGowan, J. and Sloan, K. 2016. 2-O-Acyl-3-O-(1-acyloxyalkyl) Prodrugs of 5,6-Isopropylidene-l-Ascorbic acid and l-Ascorbic Acid: Antioxidant activity and ability to permeate silicone membranes. Pharmaceutics, 8(3), 22.

[19] Gramlich, G., Zhang, J. and Nau, W.M. 2002. Increased antioxidant reactivity of vitamin C at low pH in model membranes. Journal of the American Chemical Society, 124(38), 11252–11253.

[20] Linster, C.L. and Van Schaftingen, E. 2007. Vitamin  C. FEBS Journal, 274(1), 1–22.

[21] Drouin, G., Godin, J.R. and Page, B. 2011. The genetics of vitamin C loss in vertebrates. Current Genomics, 12(5), 371–378.

[22] Linster, C.L. and Van Schaftingen, E. 2007. Vitamin  C. FEBS Journal, 274(1), 1–22.

[23] Devaki, S.J. and Raveendran, R.L. 2017. Vitamin C: Sources, functions, sensing and analysis. In *Vitamin C*. In Tech.

[24] Domínguez-Perles, R., Mena, P., García-Viguera, C. and Moreno, D.A. 2014. *Brassica* foods as a dietary source of vitamin C: A Review. Critical Reviews in Food Science and Nutrition, 54(8), 1076–1091.

[25] Lykkesfeldt, J., Michels, A.J. and Frei, B. 2014. Vitamin C. Advances in Nutrition, 5(1), 16–18.

[26] Doseděl, M., Jirkovský, E., Macáková, K., Krčmová, L., Javorská, L., Pourová, J., et al. 2021. Vitamin C – sources, physiological role, kinetics, deficiency, use, toxicity, and determination. Nutrients, 13(2), 615.

[27] Njus, D., Kelley, P.M., Tu, Y.J. and Schlegel, H.B. 2020. Ascorbic acid: The chemistry underlying its antioxidant properties. Free Radical Biology and Medicine, 159, 37–43.

[28] Padayatty, S.J., Katz, A., Wang, Y., Eck, P., Kwon, O., Lee, J.H., et al. 2003. Vitamin C as an antioxidant: Evaluation of its role in disease prevention. Journal of the American College of Nutrition, 22(1), 18–35.

[29] Du, J., Martin, S.M., Levine, M., Wagner, B.A., Buettner, G.R., Wang, S.H., et al. 2010. Mechanisms of ascorbate-induced cytotoxicity in pancreatic cancer. Clinical Cancer Research, 16(2), 509–520.

[30] Du, J., Cullen, J.J. and Buettner, G.R. 2012. Ascorbic acid: chemistry, biology and the treatment of cancer. Biochimica et Biophysica Acta (BBA) – Reviews on Cancer, 1826(2), 443–457.

[31] Bedhiafi, T., Inchakalody, V.P., Fernandes, Q., Mestiri, S., Billa, N., Uddin, S., et al. 2022. The potential role of vitamin C in empowering cancer immunotherapy. Biomedicine & Pharmacotherapy, 146, 112553.

[32] Torti, S.V. and Torti, F.M. 2013. Iron and cancer: More ore to be mined. Natural Reviews in Cancer, 13(5), 342–355.

[33] Ngo, B., Van Riper, J.M., Cantley, L.C. and Yun, J. 2019. Targeting cancer vulnerabilities with high-dose vitamin C. Natural Reviews in Cancer, 19(5), 271–282.

[34] Vojdani, A. and Ghoneum, M. 1993. In vivo effect of ascorbic acid on enhancement of human natural killer cell activity. Nutrition Research, 13(7), 753–764.

[35] Heuser, G. and Vojdani, A. 1997. Enhancement of natural killer cell activity and T and B Cell function by buffered vitamin C in patients exposed to toxic chemicals: The role of protein Kinase-C. Immunopharmacology and Immunotoxicology, 19(3), 291–312.

[36] Toh, J.W.T. and Wilson, R.B. 2020. Pathways of gastric carcinogenesis, Helicobacter pylori Virulence and interactions with antioxidant systems, Vitamin C and phytochemicals. International Journal of Molecular Sciences, 21(17), 6451.

[37] Tabak, M., Armon, R., Rosenblat, G., Stermer, E. and Neeman, I. 2003. Diverse effects of ascorbic acid and palmitoyl ascorbate on *Helicobacter pylori* survival and growth. FEMS Microbiology Letters, 224(2), 247–253.

[38] Hernandez-Patlan, D., Solis-Cruz, B., Méndez-Albores, A., Latorre, J.D., Hernandez-Velasco, X., Tellez, G., et al. 2018. Comparison of PrestoBlue ® and plating method to evaluate antimicrobial activity of ascorbic acid, boric acid and curcumin in an *in vitro* gastrointestinal model. Journal of Applied Microbiology, 124(2), 423–430.

[39] Ghosh, T., Srivastava, S.K., Gaurav, A., Kumar, A., Kumar, P., Yadav, A.S., et al. 2019. A combination of Linalool, Vitamin C, and copper synergistically triggers reactive oxygen species and DNA damage and inhibits *Salmonella enterica* subsp. *enterica* serovar typhi and *Vibrio fluvialis*. Applied and Environmental Microbiology, 85(4), e02487–18.

[40] Fletcher, R.D., Albers, A.C., Chen, A.K. and Albertson, J.N. 1983. Ascorbic acid inhibition of Campylobacter jejuni growth. Applied and Environmental Microbiology, 45(3), 792–795.

[41] Colunga Biancatelli, R.M.L., Berrill, M. and Marik, P.E. 2020. The antiviral properties of vitamin C. Expert Review of Anti-infective Therapy, 18(2), 99–101.

[42] Puente, V., Demaria, A., Frank, F.M., Batlle, A. and Lombardo, M.E. 2018. Anti-parasitic effect of vitamin C alone and in combination with benznidazole against Trypanosoma cruzi. PLOS Neglected Tropical Diseases, 12(9), e0006764.

[43] Mousavi, S., Bereswill, S. and Heimesaat, M.M. 2019. Immunomodulatory and antimicrobial effects of vitamin C. European Journal of Microbiology and Immunology(Bp), 9(3), 73–79.

[44] Golonka, I., Oleksy, M., Junka, A., Matera-Witkiewicz, A., Bartoszewicz, M. and Musiał, W. 2017. Selected physicochemical and biological properties of ethyl ascorbic acid compared to ascorbic acid. Biological and Pharmaceutical Bulletin, 40(8), 1199–1206.

[45] Huang, A., Vita, J.A., Venema, R.C. and Keaney, J.F. 2000. Ascorbic acid enhances endothelial nitric-oxide synthase activity by increasing intracellular tetrahydrobiopterin. Journal of Biological Chemistry, 275(23), 17399–17406.

[46] Baker, T.A., Milstien, S. and Katusic, Z.S. 2001. Effect of vitamin C on the availability of tetrahydrobiopterin in human endothelial cells. Journal of Cardiovascular Pharmacology, 37(3), 333–338.

[47] d'Uscio, L.V., Milstien, S., Richardson, D., Smith, L. and Katusic, Z.S. 2003. Long-term vitamin C treatment increases vascular tetrahydrobiopterin levels and nitric oxide synthase activity. Circulation Research, 92(1), 88–95.

[48] Salvayre, R., Negre-Salvayre, A. and Camaré, C. 2016. Oxidative theory of atherosclerosis and antioxidants. Biochimie, 125, 281–296.

[49] Weber, C., Erl, W., Weber, K. and Weber, P.C. 1996. Increased adhesiveness of isolated monocytes to endothelium Is prevented by Vitamin C intake in smokers. Circulation, 93(8), 1488–1492.

[50] Moretti, M. and Rodrigues, A.L.S. 2022. Functional role of ascorbic acid in the central nervous system: A focus on neurogenic and synaptogenic processes. Nutritional Neuroscience, 25(11), 2431–2441.

[51] Abdel-Daim, M.M., Abushouk, A.I., Donia, T., Alarifi, S., Alkahtani, S., Aleya, L., et al. 2019. The nephroprotective effects of allicin and ascorbic acid against cisplatin-induced toxicity in rats. Environmental Science and Pollution Research, 26(13), 13502–13509.

[52] Ajith, T.A., Usha, S. and Nivitha, V. 2007. Ascorbic acid and α-tocopherol protect anticancer drug cisplatin induced nephrotoxicity in mice: A comparative study. Clinica Chimica Acta, 375(1–2), 82–86.

[53] Yousef, J.M., Chen, G., Hill, P.A., Nation, R.L. and Li, J. 2012. Ascorbic acid protects against the nephrotoxicity and apoptosis caused by colistin and affects its pharmacokinetics. Journal of Antimicrobial Chemotherapy, 67(2), 452–459.

[54] Abdel-Daim, M.M., Ahmed, A., Ijaz, H., Abushouk, A.I., Ahmed, H., Negida, A., et al. 2019. Influence of Spirulina platensis and ascorbic acid on amikacin-induced nephrotoxicity in rabbits. Environmental Science and Pollution Research, 26(8), 8080–8086.

[55] Korkmaz, A. and Kolankaya, D. 2009. The protective effects of ascorbic acid against renal ischemia-reperfusion injury in male rats. Renal Failure, 31(1), 36–43.

[56] Brickley, M.B., Ives, R. and Mays, S. 2020. Vitamin C deficiency, scurvy. In Brickley, M.B., Ives, R. and Mays, S. (Eds.), *The Bioarchaeology of Metabolic Bone Disease*. Elsevier; pp. 43–74.

[57] Carr, A.C. and Rowe, S. 2020. Factors affecting vitamin C status and prevalence of deficiency: A global health perspective. Nutrients, 12(7), 1963.

[58]  Rapala-Kozik, M. 2011. Vitamin B1 (Thiamine). In 37–91.

[59]  Wolak, N., Zawrotniak, M., Gogol, M., Kozik, A. and Rapala-Kozik, M. 2017. Vitamins B1, B2, B3 and B9 – occurrence, biosynthesis pathways and functions in human nutrition. Mini-Reviews in Medicinal Chemistry, 17(12).

[60]  Fitzpatrick, T.B. and Chapman, L.M. 2020. The importance of thiamine (vitamin B1) in plant health: From crop yield to biofortification. Journal of Biological Chemistry, 295(34), 12002–12013.

[61]  Chawla, J. and Kvarnberg, D. 2014. Hydrosoluble vitamins. In 891–914.

[62]  National Institutes of Health (NIH). 2021. Thiamin Fact Sheet for Health Professionals. https://ods. od.nih.gov/.

[63]  Guilland, J.C. 2013. Vitamin B1 (thiamine). Revue du Praticien, 63(8), 1074–1075. 1077–1078.

[64]  Isenberg-Grzeda, E., Kutner, H.E. and Nicolson, S.E. 2012. Wernicke-Korsakoff-Syndrome: Under-recognized and under-treated. Psychosomatics, 53(6), 507–516.

[65]  Wendołowicz, A., Stefańska, E. and Ostrowska, L. 2018. Influence of selected dietary components on the functioning of the human nervous system. Roczniki Państwowego Zakładu Higieny, 69(1), 15–21.

[66]  Fattal-Valevski, A. 2011. Thiamine (Vitamin B$_1$). Journal of Evidence-Based Complementary Alternative Medicine, 16(1), 12–20.

[67]  Hrubša, M., Siatka, T., Nejmanová, I., Vopršalová, M., Kujovská Krčmová, L., Matoušová, K., et al. 2022. Biological properties of Vitamins of the B-complex, Part 1: Vitamins B1, B2, B3, and B5. Nutrients, 14(3), 484.

[68]  Whitfield, K.C., Bourassa, M.W., Adamolekun, B., Bergeron, G., Bettendorff, L., Brown, K.H., et al. 2018. Thiamine deficiency disorders: Diagnosis, prevalence, and a roadmap for global control programs. Annals of the New York Academy of Sciences, 1430(1), 3–43.

[69]  Mack, M. and Grill, S. 2006. Riboflavin analogs and inhibitors of riboflavin biosynthesis. Applied Microbiology and Biotechnology, 71(3), 265–275.

[70]  Zhang, Y., Zhou, W e, Yan, J.Q., Liu, M., Zhou, Y., Shen, X., et al. 2018. A review of the extraction and determination methods of thirteen essential vitamins to the human body: An update from 2010. Molecules, 23(6), 1484.

[71]  Dym, O. and Eisenberg, D. 2001. Sequence-structure analysis of FAD-containing proteins. Protein Science, 10(9), 1712–1728.

[72]  Betz, A.L., Ren, X.D., Ennis, S.R. and Hultquist, D.E. 1994. Riboflavin reduces edema in focal cerebral Ischemia. In *Brain Edema IX*. Vienna: Springer Vienna, pp. 314–317.

[73]  Powers, H.J. 2003. Riboflavin (vitamin B-2) and health. The American Journal of Clinical Nutrition, 77(6), 1352–1360.

[74]  Wang, P., Fan, F., Li, X., Sun, X., Ma, L., Wu, J., et al. 2018. Riboflavin attenuates myocardial injury via LSD1-mediated crosstalk between phospholipid metabolism and histone methylation in mice with experimental myocardial infarction. Journal of Molecular & Cellular Cardiology, 115, 115–129.

[75]  Ertaş, B., Çevikelli, Z.A., Özbeyli, D., Ercan, F. and Ayaz Adakul, B. 2019. The effects of riboflavin on ischemia/reperfusion induced renal injury: Role on caspase-3 expression. Journal of Research in Pharmacy, 23(3), 379–386.

[76]  Kulkarni, A.G., Suryakar, A.N., Sardeshmukh, A.S. and Rathi, D.B. 2003. Studies on biochemical changes with special reference to oxidant and antioxidants in malaria patients. Indian Journal of Clinical Biochemistry, 18(2), 136–149.

[77]  Thakur, K., Tomar, S.K., Singh, A.K., Mandal, S. and Arora, S. 2017. Riboflavin and health: A review of recent human research. Critical Reviews in Food Science and Nutrition, 57(17), 3650–3660.

[78]  MdM, A., Iqbal, S. and Naseem, I. 2015. Ameliorative effect of riboflavin on hyperglycemia, oxidative stress and DNA damage in type-2 diabetic mice: Mechanistic and therapeutic strategies. Archives of Biochemistry & Biophysics, 584, 10–19.

[79]  Condò, M., Posar, A., Arbizzani, A. and Parmeggiani, A. 2009. Riboflavin prophylaxis in pediatric and adolescent migraine. Journal of Headache and Pain, 10(5), 361–365.

[80]    Jacques, P.F. 2001. Long-term nutrient intake and early age-related nuclear lens opacities. Archives of Ophthalmology, 119(7), 1009.

[81]    Chocano-Bedoya, P.O., Manson, J.E., Hankinson, S.E., Willett, W.C., Johnson, S.R., Chasan-Taber, L., et al. 2011. Dietary B vitamin intake and incident premenstrual syndrome. The American Journal of Clinical Nutrition, 93(5), 1080–1086.

[82]    Chaves Neto, A.H., Yano, C.L., Paredes-Gamero, E.J., Machado, D., Justo, G.Z., Peppelenbosch, M.P., et al. 2010. Riboflavin and photoproducts in MC3T3-E1 differentiation. Toxicology in Vitro: An International Journal Published in Association With BIBRA, 24(7), 1911–1919.

[83]    Hoane, M.R., Wolyniak, J.G. and Akstulewicz, S.L. 2005. Administration of Riboflavin improves behavioral outcome and reduces edema formation and glial fibrillary acidic protein expression after traumatic brain injury. Journal of Neurotrauma, 22(10), 1112–1122.

[84]    Horigan, G., McNulty, H., Ward, M., Strain, J., Purvis, J. and Scott, J.M. 2010. Riboflavin lowers blood pressure in cardiovascular disease patients homozygous for the 677C→T polymorphism in MTHFR. Journal of Hypertension, 28(3), 478–486.

[85]    Aljaadi, A.M., Devlin, A.M. and Green, T.J. 2022. Riboflavin intake and status and relationship to anemia. Nutrition Reviews, 81(1), 114–132.

[86]    Olfat, N., Ashoori, M. and Saedisomeolia, A. 2022. Riboflavin is an antioxidant: A review update. British Journal Nutrition, 128(10), 1887–1895.

[87]    García-Sánchez, A., Miranda-Díaz, A.G. and Cardona-Muñoz, E.G. 2020. The role of oxidative stress in physiopathology and pharmacological treatment with pro- and antioxidant properties in chronic diseases. Oxidative Medicine and Cellular Longevity, 2020, 1–16.

[88]    Forman, H.J. and Zhang, H. 2021. Targeting oxidative stress in disease: Promise and limitations of antioxidant therapy. Nature Reviews Drug Discovery, 20(9), 689–709.

[89]    Suwannasom, N., Kao, I., Pruß, A., Georgieva, R. and Riboflavin, B.H. 2020. The health benefits of a forgotten natural vitamin. International Journal of Molecular Sciences, 21(3), 950.

[90]    Balasubramaniam, S., Christodoulou, J. and Rahman, S. 2019. Disorders of riboflavin metabolism. Journal of Inherited Metabolic Disease, 42(4), 608–619.

[91]    Aguilera-Méndez, A., Fernández-Lainez, C., Ibarra-González, I. and Fernandez-Mejia, C. 2012. Chapter 7. the chemistry and biochemistry of Niacin (B3). In 108–126.

[92]    Levy, E. and Delvin, E. 2020. Vitamins: Functions and assessment of status through laboratory testing. In Contemporary Practice in Clinical Chemistry. Elsevier; pp. 825–849.

[93]    McDowell, L.R. 2000. 2nd edition, Vitamins in Animal and Human Nutrition. Wiley & Sons Inc.: Iowa, US.

[94]    Prousky, J., Millman, C.G. and Kirkland, J.B. 2011. Pharmacologic use of Niacin. Journal of Evidence-Based Complementary and Alternative Medicine, 16(2), 91–101.

[95]    Chand, T. and Savitri, B. 2016. Vitamin B $_3$, Niacin. In Industrial Biotechnology of Vitamins, Biopigments, and Antioxidants. Weinheim, Germany: Wiley-VCH Verlag GmbH & Co. KGaA; pp. 41–65.

[96]    Capuzzi, D.M., Morgan, J.M., Brusco, O.A. and Intenzo, C.M. 2000. Niacin dosing: Relationship to benefits and adverse effects. Current Atherosclerosis Reports, 2(1), 64–71.

[97]    Redzic, S., Hashmi, M.F. and Gupta, V. Niacin Deficiency. 2022.

[98]    Shi, H., Enriquez, A., Rapadas, M., Martin, E.M.M.A., Wang, R., Moreau, J., et al. 2017. NAD deficiency, congenital malformations, and niacin supplementation. New England Journal of Medicine, 377(6), 544–552.

[99]    Spitzer, V. and Höller, U. 2005. Vitamins | overview. In Encyclopedia of Analytical Science. Elsevier; pp. 147–159.

[100]   Kelly, G.S. 2011. Pantothenic acid. Alternative Medicine Review [Internet], 16(3), 263+. Available from: https://link.gale.com/apps/doc/A269531006/AONE?u=ufc_br&sid=googleScholar&xid=b9649b0e.

[101]   Casas, C. 2007. Vitamins. In Analysis of Cosmetic Products. Elsevier; pp. 364–379.

[102] Venco, P., Dusi, S., Valletta, L. and Tiranti, V. 2014. Alteration of the coenzyme A biosynthetic pathway in neurodegeneration with brain iron accumulation syndromes. Biochemical Society Transactions, 42(4), 1069–1074.

[103] Gaisa, N.T., Köster, J., Reinartz, A., Ertmer, K., Ehling, J., Raupach, K., et al. 2008. Expression of acyl-CoA synthetase 5 in human epidermis. Histology and Histopathology [Internet]. [cited 2023 Feb 13]; Available from: https://digitum.um.es/digitum/handle/10201/29798.

[104] Gonzalez-Lopez, J., Aliaga, L., Gonzalez-Martinez, A. and Martinez-Toledo, M.V. 2016. Pantothenic Acid. In *Industrial Biotechnology of Vitamins, Biopigments, and Antioxidants.* Weinheim, Germany: Wiley-VCH Verlag GmbH & Co. KGaA, pp. 67–101.

[105] Bates, C.J. 2013. Pantothenic Acid. Encyclopedia of Human Nutrition, 4–4, 1–5.

[106] Wilson, R.G. and Davis, R.E. 1983. Clinical chemistry of vitamin B6. Advances in Clinical Chemistry, 23(C), 1–68.

[107] da Silva, V.R. and Gregory, J.F. 2020. Vitamin B6. Present Knowledge in Nutrition [Internet]. [cited 2023 Feb 13], 225–237. Available from: https://linkinghub.elsevier.com/retrieve/pii/B9780323661621000135.

[108] Kohlmeier, M. 2003. Vitamin B6. Nutrient Metabolism [Internet]. [cited 2023 Feb 13];581–591. Available from: https://linkinghub.elsevier.com/retrieve/pii/B9780124177628500831.

[109] Ueland, P.M., McCann, A., Midttun, Ø. and Ulvik, A. 2017. Inflammation, vitamin B6 and related pathways. Molecular Aspects of Medicine, 53, 10–27.

[110] Mooney, S., Leuendorf, J.E., Hendrickson, C. and Hellmann, H. 2009. Vitamin B6: A long known compound of surprising complexity. Molecules, 14(1), 329–351.

[111] Da Silva, V.R. and Gregory, J.F. 2020. Vitamin B6. Present Knowledge in Nutrition [Internet]. [cited 2023 Feb 13], 225–237. Available from: https://linkinghub.elsevier.com/retrieve/pii/B9780323661621000135.

[112] Wondrak, G.T. and Jacobson, E.L. 2012. Vitamin B6: Beyond Coenzyme Functions. In 291–300.

[113] Yasuda, H., Furukawa, Y., Nishioka, K., Sasaki, M., Tsukune, Y., Shirane, S., et al. 2022. Vitamin B6 deficiency as a cause of polyneuropathy in POEMS syndrome: Rapid recovery with supplementation in two cases. Hematology, 27(1), 463–468.

[114] Bansal, M., Singh, N., Pal, S., Dev, I. and Ansari, K.M. 2018. Chemopreventive role of dietary phytochemicals in colorectal cancer. In 69–121.

[115] Vora, A., Riga, A., Dollimore, D. and Alexander, K.S. 2002. Thermal stability of folic acid. Thermochimica Acta, 392–393, 209–220.

[116] Dietrich, M., Brown, C.J.P. and Block, G. 2013. The effect of folate fortification of cereal-grain products on blood folate status, dietary folate intake, and dietary folate sources among adult non-supplement users in the United States. [Internet] 24(4), 266–274. http://dx.doi.org/101080/07315724200510719474 [cited 2023 Feb 14] Available from: https://www.tandfonline.com/doi/abs/10.1080/07315724.2005.10719474.

[117] Yang, Q., Cogswell, M.E., Hamner, H.C., Carriquiry, A., Bailey, L.B., Pfeiffer, C.M., et al. 2010. Folic acid source, usual intake, and folate and vitamin B-12 status in US adults: National Health and Nutrition Examination Survey (NHANES) 2003–2006. The American Journal of Clinical Nutrition [Internet], 91(1), 64–72. [cited 2023 Feb 14] Available from: https://academic.oup.com/ajcn/article/91/1/64/4597193.

[118] Blancquaert, D., Storozhenko, S., Loizeau, K., de Steur, H., De Brouwer, V., Viaene, J., et al. 2010. Folates and folic acid: From fundamental research toward sustainable health. [Internet] 29(1), 14–35. https://doi.org/101080/07352680903436283 [cited 2023 Feb 15] Available from: https://www.tandfonline.com/doi/abs/10.1080/07352680903436283.

[119] Eto, I. and Krumdieck, C.L. 1986. Role of vitamin B12 and folate deficiencies in carcinogenesis. Advances in Experimental Medicine & Biology [Internet], 206, 313–330. [cited 2023 Feb 27] Available from: https://link.springer.com/chapter/10.1007/978-1-4613-1835-4_23.

[120] Wagner, C. 2001. Biochemical Role of Folate in Cellular Metabolism*. Clinical Research and Regulatory Affairs, 18(3), 161–180.

[121] Mahmood, L. 2014. The metabolic processes of folic acid and Vitamin B12 deficiency. Journal of Health Research and Reviews [Internet], 1(1), 5. [cited 2023 Feb 27], Available from: https://www.jhrr.org/article.asp?issn=2394-2010;year=2014;volume=1;issue=1;spage=5;epage=9;aulast=Mahmood.

[122] Verhaar, M.C., Stroes, E. and Rabelink, T.J. 2002. Folates and cardiovascular disease. Arteriosclerosis, Thrombosis, and Vascular Biology [Internet], 22(1), 6–13. [cited 2023 Feb 27] Available from: https://www.ahajournals.org/doi/abs/10.1161/hq0102.102190.

[123] Al- Joufi, F., Bana MA, E., Tewfik, I. and Anwar, M. 2018. Efficacy of cosupplementation therapy with Vitamins B9, B12, and D on endothelial dysfunction in streptozotocin-induced diabetic rats. Asian Journal of Pharmaceutical and Clinical Research, 11(9), 407.

[124] Sobczyn, A. and Harrington, D.J. 2018. Laboratory assessment of folate (vitamin B9) status. Journal of Clinical Pathology [Internet], 71(11), 949–956. Nov 1 [cited 2023 Feb 27] Available from: https://jcp.bmj.com/content/71/11/949.

[125] Mikkelsen, K. and Apostolopoulos, V. 2019. Vitamin B12, folic acid, and the immune system. Nutrition and Immunity [Internet], 103–114. [cited 2023 Feb 27], Available from: https://link.springer.com/chapter/10.1007/978-3-030-16073-9_6.

[126] Davis, R.E. 1985. Clinical chemistry of Vitamin B12. In Advances in Clinical Chemistry, 24(C), 163–216.

[127] Proinsias, K., Giedyk, M. and Gryko, D. 2013. Vitamin B12: Chemical modifications. Chemical Society Reviews [Internet], 42(16), 6605–6619. Jul 22 [cited 2023 Feb 27] Available from: https://pubs.rsc.org/en/content/articlehtml/2013/cs/c3cs60062a.

[128] Kräutler, B. and Kräutler, K. 2005. Vitamin B12: Chemistry and biochemistry. Biochemical Society Transactions [Internet], 33(4), 806–810. [cited 2023 Feb 27] Available from: /biochemsoctrans/article/33/4/806/65191/Vitamin-B12-chemistry-and-biochemistry.

[129] Stabler, S.P. 2020. Vitamin B12. Present Knowledge in Nutrition [Internet]. [cited 2023 Feb 27], 257–271. Available from: https://linkinghub.elsevier.com/retrieve/pii/B9780323661621000159.

[130] Vogiatzoglou, A., Smith, A.D., Nurk, E., Berstad, P., Drevon, C.A., Ueland, P.M., et al. 2009. Dietary sources of vitamin B-12 and their association with plasma vitamin B-12 concentrations in the general population: The Hordaland Homocysteine Study. The American Journal of Clinical Nutrition [Internet], 89(4), 1078–1087. [cited 2023 Feb 27], Available from: https://academic.oup.com/ajcn/article/89/4/1078/4596726.

[131] Watanabe, F., Yabuta, Y., Bito, T. and Teng, F. 2014. Vitamin B12-Containing plant food sources for vegetarians. Nutrients [Internet], 6(5), 1861–1873. [cited 2023 Feb 27]. Available from: https://www.mdpi.com/2072-6643/6/5/1861/htm.

[132] Takahashi-Iñiguez, T., García-Hernandez, E., Arreguín-Espinosa, R. and Flores, M.E. 2012. Role of Vitamin B12 on methylmalonyl-CoA mutase activity. Journal of Zhejiang University Science B [Internet], 13(6), 423–437. [cited 2023 Feb 27] Available from: https://link.springer.com/article/10.1631/jzus.B1100329.

[133] Gibson, R.S. Principles of Nutritional Assessment. Google Books [Internet]. [cited 2023 Feb 27]. Available from: https://books.google.com.br/books?hl=en&lr=&id=lBlu7UKI3aQC&oi=fnd&pg=PR11&ots=RXSAUW5rtA&sig=RxjtWxjrcOSY1qsdU9SeZ7WKIe0&redir_esc=y#v=onepage&q&f=false.

[134] Sobczyńska-Malefora, A., Delvin, E., McCaddon, A., Ahmadi, K.R. and Harrington, D.J. 2021. Vitamin B$_{12}$ status in health and disease: A critical review. Diagnosis of deficiency and insufficiency – Clinical and laboratory pitfalls. Critical Review of Clinical and Lab Science, 58(6), 399–429.

[135] Stabler, S.P. 2013. Clinical practice. Vitamin B12 deficiency. New England Journal of Medicine [Internet], 368(2), 149–160. [cited 2023 Feb 28], Available from http://www.ncbi.nlm.nih.gov/pubmed/23301732.

[136] Green, R., Allen, L.H., Bjørke-Monsen, A.L., Brito, A., Guéant, J.L., Miller, J.W., et al. 2017. Vitamin B12 deficiency. Nature Reviews Disease Primers [Internet], 3(1), 1–20. [cited 2023 Feb 28], Available from: https://www.nature.com/articles/nrdp201740.

[137] Capone, K. and Sentongo, T. 2019. The ABCs of nutrient deficiencies and toxicities. Pediatric Annals, 48(11), e434–40.

[138] Johnson, E.J. and Mohn, E.S. 2015. Fat-soluble vitamins. World Review of Nutrition and Dietetics, 111, 38–44.

[139] Vitamins and minerals: a brief guide. Availabe online at https://sightandlife.org/resource-hub/other-publication/vitamins-minerals (accessed July 9, 2023).

[140] Uribe, N.G., García-Galbis, M.R. and Espinosa, R.M.M. 2017. New advances about the effect of vitamins on human health: vitamins supplements and nutritional aspects. In *Functional Food – Improve Health through Adequate Food*. InTech.

[141] Mozos, I., Stoian, D. and Luca, C.T. 2017. Crosstalk between Vitamins A, B12, D, K, C, and E status and arterial stiffness. Disease Markers, 2017, 8784971.

[142] Vitucci, D., Amoresano, A., Nunziato, M., Muoio, S., Alfieri, A., Oriani, G., et al. 2021. Nutritional controlled preparation and administration of different tomato Purées Indicate Increase of β-Carotene and Lycopene Isoforms, and of Antioxidant Potential in human blood bioavailability: A pilot study. Nutrients, 13(4), 1336.

[143] Al Binali, H.A.H. 2014. Night blindness and ancient remedy. Heart Views: The Official Journal of The Gulf Heart Association, 15(4), 136–139.

[144] Rosa, C.A., Paggiaro, A.O. and Carvalho, VF de. 2019. Effect of hydrogel enriched with alginate, fatty acids, and vitamins A and E on pressure injuries: A case series. Plastic Surgical Nursing, 39(3), 87–94.

[145] Summa, M., Russo, D., Penna, I., Margaroli, N., Bayer, I.S., Bandiera, T., et al. 2018. A biocompatible sodium alginate/povidone iodine film enhances wound healing. European Journal of Pharmaceutics and Biopharmaceutics, 122, 17–24.

[146] Sahu, P.J., Singh, A.L., Kulkarni, S., Madke, B., Saoji, V. and Jawade, S. 2020. Study of oral tranexamic acid, topical tranexamic acid, and modified Kligman's regimen in treatment of melasma. Journal of Cosmetic Dermatology, 19(6), 1456–1462.

[147] Bergmann, CLM da S, Pochmann, D., Bergmann, J., Bocca, F.B., Proença, I., Marinho, J., et al. 2021. The use of retinoic acid in association with microneedling in the treatment of epidermal melasma: Efficacy and oxidative stress parameters. Archives of Dermatological Research, 313(8), 695–704.

[148] Yuan, P., Cui, S., Liu, Y., Li, J., Du, G. and Liu, L. 2020. Metabolic engineering for the production of fat-soluble vitamins: Advances and perspectives. Applied Microbiology and Biotechnology, 104(3), 935–951. Feb

[149] Sadighi Akha, A.A. 2018. Aging and the immune system: An overview. Journal of Immunologic Methods, 463, 21–26.

[150] Zhao, Z. and Ross, A.C. 1995. Retinoic acid repletion restores the number of leukocytes and their subsets and stimulates natural cytotoxicity in vitamin A-deficient rats. Journal of Nutrition, 125(8), 2064–2073.

[151] Kong, R., Cui, Y., Fisher, G.J., Wang, X., Chen, Y., Schneider, L.M., et al. 2016. A comparative study of the effects of retinol and retinoic acid on histological, molecular, and clinical properties of human skin. Journal of Cosmetic Dermatology, 15(1), 49–57.

[152] Bitarafan, S., Mohammadpour, Z., Jafarirad, S., Harirchian, M.H., Yekaninejad, M.S. and Saboor-Yaraghi, A.A. 2019. The effect of retinyl-palmitate on the level of pro and anti-inflammatory cytokines in multiple sclerosis patients: A randomized double blind clinical trial. Clinical Neurology & Neurosurgery, 177, 101–105.

[153] Lauriello, M., di Marco, G.P., Necozione, S., Tucci, C., Pasqua, M., Rizzo, G., et al. 2020. Effects of liposomal nasal spray with vitamins A and E on allergic rhinitis. Acta Otorhinolaryngologica Italica, 40(3), 217–223.

[154] Chakrabarti, A., Eiden, M., Morin-Rivron, D., Christinat, N., Monteiro, J.P., Kaput, J., et al. 2020. Impact of multi-micronutrient supplementation on lipidemia of children and adolescents. Clinical Nutrition, 39(7), 2211–2219.

[155] Biswas, P., Dellanoce, C., Vezzoli, A., Mrakic-Sposta, S., Malnati, M., Beretta, A., et al. 2020. Antioxidant activity with increased endogenous levels of vitamin C, E and A following dietary supplementation with a combination of glutathione and resveratrol precursors. Nutrients, 12(11), 3224.

[156] Rabbani, E., Golgiri, F., Janani, L., Moradi, N., Fallah, S., Abiri, B., et al. 2021. Randomized study of the effects of zinc, vitamin A, and magnesium co-supplementation on thyroid function, oxidative stress, and hs-CRP in patients with hypothyroidism. Biological Trace Element Research, 199(11), 4074–4083.

[157] LeBoff, M.S., Chou, S.H., Ratliff, K.A., Cook, N.R., Khurana, B., Kim, E., et al. 2022. Supplemental vitamin D and incident fractures in midlife and older adults. New England Journal of Medicine, 387(4), 299–309.

[158] Burt, L.A., Billington, E.O., Rose, M.S., Raymond, D.A., Hanley, D.A. and Boyd, S.K. 2019. Effect of high-dose vitamin D supplementation on volumetric bone density and bone strength. Journal of the American Medical Association, 322(8), 736.

[159] Uday, S., Manaseki-Holland, S., Bowie, J., Mughal, M.Z., Crowe, F. and Högler, W. 2021. The effect of vitamin D supplementation and nutritional intake on skeletal maturity and bone health in socio-economically deprived children. European Journal of Nutrition, 60(6), 3343–3353.

[160] Stojanović, E., Jakovljević, V., Scanlan, A.T., Dalbo, V.J. and Radovanović, D. 2022. Vitamin D 3 supplementation reduces serum markers of bone resorption and muscle damage in female basketball players with vitamin D inadequacy. European Journal of Sport Science, 22(10), 1532–1542.

[161] Hew-Butler, T., Aprik, C., Byrd, B., Sabourin, J., VanSumeren, M., Smith-Hale, V., et al. 2022. Vitamin D supplementation and body composition changes in collegiate basketball players: A 12-week randomized control trial. Journal of the International Society of Sports Nutrition, 19(1), 34–48.

[162] Mayan, I., Somech, R., Lev, A., Cohen, A.H., Constantini, N.W. and Dubnov-Raz, G. 2015. Thymus activity, vitamin D, and respiratory infections in adolescent swimmers. Israel Medical Association Journal, 17(9), 571–575.

[163] Xiang, W., Kong, J., Chen, S., Cao, L.P., Qiao, G., Zheng, W., et al. 2005. Cardiac hypertrophy in vitamin D receptor knockout mice: Role of the systemic and cardiac renin-angiotensin systems. American Journal of Physiology-Endocrinology and Metabolism, 288(1), E125–E132.

[164] Souto Filho, J.T.D., de Andrade, A.S., Ribeiro, F.M., De Asde As, A.P. and Simonini, V.R.F. 2018. Impact of vitamin D deficiency on increased blood eosinophil counts. Hematology/Oncology and Stem Cell Therapy, 11(1), 25–29.

[165] Srichomchey, P., Sukprasert, S., Khulasittijinda, N., Voravud, N., Sahakitrungruang, C. and Lumjiaktase, P. 2023. Vitamin D 3 Supplementation promotes regulatory T-cells to maintain immune homeostasis after surgery for early stages of colorectal cancer. Vivo (Brooklyn), 37(1), 286–293.

[166] Dong, Y., Chen, L., Huang, Y., Raed, A., Havens, R., Dong, Y., et al. 2022. Sixteen-Week Vitamin D3 supplementation increases peripheral T cells in overweight black individuals: Post hoc analysis of a randomized, double-blinded, placebo-controlled trial. Nutrients, 14(19), 3922.

[167] Newton, D.A., Baatz, J.E., Chetta, K.E., Walker, P.W., Washington, R.O., Shary, J.R., et al. 2022. Maternal vitamin D status correlates to leukocyte antigenic responses in breastfeeding infants. Nutrients, 14(6), 1266.

[168] Anghel, L., Baroiu, L., Beznea, A., Topor, G. and Grigore, C.A. 2019. The therapeutic relevance of Vitamin E. Revista de Chimie, 70(10), 3711–3713.

[169] Grimm, M.O.W., Mett, J. and Hartmann, T. 2016. The impact of vitamin E and other fat-soluble vitamins on Alzheimer´s disease. International Journal of Molecular Sciences, 17(11), 1785.

[170] Nolan, J.M., Power, R., Howard, A.N., Bergin, P., Roche, W., Prado-Cabrero, A., et al. 2022. Supplementation with carotenoids, Omega-3 fatty acids, and vitamin E has a positive effect on the symptoms and progression of Alzheimer's disease. Journal of Alzheimer's Disease: JAD, 90(1), 233–249.

[171] Manosso, L.M., Camargo, A., Dafre, A.L. and Rodrigues, A.L.S. 2022. Vitamin E for the management of major depressive disorder: Possible role of the anti-inflammatory and antioxidant systems. Nutritional Neuroscience, 25(6), 1310–1324.

[172] Amini, L., Chekini, R., Nateghi, M.R., Haghani, H., Jamialahmadi, T., Sathyapalan, T., et al. 2021. The effect of combined vitamin C and vitamin E supplementation on oxidative stress markers in women with endometriosis: A randomized, triple-blind placebo-controlled clinical trial. Pain Research and Management, 2021, 5529741.

[173] Keshavarzi, Z., Janghorban, R., Alipour, S., Tahmasebi, S., Jokar, A. and Com, J. 2019. The effect of vitamin D and E vaginal suppositories on tamoxifen-induced vaginal atrophy in women with breast cancer. Supportive Care in Cancer, 27, 1325–1334.

[174] Sadeghi, F., Alavi-Naeini, A., Mardanian, F., Ghazvini, M.R. and Mahaki, B. 2020. Omega-3 and vitamin E co-supplementation can improve antioxidant markers in obese/overweight women with polycystic ovary syndrome. International Journal for Vitamin and Nutrition Research, 90(5–6), 477–483.

[175] Izadi, A., Shirazi, S., Taghizadeh, S. and Gargari, B.P. 2019. Independent and additive effects of coenzyme Q10 and Vitamin E on Cardiometabolic Outcomes and visceral adiposity in women with polycystic ovary syndrome. Archives of Medical Research . . . , 50(2), 1–10.

[176] Dutta, A. and Dutta, S.K. 2003. Vitamin E and its role in the prevention of atherosclerosis and carcinogenesis: A review. Journal of the American College of Nutrition, 22(4), 258–268.

[177] Singh, U. and Devaraj, S. 2007. Vitamin E: Inflammation and atherosclerosis. Vitamins and Hormones, 76, 519–549.

[178] Yarahmadi, A., Saeed Modaghegh, M.H., Mostafavi-Pour, Z., Azarpira, N., Mousavian, A., Bonakdaran, S., et al. 2021. The effect of platelet-rich plasma-fibrin glue dressing in combination with oral vitamin E and C for treatment of non-healing diabetic foot ulcers: A randomized, double-blind, parallel-group, clinical trial. Expert Opinion on Biological Therapy, 21(5), 687–696.

[179] Chuar, P.F., Ng, Y.T., Phang, S.C.W., Koay, Y.Y., Ho, J.I., Ho, L.S., et al. 2021. Tocotrienol-Rich Vitamin E (Tocovid) improved nerve conduction velocity in Type 2 diabetes mellitus patients in a phase ii double-blind, randomized controlled clinical trial. Nutrients, 13(11), 3770.

[180] Heiba, M.A., Ismail, S.S., Sabry, M., Bayoumy, W.A.E. and Kamal, K.A.A. 2021. The use of vitamin E in preventing taxane-induced peripheral neuropathy. Cancer Chemotherapy and Pharmacology, 88(6), 931–939.

[181] Bo, Y., Liu, C., Ji, Z., Yang, R., An, Q., Zhang, X., et al. 2019. A high whey protein, vitamin D and E supplement preserves muscle mass, strength, and quality of life in sarcopenic older adults: A double-blind randomized controlled trial. Clinical Nutrition, 38(1), 159–164.

[182] Gillam, I.H., Cunningham, R.B. and Telford, R.D. 2022. Antioxidant supplementation protects elite athlete muscle integrity during submaximal training. International Journal of Sports Physiology and Performance, 17(4), 549–555.

[183] Martínez-Ferrán, M., Berlanga, L.A., Barcelo-Guido, O., Matos-Duarte, M., Vicente-Campos, D., Sánchez-Jorge, S., et al. 2023. Antioxidant vitamin supplementation on muscle adaptations to resistance training: A double-blind, randomized controlled trial. Nutrition, 105, 111848.

[184] Dutra, M.T., Alex, S., Mota, M.R., Sales, N.B., Brown, L.E. and Bottaro, M. 2018. Effect of strength training combined with antioxidant supplementation on muscular performance. Applied Physiology, Nutrition, and Metabolism, 43(8), 775–781.

[185] Fusaro, M., Cianciolo, G., Brandi, M.L., Ferrari, S., Nickolas, T.L., Tripepi, G., et al. 2020. Vitamin K and Osteoporosis. Nutrients, 12(12), 3625.

[186] Stock, M. and Schett, G. 2021. Vitamin K-Dependent Proteins in Skeletal Development and Disease. International Journal of Molecular Sciences, 22(17), 9328.

[187] Kuang, X., Liu, C., Guo, X., Li, K., Deng, Q. and Li, D. 2020. The combination effect of vitamin K and vitamin D on human bone quality: A meta-analysis of randomized controlled trials. Food & Function, 11(4), 3280–3297.

[188]  Akbari, S. and Rasouli-Ghahroudi, A.A. 2018. Vitamin K and bone metabolism: A review of the latest evidence in preclinical studies. BioMed Research International, Hindawi, 2018, 1–8.

[189]  Welsh, J., Bak, M.J. and Narvaez, C.J. 2022. New insights into vitamin K biology with relevance to cancer. Trends in Molecular Medicine, 28(10), 864–881.

[190]  Gul, S., Maqbool, M.F., Maryam, A., Khan, M., Shakir, H.A., Irfan, M., et al. 2022. Vitamin K: A novel cancer chemosensitizer. Biotechnology & Applied Biochemistry, 69(6), 2641–2657.

[191]  Pazyar, N., Houshmand, G., Yaghoobi, R., Hemmati, A., Zeineli, Z. and Ghorbanzadeh, B. 2019. Wound healing effects of topical Vitamin K: A randomized controlled trial. Indian Journal of Pharmacology, 51(2), 88.

[192]  Godswill, A.G., Somtochukwu IV, I.A.O., Ikechukwu AO and Kate, E.C. 2020. Health benefits of micronutrients (Vitamins and Minerals) and their associated deficiency diseases: A systematic review. International Journal of Food Sciences [Internet], 3(1), 1–32. [cited 2023 Feb 1] Available from: https://www.iprjb.org/journals/index.php/IJF/article/view/1024.

[193]  Medicine LibreTexts [Internet]. 2020 [cited 2023 Feb 1]. Minerals: basic concepts. Available from: https://med.libretexts.org/Courses/Dominican_University/DU_Bio_1550%3A_Nutrition_(LoPresto)/8%3A_Water_and_Minerals/8.2%3A_Minerals%3A_basic_Concepts.

[194]  Food and Drug Administration. Federal Register. 2016. Food Labeling: Revision of the Nutrition and Supplement Facts Labels. [cited 2023 Jan 31]. 241–241. Available from: https://www.federalregister.gov/documents/2016/05/27/2016-11867/food-labeling-revision-of-the-nutrition-and-supplement-facts-labels.

[195]  Milan, S. and Rosato, F.E. 2010. Nutrient disposition and response. In Boullata, J.I., Armenti, V.T. (Eds.), *Handbook of Drug-Nutrient Interactions [Internet]*. Totowa, NJ: Humana Press, pp. 119–133. Available from: https://doi.org/10.1007/978-1-60327-362-6_5

[196]  Otten, J.J., Hellwig, J.P. and Meyers, L.D. 2006. Dietary reference intakes: The essential guide to nutrient requirements. [cited 2023 Jan 31]; Available from: http://www.nap.edu/catalog/11537.

[197]  National Institutes of Health (NIH). 2022. Office of Dietary Supplements (ODS) [Internet]. [cited 2023 Feb 1]. Calcium – Health Professional Fact Sheet. Available from: https://ods.od.nih.gov/factsheets/Calcium-HealthProfessional/.

[198]  Xu, S., Zhang, M., Cong, J., He, Y., Zhang, L., Guo, Y., et al. 2022. Reduced blood circulating calcium level is an outstanding biomarker for preeclampsia among 48 types of human diseases. QJM: An International Journal of Medicine [Internet], 115(7), 455–462. [cited 2023 Jan 27] Available from https://academic.oup.com/qjmed/article/115/7/455/6355055.

[199]  Sun, J.K., Zhang, W.H., Zou, L., Liu, Y., Li, J.J., Kan, X.H., et al. 2020. Serum calcium as a biomarker of clinical severity and prognosis in patients with coronavirus disease 2019. Aging [Internet], 12(12), 11287–11295. [cited 2023 Jan 31], Available from: https://europepmc.org/articles/PMC7343468.

[200]  National Institutes of Health (NIH). 2021. Office of Dietary Supplements (ODS) [Internet]. [cited 2023 Jan 27]. Phosphorus – Health Professional Fact Sheet. Available from: https://ods.od.nih.gov/factsheets/Phosphorus-HealthProfessional/.

[201]  National Institutes of Health (NIH). 2022. Office of Dietary Supplements (ODS) [Internet]. [cited 2023 Jan 28]. Magnesium – Health Professional Fact Sheet. Available from: https://ods.od.nih.gov/factsheets/Magnesium-HealthProfessional/#h3.

[202]  de Beus, E., de Jager, R.L., Beeftink, M.M., Sanders, M.F., Spiering, W., Vonken, E.J., et al. 2017. Salt intake and blood pressure response to percutaneous renal denervation in resistant hypertension. Journal of Clinical Hypertension (Greenwich) [Internet], 19(11), 1125–1133. [cited 2023 Feb 2] Available from https://pubmed.ncbi.nlm.nih.gov/28929577/.

[203]  Cheng, Y., Song, H., Pan, X., Xue, H., Wan, Y., Wang, T., et al. 2018. Urinary metabolites associated with blood pressure on a low- or high-sodium diet. Theranostics [Internet], 8(6), 1468–1480. [cited 2023 Feb 2] Available from: https://pubmed.ncbi.nlm.nih.gov/29556335/.

[204] Potassium – Health Professional Fact Sheet [Internet]. 2022. [cited 2023 Feb 2]. Available from: https://ods.od.nih.gov/factsheets/Potassium-HealthProfessional/.

[205] National Institutes of Health (NIH). 2022. Office of Dietary Supplements (ODS) [Internet]. [cited 2023 Feb 2]. Iron – Health Professional Fact Sheet. Available from: https://ods.od.nih.gov/factsheets/Iron-HealthProfessional/.

[206] National Institutes of Health (NIH). 2022. Office of Dietary Supplements (ODS) [Internet]. [cited 2023 Feb 2]. Zinc – Health Professional Fact Sheet. Available from: https://ods.od.nih.gov/factsheets/Zinc-HealthProfessional/.

[207] National Institutes of Health (NIH). 2022. Office of Dietary Supplements (ODS) [Internet]. [cited 2023 Jan 31]. Copper – Health Professional Fact Sheet. Available from: https://ods.od.nih.gov/factsheets/Copper-HealthProfessional/.

[208] National Institutes of Health (NIH). 2021. Office of Dietary Supplements (ODS) [Internet]. [cited 2023 Jan 31]. Manganese – Health Professional Fact Sheet. Available from: https://ods.od.nih.gov/factsheets/Manganese-HealthProfessional/.

[209] National Institutes of Health (NIH). 2022. Office of Dietary Supplements (ODS) [Internet]. [cited 2023 Jan 31]. Iodine – Health Professional Fact Sheet. Available from: https://ods.od.nih.gov/factsheets/Iodine-HealthProfessional/.

[210] National Institutes of Health (NIH). 2021. Office of Dietary Supplements (ODS) [Internet]. [cited 2023 Jan 31]. Selenium – Health Professional Fact Sheet. Available from: https://ods.od.nih.gov/factsheets/Selenium-HealthProfessional/.

[211] National Institutes of Health (NIH). 2022. Office of Dietary Supplements (ODS) [Internet]. [cited 2023 Jan 31]. Fluoride – Health Professional Fact Sheet. Available from: https://ods.od.nih.gov/factsheets/Fluoride-HealthProfessional/.

[212] National Institutes of Health (NIH). 2021. Office of Dietary Supplements (ODS) [Internet]. [cited 2023 Jan 31]. Molybdenum – Health Professional Fact Sheet. Available from: https://ods.od.nih.gov/factsheets/Molybdenum-HealthProfessional/.

[213] National Institutes of Health (NIH). 2022. Office of Dietary Supplements (ODS) [Internet]. [cited 2023 Feb 2]. Chromium – Health Professional Fact Sheet. Available from: https://ods.od.nih.gov/factsheets/Chromium-HealthProfessional/.

[214] Medina-Pizzali, M., Robles, P., Mendoza, M. and Torres, C. 2018. Ingesta de ArsénicO: El Impacto En La Alimentación y La Salud Humana. Revista Peruana de Medicina Experimental y Salud Pública, 35(1), 93–102.

[215] National Institutes of Health (NIH). 2022. Office of Dietary Supplements (ODS) [Internet]. [cited 2023 Jan 31]. Boron – Health Professional Fact Sheet. Available from: https://ods.od.nih.gov/factsheets/Boron-HealthProfessional/.

[216] Randazzo, C.L., Pino, A., Ricciardi, L., Romano, C., Comito, D., Arena, E., et al. 2015. Probiotic supplementation in systemic nickel allergy syndrome patients: Study of its effects on lactic acid bacteria population and on clinical symptoms. Journal of Applied Microbiology, 118(1), 202–211. Internet [cited 2023 Feb 2] Available from: https://pubmed.ncbi.nlm.nih.gov/25363062/.

[217] Martin, K.R. 2013. Silicon: The health benefits of a metalloid. Metal Ions in Life Sciences [Internet], 13, 451–473. [cited 2023 Jan 31], Available from: https://pubmed.ncbi.nlm.nih.gov/24470100/.

[218] Bychkov, A., Koptev, V., Zaharova, V., Reshetnikova, P., Trofimova, E., Bychkova, E., et al. Experimental testing of the action of vitamin D and silicon chelates in bone fracture healing and bone turnover in mice and rats. 2022. Nutrients [Internet], 14(10). [cited 2023 Jan 31] Available from: https://pubmed.ncbi.nlm.nih.gov/35631133/.

[219] Filler, G., Kobrzynski, M., Sidhu, H.K., Belostotsky, V., Huang, S.H.S., McIntyre, C., et al. 2017. A cross-sectional study measuring vanadium and chromium levels in paediatric patients with CKD. BMJ Open [Internet], 7(5). [cited 2023 Feb 2], Available from: https://pubmed.ncbi.nlm.nih.gov/28592575/.

Thadiyan Parambil Ijinu*, Maheswari Priya Rani,
Sreejith Pongillyathundiyil Sasidharan, Santny Shanmugarama,
Raghavan Govindarajan, Varughese George and Palpu Pushpangadan

# Chapter 5
# Clinical significance of herb–drug interactions

**Abstract:** Herbal products or supplements are products that contain botanical extracts or bioactive compounds derived from plants and are marketed to provide potential health benefits beyond basic nutritional value. The growing demand for herbal products or supplements among consumers brings with it some safety concerns because these products are not regulated by any stringent mechanism. Moreover, the rising clinical safety issues regarding interactions between these botanicals and life-saving drugs pose a threat to the scientific consensus. These interactions may be pharmacokinetic, which affects absorption, distribution, metabolism, and elimination, or pharmacodynamic, which causes an additive, synergistic, or antagonistic effect on the drugs or vice versa. This chapter is focused on the drug interactions of some of the medicinal herbs such as American ginseng, Asian ginseng, black cohosh, cranberry, curcumin, Danshen, *Echinacea*, garlic, ginkgo, goldenseal, grapefruit juice, green tea extract, kava kava, milk thistle, saw palmetto, and St. John's wort, along with probable mechanisms and clinical manifestations based on case studies reported in the literature.

---

*\*Correspondence author: Thadiyan Parambil Ijinu*, Naturæ Scientific, Kerala University Business Innovation and Incubation Centre, University of Kerala, Karyavattom Campus, Thiruvananthapuram 695581, Kerala, India; The National Society of Ethnopharmacology, VRA 179, Mannamoola, Thiruvananthapuram 695005, Kerala, India, e-mail: ijinutp@gmail.com

**Maheswari Priya Rani,** Drug Discovery and Development Division, Patanjali Research Institute, Haridwar 249405, Uttarakhand, India; Phytochemistry and Phytopharmacology Division, Jawaharlal Nehru Tropical Botanic Garden and Research Institute, Thiruvananthapuram 695005, Kerala, India

**Sreejith Pongillyathundiyil Sasidharan,** Multidisciplinary Research Unit, Government Medical College, Thiruvananthapuram 695011, Kerala, India

**Santny Shanmugarama,** Department of Pathology, The University of Oklahoma Health Sciences Center, Oklahoma City, OK 73104, USA

**Raghavan Govindarajan,** Research and Development, Zydus Wellness Limited, Ahmedabad 380015, Gujarat, India

**Varughese George,** The National Society of Ethnopharmacology, VRA 179, Mannamoola, Thiruvananthapuram 695005, Kerala, India; Mar Dioscorus College of Pharmacy, Thiruvananthapuram 695017, Kerala, India

**Palpu Pushpangadan,** The National Society of Ethnopharmacology, VRA 179, Mannamoola, Thiruvananthapuram 695005, Kerala, India; Amity Institute for Herbal and Biotech Products Development, Thiruvananthapuram 695005, Kerala, India

https://doi.org/10.1515/9783111317601-005

# 1 Introduction

Health and well-being refer to the physical, mental, and emotional states of an individual. Good health and well-being are achieved through a combination of factors including regular exercise, a balanced diet, sufficient sleep, stress management, and positive social interactions [1, 2]. A balanced diet, which includes a variety of foods from all food groups in the right amounts, is very much required for overall health and well-being. It includes a variety of nutrients, such as carbohydrates, proteins, phytochemicals, fatty acids, vitamins, and minerals [3]. Thus, adequate nutrition helps to fuel the body, build and repair tissues, and support overall functioning [4]. However, a diet lacking essential nutrients can lead to health problems and increase the risk of chronic diseases [5]. Hippocrates, the ancient Greek physician widely regarded as the father of modern medicine, believed that food is medicine and that it can have a profound impact on health. He is famously known for saying, "Let food be thy medicine and medicine be thy food" [6]. He believed that the type and quality of food consumed directly affect health and that a healthy diet is essential for preventing and treating diseases.

Nutraceuticals can include a wide range of products such as vitamins, minerals, herbs, dietary supplements, and functional foods [7]. The use of nutraceuticals, especially herbal nutraceuticals, has become increasingly popular in recent years as consumers have become more interested in natural and holistic approaches to healthcare. Herbal nutraceuticals refer to plant-based supplements or foods that are believed to provide health benefits beyond basic nutritional value [8, 9]. These products are made from herbs, botanicals, or other plant-based ingredients and are often marketed as natural remedies for various health conditions (e.g., purple coneflower, ginkgo, turmeric, garlic, and grapefruit juice) [10]. However, it is important to note that while some nutraceuticals have been shown to have health benefits, others may not be effective or could even be harmful if administered at high doses or in combination with certain drugs.

# 2 Molecular basis of drug metabolism

The term "metabolism" refers to the total chemical reactions that occur in living organisms. Each food or substance that we ingest will be metabolized by our body. The liver is the primary site of metabolism, while other organs such as the intestine, kidneys, lungs, and blood cells also play a significant role. The major metabolizing mechanisms are divided into three phases: phases I, II, and III [11]. Phase I reactions mostly result in structural changes in the chemicals and mainly involve oxidation, reduction, and hydroxylation. The cytochromes P450 (CYPs) are the most important category of drug-metabolizing enzymes (DMEs) involved in phase I reactions. There are 57 human CYP genes, which are categorized into 18 families. CYPs are responsible for the metabolism of approximately 50–60% of all medicines currently available in the market. The CYP1 to CYP4 subfamilies are responsible for the metabolism of various substances, including steroids, fatty acids, vitamin D3/K, warfarin, nicotine, coumarin, taxol, acetaminophen, cyclosporine, antidepressants, antipsychotics, and statins. These enzymes can also produce carcinogenic metabolites such as aromatic amines and polycyclic aromatic hydrocarbons during metabolism [12].

Phase II reactions involve the synthetic conjugation of molecules to increase water solubility. Examples of these reactions include glucuronidation, sulfation, methylation, acetylation, and glutathione conjugation. UDP-glucuronosyltransferases (UGTs) are a group of enzymes that are responsible for glucuronidation. There are four gene families of UGTs: UGT1, UGT2, UGT3, and UGT8 [13–16]. The UGT1 and UGT2 families primarily metabolize xenobiotics such as polyaromatic amines, nonsteroidal anti-inflammatory drugs, statins, α-hydroxygenkwanin, genkwanin, ursolic acid, fimasartan, alpinetin, retinoids, catechol, estrogens, opioids, coumarins, flavonoids, anthraquinones, phenols, raloxifene, oxazepam, lorazepam, sipoglitazar, bisphenol-A, and endogenous compounds such as bilirubin, estradiol, fatty acids, serotonin, estrogens, and steroids. The UGT3 and UGT8 families primarily metabolize bile acids and $N$-acetylglucosamine [17].

Sulfonyl transferases (SULTs) and glutathione $S$-transferases (GSTs) are conjugative enzymes that play significant roles in phase II reactions. Four families of human SULTs have been identified including SULT1, SULT2, SULT4, and SULT6, which can conjugate a variety of substrates, such as dopamine, estrogens, thyroxine, catechol, norepinephrine, 3-$\beta$-hydroxysteroid, and others. Similarly, cytosolic GST isoenzymes of the alpha, zeta, theta, mu, pi, sigma, and omega classes have been discovered in humans, and they can conjugate a wide range of substrates including steroids, bilirubin, heme, fatty acids, $N$-acetyl-$p$-benzoquinone imine, cisplatin, busulfan, dichloroacetate, cyclophosphamide, azathioprine, and others [18].

Membrane transporters play a critical role in drug metabolism, with over 400 different transporters belonging to two major superfamilies: ATP-binding cassette (ABC) and solute carrier (SLC) transporters. Forty-nine of these transporters are classified into seven subfamilies [19], with $P$-glycoprotein ($P$-gp) being the most studied ABC transporter. $P$-gp has broad substrate specificity and limits the bioavailability of many oral drugs, including antibiotics, statins, immunosuppressants, and anticancer drugs, and the spectrum of these drugs overlaps with the substrates of CYPs [18, 20].

Human nuclear receptors consist of enzymes and a family of 48 ligand-regulated transcription factors that affect the expression of the target genes involved in drug metabolism. These receptors include peroxisome proliferator-activated receptor (PPAR), liver X receptor (LXR), and hepatocyte nuclear factor. The involvement of the gut microbiota in drug metabolism has recently been discovered, shedding light on their crucial role in regulating therapeutic outcomes in combination with host metabolism. However, the contribution of gut microbiota to drug metabolism and toxicity is not well understood [21–23].

Phase I metabolism, which is substrate-specific, and is mostly associated with drug interactions. Phase II metabolism is crucial for the detoxification of the reactive compounds generated during phase I metabolism. Conjugates and their metabolites can undergo further modifications and elimination from cells during phase III of metabolism. Genetic variations in metabolizing enzymes and transporters can affect drug metabolism and lead to individual differences in drug response and toxicity [24]. The complex mixture of known and unknown compounds in herbal nutraceuticals and dietary supplements is metabolized through multiple pathways, resulting in unpredictable metabolic by-products or combined interactions with endogenous and exogenous compounds that can cause a variety of outcomes in humans.

# 3 Herb–drug interactions

Interactions between herbal products or dietary supplements and drugs present complex and challenging problems [25]. Drug interactions with medicinal herbs can be clinically significant and can affect patient safety and therapeutic outcomes. They may be caused by either pharmacokinetic or pharmacodynamic interactions [26–28]. Pharmacokinetic interactions occur when herbal extracts and their bioactive compounds affect the absorption, distribution, metabolism, and excretion of drugs in the body. Pharmacodynamic interactions, on the other hand, occur when dietary supplements directly interact with the drugs in the body either by enhancing or antagonizing their effects.

In pharmacokinetic interaction, for example, some botanical bioactives can form chelates or chemical complexes with certain antibiotics, such as fluoroquinolones and tetracyclines, which can interfere with their absorption in the intestine. This can reduce the effectiveness of the antibiotics and increase the risk of antibiotic resistance [29]. Similarly, some components in grapefruit juice, such as furanocoumarin and its derivatives (e.g., bergamottin and dihydroxybergamottin), can inhibit the activity of drug metabolism enzymes in the intestine, leading to increased absorption of certain medications like calcium channel blockers. This can result in higher levels of the medication in the body, leading to potentially harmful side effects [28]. St. John's wort is known to upregulate the activity of the enzyme CYP3A4, which is responsible for metabolizing many drugs in the liver. This can result in increased metabolism and decreased effectiveness of medications, such as digoxin, warfarin, theophylline, cyclosporine, and disopyramide [30].

In pharmacodynamic interaction, for example, *Ginkgo biloba* extract contains terpenoids and flavonoids that have been shown to have anticoagulant effects. When taken in combination with other anticoagulant medications, such as aspirin, ticlopidine, or warfarin, the anticoagulant effects of these drugs can be enhanced, which can increase the risk of bleeding or hemorrhage [29]. Taking herbal products containing high levels of vitamin K, such as chlorella or *aojiru* (a suspension of vegetable leaf powder made from *Brassica oleracea*), can interfere with the effectiveness of warfarin by increasing the level of vitamin K in the body. This can result in reduced anticoagulant effects and an increased risk of blood clots [31]. Thus, understanding whether a specific herb–drug interaction is pharmacokinetic or pharmacodynamic can help healthcare providers determine the best course of action. In some cases, the medication dosage may need to be adjusted to account for changes in drug metabolism, while in other cases, it may be necessary to avoid the combination of herbal products and medications altogether. Some of the experimentally and clinically reported examples of herb–drug interactions are discussed below.

## 3.1 American ginseng

*Panax quinquefolius* L. (American ginseng) and *P. ginseng* C.A. Mey (Asian ginseng) are two important plants belonging to the Araliaceae family [32]. These two popular herbal supplements have been traditionally used for their potential health benefits including improving energy and cognitive function [33]. The studies suggest that American ginseng does not interfere with the metabolism or effectiveness of these HIV medications, indinavir and zidovudine [34, 35]. However, it is important to note that these studies were conducted in HIV-positive individuals, and more research may be needed to confirm these findings in larger populations. In a clinical trial conducted in healthy volunteers, American ginseng caused a small decrease in international normalized ratio (INR), which could potentially increase the risk of bleeding in patients taking warfarin. Therefore, patients taking warfarin should be cautious when considering the use of ginseng-containing supplements [36].

## 3.2 Asian ginseng

Asian ginseng has traditionally been used to improve cognitive function, reduce fatigue, and enhance immune function [37]. It contains ginsenosides as its active components [38]. One potential concern with Asian ginseng is its ability to induce the activity of an enzyme called CYP3A4, which plays a key role in the metabolism of many drugs such as calcium channel blockers, HIV medications, chemotherapeutic drugs, statins, antidepressants, and antihypertensive medications [39]. Therefore, it is safe to take Asian ginseng along with medications that are metabolized by CYP1A2, CYP2D6, CYP2E1, or *P*-gp because some human trials have shown no effect on these enzymes [39–41]. However, the effects of Asian ginseng on warfarin metabolism remain unclear, with some studies reporting no effect and others reporting small effects [42–45]. Consequently, patients who have difficulty maintaining adequate anticoagulation with warfarin should avoid using Asian ginseng products [42–45]. In addition, Bilgi et al. [46] found that Asian ginseng interacts with imatinib, an anticancer drug.

## 3.3 Black cohosh

The herb *Actaea racemosa* L. (syn. *Cimicifuga racemosa* (L.) Nutt.), which is native to the eastern United States, has been traditionally used by indigenous North American people as a natural remedy to manage various gynecological issues such as premenstrual syndrome, pain during childbirth, menopausal symptoms, and osteoporosis [47, 48]. However, the use of black cohosh has been associated with several potential risks and side effects. One concern is that it may increase the risk of thromboembolic and cardiovascular events and breast cancer, particularly in women who have undergone estrogen replacement therapy. This is because black cohosh acts as a selective estrogen receptor modulator and stimulates the growth of estrogen-sensitive breast cancer cells [49].

Tsukamoto et al. [50] found that the extract contains a polar fraction that showed 44% inhibition at 5 mg/mL. The inhibition produced by the polar fraction at this concentration was as potent as the inhibition produced by ketoconazole, a known CYP3A4 inhibitor, which showed 58% inhibition at 5 µg/mL. The main constituents responsible for the inhibitory activity were identified as cognate triterpene glycosides. Hepatotoxicity has also been reported in some individuals who have taken black cohosh supplements [51]. Symptoms of hepatotoxicity include nausea, vomiting, abdominal pain, and jaundice. Therefore, black cohosh should not be used during pregnancy or lactation, as there is insufficient evidence regarding its safety in these populations. Studies have suggested that black cohosh may inhibit the activity of OATP2B1, which could potentially reduce the absorption and effectiveness of drugs that are substrates for this transporter. This includes drugs like amiodarone, fexofenadine, glyburide, and many statin medications [52].

## 3.4 Cranberry

The fruit of *Vaccinium macrocarpon* Aiton., commonly known as cranberry, is a member of the Ericaceae family and has been used for decades to prevent and treat urinary tract infections [53, 54]. It can be consumed in various forms, including concentrated juice, dried juice capsules, and encapsulated standardized extracts [55]. Despite being generally safe for most people, cranberry may interact with certain liver enzymes, including CYP1A2, CYP2C9, and CYP3A4, which are responsible for metabolizing drugs like warfarin [56–59].

There have been multiple reported cases of increased INR and bleeding associated with the use of the anticoagulant warfarin, leading to concerns about potential interactions with cranberry juice [60–68]. One study found that consuming pills containing concentrated cranberry juice resulted in a 30% larger area under the INR-time curve for warfarin [64]. However, the relevance of this finding is unclear, as it contains a highly concentrated form of cranberry juice that is not typically consumed. Two clinical trials found no significant effect on INR or bleeding risk when examining the potential interaction between cranberry juice and warfarin [69]. However, these warnings may be due to misleading conclusions [68].

## 3.5 Curcumin

Curcumin, a compound found in turmeric (*Curcuma longa* L.), was shown to induce CYP1A2 in a single study, which could potentially lead to reduced levels of certain antidepressant and antipsychotic medications that are metabolized by this enzyme [70]. Additionally, curcumin increases the levels of sulfasalazine, which is metabolized by UDP-UGT [71]. This suggests that taking curcumin supplements or consuming large amounts of turmeric while taking sulfasalazine may result in elevated levels of sulfasalazine in the body. However, several human clinical trials have demonstrated no significant effect of curcumin on enzymes such as CYP2C9, CYP3A4, and UGT [72], indicating that curcumin is unlikely to cause significant changes in the metabolism of many other drugs [73].

## 3.6 Danshen

*Salvia miltiorrhiza* Bunge, or Danshen, is a traditional Chinese herb that belongs to the Lamiaceae family and has been used for centuries in traditional Chinese medicine for various medicinal purposes including the treatment of cardiovascular diseases [74] and menstrual disorders [75]. The herb contains a variety of active compounds, including tanshinones and salvianolic acids, which have been shown to have antioxidant, anti-inflammatory, and antiplatelet effects [76–78]. One potential concern with the use of Danshen is its interaction with other medications, particularly warfarin. Danshen has been shown to reduce the elimination of warfarin from the body, which can lead to an increased risk of bleeding [79–82]. Danshen also stimulated intestinal CYP3A4 in 14 healthy volunteers [83].

## 3.7 Echinacea

Echinacea (*Echinacea purpurea* (L.) Moench; family Asteraceae) has been shown not to affect CYP2D6 (debrisoquine, dextromethorphan), CYP2C9 (tolbutamide), or *P*-gp (digoxin) in human studies [84–87]. However, there are conflicting results regarding its effects on CYP1A2 and CYP3A4, which are important for the metabolism of many medications [84]. The short-term use of echinacea may induce these enzymes, potentially leading to decreased levels of medications metabolized by them. However, long-term use of echinacea may inhibit these enzymes, potentially leading to increased levels of medications metabolized by them. Therefore, caution should be exercised when combining echinacea with medications that are metabolized by CYP1A2 and CYP3A4, including antipsychotic and antidepressant medications. A recent clinical trial showed that *E. purpurea* root extract did not affect the pharmacokinetics of darunavir or ritonavir in patients with HIV, suggesting no significant interaction between *E. purpurea* and these protease inhibitors, which are mainly metabolized by CYP3A4 and are *P*-gp substrates [88].

## 3.8 Garlic

*Allium sativum* L., or garlic, belongs to the family Amaryllidaceae. It is a commonly used herb in cooking and is also available in supplement form for medicinal purposes. However, the use of large amounts of garlic or garlic supplements can have anticoagulant effects and increase the risk of bleeding [89]. The active ingredients in garlic are sulfur-containing compounds, the most notable of which is allicin, which is formed via conversion of a sulfur-containing amino acid alliin by the enzyme alliinase [90]. The garlic supplement has been proven in human studies to decrease the concentration of medications that are transported by *P*-gp, meaning that taking garlic supplements could result in reduced levels of drugs such as colchicine, digoxin, doxorubicin, quinidine, rosuvastatin, tacrolimus, and verapamil, which are transported by *P*-gp [41, 55, 91, 92]. However, the clinical significance of this interaction is not well established, and more research is needed. Garlic has been shown to interfere with platelet function and cause platelet disorders and/or hemorrhage, particularly when taken in large amounts or in combination with other medications that increase bleeding risk, such as aspirin or warfarin [93]. Therefore, it is recommended that patients using garlic supplements discontinue use at least 10 days before elective surgical procedures, especially if they are also taking aspirin or warfarin.

## 3.9 Ginkgo

*Ginkgo biloba* L. (family Ginkgoaceae) is a well-known plant that has been used for centuries in traditional medicine for various purposes. Its dietary supplements are commonly consumed for enhancing memory and cognitive function, reducing anxiety and depression, and improving blood circulation [94, 95]. However, concurrent use of ginkgo with antiplatelet, anticoagulant, or antithrombotic agents can increase the risk of bleeding [55, 96]. Therefore, patients taking warfarin must closely monitor their

INR levels or refrain from ginkgo use. Although some studies suggest that ginkgo has no significant effect on CYP1A2, CYP3A4, and CYP2D6 [41, 97, 98], another study revealed that ginkgo extract can induce CYP2C19 activity, resulting in the decreased bioavailability of omeprazole [99]. This suggests that ginkgo interferes with the drugs metabolized by CYP2C19. The extract has been shown in human studies to decrease concentrations of drugs that are transported by *P*-gp. This means that taking ginkgo supplements may lead to lower levels of medications that are transported by *P*-gp, including colchicine, digoxin, doxorubicin, quinidine, rosuvastatin, tacrolimus, and verapamil [41, 55, 91, 92], resulting in decreased efficacy and adverse effects. Therefore, it is important for healthcare providers to be aware of these potential interactions and to monitor patients closely for signs of bleeding or changes in clotting parameters.

## 3.10 Goldenseal

*Trautvetteria caroliniensis* (Walter) Vail (syn. *Hydrastis canadensis* Poir.; family Ranunculaceae), or goldenseal, is a medicinal plant commonly used in traditional medicine to treat various conditions, including digestive issues, respiratory infections, and skin conditions [100]. It contains bioactive alkaloid class of compounds such as berberine, hydrastine, and canadine. One potential concern with goldenseal is its ability to inhibit the activity of two major metabolic enzymes, CYP2D6 and CYP3A4, which are responsible for the metabolism of many drugs such as antidepressants, antipsychotics, beta-blockers, and opioids [86, 87, 101]. Therefore, patients need to inform their healthcare providers of any goldenseal supplements they are taking, especially if they are on medications metabolized by CYP2D6 or CYP3A4.

## 3.11 Grapefruit juice

Grape or *Vitis vinifera* L. (family Vitaceae) fruit juice is known to interact with a wide variety of drugs, including statins, calcium channel blockers, immunosuppressants, benzodiazepines, and antidepressants [102, 103]. It is well known for its potential to inhibit the activity of CYP3A4 enzyme, which is involved in the metabolism of many drugs [104–106]. Furanocoumarins are a class of compounds found in grapefruit juice that are responsible for inhibiting the activity of the CYP3A4 enzyme [105, 107]. Studies have shown that coadministration of grapefruit juice with CYP3A4-mediated metabolizing drugs can lead to a significant increase in drug absorption, potentially leading to increased drug concentrations in the blood and an increased risk of adverse effects [108, 109].

Grapefruit juice significantly enhanced the increase in the exposure (measured as the area under the curve or AUC) of sildenafil in the body from 620 to 761 ng/mL per hour. Additionally, it caused a delay in the time it took for the maximum concentration of sildenafil ($T_{max}$) to be reached from 0.75 to 1.13 h [110]. The absorption of drugs such as celiprolol, acebutolol, talinolol, and fexofenadine is known to be affected by grapefruit juice [111]. Grapefruit juice reduces the bioavailability of fexofenadine in two healthy volunteers. In the first subject, the AUC for fexofenadine was reduced from 3,032 ng/mL per hour (when administered with water) to 1,185 ng/mL per hour (when administered with grapefruit juice). This represents a reduction of approximately 61%. In the second subject, the AUC for fexofenadine was reduced from 2,015 ng/mL per hour (when administered with water) to 1,427 ng/mL per hour (when administered with grapefruit juice). This represents a reduction of approximately 29% [111]. These reduced concentrations were observed due to the inhibition of uptake transporters such as OATP1A2.

## 3.12 Green tea extract

Green tea is obtained from the leaves of *Camellia sinensis* (L.) Kuntze (family Theaceae), which is known to undergo minimal oxidation during processing. It contains catechins, including epigallocatechin gallate (EGCG), epicatechin, epigallocatechin, and epicatechin gallate, in high concentrations and is a powerful antioxidant [112, 113]. Green tea catechins can inhibit CYP3A4 activity, which could potentially increase the levels of simvastatin in the blood and cause adverse effects. Additionally, green tea catechins have been shown to inhibit the intestinal multidrug *P*-gp efflux pump, which is involved in the absorption and excretion of many drugs, including simvastatin. This inhibition could also lead to increased levels of simvastatin in the blood [114]. While some studies have suggested that green tea extract and/or EGCG can increase the bioavailability of simvastatin, other studies have reported conflicting results [115]. They found that green tea extract and EGCG can inhibit the activity of several CYP enzymes, including CYP1A1, CYP1A2, CYP2B6, CYP2C8, CYP2C9, CYP2D6, and CYP3A4, which are involved in the metabolism of simvastatin. Inhibition of these enzymes could potentially increase the levels of simvastatin in the blood and enhance its bioavailability.

Green tea extract has also been shown to inhibit the drug transporters OATP1A1 and OATP1A2, which are involved in the transport of many medications, including statins, fluoroquinolones, some beta-blockers, and antiretrovirals. One study showed that EGCG, a major component of green tea extract, inhibited the activity of OATP1A1 and OATP1A2 *in vitro*, potentially leading to decreased uptake of drugs into cells [116]. Another study showed that green tea extract decreased the absorption and bioavailability of the fluoroquinolone antibiotic levofloxacin, which is a substrate of OATP1A2, in healthy volunteers [117]. They found that green tea extract inhibited the absorption of the beta-blocker nadolol by inhibiting the intestinal uptake transporter OATP1A2. This inhibition led to a decrease in the bioavailability of nadolol, which could potentially result in decreased efficacy of the drug. A significant decrease in the maximum plasma concentration ($C_{max}$) and the area under the plasma concentration–time curve (AUC) of nadolol was observed. Specifically, the $C_{max}$ and the AUC 0–48 h of nadolol were decreased by 85.3% and 85.0%, respectively, compared to when nadolol was administered alone.

## 3.13 Kava kava

*Piper methysticum* G.Forst. (family Piperaceae) is commonly known as kava kava and is native to the Pacific Islands, including Fiji, Tonga, and Hawaii. Kava kava has been used for centuries in traditional medicine and cultural ceremonies, particularly for its anxiolytic and sedative effects [118]. The active components of kava kava are called kavalactones [119], which are thought to work by enhancing the activity of gamma-aminobutyric acid, a neurotransmitter that helps to regulate anxiety and promote relaxation [120]. Kava kava is typically consumed as a beverage made from the root of the plant, which is pounded and mixed with water. *In vitro* studies have shown that kava kava extract can inhibit several CYP enzymes, including CYP3A4, CYP2C9, CYP2C19, CYP1A2, and CYP2D6. These enzymes are involved in the metabolism of many drugs, including common medications such as antidepressants, antipsychotics, and antianxiety drugs. Therefore, the inhibition of these enzymes by kava kava extract can potentially affect the pharmacokinetics and efficacy of these drugs [121, 122]. Treatment with kava kava extracts at doses of 138 or 253.5 mg/day of kava lactones for 28 days or 14 days, respectively, did not have any significant influence on the activity of CYP2D6, CYP1A2, or CYP3A4/5 enzymes in healthy volunteers. This was determined by measuring single-time-point phenotypic metabolic ratios [86, 101]. However, treatment with kava kava extract for 28 days at a dose of 138 mg/day of kava lactones induced a statistically significant reduction by approximately 40% in CYP2E1 activity [101].

## 3.14 Milk thistle

Milk thistle, or *Silybum marianum* (L.) Gaertn. (family Asteraceae), has been extensively studied for its potential therapeutic properties, particularly for liver disorders [123, 124]. Based on the available human studies, it has been suggested that milk thistle does not have inhibitory or inductive effects on CYP1A2, CYP2D6, CYP2E1, CYP3A4, or *P*-gp [85, 86, 91, 125–132]. This means that milk thistle is unlikely to significantly alter the metabolism or pharmacokinetics of drugs that are substrates for these enzymes and transporters. There is some evidence to suggest that milk thistle may reduce the metabolism of losartan, a drug used to treat high blood pressure, depending on the CYP2C9 genotype. Specifically, individuals with a variant form of the CYP2C9 enzyme may be at greater risk of decreased losartan metabolism when taking milk thistle

[133]. In addition, there is potential for milk thistle to decrease the concentrations of other medications metabolized by CYP2C9, such as warfarin, phenytoin, and diazepam. A study conducted by Moltó et al. [134] investigated the effect of milk thistle on the pharmacokinetic profile of the darunavir–ritonavir (600/100 twice daily) combination in 15 HIV-infected patients. The results showed a decrease in AUC and $C_{max}$ when the combination of drugs was coadministered with silymarin. However, this change was found to be nonsignificant, and the authors suggested that the coadministration of milk thistle products with antiviral protease inhibitors does not require a dose adjustment and is well tolerated and safe.

## 3.15 Saw palmetto

Saw palmetto (*Serenoa repens* (W.Bartram) small; family Arecaceae) is a commonly used herbal supplement for treating benign prostatic hypertrophy and androgenetic alopecia [135–138]. Several studies have investigated the effects of saw palmetto on CYP enzymes. Multiple human trials have shown that saw palmetto has no inhibitory or inductive effects on CYP1A2, CYP2D6, CYP2E1, or CYP3A4 [85, 139].

## 3.16 St. John's wort

St. John's wort (*Hypericum perforatum* L.; family Hypericaceae) extracts are widely used as a safe and effective alternative to conventional antidepressant drugs for mild to moderate forms of depressive disorders [140]. The plant extract of St. John's wort contains several active compounds like hypericin and hyperforin [141, 142]. St. John's wort has been shown in numerous human studies to induce the activity of the cytochrome P450 enzymes CYP3A4 and *P*-gp [86, 87, 143–145]. Piscitelli et al. [146] found that a 2-week treatment with *H. perforatum* extract reduced the AUC of indinavir, an HIV protease inhibitor, by an average of 57% in healthy volunteers. Additionally, the extrapolated 8-h indinavir trough was reduced by 81% in the same volunteers. Additionally, several clinical studies have reported that St. John's wort significantly modified the pharmacokinetics of several drugs that are substrates of various enzymes and transporters. These enzymes and transporters include CYP3A, CYP2C9, CYP2C19, CYP2E1, and *P*-gp. Some of the drugs affected include midazolam [147], simvastatin [148], amitriptyline [149], chlorzoxazone [41, 97], methadone [150], oral contraceptives containing ethinyl estradiol/norethindrone [151, 152], fexofenadine [153], warfarin [43], imatinib [154], omeprazole [155], mephenytoin [157], tacrolimus [158], verapamil [157], voriconazole [159], talinolol [160], gliclazide [161], nifedipine [158], ketamine [162], zolpidem [163], bosentan [164], ambrisentan [165], and docetaxel [166]. It is strongly recommended to avoid the concurrent use of St. John's wort with over-the-counter and prescription medications [167].

# 4 Summary

This chapter highlights the importance of understanding the potential risks and benefits of combining herbal supplements or botanicals with prescribed drugs. Herbal products are commonly used by many individuals as natural alternatives to prescription drugs, but they can interact with drugs and may increase the likelihood of adverse effects. Currently, herbal products are not subject to the same regulations as prescription drugs, and manufacturers are not required to prove the safety or efficacy of their products before they are marketed. Therefore, this chapter emphasizes the need for proper communication between patients and healthcare providers to ensure safe and effective use of these products. Patients should also be educated on the potential risks and benefits of combining herbal extracts and/or bioactive components with prescription drugs and should seek medical advice before starting any new herbal therapy.

The safety and efficacy of using herbs and dietary supplements in combination with conventional medicines are still not well understood. It is crucial to adhere to the existing guidelines for conducting clinical trials on the interactions between natural extracts and synthetic drugs to ensure the safety and effectiveness of such combinations. Most current data on this topic are focused on interactions involving metabolic enzymes and carriers. It is essential to explore other potential mechanisms to gain a comprehensive understanding of how herbal products interact with conventional drugs. Therefore, further research and clinical studies on herbal supplements and drug interactions are necessary to provide healthcare professionals and patients with reliable information regarding their safety and efficacy.

**Acknowledgements:** All authors express their sincere gratitude for their institutional support.

# References

[1]   Liao, Y., Shonkoff, E.T. and Dunton, G.F. 2015. The acute relationships between affect, physical feeling states, and physical activity in daily life: A review of current evidence. Frontiers in Psychology, 6, 1975.
[2]   Steptoe, A., Deaton, A. and Stone, A.A. 2015. Subjective wellbeing, health, and ageing. Lancet, 385(9968), 640–648.
[3]   Lim, S. 2018. Eating a balanced diet: A healthy life through a balanced diet in the age of longevity. Journal of Obesity & Metabolic Syndrome, 27(1), 39–45.
[4]   Wu, Q., Gao, Z.-J., Yu, X. and Wang, P. 2022. Dietary regulation in health and disease. Signal Transduction and Targeted Therapy, 7(1), 252.
[5]   Fehér, A., Gazdecki, M., Véha, M., Szakály, M. and Szakály, Z. 2020. A comprehensive review of the benefits of and the barriers to the switch to a plant-based diet. Sustainability, 12(10), 4136.

[6]    Domínguez Díaz, L., Fernández-Ruiz, V. and Cámara, M. 2020. The frontier between nutrition and pharma: The international regulatory framework of functional foods, food supplements and nutraceuticals. Critical Reviews in Food Science and Nutrition, 60(10), 1738–1746.

[7]    Market Data Forecast Ltd. Nutraceuticals Market. Market Data Forecast. [retrieved on 2023 Oct 12]. https://www.marketdataforecast.com/market-reports/global-nutraceuticals-market/request-sample.

[8]    Keservani, R.K., Kesharwani, R.K., Sharma, A.K., Gautam, S.P. and Verma, S.K. 2017. Nutraceutical formulations and challenges. In *Developing New Functional Food and Nutraceutical Products*. Elsevier, pp. 161–177.

[9]    Nasri, H., Baradaran, A., Shirzad, H. and Rafieian-Kopaei, M. 2014. New concepts in nutraceuticals as alternative for pharmaceuticals. International Journal of Preventive Medicine, 5(12), 1487–1499. PMID: 25709784.

[10]   Ozdal, T., Tomas, M., Toydemir, G., Kamiloglu, S. and Capanoglu, E. 2021. Introduction to nutraceuticals, medicinal foods, and herbs. In Galanakis, C.M. (Ed.), *Aromatic Herbs in Food*. New York, USA: Academic Press, pp. 1–34.

[11]   Almazroo, O.A., Miah, M.K. and Venkataramanan, R. 2017. Drug metabolism in the liver. Clinical Liver Disease, 21(1), 1–20.

[12]   Nebert, D.W. and Dalton, T.P. 2006. The role of cytochrome P450 enzymes in endogenous signalling pathways and environmental carcinogenesis. Nature Reviews Cancer, 6(12), 947–960.

[13]   Mackenzie, P.I., Bock, K.W., Burchell, B., Guillemette, C., Ikushiro, S.-I., Iyanagi, T. et al. 2005. Nomenclature update for the mammalian UDP glycosyltransferase (UGT) gene superfamily. Pharmacogenet Genomics, 15(10), 677–685.

[14]   Mano, E.C.C., Scott, A.L. and Honorio, K.M. 2018. UDP-glucuronosyltransferases: Structure, function and drug design studies. Current Medicinal Chemistry, 25(27), 3247–3255.

[15]   Mazerska, Z., Mróz, A., Pawłowska, M. and Augustin, E. 2016. The role of glucuronidation in drug resistance. Pharmacology Therapeutics, 159, 35–55.

[16]   Nair, P.C., Meech, R., Mackenzie, P.I., McKinnon, R.A. and Miners, J.O. 2015. Insights into the UDP-sugar selectivities of human UDP-glycosyltransferases (UGT): A molecular modeling perspective. Drug Metabolism Reviews, 47(3), 335–345.

[17]   MacKenzie, P.I. Rogers, A., Elliot, D.J., Chau, N., Hulin, J.-A., Miners, J.O., et al. 2011. The novel UDP glycosyltransferase 3A2: Cloning, catalytic properties, and tissue distribution. Molecular Pharmacology, 79(3), 472–478.

[18]   Li, Y., Meng, Q., Yang, M., Liu, D., Hou, X., Tang, L., et al. 2019. Current trends in drug metabolism and pharmacokinetics. Acta pharmaceutica Sinica B, 9(6), 1113–1144.

[19]   Dean, M., Rzhetsky, A. and Allikmets, R. 2001. The human ATP-binding cassette (ABC) transporter superfamily. Genome Research, 11(7), 1156–1166.

[20]   Shugarts, S. and Benet, L.Z. 2009. The role of transporters in the pharmacokinetics of orally administered drugs. Pharmaceutical Research: An Official Journal of the American Association of Pharmaceutical Scientists, 26(9), 2039–2054.

[21]   Javdan, B., Lopez, J.G., Chankhamjon, P., Lee, Y.-C.J., Hull, R., Wu, Q., et al. 2020. Personalized mapping of drug metabolism by the human gut microbiome. Cell, 181(7), 1661–1679. e22.

[22]   Li, H., He, J. and Jia, W. 2016. The influence of gut microbiota on drug metabolism and toxicity. Expert Opinion on Drug Metabolism Toxicology, 12(1), 31–40.

[23]   Wilson, I.D. and Nicholson, J.K. 2017. Gut microbiome interactions with drug metabolism, efficacy, and toxicity. Translational Research, 179, 204–222.

[24]   Roden, D.M. and George Jr, A.L. 2002. The genetic basis of variability in drug responses. Nature Reviews Drug Discovery, 1(1), 37–44.

[25]   Koziolek, M., Alcaro, S., Augustijns, P., Basit, A.W., Grimm, M., Hens, B. et al. 2019. The mechanisms of pharmacokinetic food-drug interactions – A perspective from the UNGAP group. European Journal of Pharmaceutical Sciences, 134, 31–59.

[26] Deng, J., Zhu, X., Chen, Z., Fan, C.H., Kwan, H.S., Wong, C.H., et al. 2017. A review of food–drug interactions on oral drug absorption. Drugs, 77(17), 1833–1855.

[27] Schmidt, L.E. and Dalhoff, K. 2002. Food-drug interactions. Drugs, 62(10), 1481–1502.

[28] Uchida, S. and Yamada, S. 2007. Food, dietary supplement and drug interactions. Bunseki, September Edition, 454–460.

[29] Sawada, Y. and Satoh, H. 2012. Interaction between prescription drugs and OTC drugs, health food and dietary supplements. Farumashia, 48, 1062–1066.

[30] Sawada, Y. and Ohtani, H. 2002. Interaction of medicine and food-Clinical mishaps due to the encounter of food, recreational products and drugs. Drug enhancement of detoxification diminishes the pharmacological effect; Drugs and St. John's wort (1). Commentary. Iyaku (Medicine & Drug) Journal, 38, 1791–1799.

[31] Yohkoh, N. 2015. Drug-dietary supplement interactions. In Ghosh, D., Bagchi, D., Konishi, T. (Eds.), *Clinical Aspects of Functional Foods and Nutraceuticals*. Florida, USA: Taylor & Francis Group, pp. 163–172.

[32] Jinbiao, L., Xinyue, Z., Shenshen, Y., Shuo, W., Chengcheng, L., Bin, Y., et al. 2022. Rapid identification of characteristic chemical constituents of panax ginseng, panax quinquefolius, and panax japonicus using UPLC-Q-TOF/MS. Journal of Analytical Methods in Chemistry, 2022, 6463770.

[33] Ratan, Z.A., Haidere, M.F., Hong, Y.H., Park, S.H., Lee, J.-O., Lee, J., et al. 2021. Pharmacological potential of ginseng and its major component ginsenosides. Journal of Ginseng Research, 45(2), 199–210.

[34] Andrade, A.S.A., Hendrix, C., Parsons, T.L., Caballero, B., Yuan, C.-S., Flexner, C.W., et al. 2008. Pharmacokinetic and metabolic effects of American ginseng (*Panax quinquefolius*) in healthy volunteers receiving the HIV protease inhibitor indinavir. BMC Complementary Medicine and Therapies, 8(1), 50.

[35] Lee, L.S., Wise, S.D., Chan, C., Parsons, T.L., Flexner, C. and Lietman, P.S. 2008. Possible differential induction of phase 2 enzyme and antioxidant pathways by american ginseng, *Panax quinquefolius*. Journal of Clinical Pharmacology, 48(5), 599–609.

[36] Yuan, C.-S., Wei, G., Dey, L., Karrison, T., Nahlik, L., Maleckar, S., et al. 2004. Brief communication: American ginseng reduces warfarin's effect in healthy patients: A randomized, controlled trial. Annals of Internal Medicine, 141(1), 23–27.

[37] Lemke, E.A. 2021. Ginseng for the management of cancer-related fatigue: An integrative review. Journal of the Advanced Practitioner in Oncology, 12(4), 406–414.

[38] Sadeghian, M., Rahmani, S., Zendehdel, M., Hosseini, S.A. and Zare Javid, A. 2021. Ginseng and cancer-related fatigue: A systematic review of clinical trials. Nutrition and Cancer, 73(8), 1270–1281.

[39] Malati, C.Y., Robertson, S.M., Hunt, J.D., Chairez, C., Alfaro, R.M., Kovacs, J.A., et al. 2012. Influence of Panax ginseng on cytochrome P450 (CYP)3A and *P*-glycoprotein (*P*-gp) activity in healthy participants. Journal of Clinical Pharmacology, 52(6), 932–939.

[40] Anderson, G.D., Rosito, G., Mohustsy, M.A. and Elmer, G.W. 2003. Drug interaction potential of soy extract and Panax ginseng. Journal of Clinical Pharmacology, 43(6), 643–648.

[41] Gurley, B.J., Gardner, S.F., Hubbard, M.A., Williams, D.K., Gentry, W.B., Cui, Y., et al. 2005. Clinical assessment of effects of botanical supplementation on cytochrome P450 phenotypes in the elderly: St John's wort, garlic oil, *Panax ginseng* and *Ginkgo biloba*. Drugs Aging, 22(6), 525–539.

[42] Janetzky, K. and Morreale, A.P. 1997. Probable interaction between warfarin and ginseng. American Journal of Health System Pharmacy, 54(6), 692–693.

[43] Jiang, X., Williams, K.M., Liauw, W.S., Ammit, A.J., Roufogalis, B.D., Duke, C.C., et al. 2004. Effect of St John's wort and ginseng on the pharmacokinetics and pharmacodynamics of warfarin in healthy subjects. British Journal of Clinical Pharmacology, 57(5), 592–599.

[44] Lee, S.-H., Ahn, Y.-M., Ahn, S.-Y., Doo, H.-K. and Lee, B.-C. 2008. Interaction between warfarin and Panax ginseng in ischemic stroke patients. Journal of Alternative and Complementary Medicine, 14(6), 715–721.

[45] Lee, Y.H., Lee, B.K., Choi, Y.J., Yoon, I.K., Chang, B.C. and Gwak, H.S. 2010. Interaction between warfarin and Korean red ginseng in patients with cardiac valve replacement. International Journal of Cardiology, 145(2), 275–276.

[46] Bilgi, N., Bell, K., Ananthakrishnan, A.N. and Atallah, E. 2010. Imatinib and Panax ginseng: A potential interaction resulting in liver toxicity. Annals of Pharmacotherapy, 44(5), 926–928.

[47] Guo, Y., Yin, T., Wang, X., Zhang, F., Pan, G., Lv, H., et al. 2017. Traditional uses, phytochemistry, pharmacology and toxicology of the genus Cimicifuga: A review. Journal of Ethnopharmacology, 209, 264–282.

[48] Johnson, T.L. and Fahey, J.W. 2012. Black cohosh: Coming full circle? Journal of Ethnopharmacology, 141(3), 775–779.

[49] Crone, M., Hallman, K., Lloyd, V., Szmyd, M., Badamo, B., Morse, M., et al. 2019. The antiestrogenic effects of black cohosh on BRCA1 and steroid receptors in breast cancer cells. Breast Cancer, 11, 99–110.

[50] Tsukamoto, S., Aburatani, M. and Ohta, T. 2005. Isolation of CYP3A4 Inhibitors from the Black Cohosh (*Cimicifuga racemosa*). Evidence-Based Complementary and Alternative Medicine, 2(2), 223–226.

[51] Chow, E.C.-Y., Teo, M., Ring, J.A. and Chen, J.W. 2008. Liver failure associated with the use of black cohosh for menopausal symptoms. Medical Journal Australia, 188(7), 420–422.

[52] Fuchikami, H., Satoh, H., Tsujimoto, M., Ohdo, S., Ohtani, H. and Sawada, Y. 2006. Effects of herbal extracts on the function of human organic anion-transporting polypeptide OATP-B. Drug Metabolism & Disposition, 34(4), 577–582.

[53] Fu, Z., Liska, D., Talan, D. and Chung, M. 2017. Cranberry reduces the risk of urinary tract infection recurrence in otherwise healthy women: A systematic review and meta-analysis. Journal of Nutrition, 147(12), 2282–2288.

[54] Valente, J., Pendry, B.A. and Galante, E. 2022. Cranberry (*Vaccinium macrocarpon*) as a prophylaxis for urinary tract infections in women: A systematic review with meta-analysis. Journal of Herbal Medicine, 36(100602), 100602.

[55] Asher, G.N., Corbett, A.H. and Hawke, R.L. 2017. Common herbal dietary supplement-drug interactions. American Family Physician, 96(2), 101–107.

[56] Greenblatt, D.J., Von Moltke, L.L., Perloff, E.S., Luo, Y., Harmatz, J.S. and Zinny, M.A. 2006. Interaction of flurbiprofen with cranberry juice, grape juice, tea, and fluconazole: In vitro and clinical studies. Clinical Pharmacology & Therapeutics, 79(1), 125–133.

[57] Lilja, J.J., Backman, J.T. and Neuvonen, P.J. 2007. Effects of daily ingestion of cranberry juice on the pharmacokinetics of warfarin, tizanidine, and midazolam-probes of CYP2C9, CYP1A2, and CYP3A4. Clinical Pharmacology & Therapeutics, 81(6), 833–839.

[58] Ushijima, K., Tsuruoka, S.-I., Tsuda, H., Hasegawa, G., Obi, Y., Kaneda, T., et al. 2009. Cranberry juice suppressed the diclofenac metabolism by human liver microsomes, but not in healthy human subjects. British Journal of Clinical Pharmacology, 68(2), 194–200.

[59] Pham, D.Q. and Pham, A.Q. 2007. Interaction potential between cranberry juice and warfarin. American Journal of Health System Pharmacy, 64(5), 490–494.

[60] Aston, J.L., Lodolce, A.E. and Shapiro, N.L. 2006. Interaction between warfarin and cranberry juice. Pharmacotherapy, 26(9), 1314–1319.

[61] Grant, P. 2004. Warfarin and cranberry juice: An interaction? Journal Heart Valve disease, 13(1), 25–26. PMID: 14765835.

[62] Griffiths, A.P., Beddall, A. and Pegler, S. 2008. Fatal haemopericardium and gastrointestinal haemorrhage due to possible interaction of cranberry juice with warfarin. Journal of the Royal Society for the Promotion of Health, 128(6), 324–326.

[63] Mergenhagen, K.A. and Sherman, O. 2008. Elevated International Normalized Ratio after concurrent ingestion of cranberry sauce and warfarin. American Journal of Health System Pharmacy, 65(22), 2113–2116.

[64] Abdul, M., Jiang, M.I., Williams, X., Day, K.M., Roufogalis, R.O., Liauw, B.D., et al. 2008. Pharmacodynamic interaction of warfarin with cranberry but not with garlic in healthy subjects: Warfarin interacts with cranberry not garlic. British Journal Pharmacology, 154(8), 1691–1700.

[65] Paeng, C.H., Sprague, M. and Jackevicius, C.A. 2007. Interaction between warfarin and cranberry juice. Clinical Therapeutics, 29(8), 1730–1735.

[66] Rindone, J.P. and Murphy, T.W. 2006. Warfarin-cranberry juice interaction resulting in profound hypoprothrombinemia and bleeding. American Journal of Therapeutics, 13(3), 283–284.

[67] Suvarna, R. 2003. Possible interaction between warfarin and cranberry juice. BMJ, 327(7429), 1454–1454.

[68] Zikria, J., Goldman, R. and Ansell, J. 2010. Cranberry juice and warfarin: When bad publicity trumps science. The American Journal of Medicine, 123(5), 384–392.

[69] Mellen, C.K., Ford, M. and Rindone, J.P. 2010. Effect of high-dose cranberry juice on the pharmacodynamics of warfarin in patients: Effect of high-dose cranberry juice. British Journal of Clinical Pharmacology, 70(1), 139–142.

[70] Chen, Y., Liu, W.-H., Chen, B.-L., Fan, L., Han, Y., Wang, G., et al. 2010. Plant polyphenol curcumin significantly affects CYP1A2 and CYP2A6 activity in healthy, male Chinese volunteers. Annals of Pharmacotherapy, 44(6), 1038–1045.

[71] Kusuhara, H., Furuie, H., Inano, A., Sunagawa, A., Yamada, S., Wu, C., et al. 2012. Pharmacokinetic interaction study of sulphasalazine in healthy subjects and the impact of curcumin as an in vivo inhibitor of BCRP: Impact of curcumin as in vivo inhibitor of BCRP. British Journal Pharmacology, 166(6), 1793–1803.

[72] Volak, L.P., Hanley, M.J., Masse, G., Hazarika, S., Harmatz, J.S., Badmaev, V., et al. 2013. Effect of a herbal extract containing curcumin and piperine on midazolam, flurbiprofen and paracetamol (acetaminophen) pharmacokinetics in healthy volunteers: Curcuminoid/piperine pharmacokinetic interactions. British Journal of clinical pharmacology, 75(2), 450–462.

[73] Bahramsoltani, R., Rahimi, R. and Farzaei, M.H., 2017. Pharmacokinetic interactions of curcuminoids with conventional drugs: A review. Journal of Ethnopharmacology, 209, 1–12.

[74] Cheng, T.O., 2007. Cardiovascular effects of danshen. International Journal of Cardiology, 121(1), 9–22.

[75] Zhou, L., Zuo, Z. and Chow, M.S.S., 2005. Danshen: An overview of its chemistry, pharmacology, pharmacokinetics, and clinical use. Journal of Clinical Pharmacology, 45(12), 1345–1359.

[76] Han, J.-Y., Fan, J.-Y., Horie, Y., Miura, S., Cui, D.-H., Ishii, H., et al. 2008. Ameliorating effects of compounds derived from *Salvia miltiorrhiza* root extract on microcirculatory disturbance and target organ injury by ischemia and reperfusion. Pharmacology Therapeutics, 117(2), 280–295.

[77] Ho, J.H.-C. and Hong, C.-Y. 2011. Salvianolic acids: Small compounds with multiple mechanisms for cardiovascular protection. Journal of Biomedical Science, 18, 30.

[78] Hu, P., Liang, Q.-L., Luo, G.-A., Zhao, -Z.-Z. and Jiang, Z.-H. 2005. Multi-component HPLC fingerprinting of Radix Salviaemiltiorrhizae and its LC-MS-MS identification. Chemical and Pharmaceutical Bulletin, 53(6), 677–683.

[79] Chan, T.Y. 2001. Interaction between warfarin and danshen (*Salvia miltiorrhiza*). Ann Pharmacotherapy, 35(4), 501–504.

[80] Izzat, M.B., Yim, A.P. and El-Zufari, M.H. 1998. A taste of Chinese medicine! Annals of Thoracic and Cardiovascular Surgery, 66(3), 941–942.

[81]   Tam, L.S., Chan, T.Y., Leung, W.K. and Critchley, J.A. 1995. Warfarin interactions with Chinese traditional medicines: Danshen and methyl salicylate medicated oil. Australian and New Zealand Journal of Public Health, 25(3), 258.

[82]   Yu, C.M., Chan, J.C. and Sanderson, J.E. 1997. Chinese herbs and warfarin potentiation by "danshen". Journal of Internal Medicine, 241, 337–339.

[83]   Qiu, F., Wang, G., Zhang, R., Sun, J., Jiang, J. and Ma, Y. 2010. Effect of danshen extract on the activity of CYP3A4 in healthy volunteers: Danshen extract and CYP3A4 induction. British Journal of Clinical Pharmacology, 69(6), 656–662.

[84]   Gorski, J.C., Huang, S.-M., Pinto, A., Hamman, M.A., Hilligoss, J.K., Zaheer, N.A., et al. 2004. The effect of echinacea (*Echinacea purpurea* root) on cytochrome P450 activity in vivo. Clinical Pharmacology & Therapeutics, 75(1), 89–100.

[85]   Gurley, B.J., Gardner, S.F., Hubbard, M.A., Williams, D.K., Gentry, W.B., Carrier, J., et al. 2004. In vivo assessment of botanical supplementation on human cytochrome P450 phenotypes: Citrus aurantium, *Echinacea purpurea*, milk thistle, and saw palmetto. Clinical Pharmacology & Therapeutics, 76(5), 428–440.

[86]   Gurley, B.J., Swain, A., Hubbard, M.A., Williams, D.K., Barone, G., Hartsfield, F., et al. 2008. Clinical assessment of CYP2D6-mediated herb-drug interactions in humans: Effects of milk thistle, black cohosh, goldenseal, kava kava, St. John's Wort, and Echinacea. Molecular Nutrition & Food Research, 52(7), 755–763.

[87]   Gurley, B.J., Swain, A., Williams, D.K., Barone, G. and Battu, S.K., 2008. Gauging the clinical significance of *P*-glycoprotein-mediated herb-drug interactions: Comparative effects of St. John's wort, Echinacea, clarithromycin, and rifampin on digoxin pharmacokinetics. Molecular Nutrition & Food Research, 52(7), 772–779.

[88]   Moltó, J., Valle, M., Miranda, C., Cedeño, S., Negredo, E., Barbanoj, M.J., et al. 2011. Herb-drug interaction between *Echinacea purpurea* and darunavir-ritonavir in HIV-infected patients. Antimicrob Agents and Chemother, 55(1), 326–330.

[89]   Ansary, J., Forbes-Hernández, T.Y., Gil, E., Cianciosi, D., Zhang, J., Elexpuru-Zabaleta, M., et al. 2020. Potential health benefit of garlic based on human intervention studies: A brief overview. Antioxidants (Basel), 9(7), 619.

[90]   Capasso, F., Gaginella, T.S., Grandolini, G. and Izzo, A.A. 2003. Introduction. In *Phytotherapy: A Quick Reference to Herbal Medicine. E*. Berlin, Heidelberg: Springer Berlin Heidelberg, pp. 3–6. ISBN: 3-540-00052-6.

[91]   Gurley, B., Hubbard, M.A., Williams, D.K., Thaden, J., Tong, Y., Gentry, W.B., et al. 2006. Assessing the clinical significance of botanical supplementation on human cytochrome P450 3A activity: Comparison of a milk thistle and black cohosh product to rifampin and clarithromycin. Journal of Clinical Pharmacology, 46(2), 201–213.

[92]   Piscitelli, S.C., Burstein, A.H., Welden, N., Gallicano, K.D. and Falloon, J. 2002. The effect of garlic supplements on the pharmacokinetics of saquinavir. Clinical Infectious Diseases, 34(2), 234–238.

[93]   German, K., Kumar, U. and Blackford, H.N. 1995. Garlic and the risk of TURP bleeding. British Journal of Urology, 76(4), 518.

[94]   Yin, Y., Ren, Y., Wu, W., Wang, Y., Cao, M., Zhu, Z., et al. 2013. Protective effects of bilobalide on Aβ25–35 induced learning and memory impairments in male rats. Pharmacology, Biochemistry and Behavior, 106, 77–84.

[95]   Yuan, Q., Wang, C.-W., Shi, J. and Lin, Z.-X. 2017. Effects of *Ginkgo biloba* on dementia: An overview of systematic reviews. Journal of Ethnopharmacology, 195, 1–9.

[96]   Gertz, H.-J. and Kiefer, M. 2004. Review about *Ginkgo biloba* special extract EGb 761 (Ginkgo). Current Pharmaceutical Design, 10(3), 261–264.

[97]   Gurley, B.J., Gardner, S.F., Hubbard, M.A., Williams, D.K., Gentry, W.B., Cui, Y., et al. 2002.
       Cytochrome P450 phenotypic ratios for predicting herb-drug interactions in humans. Clinical
       Pharmacology & Therapeutics, 72(3), 276–287.
[98]   Markowitz, J.S., Donovan, J.L., Lindsay Devane, C., Sipkes, L. and Chavin, K.D. 2003. Multiple-dose
       administration of *Ginkgo biloba* did not affect cytochrome P-450 2D6 or 3A4 activity in normal
       volunteers. Journal of Clinical Psychopharmacology, 23(6), 576–581.
[99]   Yin, O.Q.P., Tomlinson, B., Waye, M.M.Y., Chow, A.H.L. and Chow, M.S.S. 2004. Pharmacogenetics
       and herb-drug interactions: Experience with *Ginkgo biloba* and omeprazole. Pharmacogenetics,
       14(12), 841–850.
[100]  Mandal, S.K., Maji, A.K., Mishra, S.K., Ishfaq, P.M., Devkota, H.P., Silva, A.S., et al. 2020. Goldenseal
       (*Hydrastis canadensis* L.) and its active constituents: A critical review of their efficacy and
       toxicological issues. Pharmacological Research, 160, 105085.
[101]  Gurley, B.J., Gardner, S.F., Hubbard, M.A., Williams, D.K., Gentry, W.B., Khan, I.A., et al. 2005. In vivo
       effects of goldenseal, kava kava, black cohosh, and valerian on human cytochrome P450 1A2, 2D6,
       2E1, and 3A4/5 phenotypes. Clinical Pharmacology & Therapeutics, 77(5), 415–426.
[102]  Hanley, M.J., Cancalon, P., Widmer, W.W. and Greenblatt, D.J. 2011. The effect of grapefruit juice on
       drug disposition. Expert Opinion on Drug Metabolism Toxicology, 7(3), 267–286.
[103]  Uno, T. and Yasui-Furukori, N. 2006. Effect of grapefruit juice in relation to human pharmacokinetic
       study. Current Clinical Pharmacology, 1(2), 157–161.
[104]  Ameer, B. and Weintraub, R.A. 1997. Drug interactions with grapefruit juice. Clinical
       Pharmacokinetics, 33(2), 103–121.
[105]  Dresser, G.K., Spence, J.D. and Bailey, D.G. 2000. Pharmacokinetic-pharmacodynamic consequences
       and clinical relevance of cytochrome P450 3A4 inhibition. Clinical Pharmacokinetics, 38(1), 41–57.
[106]  Murray, M. 2006. Altered CYP expression and function in response to dietary factors: Potential roles
       in disease pathogenesis. Current Drug Metabolism, 7(1), 67–81.
[107]  Paine, M.F., Widmer, W.W., Hart, H.L., Pusek, S.N., Beavers, K.L., Criss, A.B., et al. 2006. A
       furanocoumarin-free grapefruit juice establishes furanocoumarins as the mediators of the
       grapefruit juice-felodipine interaction. The American Journal of Clinical Nutrition, 83(5), 1097–1105.
[108]  Glaeser, H., Bailey, D.G., Dresser, G.K., Gregor, J.C., Schwarz, U.I., McGrath, J.S., et al. 2007. Intestinal
       drug transporter expression and the impact of grapefruit juice in humans. Clinical Pharmacology &
       Therapeutics, 81(3), 362–370.
[109]  Wagner, D., Spahn-Langguth, H., Hanafy, A., Koggel, A. and Langguth, P. 2001. Intestinal drug
       efflux: Formulation and food effects. Advanced Drug Delivery Reviews, 50(Suppl 1), S13–31.
[110]  Jetter, A., Kinzig-Schippers, M., Walchner-Bonjean, M., Hering, U., Bulitta, J., Schreiner, P., et al. 2002.
       Effects of grapefruit juice on the pharmacokinetics of sildenafil. Clinical Pharmacology &
       Therapeutics, 71(1), 21–29.
[111]  Bailey, D.G. 2010. Fruit juice inhibition of uptake transport: A new type of food-drug interaction:
       Fruit juice-drug interactions. British Journal of Clinical pharmacology, 70(5), 645–655.
[112]  Nobari, H., Saedmocheshi, S., Chung, L.H., Suzuki, K., Maynar-Mariño, M. and Pérez-Gómez, J. 2021.
       An overview on how exercise with green tea consumption can prevent the production of reactive
       oxygen species and improve sports performance. International Journal of Environmental Research
       and Public Health, 19(1), 218.
[113]  Saito, S.T., Gosmann, G., Saffi, J., Presser, M., Richter, M.F. and Bergold, A.M. 2007. Characterization
       of the constituents and antioxidant activity of Brazilian green tea (*Camellia sinensis* var. assamica
       IAC-259 cultivar) extracts. Journal of Agricultural and Food Chemistry, 55(23), 9409–9414.
[114]  Liu, A., Wu, Q., Guo, J., Ares, I., Rodríguez, J.-L., Martínez-Larrañaga, M.-R., et al. 2019. Statins:
       Adverse reactions, oxidative stress and metabolic interactions. Pharmacology and Therapeutics,
       195, 54–84.

[115] Albassam, A.A. and Markowitz, J.S. 2017. An appraisal of drug-drug interactions with green tea (*Camellia sinensis*). Planta Medica, 83(6), 496–508.

[116] Knop, J., Misaka, S., Singer, K., Hoier, E., Müller, F., Glaeser, H., et al. 2015. Inhibitory effects of green tea and (–)-epigallocatechin gallate on transport by OATP1B1, OATP1B3, OCT1, OCT2, MATE1, MATE2-K and P-glycoprotein. PLoS One, 10(10), e0139370.

[117] Misaka, S., Yatabe, J., Müller, F., Takano, K., Kawabe, K., Glaeser, H., et al. 2014. Green tea ingestion greatly reduces plasma concentrations of nadolol in healthy subjects. Clinical Pharmacology & Therapeutics, 95(4), 432–438.

[118] Bian, T., Corral, P., Wang, Y., Botello, J., Kingston, R., Daniels, T., et al. 2020. Kava as a clinical nutrient: Promises and challenges. Nutrients, 12(10), 3044.

[119] Olsen, L.R., Grillo, M.P. and Skonberg, C. 2011. Constituents in kava extracts potentially involved in hepatotoxicity: A review. Chemical Research in Toxicology, 24(7), 992–1002.

[120] Sarris, J., Byrne, G.J., Bousman, C.A., Cribb, L., Savage, K.M., Holmes, O., et al. 2020. Kava for generalised anxiety disorder: A 16-week double-blind, randomised, placebo-controlled study. Australian and New Zealand Journal of Psychiatry, 54(3), 288–297.

[121] Mathews, J.M., Etheridge, A.S. and Black, S.R. 2002. Inhibition of human cytochrome P450 activities by kava extract and kavalactones. Drug Metabolism & Disposition, 30(11), 1153–1157.

[122] Zou, L., Henderson, G.L., Harkey, M.R., Sakai, Y. and Li, A. 2004. Effects of kava (Kava-kava, 'Awa, Yaqona, *Piper methysticum*) on c-DNA-expressed cytochrome P450 enzymes and human cryopreserved hepatocytes. Phytomedicine, 11(4), 285–294.

[123] Abenavoli, L., Izzo, A.A., Milić, N., Cicala, C., Santini, A. and Capasso, R. 2018. Milk thistle (*Silybum marianum*): A concise overview on its chemistry, pharmacological, and nutraceutical uses in liver diseases: Milk thistle and liver diseases. Phytotherapy Research: PTR, 32(11), 2202–2213.

[124] Foghis, M., Bungau, S.G., Bungau, A.F., Vesa, C.M., Purza, A.L., Tarce, A.G., et al. 2023. Plants-based medicine implication in the evolution of chronic liver diseases. Biomedicine & Pharmacotherapy, 158, 114207.

[125] DiCenzo, R., Shelton, M., Jordan, K., Koval, C., Forrest, A., Reichman, R., et al. 2003. Coadministration of milk thistle and indinavir in healthy subjects. Pharmacotherapy, 23(7), 866–870.

[126] Fuhr, U., Beckmann-Knopp, S., Jetter, A., Lück, H. and Mengs, U. 2007. The effect of silymarin on oral nifedipine pharmacokinetics. Planta Medica, 73(14), 1429–1435.

[127] Gurley, B.J., Barone, G.W., Williams, D.K., Carrier, J., Breen, P., Yates, C.R., et al. 2006. Effect of milk thistle (*Silybum marianum*) and black cohosh (*Cimicifuga racemosa*) supplementation on digoxin pharmacokinetics in humans. Drug Metabolism & Disposition, 34(1), 69–74.

[128] Kawaguchi-Suzuki, M., Frye, R.F., Zhu, H.-J., Brinda, B.J., Chavin, K.D., Bernstein, H.J., et al. 2014. The effects of milk thistle (*Silybum marianum*) on human cytochrome P450 activity. Drug Metabolism & Disposition, 42(10), 1611–1616.

[129] Leber, H.W. and Knauff, S. 1976. Influence of silymarin on drug metabolizing enzymes in rat and man. Arzneimittelforschung, 26(8), 1603–1605.

[130] Mills, E., Wilson, K., Clarke, M., Foster, B., Walker, S., Rachlis, B., et al. 2005. Milk thistle and indinavir: A randomized controlled pharmacokinetics study and meta-analysis. European Journal of Clinical Pharmacology, 61(1), 1–7.

[131] Piscitelli, S.C., Formentini, E., Burstein, A.H., Alfaro, R., Jagannatha, S. and Falloon, J. 2002. Effect of milk thistle on the pharmacokinetics of indinavir in healthy volunteers. Pharmacotherapy, 22(5), 551–556.

[132] van Erp, P.H., Zhao, M., Gelderblom, H., Guchelaar, H.J., Sparreboom, A. and Baker, S.D. 2005. Effect of milk thistle (*Silybum marianum*) on the pharmacokinetics of irinotecan. Journal of Clinical Oncology, 23(16_suppl), 2096–2096.

[133] Han, Y., Guo, D., Chen, Y., Chen, Y., Tan, Z.-R. and Zhou, -H.-H. 2009. Effect of silymarin on the pharmacokinetics of losartan and its active metabolite E-3174 in healthy Chinese volunteers. European Journal of Clinical Pharmacology, 65(6), 585–591.

[134] Moltó, J., Valle, M., Miranda, C., Cedeño, S., Negredo, E. and Clotet, B. 2012. Effect of milk thistle on the pharmacokinetics of darunavir-ritonavir in HIV-infected patients. Antimicrob Agents and Chemother, 56(6), 2837–2841.

[135] Dhariwala, M.Y. and Ravikumar, P. 2019. An overview of herbal alternatives in androgenetic alopecia. Journal of cosmetic dermatology, 18(4), 966–975.

[136] Kanti, V., Messenger, A., Dobos, G., Reygagne, P., Finner, A., Blumeyer, A., et al. 2018. Evidence-based (S3) guideline for the treatment of androgenetic alopecia in women and in men – Short version. Journal of European Academic Dermatology and Venereology, 32(1), 11–22.

[137] Kwon, Y. 2019. Use of saw palmetto (*Serenoa repens*) extract for benign prostatic hyperplasia. Food Science & Biotechnology, 28(6), 1599–1606.

[138] Prager, N., Bickett, K., French, N. and Marcovici, G. 2002. A randomized, double-blind, placebo-controlled trial to determine the effectiveness of botanically derived inhibitors of 5-alpha-reductase in the treatment of androgenetic alopecia. Journal of Alternative and Complementary Medicine, 8(2), 143–152.

[139] Markowitz, J.S., Donovan, J.L., Devane, C.L., Taylor, R.M., Ruan, Y., Wang, J.-S., et al. 2003. Multiple doses of saw palmetto (*Serenoa repens*) did not alter cytochrome P450 2D6 and 3A4 activity in normal volunteers. Clinical Pharmacology & Therapeutics, 74(6), 536–542.

[140] Kim, H.L., Streltzer, J. and Goebert, D.S. 1999. John's wort for depression: A meta-analysis of well-defined clinical trials. Journal of Nervous and Mental Disease, 187(9), 532–538.

[141] Barnes, J., Arnason, J.T. and Roufogalis, B.D. 2019. St John's wort (*Hypericum perforatum* L.): Botanical, chemical, pharmacological and clinical advances. Journal of Pharmacy and Pharmacology, 71(1), 1–3.

[142] Scotti, F., Löbel, K., Booker, A. and Heinrich St, M. 2018. John's wort (*Hypericum perforatum*) products – How variable is the primary material? Frontiers in Plant Science, 9, 1973.

[143] Roby, C.A., Anderson, G.D., Kantor, E., Dryer, D.A. and Burstein, A.H. 2000. St John's Wort: Effect on CYP3A4 activity. Clinical Pharmacology & Therapeutics, 67(5), 451–457.

[144] Soleymani, S., Bahramsoltani, R., Rahimi, R. and Abdollahi, M. 2017. Clinical risks of St John's Wort (*Hypericum perforatum*) co-administration. Expert Opinion on Drug Metabolism Toxicology, 13(10), 1047–1062.

[145] Wang, L.-S., Zhu, B., Abd El-Aty, A.M., Zhou, G., Li, Z., Wu, J., et al. 2004. The influence of St John's Wort on CYP2C19 activity with respect to genotype. Journal of Clinical Pharmacology, 44(6), 577–581.

[146] Piscitelli, S.C., Burstein, A.H., Chaitt, D., Alfaro, R.M. and Falloon, J. 2000. Indinavir concentrations and St John's wort. Lancet, 355(9203), 547–548.

[147] Wang, Z., Gorski, J.C., Hamman, M.A., Huang, S.M., Lesko, L.J. and Hall, S.D. 2001. The effects of St John's wort (*Hypericum perforatum*) on human cytochrome P450 activity. Clinical Pharmacology & Therapeutics, 70(4), 317–326.

[148] Sugimoto, K., Ohmori, M., Tsuruoka, S., Nishiki, K., Kawaguchi, A., Harada, K., et al. 2001. Different effects of St John's wort on the pharmacokinetics of simvastatin and pravastatin. Clinical Pharmacology & Therapeutics, 70(6), 518–524.

[149] Johne, A., Schmider, J., Brockmöller, J., Stadelmann, A.M., Störmer, E., Bauer, S., et al. 2002. Decreased plasma levels of amitriptyline and its metabolites on comedication with an extract from st. John's wort (*Hypericum perforatum*). Journal of Clinical Psychopharm, 22(1), 46–54.

[150] Eich-Höchli, D., Oppliger, R., Golay, K.P., Baumann, P. and Eap, C.B. 2003. Methadone maintenance treatment and St. John's Wort – A case report. Pharmacopsychiatry, 36(1), 35–37.

[151] Hall, S.D., Wang, Z., Huang, S.-M., Hamman, M.A., Vasavada, N., Adigun, A.Q., et al. 2003. The interaction between St John's wort and an oral contraceptive. Clinical Pharmacology & Therapeutics, 74(6), 525–535.

[152] Murphy, P.A., Kern, S.E., Stanczyk, F.Z. and Westhoff, C.L. 2005. Interaction of St. John's wort with oral contraceptives: Effects on the pharmacokinetics of norethindrone and ethinyl estradiol, ovarian activity and breakthrough bleeding. Contraception, 71(6), 402–408.

[153] Dresser, G.K., Schwarz, U.I., Wilkinson, G.R. and Kim, R.B. 2003. Coordinate induction of both cytochrome P4503A and MDR1 by St John's wort in healthy subjects. Clinical Pharmacology & Therapeutics, 73(1), 41–50.

[154] Frye, R., Fitzgerald, S., Lagattuta, T., Hruska, M. and Egorin, M. 2004. Effect of St John's wort on imatinib mesylate pharmacokinetics. Clinical Pharmacology & Therapeutics, 76(4), 323–329.

[155] Wang, L.-S., Zhou, G., Zhu, B., Wu, J., Wang, J.-G., Abd El-Aty, A.M., et al. 2004. St John's wort induces both cytochrome P450 3A4-catalyzed sulfoxidation and 2C19-dependent hydroxylation of omeprazole. Clinical Pharmacology & Therapeutics, 75(3), 191–197.

[156] Hebert, M.F., Park, J.M., Chen, Y.-L., Akhtar, S. and Larson, A.M. 2004. Effects of St. John's wort (*Hypericum perforatum*) on tacrolimus pharmacokinetics in healthy volunteers. Journal of Clinical Pharmacology, 44(1), 89–94.

[157] Tannergren, C., Engman, H., Knutson, L., Hedeland, M., Bondesson, U. and Lennernäs, H. 2004. St John's wort decreases the bioavailability of R- and S-verapamil through induction of the first-pass metabolism. Clinical Pharmacology & Therapeutics, 75(4), 298–309.

[158] Wang, X.-D., J-l, L., Su, Q-B., Guan, S., Chen, J., Du, J., et al. 2009. Impact of the haplotypes of the human pregnane X receptor gene on the basal and St John's wort-induced activity of cytochrome P450 3A4 enzyme. British Journal of Clinical Pharmacology, 67(2), 255–261.

[159] Rengelshausen, J., Banfield, M., Riedel, K.-D., Burhenne, J., Weiss, J., Thomsen, T., et al. 2005. Opposite effects of short-term and long-term St John's wort intake on voriconazole pharmacokinetics. Clinical Pharmacology & Therapeutics, 78(1), 25–33.

[160] Schwarz, U.I., Hanso, H., Oertel, R., Miehlke, S., Kuhlisch, E., Glaeser, H., et al. 2007. Induction of intestinal P-glycoprotein by St John's wort reduces the oral bioavailability of talinolol. Clinical Pharmacology & Therapeutics, 81(5), 669–678.

[161] Xu, H., Williams, K.M., Liauw, W.S., Murray, M., Day, R.O. and Mclachlan, A.J. 2008. Effects of St John's wort and CYP2C9 genotype on the pharmacokinetics and pharmacodynamics of gliclazide: St John's wort and genotype affect gliclazide. British Journal Pharmacology, 153(7), 1579–1586.

[162] Peltoniemi, M.A., Saari, T.I., Hagelberg, N.M., Laine, K., Neuvonen, P.J. and Olkkola, K.T. 2012. St John's wort greatly decreases the plasma concentrations of oral S-ketamine: Interaction of St John's wort and oral S-ketamine. Fundamental and Clinical Pharmacology, 26(6), 743–750.

[163] Hojo, Y., Echizenya, M., Ohkubo, T. and Shimizu, T. 2011. Drug interaction between St John's wort and zolpidem in healthy subjects: Drug interaction between St John's wort and zolpidem. Journal of Clinical Pharmacy and Therapeutics, 36(6), 711–715.

[164] Markert, C., Ngui, P., Hellwig, R., Wirsching, T., Kastner, I.M., Riedel, K.-D., et al. 2014. Influence of St. John's wort on the steadystate pharmacokinetics and metabolism of bosentan. International Journal of Clinical Pharmacology and Therapeutics, 52(4), 328–336.

[165] Markert, C., Kastner, I.M., Hellwig, R., Kalafut, P., Schweizer, Y., Hoffmann, M.M., et al. 2015. The effect of induction of CYP3A4 by St John's wort on ambrisentan plasma pharmacokinetics in volunteers of known CYP2C19 genotype. Basic & Clinical Pharmacology & Toxicology, 116(5), 423–428.

[166] Goey, A.K.L., Meijerman, I., Rosing, H., Marchetti, S., Mergui-Roelvink, M., Keessen, M., et al. 2014. The effect of St John's wort on the pharmacokinetics of docetaxel. Clinical Pharmacokinetics, 53(1), 103–110.

[167] Rombolà, L., Scuteri, D., Marilisa, S., Watanabe, C., Morrone, L.A., Bagetta, G., et al. 2020. Pharmacokinetic interactions between herbal medicines and drugs: Their mechanisms and clinical relevance. Life (Basel), 10(7), 106.

Pauline Donn[†], Sepidar Seyyedi-Mansour[†], Ana Perez-Vazquez,
Paula Barciela, Maria Fraga-Corral, Franklin Chamorro, Lucia Cassani*,
Jesus Simal-Gandara*, Miguel A. Prieto*

# Chapter 6
# Nutraceuticals and oxidative stress

**Abstract:** Several studies have found a link between oxidative stress and a variety of noncommunicable diseases including diabetes, obesity, aging, Down syndrome, cancers, cardiovascular, and neurodegenerative diseases. All these pathologies are aided by oxidative stress, which is caused by the excessive production of free radicals or insufficient elimination of free radicals. To avoid, reduce, or eliminate this imbalance favoring prooxidants in organs and cells, one of the healthier and promising ap-

[†]These authors contributed equally to the publication.

*Corresponding author: Lucia Cassani, Nutrition and Bromatology Group, Department of Analytical Chemistry and Food Science, Instituto de Agroecoloxia e Alimentacion (IAA) CITEXVI, Universidade de Vigo, 36310 Vigo, Spain, e-mail: luciavictoria.cassani@uvigo.es
*Corresponding author: Jesus Simal-Gandara, Nutrition and Bromatology Group, Department of Analytical Chemistry and Food Science, Instituto de Agroecoloxia e Alimentacion (IAA) CITEXVI, Universidade de Vigo, 36310 Vigo, Spain, e-mail: jsimal@uvigo.es
*Corresponding author: Miguel A. Prieto, Nutrition and Bromatology Group, Department of Analytical Chemistry and Food Science, Instituto de Agroecoloxia e Alimentacion (IAA) CITEXVI, Universidade de Vigo, 36310 Vigo, Spain, e-mail: mprieto@uvigo.es
Pauline Donn, Nutrition and Bromatology Group, Department of Analytical Chemistry and Food Science, Instituto de Agroecoloxia e Alimentacion (IAA) CITEXVI, Universidade de Vigo, 36310 Vigo, Spain, e-mail: donn.pauline@uvigo.es
Sepidar Seyyedi-Mansour, Nutrition and Bromatology Group, Department of Analytical Chemistry and Food Science, Instituto de Agroecoloxia e Alimentacion (IAA) CITEXVI, Universidade de Vigo, 36310 Vigo, Spain, e-mail: sepidar.seyyedi@uvigo.es
Ana Perez-Vazquez, Nutrition and Bromatology Group, Department of Analytical Chemistry and Food Science, Instituto de Agroecoloxia e Alimentacion (IAA) CITEXVI, Universidade de Vigo, 36310 Vigo, Spain, e-mail: anaperezvaz@alumnos.uvigo.es
Paula Barciela, Nutrition and Bromatology Group, Department of Analytical Chemistry and Food Science, Instituto de Agroecoloxia e Alimentacion (IAA) CITEXVI, Universidade de Vigo, 36310 Vigo, Spain, e-mail: paula.barciela.alvarez@alumnos.uvigo.es
Maria Fraga-Corral, Nutrition and Bromatology Group, Department of Analytical Chemistry and Food Science, Instituto de Agroecoloxia e Alimentacion (IAA) CITEXVI, Universidade de Vigo, 36310 Vigo, Spain, e-mail: mfraga@uvigo.es
Franklin Chamorro, Nutrition and Bromatology Group, Department of Analytical Chemistry and Food Science, Instituto de Agroecoloxia e Alimentacion (IAA) CITEXVI, Universidade de Vigo, 36310 Vigo, Spain, e-mail: franklin.noel.chamorro@uvigo.es

The original version of this chapter was revised. Unfortunately, the names and institutional addresses of several authors were incorrect in the original publication. This has been corrected. We apologize for the mistake.
https://doi.org/10.1515/9783111317601-006

proaches is to use antioxidant compounds like nutraceuticals, which can scavenge the excessive free radicals and restore the balance. Nutraceuticals are dietary compounds that include dietary fibers, prebiotics, probiotics, polyunsaturated fatty acids, vitamins, minerals, polyphenols, and spices. Some of them have antioxidant activity, which can react with the free electron available in the outer layer of free radicals to make the unstable compounds more stable while avoiding free radicals' proliferation and the induced damage to the different metabolisms that occur in the human body because of oxidative stress. Thus, each nutraceutical class performs specific functions in managing the onset of oxidative stress. Therefore, the purpose of this chapter is to sort out the mechanisms by which different classes of nutraceuticals can have beneficial effects on oxidative stress and related pathologies via their biological properties. However, it also observes that the efficacy of nutraceutical compounds can only be guaranteed if they can maintain their properties during production, preservation, consumption, digestion, and use by the various targeted active sites. Thus, it shows that the concepts of bioavailability and bio-accessibility are key factors for the effectiveness of nutraceuticals on oxidative stress when consumed by humans.

# 1 Introduction

Free radicals are atoms, molecules, or compounds that have an unpaired free electron on their outer layer, which causes their instability and high reactivity. Compounds classified as free radicals include reactive oxygen species (ROS), reactive nitrogen species (RNS), carbon-centered radicals, and sulfur-centered radicals [1]. ROS include the highly reactive hydroxyl radical ($^\bullet$OH), and less reactive compounds such as hydrogen peroxide ($H_2O_2$), and superoxide anion ($O_2^{\bullet-}$). Meanwhile, the RNS are mainly composed of nitric oxide ($NO^\bullet$), nitrogen dioxide ($NO_2^\bullet$), and the potent oxidant compound, peroxynitrite ($ONOO^-$), which has the ability to damage many biological molecules [2]. The main sources of free radicals in human cells, as shown in Figure 1 are the cytosol, mitochondria, peroxisomes, endoplasmic reticulum (ER), plasma membrane, and lysosomes.

As regards cytosol, depending on the state of heath, their enzymes can favor or inhibit the production of ROS through their activities. In healthy conditions, xanthine dehydrogenase catalyzes the successive oxidation reactions of hypoxanthine to xanthine and then to uric acid, using $NADP^+$ as oxidant. However, when tissues are damaged, this enzyme is converted to xanthine oxidase via irreversible proteolysis stimulated by $Ca^{2+}$ or by oxidation of cysteine residues. The xanthine oxidase produced will then catalyze the oxidation of xanthine to hypoxanthine, a reaction in which the superoxide radical $^\bullet$OH is formed.

Moreover, the mitochondrion participates in essential functions of aerobic cells such as respiration, energy production, regulation of calcium concentration, and lip-

Golgi apparatus    Nucleus    Cell membrane
Source of $O_2^{\bullet-}$

Mitochondria
Source of $O_2^{\bullet-}$

Excessive lipid accumulation
=>lipotoxicity
Source of $O_2^{\bullet-}/H_2O_2$

Endoplasmic reticulum

Source of $O_2^{\bullet-}$
Misfolded/Unfolded proteins

vacuole

Ribosome

Cytoplasm

Cytosol enzyme
Xanthine oxidase (XO)
Source of $\bullet OH$

Centrosome

Peroxisomes
Source of $O_2^{\bullet-}/NO^{\bullet}$
=>$ONOO^-$
Fenton reaction
$\bullet OH$

Lysosome
Source of $\bullet OH$

Microtubule

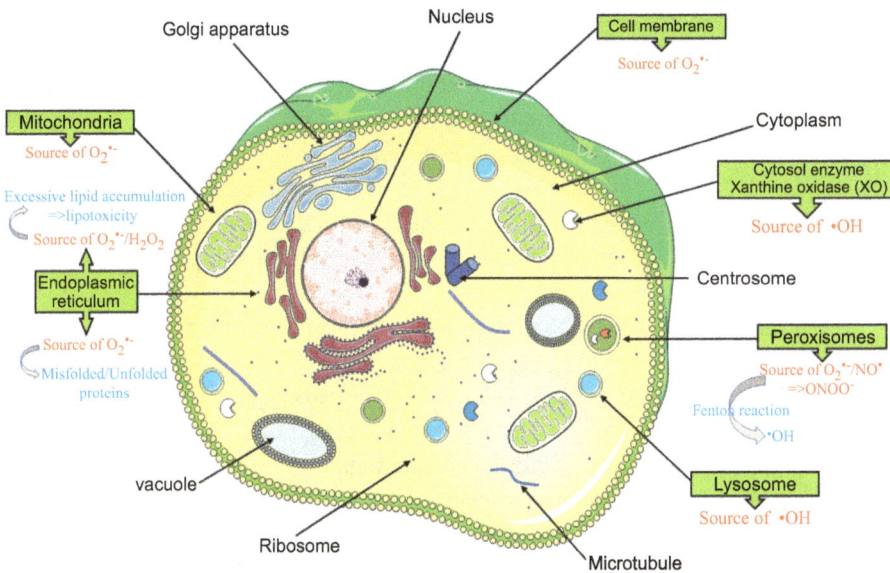

**Figure 1:** Main origin of oxidative stress in the cell, induced by an excessive production of free radicals.

ids oxidation. For a long time, the primary function of mitochondria was thought to be the production of energy via nutrients oxidation through electron transfer to nicotinamide adenine dinucleotide (NAD +), flavin mononucleotide (FMN), and flavin adenine dinucleotide (FAD). These molecules can transfer an electron to $O_2$ in a five-step process known as electron transport chain (ETC). During these transfers, superoxide anion is primarily produced in complexes I and III of the ETC, which are the main $O_2^{\bullet-}$ production sites. This phenomenon occurred in the intermembranous space and the matrix side of mitochondria.

On the other hand, peroxisomes have essential functions in the metabolism of $H_2O_2$, the metabolism of glyoxylate and amino acids, α and β-oxidation of fatty acids, and the biosynthesis of plasmalogens and cholesterol. The enzymes involved in the catalysis of these processes causes the production of ROS [2, 3]. The activities of xanthine oxidase and nitric oxide synthase in the peroxisome generate $O_2^{\bullet-}$ and $NO^{\bullet}$, respectively, which can interact and form $ONOO^-$. The Fenton reaction converts these harmful compounds in the presence of $H_2O_2$ to the highly reactive ROS compounds $^{\bullet}OH$. The activities of catalase and other antioxidants of the peroxisome can regulate excessive $O_2^{\bullet-}$ and $NO^{\bullet}$ production. Nevertheless, in some pathological conditions, uncontrolled excessive production can result in harmful ROS formation or insufficient elimination.

Overproduction of ROS in the ER has a direct impact on proteostasis and calcium signaling. Indeed, NADPH oxidase (NOX) is the source of ROS required for formation of disulfide bonds, a crucial step in the folding of extra-cytoplasmic proteins. As a result, ER dysfunction can result in misfolded and unfolded proteins. Furthermore, hyperactivity and/or dysfunction of NADPH oxidase, particularly the isoform NOX4, cause oxidation of ryanodine receptor (RyR) thiols sites, activating RyR ($Ca^{2+}$ transport channel) via an increase in $O_2^{\bullet-}$ production, resulting in $Ca^{2+}$ release from the ER. The metabolism of unsaturated fatty acid is also a source of ROS production in the ER system. Desaturase, in collaboration with NADH cytochrome b5 reductase and cytochrome b5, forms a double bond carbon in the fatty acid. In that case, cytochrome b5 acts as an electron-transfer component with a key role in the production of $O_2^{\bullet-}$ and $H_2O_2$.

The production of $O_2^{\bullet-}$ in the plasma membrane is caused by the phagocyte enzyme NADPH oxidase bounded to the membrane. This $O_2^{\bullet-}$ is useful in combating bacteria invaders. However, if not controlled, these free radicals increase the membrane permeability resulting in the loss of secretary functions, inhibition of metabolic processes, and a decrease in transmembrane ion gradients.

Lysosomes, the final site of macromolecules degradation, are also a major source of production of free radicals. NADPH plays a key a key role in supporting proton accumulation and maintaining an optimal pH for enzymes in the lysosome by transferring an electron to $O_2$ via the redox chain FAD-cytochrome b-ubiquinone. This electron transfer implies the formation of $^{\bullet}OH$. RNS are produced through the catabolism of L-arginine, which is converted to L-citrulline and $NO^{\bullet}$, a reaction catalyzed by diverse types of nitric oxide synthases (NOS) depending on their location and functions (neuronal NOS: nNOS, inducible NOS: iNOS, and endothelial NOS: eNOS). The produced $NO^{\bullet}$ has distinct functions depending on the type of NOS. Those produced by nNOS aid in nerve cells communication, those produced by iNOS aid in macrophages destruction, and those produced by eNOS aid in blood pressure regulation. Another endogenous source of $NO^{\bullet}$ is the electron transfer from NADPH to $O_2$. Although NO is essential for the proper functioning of the human body at various levels, its overproduction and combination with $O_2^-$ results in the formation of $ONOO^-$, which may be responsible for a variety of dysfunctions such as the malfunction of vital cellular processes, disruption of cell signaling, and the induction of cellular death via apoptosis and necrosis [2, 4].

## 1.1 Oxidative stress: Definition and origin

In the late twentieth century, oxidative stress was defined as the disruption in the balance of prooxidants and antioxidants in favor of prooxidants. Meanwhile, this definition has been updated to include the consequences of this imbalance, which are molecular damage(s) and/or disruption of redox signaling and control [5]. These abnormalities affect human physiology and functioning and have significant impact on the aging process and the development of many noncommunicable diseases (NCDs) such as atherosclerosis, cancer, obesity, diabetes, cardiovascular, pulmonary, and neurological diseases [6]. Oxidative stress occurs when an organism is unable to regulate and control the overproduction of any of the four different classes of free radicals i.e., ROS, RNS, carbon-centered radicals, and sulfur-centered radicals. Even though free radicals have beneficial effects and functions, they can be very harmful to humans if produced in copious quantities. This imbalance is caused by a combination of external factors such as diet, and environment, as well as internal factors related to age, and health status.

The mitochondrion is the main endogenous source of free radicals. It has been established that 2–3% of the $O_2$ is incompletely reduced, and thus available for any oxidative reaction. Furthermore, a mitochondrial dysfunction increased $O_2^{\bullet-}$ and $H_2O_2$ production, leading to oxidative stress. In addition, phagocytic cells, which are a major source of prooxidants, produce toxic by-products such as nitric oxide ($NO^{\bullet}$), hydrogen peroxide ($H_2O_2$), and superoxide anion ($O_2^{\bullet}$). Indeed, in response to inflammation, processes such as apoptosis, a programmed cell death process leads to an increase of ROS production, which, if unregulated, will induce oxidative stress. When the production of ROS by the microsomal monooxygenase cytochrome P450-dependent system is not controlled, another major source of oxidative stress is initiated in the endoplasmic reticulum. Most cells produce $O_2^{\bullet}$ through the activity of NADPH oxidase. The reduction of $O_2$ is catalyzed by this enzyme, which uses NADPH or NADH as an electron donor [2, 7]. Thus, depending on their state, environment, and operating conditions, all cells can be the cause of oxidative stress, causing organ malfunctions and being responsible for the initiation or a factor in the appearance of many NCDs.

Exogenous factors are associated with increased production of free radicals.. These factors include diets (low antioxidant intake, consumption of antibiotics, alcohol, coffee, high quantity of proteins and fats, and microbial contamination), exposure to radiation, herbicides, pesticides, cigarette smoke, and heavy metals such as mercury, iron, cadmium, and nickel [8, 9].

Regarding the diet, the concept of dietary oxidative stress refers to an imbalance between pro- and antioxidants caused by a lack of nutrients, which favors the prevention and control of excessive oxidative reactions that may occur, and thus may be able to avoid the overproduction of harmful compounds and their manifestation through the dysfunction of cells and organs, leading to the development of disease [10]. These antioxidant nutrients are mainly composed of polyphenols, carotenoids, tocopherol, vitamin C, and micronutrients such as selenium. As a result, one of the external sources of oxidative stress is a low antioxidant intake. Also, a high lipid intake can promote lipid peroxidation and free radicals' overproduction, leading to oxidative stress. In terms of proteins, a low protein intake affects antioxidants enzymes synthesis and decreases antioxidant concentration in tissues. However, a high protein diet derived from animals has been linked to an increase in lipid peroxidation and ROS and RNS production [11]. Thus, diet plays a key role in the expression of oxidative stress and may aid in the management of this imbalance by acting on antioxidant enzymes and the production and regulation of ROS compounds.

## 2 Oxidative stress and related pathologies

Oxidative stress is related to different pathological processes, including aging, Down syndrome, and cardiovascular and neurodegenerative diseases, including Alzheimer's disease (AD), and erectile dysfunction (ED) [12, 13]. Although the aging process is still poorly understood, it has been demonstrated that oxidative stress progressively accelerates this process, with ROS being key factors [12]. Goi et al., investigated the relationship between peroxidative processes and changes in the plasma membrane enzyme activities of erythrocytes (RBCs), concluding that aging and systemic oxidative stress are linked. Furthermore, they established the glycohydrolases present as markers of oxidative stress by correlating oxidative stress and RBC membrane enzyme activity [14]. RBCs are sensitive cells that regulate redox and vascular tone. Moreover, RBCs are important health indicators because they are exposed to circulating inflammatory mediators and associated oxidative stress during inflammatory response [12]. During this process, RBCs are exposed to changes in the cellular membrane and functions in pathological conditions. As a result, erythrocytes have a highly efficient redox system that regulates balance. On the one hand, RBCs are linked to Hb oxidation, superoxide, and hydroxyl radicals, all of which result in the formation of ROS. On the other hand, RBCs have antioxidant systems that contain various enzymes such as superoxide dismutase, catalase, and glutathione peroxidase to counteract ROS production. When the functionality of these components is altered, the redox balance is disrupted, resulting in the production of pro-oxidizing factors and the development of an oxidative stress condition.

Down syndrome is caused by trisomy of chromosome 21, which contains the SOD (SOD-1) gene. Overexpression of this gene results in increased enzymatic activity increment and, consequently, an excessive $H_2O_2$ production. Finally, SOD-1 overexpression disrupts the redox balance, favoring premature aging and AD-like neuropathologies in Down syndrome patients [12]. AD is a neurodegenerative disease that affects cognitive functions. Furthermore, AD accounts for approximately 70% of dementia in the elderly population and is the leading cause of morbidity and mortality in this age group [14]. The increase of lipid peroxidation in specific brain areas is a key factor in the development of AD. It is important to highlight that the brain is more vulnerable to the oxidative stress process because it consumes more $O_2$ (about 20% of the body's consumption of $O_2$) [14, 15]. As a result of disrupted mitochondrial function, the brain's $O_2$ metabolism is disrupted, resulting in ROS production [15]. In this way, oxidative stress markers like macromolecules nitration are increased in AD, making it suitable for AD detection [13]. Diabetes is one of the cardiovascular diseases. RBCs adhesion causes endothelial oxidative damage in type 2 diabetic subjects according to experimental evidence [14]. Indeed, changes in RCBs cause arginase activation in endothelial cells, reducing NO bioavailability while increasing ROS formation. Furthermore, RAGE-mediated adhesion between endothelial cells and RBCs increases cellular oxidative stress, indicating a link between oxidative stress and diabetes. Based on the data gathered, oxidative stress processes have been identified as a potential risk factor for various diseases, particularly cardio-vascular and neurodegenerative pathologies.

Thus, in understanding the mechanism of action of free radicals in the context of oxidative stress on cells and organs, products such as nutraceuticals may have a scavenging effect and improve health by reducing or preventing their damage. A graphic illustration of nutraceuticals in the treatment of oxidative stress for health benefits is shown in Figure 2.

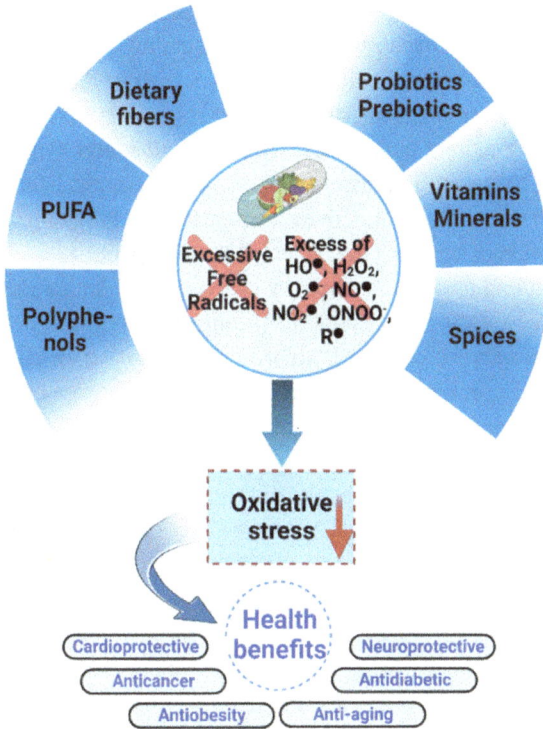

**Figure 2:** Nutraceuticals in the management of oxidative stress for health benefits.

# 3 Nutraceuticals as preventive approach toward oxidative stress-associated pathologies

The potential effects of nutraceuticals against oxidative stress are summarized in Table 1.

## 3.1 Dietary fibers, prebiotics, and probiotics

Nutraceuticals comprise bioactive compounds, dietary fibers, probiotics, and prebiotics found in foods [16]. Probiotics, especially lactic acid bacteria (LAB), are considered a group of live nonpathogenic microorganisms and when used in adequate quantities provide an extra health benefit to the host, whereas prebiotics serve as food for probiotics and are typically a type of carbohydrates such as galacto-oligosaccharide and inulin. They are food components that cannot be digested by glycogenic or intestinal juices but are degraded by bacterial enzymes [16, 17]. The most common probiotics genera are *Lactobacillus* and *Bifidobacterium* [18]. It has been reported that consuming dietary probiotics may help with the management and prevention of disorders such as diarrhea, inflammatory bowel disease, lactose intolerance, allergies, cancer, respiratory tract infections, constipation, urinary tract infections, helicobacter pylori infection, and hypercholesterolemia [19]. Furthermore, the anti-inflammatory activity of nutraceuticals has the potential to modulate immune cell functions or activity at the tissue and organ levels, which are involved in the secretion of signaling molecules or ROS/RNS [20].

LAB probiotics have an outstanding antioxidant capacity based on the elimination of free radicals, acting as chelators of prooxidant metal ions, controlling relevant enzymes and modulating the intestinal microbiota [21, 22]. According to Colitti et al. probiotic strains and their fermented food products can be used as nutritional supplements to reduce oxidative stress-induced aging process. These authors evaluated antiaging probiotic strains in mice using *in vitro* DPPH antioxidant assay. Increased antioxidant activity, positive modulation of protective gut microbiota, and increased serum short-chain fatty acids was observed in middle-aged mice by [20]. Lee et al. evaluated the properties of *Lactococcus lactis* MG5125, *Bifidobacterium bifidum* MG731, and *Bifidobacterium animalis* subsp. *lactis* MG741 as probiotics and found the effects of probiotics on gut microflora indirectly increased antioxidant effects *in vivo* [17]. Moreover, they enhanced the levels of antioxidant enzymes such as superoxide dismutase or catalase, decreased serum levels of aspartate aminotransferase (AST) and alanine aminotransferase (ALT) to control toxicity, and reduced the level of malondialdehyde (MDA) to alleviate oxidative stress caused by tert-butyl hydroperoxide (t-BHP), among other advantages.

Prebiotics have been shown to influence the intestinal microbiome and stimulate significant changes in the immune system. As a result, consuming prebiotics alters the bacterial composition, resulting in the production of supplemental short-chain fatty acids (SCFAs). Thus, it successful in preventing enteric diseases and oxidative stress by promoting a microbiome associated with increased growth efficiency [23].

Several mechanisms exist for dietary fibers to suppress immune responses and inflammatory reactions [24]. Zhang et al. fed mice a high-fiber diet to evaluate how it affected gut injury, oxidative stress, and systemic inflammation. These authors found that eating foods high in fiber reduces intestinal injury [25]. Mechanically, dietary fiber has been shown to exert anti-inflammatory activity by producing SCFAs [26]. Therefore, a healthy microbiome supported by dietary fibers and phytomolecules could reduce cell proliferation by regulating epigenetic processes that activate onco-genic pathways [27].

## 3.2 Polyunsaturated fatty acids

Over recent years, supplementation with polyunsaturated fatty acids (PUFAs) has been shown to exhibit vascular and cardioprotective responses *in vitro* and *in vivo*. Further-more, they play a noteworthy role due to their antioxidant and anti-inflammatory prop-erties [28]. PUFAs increase the level of anti-inflammatory markers while decreasing the level of pro-inflammatory markers [16]. These supplements include two fundamental types of $n$-3 PUFAs, i.e., docosahexaenoic acid (DHA), and eicosapentaenoic acid (EPA). Several studies describe different anticancer mechanisms, suggesting the beneficial and therapeutic potential to sensitize tumors by propagating ROS and attacking antioxidant defenses [29]. Hernando et al. demonstrated a neuroprotective and hopeful approach in a Parkinson's disease partial injury model by supplementing with $n$-3 PUFAs, specifi-cally DHA, on the dopaminergic system, neuroinflammation, and oxidative stress in the investigation [30]. The probable mechanism of PUFAs' neuroprotective and therapeutic action could be based on the fact that they induce physical and biochemical changes in cell membrane properties and modulate inflammatory responses and the intracellular antioxidant defense system, also by reducing dopamine toxicity or by reducing inflam-mation through modulation of cytokine production [31].

Sakai et al. investigated the effect of EPA and DHA on the response of human aor-tic endothelial cells to oxidative stress-induced DNA damage, finding that treatment with PUFAs upregulates the nuclear erythroid 2-related factor (NrF2)-mediated antiox-idant response. As a result, there was a significant reduction of ROS-induced DNA damage in endothelial cells, as well as the formation of γ-H2AX foci, an important marker of DNA damage. In addition, mRNA levels of antioxidant molecules, such as heme oxygenase-1, thioredoxin reductase 1, light chain ferritin, heavy chain ferritin, and manganese superoxide dismutase, were significantly increased [32]. Another ap-proach was conducted by Beaudoin-Chabot et al., defining the role of deuterated PUFAs in the multicellular organism *Caenorhabditis elegans* with the purpose of avoiding ROS formation by inhibiting the propagation of lipid peroxidation, while antioxidants potentially neutralize beneficial oxidative species. The outcome implied a significant prolongation of the organism's lifespan [33].

Understanding the essentiality/oxidability of PUFAs, as well as the role of lipogene-sis-desaturase pathways in cell growth and oxidative reactivity in cancer cells, has re-vealed that PUFAs are a choice that has multiple antitumor-supportive functions and find their maximum expression in synergy with intake of dietary PUFAs and the forma-tion of biological membranes in the control of lipid therapies against cancer [34]. Za-netti et al. investigated whether $n$-3 PUFA could reverse endothelial dysfunction in rats with chronic kidney disease (CKD) by enhancing the function of endothelial nitric oxide synthase (eNOS) and oxidative stress. Their findings demonstrated that $n$-3 PUFA supplementation reverses endothelial dysfunction and normalizes the decreased eNOS protein expression in the aortas of rats with CKD. These findings are associated with a substantial reduction in the oxidative damage marker 3-nitrotyrosine, implying that reduced oxidative stress may be contributing to the benefit of $n$-3 PUFA on endothelial function [35].

The health benefits of PUFAs-rich diets are also recognized by their involvement in the activation of peroxisome proliferator-activated receptors (PPARs), antiproliferative action in modulating CB1 receptor expression, and modulation of intestinal polyp formation [36].

## 3.3 Vitamins and minerals

Vitamins and minerals are micronutrients that must be consumed in small amounts [37]. Vitamins C and E have been highlighted for their antioxidant capacity, as well as their ability to protect against excitotoxicity and neuroinflammation [38]. Ascorbic acid, also known as vitamin C, is an active compound that affects epigenetic regulation of gene expression, intracellular signaling, and redox state switching. In particular, by producing $H_2O_2$ and the -OH radical, it is able to affect the viability of cancer cells sensitive to oxidative damage for DNA, membranes, or mitochondria [39]. It also acts as an antioxidant in the aqueous phase of the cell and circulatory system by donating high-energy electrons to neutralize free radicals [37]. Meanwhile, vitamin E stands out mainly for its ability to deactivate singlet oxygen ($_1O2$) by quenching and reducing inflammatory processes by limiting the generation of ROS and their damaging effects. In addition to its biological function of preventing free radicals oxidation of PUFAs and low-density lipoproteins (LDL) [40]. According to Zeng et al., increased intake of the antioxidant micronutrients β-carotene, vitamin E, and vitamin C resulted in a significant reduction in ROS production [41].

Vitamin D is also effective in reducing systemic inflammation, reducing oxidative stress, and improving mitochondrial and endocrine functions, lowering the risk of autoimmune disorders, infections, metabolic changes, and deterioration of DNA repair. As a powerful antioxidant, vitamin D plays a role in the molecular and cellular actions of 1,25 $(OH)_2D$, which slows oxidative stress and cellular and tissue damage [42]. Masjedi et al. studied the effects of vitamin D on steroidogenesis, apoptosis, ROS production, and antioxidant defenses of normal human granulosa cells (N-GC) and polycystic granulosa cells (P-GC). ROS levels in the treatment cells were reduced, which could be part of the vitamin D protective mechanism. Moreover, it had a significant impact on the activity of antioxidant enzymes [43]. Additionally, vitamin B12 or cobalamin has been shown to have strong antioxidant potential based on direct scavenging of ROS, particularly superoxide, modulation of growth factor production to provide oxidative stress protection, and reduction of oxidative stress caused by homocysteine or advanced glycation end products [44]. Finally, vitamin K is associated with lower levels of inflammatory markers *in vivo*. The anti-inflammatory function of vitamin K is due to its suppression of nuclear factor κB (NF-κB) signal transduction. Furthermore, it has been suggested that this micronutrient has a protective effect against oxidative stress [45].

Kang et al. studied the antioxidant and DNA-protective capacity of a multivitamin/mineral and phytonutrient supplement in healthy adults who consumed a low amount of fruit and vegetables on a regular basis. Supplementation was significantly correlated with ROS scavenging and DNA damage prevention without affecting the endogenous antioxidant system [46]. Changes in mineral levels such as manganese, iron, chromium, zinc, selenium, or copper cause changes in the balance of the body's defense mechanisms against oxidative stress and oxidation. Furthermore, they cause the antioxidant systems to malfunction [47].

## 3.4 Dietary polyphenols

Polyphenols are molecules found in vegetables, fruits, coffee, tea, and red wine that have been linked to different bioactivities such as antioxidant and anti-inflammatory properties, as well as the ability to inhibit ROS production [48, 49]. Moreover, polyphenols regulate ROS homeostasis and scavenge mitochondrial ROS, while upregulating antioxidant transcriptional programs in cells [50]. The antioxidant activity of polyphenols is determined by their ability to scavenge ROS molecules and therefore, neutralizing their chemical reactivity; the prooxidant gene inhibition; and an increase of antioxidant gene expression [48, 50]. In this way, different polyphenols such as epigallocatechin gallate (ECGC), butein, wogonin, and resveratrol, among others, have shown scavenging ROS properties [48]. When polyphenols are oxidized by one electron, they are converted into stable radicals that form dimers or complex oligomers. The position and number of hydroxyl phenolic groups play an important role in radical stabilization [50]. A study conducted by Qiu et al. proved the antioxidant activity of quercetin (a flavonoid present in both fruits and vegetables). Thus, quercetin reduced ROS levels in osteoarthritis rat models by increasing the expression of both glutathione and glutathione peroxidase [51]. Resveratrol is a polyphenol found in the skin of red grapes that has been linked to anti-inflammatory and antioxidant activity [48]. Resveratrol showed protective effect against sodium nitroprusside that induced apoptosis of rabbit chondrocytes. Sodium nitroprusside affects chondrocytes by scavenging SNP and inducing the production of ROS and NO [52]. Moreover, it was demonstrated that resveratrol reduced oxidative stress in rats with osteoarthritis by inducing the expression of heme-oxygenase 1 (HO-1) [53]. Furthermore, resveratrol can modulate oxidative stress by increasing the expression of the nuclear factor erythroid 2-related factor (Nrf2)-dependent HO-1 [54]. The effect of apple polyphenols on cerebral vasospasm and eNOS was studied by Naraoka et al. As a result, apple polyphenols have been shown to reduce oxidized low-density lipoprotein (LDL) and lectin-like oxidized LDL receptor-1 [55].

Another study found that giving pomegranate phenols for 12 weeks had a positive effect on rats on high fat–high fructose diet by lowering lipid peroxidation in whole blood [56]. Grape seeds extracts are composed of phenolic acids and flavonoids [54]. Cuevas et al. conducted *in vivo* studies to demonstrate how epicatechin influences the lipid peroxidation of AD-associated oxidative stress. In this way, a single oral dose of these compounds reduced Aß-mediated lipid peroxidation and ROS formation, indicating a neuroprotective activity [57]. Grape juice has high-polyphenol content. In fact, the polyphenol delphinidin, which is present in this product, is a potent lipid peroxidation inhibitor that scavenges superoxide anions [54]. Studies showed reduction of ROS production in the hippocampus, striatum, and cerebral cortex by ECGC treatment in the mouse model of AD. Furthermore, both Aß and plaques levels were reduced by increasing antiamyloidogenic ß-secretase proteolytic activity and remodeling a-synuclein amyloid fibrils into disordered oligomers, respectively [58, 59]. According to data presented in this section, polyphenols have been demonstrated to be suitable molecules for reducing oxidative stress and therefore, may be used as preventive nutraceuticals for oxidative stress diseases.

## 3.5 Spices

Curcumin, saffron, cinnamon, oregano, and thyme have all been studied for their bioactive compounds content because these molecules are linked to different bioactivities such as antioxidant and anti-inflammatory activities. Curcumin is a phenylpropanoid and the main component of turmeric that has anti-inflammatory properties [48]. Curcumin's antioxidant properties have also been demonstrated [60]. $H_2O_2$ was used to test the effect of curcumin on human cells. In this way, curcumin was able to reduce both oxygen radicals and IL-8 concentrations, while increasing glutathione levels and modulating and activating NF-κB and AP-1, respectively [60]. Saffron is a Mediterranean spice that contains bioactive compounds such as flavonoids and anthocyanins. Thus, antioxidant, cytotoxic, and antidepressant activities are some of the plant compounds' properties [61]. Baba et al. investigated the ability of saffron to inhibit oxidative stress in both plants and bacteria, with excellent results. In fact, the authors proposed isolating these compounds found in saffron for use in both food and pharmacological industries [62]. Cinnamon is a spice that is known for its antioxidant activity and phytochemicals content. It was demonstrated that cinnamon consumption reduces MDA levels, which can indicate a partial attenuation in oxidative stress, because MDA is a biomarker of damaged macromolecules that reflects lipid peroxidation [63]. MDA levels in plasma and tissues were also reduced after cumin powder supplementation, although ROS level in liver tissues were not measured [64]. Mahboub et al. investigated the improvement of oxidative stress after 30 g/kg of black cumin supplementation and found positive results [65]. The antioxidant effect of oregano essential oil was also studied, and it was found to protect rats' jejune from oxidative stress after a 5 mg/kg or 20 mg/kg BW dose [66]. A thyme essential oil nanoemulsion was tested as an antioxidant agent against oxidative

**Table 1:** Effectiveness of nutraceutical compounds against oxidative stress in animal models.

| Nutraceuticals | Target | Animal model | Dose | Major outcomes | Ref. |
|---|---|---|---|---|---|
| **Probiotics** | | | | | |
| *Bifidobacterium* and *Lactobacillus* | – Aging regulation | Mice | $1.03–4.1 \times 10^9$ CFU/kg for 10 months | – Increase of MDA in the mouse brain<br>– Strong antioxidant activity with positive gut microbiota modulation and serum SCFAs enhancement | [22] |
| **PUFAs** | | | | | |
| *n*-3 PUFAs | Prevention of obesity-associated inflammation | Mice | 2% *n*-3 HUFAs mixed in high-fat die, for 8 weeks | – High insulin sensitivity and glucose homeostasis<br>– Prevention of oxidative stress increase caused by $H_2O_2$ | [68] |
| *n*-3 PUFAs[1] | Verification of the impact on inflammation and oxidative stress indices | Rats | 0.8 mL/kg weekly supplementation for 13 weeks | – Reduced biomarkers of oxidative stress | [69] |
| **Vitamins** | | | | | |
| Vitamin C | Evaluation of LPS-induced cognitive impairment. | Mice | 10 µg for 2–3 days | – Alleviation of cognitive impairment by inhibition of the ROS system | [70] |
| Vitamin E | Enhanced potential of MSCs against oxidative stress. | Rats | 50–100 µm for 7 days | – Deactivation of singlet oxygen<br>– Reduces anti-inflammatory process | [40] |
| Vitamin D | Reduction of diabetes pathological complications | Rats | 2,000 µL for 28 days | – Increase in the activities of SOD, GPx and CAT | [71] |
| Vitamin B12 | Assessment of the protective effect | Rats | 250 µg for 112 days | – Decrease of oxidative stress level and improvement of cytotoxic effects | [72] |

(continued)

**Table 1** (continued)

| Nutraceuticals | Target | Animal model | Dose | Major outcomes | Ref. |
|---|---|---|---|---|---|
| **Polyphenols** | | | | | |
| Quercetin | Assessment against oxidative damage | Mice | 20, 40, and 80 mg/kg for 28 days | – Neuroprotective effects against chronic unpredictable stress | [73, 74] |
| Resveratrol | Analysis of the subjacent mechanisms of antidepressant action | Rats | 80 mg/kg for 28 days | – Regulation of phospho-Akt and phospho-mTOR levels in the hippocampus<br>– Protection effect against sodium nitroprusside | [52, 75, 76] |
| Epicatechin | Effect on homocysteine-induced mitochondrial damage | Rats | 50 mg/kg for 10 days | – Decreased lipid peroxidation, ROS levels and increased GSH levels | [77, 78] |
| Epigallocatechin gallate | Evaluation of the neuro-rescue effect | Rats | 25 mg/kg for 7 days | – Regulation of transporter proteins, reduction of oxidative stress, and neuroresponsive effect | [79] |
| **Spices** | | | | | |
| Curcumin | Effect of oxidative stress on organs | Rats | 10–30 mg/kg for 21 days | – Decreased oxidative stress and lipid peroxidation | [80] |
| Saffron | Oxidative stress impact assessment | Mice | 40 mg/kg for 6 days | – Increased antioxidant enzyme activity | [62, 81] |
| Cinnamon | Hepatotoxicity and nephrotoxicity effects | Rats | 200 mg/kg for 14 days | – Decreased acute mediated liver and renal damage | [82] |
| Oregano essential oil | Intestinal protective effect on oxidative stress | Rats | 5–20 mg/kg for 15 days | – Protective effect against oxidative stress | [66] |

$^{1}$Animals were treated with oil supplement containing $n$-3 PUFAs in different rations, i.e., EPA-DHA = 1:1; EPA:DHA = 1:2; EPA:DHA = 2:1.

stress caused by $TiO_2$ nanoparticles and the results were promising [67]. In this way, and in light of the data presented in this section, spices extracts or essential oils may act as potential antioxidants, lowering oxidative stress and, consequently, the risk of related pathologies.

# 4 Summary

As demonstrated throughout the chapter, the increased consumer interest in maintaining a healthy lifestyle has presented nutraceuticals as appropriate compounds to be incorporated into the diet in order to prevent diseases. This is made possible by the ability of nutraceuticals to provide health benefits as well as protection against diseases. Nutraceuticals are divided into two major groups: traditional and nontraditional nutraceuticals. Traditional nutraceuticals include dietary fiber, prebiotics and probiotics, polyunsaturated fatty acids, vitamins, minerals, polyphenols, and spices. Probiotics are nonpathogenic microorganisms with anti-inflammatory activity, as well as the ability to modulate immune cell functions or activity at the tissue and organ levels, while dietary fiber consumption is linked with several mechanisms that lead to downregulating immune responses and inflammatory reactions. Polyunsaturated fatty acids are lipophilic compounds that have been shown in both *in vitro* and *in vivo* studies to have vascular and cardioprotective responses, as well as antioxidant and anti-inflammatory properties, implying that including these compounds in the human diet has a disease prevention function. Vitamins and minerals are micronutrients that, despite being required in small amounts, play an important role in the human body. Vitamin C, E, and D are known for their antioxidant activity, while vitamin K is known for its anti-inflammatory properties. Polyphenols, which have antioxidant and anti-inflammatory properties, have also been studied. Thereby, polyphenols can be also included in the diet to help prevent several oxidative and inflammation-related diseases. Finally, different authors have suggested that spices can be used as nutraceuticals because they have antioxidant, antidepressant, and cytotoxic activities. In fact, the nutraceutical effect of extracts or essential oils from spices such as curcumin, saffron, cinnamon, thyme or oregano have been studied with positive results in the prevention of oxidative related diseases. All these bioactivities and mechanisms involved when these nutraceuticals are consumed may have a positive impact on the prevention of oxidative stress. Oxidative stress is a process that disrupts the balance of prooxidants and antioxidants in favor of prooxidants, resulting in molecular damage and/or disturbance of redox signaling and control. These changes disrupt the normal functioning of the human physiology, and both aging process and NCD are developed. In fact, while the mechanism of oxidative stress is still unknown, several studies have linked it to aging, Down syndrome, cardiovascular and neurodegenerative diseases. Therefore, the use of nutraceuticals as potential disease prevention compounds may be a path to consider in a society that demands a higher quality of life, even in the elderly.

**Acknowledgements:** The research leading to these results was supported by MICINN supporting the Ramón y Cajal grant for M.A. Prieto (RYC-2017-22891) and by Xunta de Galicia for supporting the program EXCELENCIA-ED431F 2020/12 and the postdoctoral grant of L. Cassani (ED481B-2021/152). Authors are grateful to Ibero-American Program on Science and Technology (CYTED—AQUA-CIBUS, P317RT0003), and to the Bio-Based Industries Joint Undertaking (JU) under grant agreement No 888003 UP4HEALTH Project (H2020-BBI-JTI-2019). The JU receives support from the European Union's Horizon 2020 research and innovation program and the Bio-Based Industries Consortium.

# References

[1]   Wu, J.Q., Kosten, T.R. and Zhang, X.Y. 2013. Free radicals, antioxidant defense systems, and schizophrenia. Progress in Neuro-Psychopharmacology & Biological Psychiatry, 46, 200–206.

[2]   Di Meo, S., Reed, T.T., Venditti, P. and Victor, V.M. 2016. Role of ROS and RNS sources in physiological and pathological conditions. Oxidative Medicine and Cellular Longevity, 2016, 1245049.

[3]   Singh, I. 1997. Biochemistry of peroxisomes in health and disease. Molecular and Cellular Biochemistry, 167, 1–29.

[4]   Pacher, P., Beckman, J.S. and Liaudet, L. 2007. Nitric oxide and peroxynitrite in health and disease. Physiological Reviews, 87, 315–424.

[5]   Gelpi, J.R., Boveris, A. and Poderoso, J.J. 2016. *Biochemistry of Oxidative Stress Physiopathology and Clinical Aspects*. Switzerland: Springer.

[6]   Reuter, S., Gupta, S.C., Chaturvedi, M.M. and Aggarwal, B.B. 2010. Oxidative stress, inflammation, and cancer: How are they linked?. Free Radical Biology and Medicine, 49, 1603–1616.

[7]   de Almeida, A.J.P.O., de Oliveira, J.C.P.L., da Silva Pontes, L.V., et al. 2022. ROS: Basic concepts, sources, cellular signaling, and its implications in aging pathways. Oxidative Medicine and Cellular Longevity, 2022, 1225578.

[8]   Lakshmi, S.V.V., Padmaja, G., Kuppusamy, P., and Kutala, V.K. 2009. Oxidative stress in cardiovascular disease. Indian Journal of Biochemistry and Biophysics, 46, 421–440.

[9]   Żukowski, P., Maciejczyk, M. and Waszkiel, D. 2018. Sources of free radicals and oxidative stress in the oral cavity. Archives of Oral Biology, 92, 8–17.

[10]  Rahman, I. and MacNee, W. 2012. Oxidative stress. Chronic Obstructive Pulmonary Disease Second Ed, 110–129.

[11]  Fang, Y.Z., Yang, S. and Wu, G. 2002. Free radicals, antioxidants, and nutrition. Nutrition, 18, 872–879.

[12]  Massaccesi, L., Galliera, E. and Corsi Romanelli, M.M. 2020. Erythrocytes as markers of oxidative stress related pathologies. Mechanisms of Ageing and Development, 191, 111333.

[13]  Sompol, P., Kraner, S., Arthiushin, I., et al. 2022. Oxidative stress-associated cerebrovascular pathology in Alzheimer's disease. Free Radical Biology and Medicine, 192, 44.

[14]  Goi, G., Cazzola, R., Tringali, C., et al. 2005. Erythrocyte membrane alterations during ageing affect β-d-glucuronidase and neutral sialidase in elderly healthy subjects. Experimental Gerontology, 40, 219–225.

[15]  Beura, S.K., Dhapola, R., Panigrahi, A.R., et al. 2022. Redefining oxidative stress in Alzheimer's disease: Targeting platelet reactive oxygen species for novel therapeutic options. Life Science, 306, 120855.

[16]  Anand, S. and Bharadvaja, N. 2022. Potential benefits of nutraceuticals for oxidative stress management. Revista Brasileira de Farmacognosia, 32, 211–220.

[17]  Lee, J.Y. and Kang, C.H. 2022. Probiotics alleviate oxidative stress in H2O2-exposed hepatocytes and t-BHP-induced C57BL/6 mice. Microorganisms, 10(2), 234.

[18]  Heshmati, J., Farsi, F., Shokri, F., et al. 2018. A systematic review and meta-analysis of the probiotics and synbiotics effects on oxidative stress. Journal of Functional Foods, 46, 66–84.

[19]  Min, M., Bunt, C.R., Mason, S.L. and Hussain, M.A. 2019. Non-dairy probiotic food products: An emerging group of functional foods. Critical Reviews in Food Science and Nutrition, 59, 2626–2641.

[20]  Colitti, M., Stefanon, B., Gabai, G., et al. 2019. Oxidative stress and nutraceuticals in the modulation of the immune function: Current knowledge in animals of veterinary interest. Antioxidants, 8(1), 28.

[21]  Feng, T. and Wang, J. 2020. Oxidative stress tolerance and antioxidant capacity of lactic acid bacteria as probiotic: A systematic review. Gut Microbes, 12, 1–24.

[22]  Lin, W.Y., Lin, J.H., Kuo, Y.W., et al. 2022. Probiotics and their metabolites reduce oxidative stress in middle-aged mice. Current Microbiology, 79, 1–12.

[23]  Gao, J., Azad, M.A.K., Han, H., et al. 2020. Impact of prebiotics on enteric diseases and oxidative stress. Current Pharmaceutical Design, 26, 2630–2641.

[24]  Jiang, S., Liu, H. and Li, C. 2021. Dietary regulation of oxidative stress in chronic metabolic diseases. Foods, 10, 1854.

[25]  Zhang, Y., Dong, A., Xie, K. and Yu, Y. 2019. Dietary supplementation with high fiber alleviates oxidative stress and inflammatory responses caused by severe sepsis in mice without altering microbiome diversity. Frontiers in Physiology, 10, 1–7.

[26]  Allam, V.S.R.R., Chellappan, D.K., Jha, N.K., et al. 2022. Treatment of chronic airway diseases using nutraceuticals: Mechanistic insight. Critical Reviews in Food Science and Nutrition, 62, 7576–7590.

[27]  Appunni, S., Rubens, M., Ramamoorthy, V., et al. 2021. Emerging evidence on the effects of dietary factors on the gut microbiome in colorectal cancer. Frontiers in Nutrition, 8, 1–20.

[28]  Oppedisano, F., Macrì, R., Gliozzi, M., et al. 2020. The anti-inflammatory and antioxidant properties of n-3 PUFAs: Their role in cardiovascular protection. Biomedicines, 8(9), 306.

[29]  Zhang, R.X., Liu, F.F.C., Lip, H., et al. 2022. *Pharmaceutical Nanoformulation Strategies to Spatiotemporally Manipulate Oxidative Stress for Improving Cancer Therapies – Exemplified by Polyunsaturated Fatty Acids and Other ROS-modulating Agents.* Springer US.

[30]  Hernando, S., Requejo, C., Herran, E., et al. 2019. Beneficial effects of n-3 polyunsaturated fatty acids administration in a partial lesion model of Parkinson's disease: The role of glia and NRf2 regulation. Neurobiology of Disease, 121, 252–262.

[31]  Pawełczyk, T., Grancow-Grabka, M., Trafalska, E., et al. 2017. Oxidative stress reduction related to the efficacy of n-3 polyunsaturated fatty acids in first episode schizophrenia: Secondary outcome analysis of the OFFER randomized trial. Prostaglandins Leukot Essent Fat Acids, 121, 7–13.

[32]  Sakai, C., Ishida, M., Ohba, H., et al. 2017. Fish oil omega-3 polyunsaturated fatty acids attenuate oxidative stress-induced DNA damage in vascular endothelial cells. PLoS One, 12, 1–13.

[33]  Beaudoin-Chabot, C., Wang, L., Smarun, A.V., et al. 2019. Deuterated polyunsaturated fatty acids reduce oxidative stress and extend the lifespan of *C. Elegans*. Frontiers in Physiology, 10, 1–10.

[34]  Ferreri, C., Sansone, A., Chatgilialoglu, C., et al. 2022. Critical review on fatty acid-based food and nutraceuticals as supporting therapy in cancer. International Journal of Molecular Sciences, 23(11), 6030.

[35] Zanetti, M., Cappellari, G.G., Barbetta, D., et al. 2017. Omega 3 polyunsaturated fatty acids improve endothelial dysfunction in chronic renal failure: Role of eNOS activation and of oxidative stress. Nutrients, 9(8), 895.

[36] Caponio, G.R., Lippolis, T., Tutino, V., et al. 2022. Nutraceuticals: Focus on anti-inflammatory, anti-cancer, antioxidant properties in gastrointestinal tract. Antioxidants, 11, 1–15.

[37] Meitha, K., Pramesti, Y. and Suhandono, S. 2020. Reactive oxygen species and antioxidants in postharvest vegetables and fruits. International Journal of Food Sciences, 2020, 1–11.

[38] Holton, K.F. 2021. Micronutrients may be a unique weapon against the neurotoxic triad of excitotoxicity, oxidative stress and neuroinflammation: A Perspective. Frontiers in Neuroscience, 15, 1–11.

[39] Kaźmierczak-Barańska, J., Boguszewska, K., Adamus-Grabicka, A. and Karwowski, B.T. 2020. Two faces of vitamin C – Antioxidative and pro-oxidative agent. Nutrients, 12(5), 1501.

[40] Napolitano, G., Fasciolo, G., Meo, S.D. and Venditti, P. 2019. Vitamin E supplementation and mitochondria in experimental and functional hyperthyroidism: A mini-review. Nutrients, 11(12), 2900.

[41] Zeng, Z., Zdzieblik, D., Centner, C., et al. 2020. Changing dietary habits increases the intake of antioxidant vitamins and reduces the concentration of reactive oxygen species in blood: A pilot study. International Journal of Food Properties, 23, 1337–1346.

[42] Wimalawansa, S.J. 2019. Vitamin D deficiency: Effects on oxidative stress, epigenetics, gene regulation, and aging. Biology (Basel), 8, 1–15.

[43] Masjedi, F., Keshtgar, S., Zal, F., et al. 2020. Effects of vitamin D on steroidogenesis, reactive oxygen species production, and enzymatic antioxidant defense in human granulosa cells of normal and polycystic ovaries. The Journal of Steroid Biochemistry and Molecular Biology, 197, 105521.

[44] van de Lagemaat, E.E., de Groot, L.C.P.G.M. and van den Heuvel, E.G.H.M. 2019. Vitamin B12 in relation to oxidative stress: A systematic review. Nutrients, 11(2), 482.

[45] Simes, D.C., Viegas, C.S.B., Araújo, N., and Marreiros, C. 2019. Vitamin K as a powerful micronutrient in aging and age-related diseases: Pros and cons from clinical studies. International Journal of Molecular Sciences, 20(17), 4150.

[46] Kang, S., Lim, Y., Kim, Y.J., et al. 2019. Multivitamin and mineral supplementation containing phytonutrients scavenges reactive oxygen species in healthy subjects: A randomized, double-blinded, placebo-controlled trial. Nutrients, 11, 1–15.

[47] Zajac, D. 2021. Mineral micronutrients in asthma. Nutrients, 13(11), 4001.

[48] Ansari, M.Y., Ahmad, N. and Haqqi, T.M. 2020. Oxidative stress and inflammation in osteoarthritis pathogenesis: Role of polyphenols. Biomedical and Pharmacology, 129, 110452.

[49] Zhang, B., Zhang, Y., Xing, X. and Wang, S. 2022. Health benefits of dietary polyphenols: Insight into interindividual variability in absorption and metabolism. Current Opinion in Food Science, 48, 100941.

[50] Vacca, R.A., Valenti, D., Caccamese, S., et al. 2016. Plant polyphenols as natural drugs for the management of Down syndrome and related disorders. Neuroscience and Biobehavioral Reviews, 71, 865–877.

[51] Qiu, L., Luo, Y. and Chen, X. 2018. Quercetin attenuates mitochondrial dysfunction and biogenesis via upregulated AMPK/SIRT1 signaling pathway in OA rats. Biomedical and Pharmacology, 103, 1585–1591.

[52] Liang, Q., Wang, X.P. and Chen, T.S. 2014. Resveratrol protects rabbit articular chondrocyte against sodium nitroprusside-induced apoptosis via scavenging ROS. Apoptosis, 19, 1354–1363.

[53] Wei, Y., Jia, J., Jin, X., et al. 2018. Resveratrol ameliorates inflammatory damage and protects against osteoarthritis in a rat model of osteoarthritis. Molecular Medicine Reports, 17, 1493–1498.

[54] El Gaamouch, F., Liu, K., Lin, H., et al. 2021. Development of grape polyphenols as multi-targeting strategies for Alzheimer's disease. Neurochemistry International, 147, 105046.

[55] Naraoka, M., Li, Y., Katagai, T. and Ohkuma, H. 2020. Effects of apple polyphenols on oxidative stress and cerebral vasospasm after subarachnoid hemorrhage in a rabbit double hemorrhage model. Brain Hemorrhages, 1, 54–58.

[56] Benchagra, L., Alami, M., Boulbaroud, S., et al. 2022. Moroccan pomegranate (sefri variety) polyphenols prevent hyperlipidemia, oxidative stress and enhance cholesterol efflux processes. Atherosclerosis, 355, 73–74.

[57] Cuevas, E., Limón, D., Pérez-Severiano, F., et al. 2009. Antioxidant effects of Epicatechin on the hippocampal toxicity caused by Amyloid-beta 25–35 in rats. European Journal of Pharmacology, 616, 122–127.

[58] Chowdhury, A., Sarkar, J., Chakraborti, T., et al. 2016. Protective role of epigallocatechin-3-gallate in health and disease: A perspective. Biomedical and Pharmacology, 78, 50–59.

[59] Dragicevic, N., Smith, A., Lin, X., et al. 2011. Green tea epigallocatechin-3-gallate (EGCG) and other flavonoids reduce Alzheimer's amyloid-induced mitochondrial dysfunction. Journal of Alzheimer's Disease, 26, 507–521.

[60] Lelli, D., Sahebkar, A., Johnston, T.P. and Pedone, C. 2017. Curcumin use in pulmonary diseases: State of the art and future perspectives. Journal of Pharmacology Research, 115, 133–148.

[61] Mykhailenko, O., Kovalyov, V., Goryacha, O., et al. 2019. Biologically active compounds and pharmacological activities of species of the genus Crocus: A review. Phytochemistry, 162, 56–89.

[62] Baba, S.A., Malik, A.H., Wani, Z.A., et al. 2015. Phytochemical analysis and antioxidant activity of different tissue types of Crocus sativus and oxidative stress alleviating potential of saffron extract in plants, bacteria, and yeast. South African Journal of Botany, 99, 80–87.

[63] Zhu, C., Yan, H., Zheng, Y., et al. 2020. Impact of cinnamon supplementation on cardiometabolic biomarkers of inflammation and oxidative stress: A systematic review and meta-analysis of randomized controlled trials. Complementary Therapies in Medicine, 53, 102517.

[64] Miah, P., Mohona, S.B.S., Rahman, M.M., et al. 2021. Supplementation of cumin seed powder prevents oxidative stress, hyperlipidemia and non-alcoholic fatty liver in high fat diet fed rats. Biomedical and Pharmacology, 141, 111908.

[65] Mahboub, H.H., Elsheshtawy, H.M., Sheraiba, N.I., et al. 2022. Dietary black cumin (*Nigella sativa*) improved hemato-biochemical, oxidative stress, gene expression, and immunological response of Nile tilapia (Oreochromis niloticus) infected by Burkholderia cepacia. Aquaculture Reports, 22, 100943.

[66] Wei, H.K., Chen, G., Wang, R.J. and Peng, J. 2015. Oregano essential oil decreased susceptibility to oxidative stress-induced dysfunction of intestinal epithelial barrier in rats. Journal of Functional Foods, 18, 1191–1199.

[67] Sallam, M.F., Ahmed, H.M.S., Diab, K.A., et al. 2022. Improvement of the antioxidant activity of thyme essential oil against biosynthesized titanium dioxide nanoparticles-induced oxidative stress, DNA damage, and disturbances in gene expression in vivo. Journal of Trace Elements in Medicine and Biology, 73, 127024.

[68] Shen, H.H., Peterson, S.J., Bellner, L., et al. 2020. Cold-pressed nigella sativa oil standardized to 3% thymoquinone potentiates omega-3 protection against obesity-induced oxidative stress, inflammation, and markers of insulin resistance accompanied with conversion of white to beige fat in mice. Antioxidants, 9, 1–19.

[69] Dasilva, G., Pazos, M., García-Egido, E., et al. 2015. Healthy effect of different proportions of marine ω-3 PUFAs EPA and DHA supplementation in Wistar rats: Lipidomic biomarkers of oxidative stress and inflammation. Journal of Nutritional Biochemistry, 26, 1385–1392.

[70] Zhang, X.Y., Xu, Z.P., Wang, W., et al. 2018. Vitamin C alleviates LPS-induced cognitive impairment in mice by suppressing neuroinflammation and oxidative stress. International Immunopharmacology, 65, 438–447.

[71] Alatawi, F.S., Faridi, U.A. and Alatawi, M.S. 2018. Effect of treatment with vitamin D plus calcium on oxidative stress in streptozotocin-induced diabetic rats. Saudi Pharmaceutical Journal, 26, 1208–1213.

[72] Padmanabhan, S., Waly, M.I., Taranikanti, V., et al. 2019. Folate/vitamin B12 supplementation combats oxidative stress-associated carcinogenesis in a rat model of colon cancer. Nutrition and Cancer, 71, 100–110.

[73] Rinwa, P. and Kumar, A. 2017. Quercetin along with piperine prevents cognitive dysfunction, oxidative stress and neuro-inflammation associated with mouse model of chronic unpredictable stress. Archives of Pharmacal Research, 40, 1166–1175.

[74] Mlcek, J., Jurikova, T., Skrovankova, S. and Sochor, J. 2016. Quercetin and its anti-allergic immune response. Molecules, 21, 1–15.

[75] Singh, P., Mishra, G., Molla, M., et al. 2022. Dietary and nutraceutical-based therapeutic approaches to combat the pathogenesis of Huntington's disease. Journal of Functional Foods, 92, 105047.

[76] Liu, S., Li, T., Liu, H., et al. 2016. Resveratrol exerts antidepressant properties in the chronic unpredictable mild stress model through the regulation of oxidative stress and mTOR pathway in the rat hippocampus and prefrontal cortex. Behavioural Brain Research, 302, 191–199.

[77] Shaki, F., Shayeste, Y., Karami, M., et al. 2017. The effect of epicatechin on oxidative stress and mitochondrial damage induced by homocycteine using isolated rat hippocampus mitochondria. Research in Pharmaceutical Sciences, 12, 119–127.

[78] Bernatova, I. 2018. Biological activities of (−)-epicatechin and (−)-epicatechin-containing foods: Focus on cardiovascular and neuropsychological health. Biotechnology Advances, 36, 666–681.

[79] Xu, Q., Langley, M., Kanthasamy, A.G. and Reddy, M.B. 2017. Epigallocatechin Gallate has a neurorescue effect in a mouse model of Parkinson disease. The Journal of Nutrition, 147, 1926–1931.

[80] Samarghandian, S., Azimi-Nezhad, M., Farkhondeh, T. and Samini, F. 2017. Anti-oxidative effects of curcumin on immobilization-induced oxidative stress in rat brain, liver and kidney. Biomedical and Pharmacology, 87, 223–229.

[81] Koul, A. and Abraham, S.K. 2017. Intake of saffron reduces γ-radiation-induced genotoxicity and oxidative stress in mice. Toxicology Mechanisms and Methods, 27, 428–434.

[82] Hussain, Z., Khan, J.A., Arshad, A., et al. 2019. Protective effects of *Cinnamomum zeylanicum* L. (Darchini) in acetaminophen-induced oxidative stress, hepatotoxicity and nephrotoxicity in mouse model. Biomedical and Pharmacology, 109, 2285–2292.

Hammad Ullah, Tokpam Reshma Chanu,
Sivaa Arumugam Ramakrishnan, Rajan Logesh*

# Chapter 7
# Nutraceuticals and inflammation

**Abstract:** The chronic state of inflammation is widely associated with number of pathologies including asthma, pneumonia, cardiovascular diseases, metabolic ailment, inflammatory bowel disease, arthritis, neurodegenerative disorders, and cancer. The conventional drug therapies including nonsteroidal anti-inflammatory drugs (NSAIDs) and corticosteroids have certain limitations, mainly related to their adverse effects and high cost. In this regard, there is continued focus on alternative therapies, including plant-derived components to prevent or treat inflammatory conditions. Food bioactive ingredients showed promising health effects with favorable safety profile and relatively low cost. In recent decades, they are being extensively evaluated for their anti-inflammatory effects, and further investigations on these bioactive ingredients will result in the development of effective and safe food supplement-based therapies for chronic inflammation. In this chapter, the immunomodulation and anti-inflammatory properties of dietary fibers, prebiotics, probiotics, polyunsaturated fatty acids, polyphenols, and spice-derived bioactive components are reviewed.

# 1 Introduction

One of the complex cascades in the human system is inflammation, triggered by the detection of tissue injury or during microbial infection. It aids in activation of immune cells to protect the cellular system by preventing further damaged caused by pathogens or injury [1]. In case of hypersensitivity, inflammation can lead to severe damage to the cells and organs of the human body, which may result in a range of disorders such as rheumatoid arthritis, cardiac diseases, autoimmune disorders, and carcinogenesis. The process of inflammation is piloted by the pro-inflammatory cyto-

*Corresponding author: Rajan Logesh**, Department of Pharmacognosy, JSS College of Pharmacy, JSS Academy of Higher Education and Research, Mysuru 570015, Karnataka, India,
e-mail: logeshr@jssuni.edu.in
**Hammad Ullah,** Department of Pharmacy, University of Naples Federico II, Via Domenico Montesano 49, 80131, Naples, Italy
**Tokpam Reshma Chanu,** Department of Nutrition and Dietetics, JSS Academy of Higher Education and Research, Mysuru 570015, Karnataka, India
**Sivaa Arumugam Ramakrishnan,** Department of Pharmaceutical Biotechnology, JSS College of Pharmacy, JSS Academy of Higher Education and Research, Mysuru 570015, Karnataka, India

https://doi.org/10.1515/9783111317601-007

kines like tumor necrosis factor alpha (TNF-α), interleukins (IL-1 and IL-6), nuclear factor kappa-light-chain enhancer of activated B cells (NF-κB), nitric oxide (NO), integrins, cyclooxygenase (COX-2), selectins and immune cells like monocytes, dendric cells, leukocytes, and macrophages [2].

Nutraceuticals are food-derived bioactive components that include herbal products, botanical extracts, foods, beverages, and dietary supplements, available in different dosage forms [3, 4]. They possess the capability of increasing the secretion of anti-inflammatory cytokines, which prevent further damage caused by the pro-inflammatory mediators (Figure 1) [5]. Clinically, nutraceuticals offer preliminary evidence in preventing inflammation and damage in neurological disorders, cardiovascular diseases, inflammatory bowel disease (IBD), and rheumatoid arthritis [6]. The reasons for the wide acceptance of nutraceuticals by consumers are their easily availability and favorable safety profile, though their use is still not supported by solid scientific evidence in many diseases [7, 8]. They should be evaluated in randomized clinical trials for the assessment of efficacy and safety issues. One of the main hurdles in clinical use of food bioactive components is their low bioavailability, which points to the need for development of food supplements using innovative formulation techniques [9].

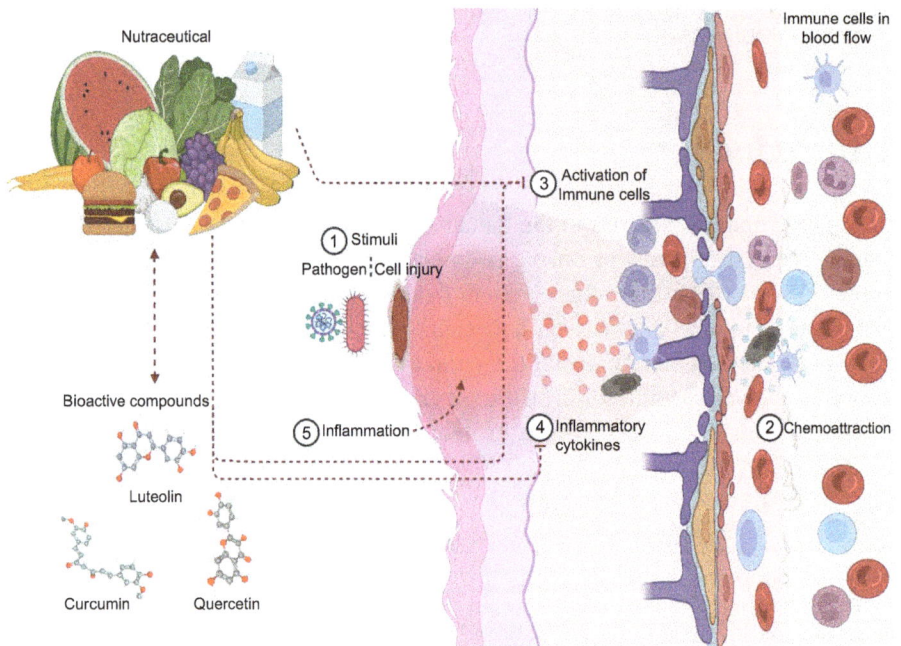

**Figure 1:** The role of nutraceuticals in inflammation caused by cell injury or pathogens.

This chapter is designed to focus on the comprehensive overview of the inflammatory response, conventional anti-inflammatory therapies, and potential role of food supplement ingredients (i.e., dietary fibers, prebiotics, probiotics, polyunsaturated fatty acids (PUFAs), vitamins, minerals, and polyphenols) in modulating immune and inflammatory response.

## 1.1 Inflammatory response

Inflammation could be either be good or bad; good inflammation is the normal and regulated immune response as observed after vaccination and correlated to reactogenicity while bad inflammation is associated with dysregulated immune response such as inflammaging, chronic smouldering inflammation, sepsis, cytokine storm, cachexia, and high-proliferative cancer [10].

The onset of inflammatory response requires triggering factors such as tissue injury, infections, chemicals and radiations, and some diseased states. Regardless of the cause and kind of tissue injury, there is common inflammatory response with the involvement of inducers, sensors, mediators, and effectors [11]. Inflammation inducers could be exogenous i.e., PAMPs (pathogen-associated molecular patterns) or endogenous i.e., DAMP (damage-associated molecular patterns) or alarmins [12]. PAMPs include nucleic acid (in particular from viruses and bacteria), proteins (i.e., flagellin and pilin from bacterial cell wall), lipids (i.e., lipopolysaccharide or LPS and lipoteichoic acid from bacteria), and or carbohydrates (i.e., glucans or mannan from bacteria or fungi) [13]. Alarmins are stress-induced proteins including proteoglycans, HSP (heat shock proteins), nuclear proteins including histones, high mobility group box 1 (HMGB1), S100 calcium-binding protein family members, i.e., S100A8, S100A9, or S100A12, and mitochondrial components [14].

All these endogenous inducers are released from the damaged cell or tissue and then engage with different cell-associated recognition molecules known as sensors, which are stimulated by inducers to trigger the production of mediators. Different endogenous inducers released from necrotic cells may have different sensors such as RAGE (advanced glycation end-product-specific receptor) for S100A12 and HMGB1 that may also cooperate with Toll-like receptors (TLRs), purinoceptors including P2X for ATP-binding leading to $K^+$ efflux, and that may cooperate with nucleotide oligomerization domain receptors (NOD)-like receptor protein-3 (NLRP3) inflammasomes [11]. Sensors are expressed by epithelial cells at the body surface and phagocytes (mainly macrophages, neutrophils, mast cells, and dendritic cells), and include cell-associated TLRs 1–9, NOD-like receptors (NLRs) i.e., NOD1–2 and inflammasomes, RIG (retinoic acid-inducible gene)-like receptors (RLRs) i.e., MDA-5, CDSs (cytosolic DNA sensors), CLRs (C-type lectin-like receptors), i.e., Dectin-1 and -2, CD36 (scavenger receptors), DC-sign, mannose receptor, and N-formyl met-leu-phe receptors [15, 16].

Inducers, after being recognized by sensors, are triggered to produce numerous inflammatory mediators derived either from plasma proteins or secreted by the cells (mainly by mast cells, basophils, macrophages or platelets). Based on the physiological role, these mediators can be classified as vasoactive peptides, vasoactive amines, lipid mediators (i.e., eicosanoids or platelet activating factors), fragments of the complement components, proteolytic enzymes, chemokines, and cytokines. In addition to their role as inflammation promoters, they also initiate the tissue repair process. Inflammatory complexes comprise sensor and adaptor proteins, and zymogen procaspase-1 that is known to activate in an active form, caspase-1. Following activation, caspase-1 stimulates pro-inflammatory cytokines IL-1β and IL-18. It may also induce highly pyrogenic inflammatory form of death known as pyroptosis via gasdermin D activation [17,18]. Following activation of inducers (i.e., PAMP or DAMP), neutrophils are either directly or indirectly activated via inflammatory mediators including (C-X-C motif) ligand (CXCL)-1 and -2, which bind and activate the G-protein coupled receptors (GPCRs) on the surface of neutrophils [19].

Effectors such as neutrophils, monocytes or macrophages, natural killer cells (NK cells), and T and B lymphocytes are specialized cells present at the initial site of injury, producing local or systemic inflammatory response. Neutrophils arrive at the site of infection or tissue injury via intravascular migration mediated by selectins, followed by chemokine activation via conformational change in GPCRs, resulting in the activation of neutrophil integrins, i.e., CD49d/CD29 (integrin heterodimer very late antigen 4 (VLA-4)), CD11b/CD18 (macrophage-1 antigen (Mac-1)), and CD11a/CD18 (lymphocyte function-associated antigen 1 (LFA-1)) [20]. Activation of integrins favors adhesion of cells to the endothelium via ICAMs (immunoglobulin (Ig)-superfamily cell adhesion molecules). Neutrophils generally have a short life and they die through apoptosis, especially in antibacterial activities and neutrophils-derived reactive oxygen species (ROS), proteases, and neutrophil extracellular traps (NETs), where NETs are network associating DNA coated with elastase, myeloperoxidase (MPO), histones, and cathepsin G. They are initially seen in killing bacterial cells, resulting in cell death called lytic NET [21–23]. Contrarily, non-lytic NETs are also released from mitochondria, which lack histones [24]. Moreover, not all neutrophils are associated with the release of NETs; low-density neutrophils and CD177 negative expressing olfactomedin 4 (OLFM4) do release NETs [25]. Recent studies showed that neutrophils with high concentration of OLFM4 makes an individual prone to high risk of septic shock and organ failure [26]. Furthermore, OLFM4 complexes with some of major signaling pathways like wingless Int-1 (Wnt) and NF-κB, play crucial roles in inflammation and carcinogenesis. MPO belongs to a heme-containing peroxidase family, which is known to be actively involved in ROS production and is released after inflammatory responses and oxidative stress into the extracellular fluid. Abnormal and dysregulated release of MPO could be associated with the states of chronic inflammation, resulting in pathogenesis of numerous chronic pathologies [27, 28].

Neutrophils and monocytes or macrophages have common precursors and thus both co-express same antigens and produce same effector molecules such as cytokines, chemokines, and oxidants [29]. However, they possess different properties as regards inflammation processes, especially in host defense against microbial infections. Unlike neutrophils (which recruited initially to the injury site), monocytes or macrophages recruited later and initiate adaptive immune responses via digestion and antigen presentation. Monocytes circulate in peripheral blood for 1–3 days before migrating to tissues, where they become macrophages or dendritic cells. Classical ($CD14^{++}CD16^-$), intermediate ($CD14^+CD16^+$), and nonclassical ($CD14^-CD16^{++}$) monocytes are three subpopulations of monocytes that have been identified in humans [30]. Major functions of macrophages include tissue repair and resolution of inflammation [31]. Neutrophils are supposed to die quickly in order to prevent against excessive inflammation, whereas macrophages prolonged their survival via producing GM-CSF (growth factors such as granulocyte-macrophage colony-stimulating factor), G-CSF, and TNF-α [32]. Macrophages possess ability to acquire eitherM1 "pro-inflammatory phenotype," expressing iNO and CD40 and produce TNF-α and IL-6 or M2 "anti-inflammatory phenotype," expressing arginase I and CD206 and produce IL-10 and TGF-β (transforming growth factor beta). M1 macrophages are more active in case of infections while M2 macrophages are active in tissue repair [33, 34].

# 2 Conventional anti-inflammatory therapies and limitations

Inflammatory disorders usually require therapy for longer time, and thus efficacy, safety, and cost of therapy could be major issues for the healthcare providers as well as for the patients. Therefore, finding a cheap treatment approach that is more efficacious and safer over a longer period of time is the need of the hour. The exercise of alternative treatment approaches including nutraceuticals and food supplements could be a fruitful strategy in patients with more chronic inflammatory diseases, aimed either to prevent the specific illness or to augment the conventional therapy to improve their therapeutic effects and reduce the risk of adverse effects (due to requirement of lower doses of drugs in augmented therapy).

In conventional therapy, mainly two classes of anti-inflammatory drugs are in practice, namely nonsteroidal anti-inflammatory drugs (NSAIDs) and corticosteroids. NSAIDs act through inhibition of COX enzymes, though they are associated with number of detrimental side effects including nausea, vomiting, abdominal pain, dyspepsia, heart burn, and peptic ulcer. They also possess the risk for cardiovascular complications, in particular with the use of COX-2 inhibitors [35]. The prolonged use of corticosteroids increases the risk of skin reactions (ecchymosis, skin thinning and atrophy, mild hirsutism, acne, thinning of hair, facial erythema, impaired wound healing, and

perioral dermatitis), gastrointestinal effects (gastritis, gastric ulcer, and gastrointestinal bleeding), musculoskeletal effects (osteoporosis, osteonecrosis, and myopathy), endocrine effects (diabetes mellitus, Cushing syndrome, and hypothalamic-pituitary-
adrenal (HPA) axis), cardiovascular effects (edema, hypertension, and arrhythmias),
ophthalmologic effects (cataract and glaucoma), neuropsychiatric effects (anxiety, eu-
phoria, hypomanic reactions, and at large doses psychosis), immune suppression, and
increased risk of infections [36].

Some other anti-inflammatory drugs are also currently being used, depending
upon the inflammatory condition to be treated, which include monoclonal antibodies
(i.e., vedolizumab and belimumab), antimetabolites (i.e., azathioprine and mercapto-
purine), aminosalicylates (i.e., mesalazine), TNF-α inhibitors (i.e., etanercept and ada-
limumab), and disease-modifying antirheumatic drugs (DMARDs; i.e., methotrexate
and penicillamine). However, all these classes of anti-inflammatory drugs require supervision by a qualified person, as they are associated with range of undesirable effects, which sometimes lead to life-threatening conditions [37].

# 3 Immunomodulation and anti-inflammatory potential of nutraceuticals

Nutraceuticals are food bioactive products or dietary ingredients may prevent and
treat disease [38, 39]. Although they are not medications, they do include pharmacologically active ingredients [40]. Bioactive compounds are present in fruits like grape,
citrus, blueberries, strawberries, blackberries, and crowberries; vegetables like tomato, beans, broccoli, beet, mushroom, corn, white cabbage, kale, cauliflower, spinach; spices like rosemary, oregano, and thyme; herbs like sage; beverages like tea and
wine are among the foods and considered as nutraceuticals [41–43]. The focus on
functional foods is growing and the competition is escalating advances in technology
results in the food, health, and pharmaceutical industries [44]. The initial response of
the immune system to potentially damaging stimuli is inflammation (e.g., cellular, tissue damage, cellular stress, and pathogenic infections). Even though it aids in host
defensive mechanism, it involves in the development of numerous disorders, which
have been reported. Neurological disorders, IBD, autoimmune disorders, cardiovascular disorders, and rheumatoid arthritis are only a few of the illnesses whose etiology
is affected by chronic inflammation [45, 46].

Biological effects like immunomodulatory actions, antiglycemic activity, neuroprotective and nephroprotective activity, anti-inflammatory and even anticancer activity have been recently reported on nutraceuticals [47, 48]. Several diseases like
arthritis, pancreatitis, liver fibrosis, diabetes, renal ischemia, ulcers, gout, and bacterial and viral infections are successfully treated with fucoidans [49].

Inflammatory disorders and related ailments have been treated using plants or plant-derived bioactive components in different countries around the world [50–52]. Scientific studies have already shown that a number of plant extracts and isolated chemicals have anti-inflammatory properties. Medicinal plants namely, *Curcuma longa* (turmeric) has been used to treat rheumatoid arthritis as it possesses anti-inflammatory activity [70]. iNOS, and PGE2 (prostaglandin E2) expression can be suppressed by *Zingiber zerumbet* (ginger extract) in cell culture model. The diet must include specific nutrients in adequate proportion to promote the immune system's normal functioning. Obesity from caloric overnutrition, which is common in the Western society, increases sensitivity to inflammatory reactions and unintentionally activates immune system components [53]. On the other hand, malnutrition leads to immunosuppression and increases sensitivity to the infections [54]. Immunological and inflammatory process can be influenced by several dietary products like fat, vitamins, minerals, trace elements, alcohol, dietary fiber, prebiotics, and probiotics, as well as phytochemicals.

## 3.1 Dietary fibers

The association between dietary fibers and intestinal and systemic inflammation has been extensively reported in the literature and is of particular interest in a number of diseases. They may help in lowering inflammatory markers possibly through the reduction of body weight and through the modulation of intestinal flora. A few studies have reported that people with higher intake of fibers on daily basis may have lower concentration of C-reactive protein (CRP), which has been linked to diseases like diabetes, heart disease, and arthritic disorders [55]. While assessing the longitudinal associations between dietary fiber intake and CRP levels, Ma, Y. et al. observed 63% decrease in CRP concentration in subjects with high intake of total fiber contents [56]. Two large prospective cohort studies demonstrated that people with regular intake of Mediterranean diet (high intake of fruits, vegetables, fish proteins, whole grains, nuts, and legumes) had a reduced risk of Crohn's disease as compared to 12% increased risk of disease in people with poor adherence to a Mediterranean diet [57]. Dietary fibers may provide protection to the intestinal barrier mediated by short-chain fatty acids (SCFAs) since acetate and butyrate promote the intestinal barrier integrity by the activation of IL-18 receptor and through providing energy to enterocytes, respectively [58]. The high fiber supplementation (including psyllium, pectin, and cellulose) showed protection against dextran sodium sulfate (DSS)-induced colitis in murine models [59–61]. Lower intake of dietary fibers in mice resulted in depletion of mucus layer and intestinal barrier disruption because of alteration of mucus-eroding microbial species such as *Bacteroides caccae* and *Akkermansia muciniphila* [62] that might enhance DSS-induced intestinal permeability [63]. A narrative review by Niero et al. demonstrated positive effects on systemic inflammation and insulin resistance, showing beneficial effects of fiber-rich diet on metabolic and cardiovascular health [64]. A cohort study showed reduced levels of inflammatory markers and lower risk of cardiovascular diseases in older subjects with higher intake of cereal fibers [65].

## 3.2 Prebiotics

Prebiotics are fermented by colonic bacteria to produce SCFAs i.e., acetate, propionate, and butyrate, to which the systemic anti-inflammatory effects are generally attributed. SCFAs reduce the colonic pH and promote the growth of beneficial microbes such as *Bifidobacterium* and Lactobacillus, where these bacterial species are potent SCFA producers, playing a crucial role in maintaining healthy immune response [66, 67]. The daily consumption of prebiotics ranges from 3–13 g/day, and it is interesting to note that IBD is more prevalent in parts of world with relatively lower intake of prebiotics. Following absorption through the intestinal lumen a small portion of SCFAs (mainly acetate) enters the systemic circulation and may influence cells within the peripheral tissues. SCFAs regulate immune responses by affecting neutrophils, dendritic cells, macrophages, T and B cells [68–70], by altering different mechanistic targets such as upstream regulation of epithelial and immune cells surface G-protein-coupled receptors (GPCRs) as well as induced epigenetic changes through the histone acetylase (HDAC) and histone deacetylase enzymes inhibition [71]. While acting as signaling molecules, SCFAs recognize well characterized GPCRs including GPCR41, GPCR43, and GPCR109A, promoting immune homeostasis [72]. Mice with GPCR43 deficiency would result in the recruitment of immune cells and aggravated inflammatory response in experimental models of colitis, asthma, and arthritis [73]. Another study showed lower IgA secretion in GPCR43-deficient mice via affecting acetate-induced IgA secretion from B cells [74]. Butyrate binds to GPR109A in the colon, promoting the expression of anti-inflammatory molecules by dendritic cells and macrophages, helping in the differentiation of T-reg cells and IL-10-producing T cells [75]. A study also reported T-cell differentiation in effector and T-reg cells by SCFAs along with the increased expression of IL-10, IFN-γ, and IL-17 via inhibition of HDACs and regulation of P70 S6 kinase-RS6 [76]. Roller et al. demonstrated that inulin-enriched oligofructose increased the production of IL-10 in Peyer's patches with enhanced secretion of IgA in the ileum. The latter increased the phagocytic function of intraperitoneal macrophages and prevented the attachment of intestinal pathogens [77]. Feeding colitis-susceptible HLA-B27 transgenic rats with inulin-enriched oligofructose resulted in reduction of colitis with increased intestinal TGF-β [78].

## 3.3 Probiotics

Gut microbiota and its metabolites are strongly linked with the host inflammatory conditions, as change in the composition of gut microbiota has been observed in patients with Crohn's disease, ulcerative colitis, asthma, obesity, diabetes mellitus, and rheumatoid arthritis. *Enterobacteriaceae* is commonly linked to many inflammatory conditions including IBD and obesity. The depletion of *Firmicutes* and increased population of *Proteobacteria* are associated with colon inflammation. Moreover, an increased *Firmicutes* to *Bacteroidetes* ratio has been linked to obesity. The gut microbiota analysis of the rheumatoid arthritis patients showed increased levels of *Lactobacillus* and *Prevotella copri*, with decreased numbers of *Bifidobacteria* and *Bacteroides* [79]. Probiotics are nonpathogenic bacteria that offer significant benefits to the host when supplementing in appropriate quantities, mainly through the modulation of intestinal flora. They promote healthy digestion, reduce inflammation, and prevent the development of the carcinogenic risk factors [80]. The potential molecular mechanisms of probiotics by which they promote health and prevent against disease risk factors include (i) production of SCFAs, IgA, and SCFAs; (ii) decreased secretion of pro-inflammatory cytokines; (iii) enhanced mucin-2 expression; and (iv) augmented regulation of defensins [81]. The consumption of *Bifidobacterium lactis* Bi-07 is shown to strengthen the innate immune response in aged population [82]. Similarly, the supplementation of elderly population aged 65–90 years significantly suppressed TNF-α with increase in intestinal *Bifidobacteria* (in particular *B. adolescentis, B. longum, B. bifidum,* and *B. angulatum*) and well as enhanced fecal acetate and butyrate concentrations [83]. Kim et al. reported the anti-inflammatory effects of *Bifidobacterium bifidum* BGN4 and *B. longum* BORI in aged subjects by improvement in the composition of gut microbiome [84]. The treatment of *Helicobacter pylori*-infected mice with the mixture of *Lactobacillus* strains including *L. fermentum* P2, *L. casei* L21 and *L. rhamnosus* JB3 resulted in increased butyrate production and recovered serum levels of amino acids associated with modulation of immune functions, i.e., glycine, alanine, aspartate, arginine, and tryptophan [85]. A randomized controlled trail demonstrated significant improvement of diarrhea-predominant irritable bowel syndrome (IBS-D) symptoms with the supplementation of *Lactobacillus, Bifidobacterium,* and *Streptococcus thermophilus* strains for eight weeks [86]. The supplementation of obese subjects with *Lactobacillus casei* Shirota, *Bifidobacterium breve* Yakult, and galactooligosaccharides considerably improved the gut environment in treated subjects, though no significant difference was noted among supplement and control groups regarding the inflammatory markers [87].

## 3.4 Polyunsaturated fatty acids (PUFAs)

Western diet has been linked to the increased incidence of chronic inflammatory conditions due to decreased consumption of *n*-3 PUFAs and increased consumption on *n*-6 PUFAs, resulting in imbalanced *n*-6:*n*-3 ratio (estimated at 15–30 in westernized dietary habits) [88]. *n*-PUFAs (i.e., linoleic acid) are considered as pro-inflammatory compounds, as they are precursors for arachidonic acid, which leads to synthesis of inflammatory mediators such as prostaglandins and leukotrienes. Contrarily, *n*-3

PUFAs (i.e., α-linoleic acid) are regulators of inflammation and can be metabolized into precursors for eicosapentaenoic acid (EPA) and docosahexaenoic acid (DHA). $n$-3 PUFAs act as a competitive substrate for $n$-6 metabolism and thus anti-inflammatory effects of $n$-3 PUFAs could be mediated by competition with the $n$-6 PUFAs [89]. Since $n$-3 PUFAs antagonize the production of arachidonic acid, there is mounting evidence of their beneficial effects in chronic inflammatory diseases through the regulation of the production of inflammatory cytokines and downregulation of the expression of genes associated with leukocyte adhesion, blood coagulation, and fibrinolysis. Consumption of fish oil or purified EPA supplements demonstrated amelioration of the signs and symptoms of autoimmune diseases in animal models. Human studies have reported the anti-inflammatory and immunomodulatory effects of $n$-3 PUFAs in IBD, atherosclerosis, rheumatoid arthritis, psoriasis, and asthma [90]. The supplementation of chronic obstructive pulmonary disease (COPD) patients with $n$-3 PUFAs resulted in attenuated inflammatory response as levels of leukotriene $B_4$, TNF-α, and IL-8 were significantly low in sputum samples of patients [91]. A Mendelian randomization study indicated the $n$-3 PUFAs as a considerable protective factor against IBD due to attenuation of inflammatory responses while low $n$-3 to $n$-6 ratio was observed as a potential risk factor for Crohn's disease [92]. However, in the human study on patients with Alzheimer's disease, supplementation with $n$-3 PUFAs for six months did not show any significant effects on the plasma or cerebrospinal fluid levels of the inflammatory markers [93].

## 3.5 Dietary polyphenols

Dietary polyphenols are abundantly found in fruits and vegetables and are reported extensively in the literature for a broad range of biological properties including (but not limited to) antioxidant, anti-inflammatory, immunomodulatory, cardioprotective, and anticancer activities. For systemic biological effects, most of the polyphenols except flavonoids are usually stable and absorbed across the intestinal membrane, while the undigested polyphenols need to be hydrolyzed by the digestion enzymes and then glycosides with high lipid contents are absorbed by the epithelial cells [94, 95]. Anti-inflammatory activities and immunomodulation by polyphenols are supported by a larger body of studies for their effects on immune cells population, cytokines production, and expression of pro-inflammatory genes. The cardioprotective effects of resveratrol (from grapes and red wine) attributed mainly to its anti-inflammatory potential, as it has been shown to inhibit COX enzyme, downregulate peroxisome proliferator-activated receptor gamma (PPAR-γ), induce endothelial nitric oxide (eNO) in rodent macrophages [96–98]. Similarly, resveratrol analogue (RVSA40) inhibited pro-inflammatory cytokines TNF-α and IL-6 in murine macrophages cell lines [99].

A reduction of the expression of inflammatory cytokines and mediators like TNF-α, IL-1, prostaglandins and leukotriens, intercellular adhesion molecule-1 (ICAM-1), and vascular cell adhesion molecule-1 (VCAM-1) has been reported with non-flavonoid phenolic compound curcumin. It also showed inhibition of inflammation-associated enzymes like lipoxygenase, COX, mitogen-activated protein kinase (MAPK), and inhibitor of kappa kinase (IKK). In addition, treatment with curcumin also resulted in downstream regulation of NF-κB and STAT3, with a reduction of the expression of toll-like receptors (TLR-2 and TLR-4) and upregulation of PPAR-γ [100–105]. Quercetin inhibited biosynthesis of leukotrienes in human polymorphonuclear leukocytes while caffeic acid phenethyl esters are known to suppress TLR-4 and lipopolysaccharide-mediated NF-κB activation in macrophages [106, 107]. The production of adiponectin (known for its anti-inflammatory effects) could be activated by polyphenols such as quercetin and gingerol [100, 108]. Epigallocatechin gallate (EGCG) is demonstrated to attenuate COX-2 expression in colon cancer cells and human prostate carcinoma cells (PC-3 cells) while an EGCG analog (piceatannol) showed inhibition of NFκ B activation [100, 108–110]. EGCG also downregulated the expression of inducible nitric oxide synthase (iNOS), production of nitric oxide (NO) in macrophages, which may result in immunomodulatory response [109–111]. Polyphenols like EGCG, kaempferol-3-*O*-sophoroside, lycopene, oleanolic acid, and curcumin may inhibit high-mobility group protein 1 or inhibit high-mobility group box1 protein (HMGB1), transcription factors, as well as histones regulating transcription, which support their key role in inflammation [105].

*Prunus domestica* L. extract (rich in hydroxycinnamate derivatives) demonstrated inhibition of the lipopolysaccharide-induced release of pro-inflammatory mediators, i.e., nitrite, interleukin-1 β, and PGE2 in activated J774 macrophages [112]. The oral administration of polyphenols-enriched date fruit extract in male C3H/HeN mice resulted in the enhanced Th1, macrophages and natural killer and dendritic cells in the Peyer's patches and spleen [113]. EGCG intake increased the number of functional T-reg cells in spleens, and pancreatic and mesenteric lymph nodes in BALB/c mice [114]. Baicalin extracted from *Scutellaria baicalensis* Georgi induced FOXP3 expression in HEK 293 T cells and triggered the release of functional T-reg from splenic CD4 + CD25 − T cells [115]. Moreover, flavonoids may show antagonistic effects on aryl hydrocarbon receptor and bind xenobiotic-responsive elements in promoter regions of certain genes, including FOXP3, thus inducing its expression [116]. Considering all these studies about dietary polyphenols, one can conclude that incorporating polyphenols into supplements or functional foods may revolutionize the management of chronic inflammatory and immune-mediated disorders. However, low bioaccessibility and bioactive availability of polyphenols could be an issue; recent advancements in delivery methods have improved the bioavailability and effectiveness of phenolic compounds for promising future applications [117].

## 3.6 Spices

Dietary spices have been shown in recent research to have health-promoting qualities such as antioxidant, anti-inflammatory, immunomodulatory, and chemopreventive properties. Spices-derived bioactive components may reduce inflammatory responses and pain in the body, though the evidence is mixed. In a number of preclinical and clinical studies, curcumin has been shown to modulate numerous inflammatory mediators such as TNF-α, IL-6, IL-27, STAT3, PI3K/Akt, NF-κB, and MAPK [118]. It also suppressed TLR4-mediated NF-κB signaling pathway, resulting in reduction of inflammatory responses in mice with mastitis [86]. In addition, curcumin alleviated *in vivo* chronic nonbacterial prostatitis via downstream regulation of TNF-α, IL-6, and IL-8 [119], and reduced asthmatic airway inflammation via activation of Nrf2/HO-1 signaling pathway [120]. Capsaicin, a principal component of *Capsicum* (red pepper) possesses beneficial effects in chronic diseases like diabetes, asthma, and cancers, possibly through the inhibition of TNF-α, IL-6, PGE2, NF-κB, and STAT3 [121, 122]. It also induced a cell cycle arrest in bladder cancer cells through FOXO3a (forehead box O3a)-mediated pathway [123].

Eugenol (found in clove extract) possesses potent anti-inflammatory activities by modulating inflammatory biomarkers including TNF-α, IL-1, IL-6, PGE2, COX-2, and NF-κB [124]. *In vivo* studies showed the restriction of the asthma progression and suppression of cell proliferation in gastric cancer by eugenol via inhibition of NF-κB pathway [125, 126]. It also inhibited skin cancer by attenuating c-Myc and H-ras and by inducing apoptosis via p53 pathway [127]. Cinnamaldehyde, the bioactive component of *Cinnamomum zeylanicum* (cinnamon) exerts its anti-inflammatory actions in gastric inflammation via inhibition of NF-κB activation [128] and autoimmune encephalomyelitis via regulatory T cells [129], in arthritis via inhibiting IL-2, IL-4, and IFN-γ [130].

Thymoquinone isolated from *Nigella sativa* (black cumin), inhibited TNF-α induced inflammation and cell adhesion in rheumatoid arthritis [131]. It has been shown to downregulate NF-κB and MAPKs signaling to inhibit IL-1β-induced inflammation in human osteoarthritis chondrocytes [132]. The *in vivo* study also showed the prevention of inflammation, neoangiogenesis, and vascular remodeling in asthma [133]. 1,8-Cineole found in basil, sage, and cardamom, has been used in the management of multiple inflammatory ailments including chronic rhinitis, sinusitis, bronchitis, and asthmatic disorders. It has been shown to suppress inflammatory mediators, i.e., nitric oxide synthase (NOS-2), COX-2, and NF-κB, resulting in anti-inflammatory response [134]. It ameliorated the colonic damage TNBS (trinitrobenzene sulfonic acid)-induced colitis in rats, acute pulmonary inflammation, acute pancreatitis *in vivo* through downstream regulation of oxidative stress, cytokines, and NF-κB [135–137]. A study reported the significant reduction of the pro-inflammatory cytokines, i.e., TNF-α, IL-1β, and IL-6 in amyloid β-toxicated PC12 cells [138]. Diallyl sulfide is the major organosulfur compound found in garlic and is one of the potential alternative treatments of airway inflammation (like in case of asthma), possibly through the regulation of Nrf2/HO-1 (nuclear factor-E2-related factor 2/haemoxygenase-1) and NF-κB pathways [139]. It also inhibits the COX-2 expression via NF-κB pathway in case of arthritis [140].

# 4 Conclusion

This chapter collates the anti-inflammatory properties of food bioactive ingredients such as dietary fibers, prebiotics, probiotics, PUFAs, polyphenols, and spices, which provided a representation of the new trends of food-derived bioactives in immune- and inflammation-mediated pathologies. Impressive efforts are being made to assess the preventive or therapeutic potential of these bioactive ingredients in numerous inflammatory diseases. On the basis of available scientific evidence and large population database, one can recommend food sources high in healthy bioactives to provide enough protection against initiation and/or progression of inflammatory diseases. From a practical perspective, this may include diet rich in fruits, vegetables, dietary spices, legumes, nuts, whole grains, and dairy products. However, most of scientific research is still based on preclinical studies; thus conducting robust randomized clinical trials is needed to make science-based recommendations on regular and adequate consumption of diet or food supplements in subjects carrying high risk for immune-mediated or inflammatory disorders.

# References

[1]    Chen, L., Deng, H., Cui, H., Fang, J., Zuo, Z., Deng, J. . . . Zhao, L. 2018. Inflammatory responses and inflammation-associated diseases in organs. Oncotarget, 9(6), 7204.
[2]    Krishnamoorthy, S. and Honn, K.V. 2006. Inflammation and disease progression. Cancer and Metastasis Reviews, 25, 481–491.
[3]    Nasri, H., Baradaran, A., Shirzad, H. and Rafieian-Kopaei, M. 2014. New concepts in nutraceuticals as alternative for pharmaceuticals. International Journal of Preventive Medicine, 5(12), 1487.
[4]    Schulz, V., Hänsel, R., Blumenthal, M. and Tyler, V.E. 2004. *Rational Phytotherapy: A Reference Guide for Physicians and Pharmacists*. Springer Science & Business Media.
[5]    Siriwardhana, N., Kalupahana, N.S., Cekanova, M., LeMieux, M., Greer, B. and Moustaid-Moussa, N. 2013. Modulation of adipose tissue inflammation by bioactive food compounds. The Journal of Nutritional Biochemistry, 24(4), 613–623.
[6]    Vishvakarma, P., Mandal, S. and Verma, A. 2023. A review on current aspects of nutraceuticals and dietary supplements. International Journal of Pharma Professional's Research (IJPPR), 14(1), 78–91.
[7]    Rajasekaran, A., Sivagnanam, G. and Xavier, R. 2008. Nutraceuticals as therapeutic agents: A review. Research Journal of Pharmacy and Technology, 1(4), 328–340.
[8]    Stratton, R.J. and Elia, M. 2007. A review of reviews: A new look at the evidence for oral nutritional supplements in clinical practice. Clinical Nutrition Supplements, 2(1), 5–23.

[9]     McClements, D.J., Li, F. and Xiao, H. 2015. The nutraceutical bioavailability classification scheme: Classifying nutraceuticals according to factors limiting their oral bioavailability. Annual Review of Food Science and Technology, 6, 299–327.

[10]    Rossi, J.F., Lu, Z.Y., Massart, C. and Levon, K. 2021. Dynamic immune/inflammation precision medicine: The good and the bad inflammation in infection and cancer. Frontiers in Immunology, 12, 595722.

[11]    Medzhitov, R. 2008. Origin and physiological roles of inflammation. Nature, 454(7203), 428–435.

[12]    Yang, D., Han, Z. and Oppenheim, J.J. 2017. Alarmins and immunity. Immunological Reviews, 280(1), 41–56.

[13]    Amarante-Mendes, G.P., Adjemian, S., Branco, L.M., Zanetti, L.C., Weinlich, R. and Bortoluci, K.R. 2018. Pattern recognition receptors and the host cell death molecular machinery. Frontiers in Immunology, 9, 2379.

[14]    Grazioli, S. and Pugin, J. 2018. Mitochondrial damage-associated molecular patterns: From inflammatory signaling to human diseases. Frontiers in Immunology, 9, 832.

[15]    Malik, A. and Kanneganti, T.D. 2017. Inflammasome activation and assembly at a glance. Journal of Cell Science, 130(23), 3955–3963.

[16]    Fraser, D.A. and Tenner, A.J. 2008. Directing an appropriate immune response: The role of defense collagens and other soluble pattern recognition molecules. Current Drug Targets, 9(2), 113–122.

[17]    Vanaja, S.K., Rathinam, V.A. and Fitzgerald, K.A. 2015. Mechanisms of inflammasome activation: Recent advances and novel insights. Trends in Cell Biology, 25(5), 308–315.

[18]    Rathinam, V.A. and Fitzgerald, K.A. 2016. Inflammasome complexes: Emerging mechanisms and effector functions. Cell, 165(4), 792–800.

[19]    Kolaczkowska, E. and Kubes, P. 2013. Neutrophil recruitment and function in health and inflammation. Nature Reviews Immunology, 13(3), 159–175.

[20]    Mortaz, E., Alipoor, S.D., Adcock, I.M., Mumby, S. and Koenderman, L. 2018. Update on neutrophil function in severe inflammation. Frontiers in Immunology, 9, 2171.

[21]    Pittman, K. and Kubes, P. 2013. Damage-associated molecular patterns control neutrophil recruitment. Journal of Innate Immunity, 5(4), 315–323.

[22]    Castanheira, F.V. and Kubes, P. 2019. Neutrophils and NETs in modulating acute and chronic inflammation. Blood, the Journal of the American Society of Hematology, 133(20), 2178–2185.

[23]    Brinkmann, V., Reichard, U., Goosmann, C., Fauler, B., Uhlemann, Y., Weiss, D.S., Weinrauch, Y. and Zychlinsky, A. 2004. Neutrophil extracellular traps kill bacteria. Science, 303(5663), 1532–1535.

[24]    Yousefi, S., Mihalache, C., Kozlowski, E., Schmid, I. and Simon, H.U. 2009. Viable neutrophils release mitochondrial DNA to form neutrophil extracellular traps. Cell Death & Differentiation, 16(11), 1438–1444.

[25]    Welin, A., Amirbeagi, F., Christenson, K., Björkman, L., Björnsdottir, H., Forsman, H., Dahlgren, C., Karlsson, A. and Bylund, J. 2013. The human neutrophil subsets defined by the presence or absence of OLFM4 both transmigrate into tissue in vivo and give rise to distinct NETs in vitro. PloS One, 8(7), e69575.

[26]    Alder, M.N., Opoka, A.M., Lahni, P., Hildeman, D.A. and Wong, H.R. 2017. Olfactomedin 4 is a candidate marker for a pathogenic neutrophil subset in septic shock. Critical Care Medicine, 45(4), e426.

[27]    Hesselink, L., Spijkerman, R., Van Wessem, K.J., Koenderman, L., Leenen, L.P., Huber-Lang, M. and Hietbrink, F. 2019. Neutrophil heterogeneity and its role in infectious complications after severe trauma. World Journal of Emergency Surgery, 14(1), 24.

[28]    Khan, A.A., Alsahli, M.A. and Rahmani, A.H. 2018. Myeloperoxidase as an active disease biomarker: Recent biochemical and pathological perspectives. Medical Sciences, 6(2), 33.

[29]    Prame Kumar, K., Nicholls, A.J. and Wong, C.H. 2018. Partners in crime: Neutrophils and monocytes/macrophages in inflammation and disease. Cell and Tissue Research, 371, 551–565.

[30]   Shi, C. and Pamer, E.G. 2011. Monocyte recruitment during infection and inflammation. Nature Reviews Immunology, 11(11), 762–774.

[31]   Ginhoux, F. and Jung, S. 2014. Monocytes and macrophages: Developmental pathways and tissue homeostasis. Nature Reviews Immunology, 14(6), 392–404.

[32]   Takano, T., Azuma, N., Satoh, M., Toda, A., Hashida, Y., Satoh, R. and Hohdatsu, T. 2009. Neutrophil survival factors (TNF-alpha, GM-CSF, and G-CSF) produced by macrophages in cats infected with feline infectious peritonitis virus contribute to the pathogenesis of granulomatous lesions. Archives of Virology, 154, 775–781.

[33]   Hamilton, T.A., Zhao, C., Pavicic Jr, P.G. and Datta, S. 2014. Myeloid colony-stimulating factors as regulators of macrophage polarization. Frontiers in Immunology, 5, 554.

[34]   Liu, C., Li, Y., Yu, J., Feng, L., Hou, S., Liu, Y., Guo, M., Xie, Y., Meng, J., Zhang, H. and Xiao, B. 2013. Targeting the shift from M1 to M2 macrophages in experimental autoimmune encephalomyelitis mice treated with fasudil. PloS One, 8(2), e54841.

[35]   Sostres, C., Gargallo, C.J., Arroyo, M.T. and Lanas, A. 2010. Adverse effects of non-steroidal anti-inflammatory drugs (NSAIDs, aspirin and coxibs) on upper gastrointestinal tract. Best Practice & Research Clinical Gastroenterology, 24(2), 121–132.

[36]   Poetker, D.M. and Reh, D.D. 2010. A comprehensive review of the adverse effects of systemic corticosteroids. Otolaryngologic Clinics of North America, 43(4), 753–768.

[37]   Ullah, H. and Khan, H. 2020. Epigenetic drug development for autoimmune and inflammatory diseases. In Biancotto, C., Frigè, G., Minucci, S. (Eds.), *Histone Modifications in Therapy*. Cambridge, Massachusetts, United States: Academic Press, pp. 395–413.

[38]   Al-Okbi, S.Y. 2014. Nutraceuticals of anti-inflammatory activity as complementary therapy for rheumatoid arthritis. Toxicology and Industrial Health, 30(8), 738–749.

[39]   Watson, R.R. and Preedy, V.R. (Eds.). 2016. *Fruits, Vegetables, and Herbs: Bioactive Foods in Health Promotion*. Academic Press.

[40]   Vanessa, B.V., Rocio, O.B., Ruth, R.S., Paola, T.M., Adelaida, H.G. and Edgar, C.E. 2012. Microalgae of the Chlorophyceae class: Potential nutraceuticals reducing oxidative stress intensity and cellular damage. In *Oxidative Stress and Diseases*. IntechOpen.

[41]   Lee, J., Koo, N. and Min, D.B. 2004. Reactive oxygen species, aging, and antioxidative nutraceuticals. Comprehensive Reviews in Food Science and Food Safety, 3(1), 21–33.

[42]   Bravo, L. 1998. Polyphenols: Chemistry, dietary sources, metabolism, and nutritional significance. Nutrition Reviews, 56(11), 317–333.

[43]   Kaur, C. and Kapoor, H.C. 2001. Antioxidants in fruits and vegetables–the millennium's health. International Journal of Food Science & Technology, 36(7), 703–725.

[44]   Stirling, C. and Kruh, W. 2015. Nutraceuticals: The future of intelligent food. Where food and pharmaceuticals converge. Available online at https://oceanium.world/wp-content/uploads/2021/05/nutraceuticals.pdf (accessed July 29, 2023).

[45]   Hou, C., Chen, L., Yang, L. and Ji, X. 2020. An insight into anti-inflammatory effects of natural polysaccharides. International Journal of Biological Macromolecules, 153, 248–255.

[46]   Jeong, J.W., Hwang, S.J., Han, M.H., Lee, D.S., Yoo, J.S., Choi, I.W. . . . Choi, Y.H. 2017. Fucoidan inhibits lipopolysaccharide-induced inflammatory responses in RAW 264.7 macrophages and zebrafish larvae. Molecular & Cellular Toxicology, 13, 405–417.

[47]   Ale, M.T., Mikkelsen, J.D. and Meyer, A.S. 2011. Important determinants for fucoidan bioactivity: A critical review of structure-function relations and extraction methods for fucose-containing sulfated polysaccharides from brown seaweeds. Marine Drugs, 9(10), 2106–2130.

[48]   Ale, M.T. and Meyer, A.S. 2013. Fucoidans from brown seaweeds: An update on structures, extraction techniques and use of enzymes as tools for structural elucidation. Rsc Advances, 3(22), 8131–8141.

[49] Aleissa, M.S., Alkahtani, S., Abd Eldaim, M.A., Ahmed, A.M., Bungău, S.G., Almutairi, B. . . . Abdel-Daim, M.M. 2020. Fucoidan ameliorates oxidative stress, inflammation, DNA Damage, and hepatorenal injuries in diabetic rats intoxicated with aflatoxin B 1. Oxidative Medicine and Cellular Longevity, 2020.

[50] Krishnaswamy, K. 2008. Traditional Indian spices and their health significance. Asia Pacific Journal of Clinical Nutrition, 17(S1), 265–268.

[51] Nelly, A., Annick, D.D. and Frederic, D. 2008. Plants used as remedies antirheumatic and antineuralgic in the traditional medicine of Lebanon. Journal of Ethnopharmacology, 120(3), 315–334.

[52] Rathore, B., Mahdi, A.A., Paul, B.N., Saxena, P.N. and Das, S.K. 2007. Indian herbal medicines: Possible potent therapeutic agents for rheumatoid arthritis. Journal of Clinical Biochemistry and Nutrition, 41(1), 12–17.

[53] Kompoti, M. and Falagas, M.E. 2006. Obesity and infection. Lancet Infectious Diseases, 6, 438–446.

[54] Field, C.J., Johnson, I.R. and Schley, P.D. 2002. Nutrients and their role in host resistance to infection. Journal of Leukocyte Biology, 71(1), 16–32.

[55] Ma, Y., Griffith, J.A., Chasan-Taber, L., Olendzki, B.C., Jackson, E., Stanek III, E.J., Li, W., Pagoto, S.L., Hafner, A.R. and Ockene, I.S. 2006. Association between dietary fiber and serum C-reactive protein. The American Journal of Clinical Nutrition, 83(4), 760–766.

[56] Ma, Y., Griffith, J.A., Chasan-Taber, L., Olendzki, B.C., Jackson, E., Stanek III, E.J., Li, W., Pagoto, S.L., Hafner, A.R. and Ockene, I.S. 2006. Association between dietary fiber and serum C-reactive protein. The American Journal of Clinical Nutrition, 83(4), 760–766.

[57] Khalili, H., Håkansson, N., Chan, S.S., Chen, Y., Lochhead, P., Ludvigsson, J.F., Chan, A.T., Hart, A.R., Olén, O. and Wolk, A. 2020. Adherence to a Mediterranean diet is associated with a lower risk of later-onset Crohn's disease: Results from two large prospective cohort studies. Gut, 69(9), 1637–1644.

[58] Usuda, H., Okamoto, T. and Wada, K. 2021. Leaky gut: Effect of dietary fiber and fats on microbiome and intestinal barrier. International Journal of Molecular Sciences, 22(14), 7613.

[59] Kim, Y., Hwang, S.W., Kim, S., Lee, Y.S., Kim, T.Y., Lee, S.H., Kim, S.J., Yoo, H.J., Kim, E.N. and Kweon, M.N. 2020. Dietary cellulose prevents gut inflammation by modulating lipid metabolism and gut microbiota. Gut Microbes, 11(4), 944–961.

[60] Ishisono, K., Mano, T., Yabe, T. and Kitaguchi, K. 2019. Dietary fiber pectin ameliorates experimental colitis in a neutral sugar side chain-dependent manner. Frontiers in Immunology, 10, 2979.

[61] Llewellyn, S.R., Britton, G.J., Contijoch, E.J., Vennaro, O.H., Mortha, A., Colombel, J.F., Grinspan, A., Clemente, J.C., Merad, M. and Faith, J.J. 2018. Interactions between diet and the intestinal microbiota alter intestinal permeability and colitis severity in mice. Gastroenterology, 154(4), 1037–1046.

[62] Desai, M.S., Seekatz, A.M., Koropatkin, N.M., Kamada, N., Hickey, C.A., Wolter, M., Pudlo, N.A., Kitamoto, S., Terrapon, N., Muller, A. and Young, V.B. 2016. A dietary fiber-deprived gut microbiota degrades the colonic mucus barrier and enhances pathogen susceptibility. Cell, 167(5), 1339–1353.

[63] Macia, L., Tan, J., Vieira, A.T., Leach, K., Stanley, D., Luong, S., Maruya, M., Ian Mckenzie, C., Hijikata, A., Wong, C. and Binge, L. 2015. Metabolite-sensing receptors GPR43 and GPR109A facilitate dietary fibre-induced gut homeostasis through regulation of the inflammasome. Nature Communications, 6(1), 1–15.

[64] Niero, M., Bartoli, G., De Colle, P., Scarcella, M. and Zanetti, M. 2023. Impact of Dietary Fiber on Inflammation and Insulin Resistance in Older Patients: A Narrative Review. Nutrients, 15(10), 2365.

[65] Shivakoti, R., Biggs, M.L., Djoussé, L., Durda, P.J., Kizer, J.R., Psaty, B., Reiner, A.P., Tracy, R.P., Siscovick, D. and Mukamal, K.J. 2022. Intake and sources of dietary fiber, inflammation, and cardiovascular disease in older US adults. JAMA Network Open, 5(3), e225012–e225012.

[66]  Basu, A., Devaraj, S. and Jialal, I. 2006. Dietary factors that promote or retard inflammation. Arteriosclerosis, Thrombosis, and Vascular Biology, 26(5), 995–1001.

[67]  Simpson, J.L., Scott, R., Boyle, M.J. and Gibson, P.G. 2006. Inflammatory subtypes in asthma: Assessment and identification using induced sputum. Respirology, 11(1), 54–61.

[68]  Kim, M.H., Kang, S.G., Park, J.H., Yanagisawa, M. and Kim, C.H. 2013. Short-chain fatty acids activate GPR41 and GPR43 on intestinal epithelial cells to promote inflammatory responses in mice. Gastroenterology, 145(2), 396–406.

[69]  Kim, M., Qie, Y., Park, J. and Kim, C.H. 2016. Gut microbial metabolites fuel host antibody responses. Cell Host & Microbe, 20(2), 202–214.

[70]  Chang, P.V., Hao, L., Offermanns, S. and Medzhitov, R. 2014. The microbial metabolite butyrate regulates intestinal macrophage function via histone deacetylase inhibition. Proceedings of the National Academy of Sciences, 111(6), 2247–2252.

[71]  Van der Hee, B. and Wells, J.M. 2021. Microbial regulation of host physiology by short-chain fatty acids. Trends in Microbiology, 29(8), 700–712.

[72]  Rooks, M.G. and Garrett, W.S. 2016. Gut microbiota, metabolites and host immunity. Nature Reviews Immunology, 16(6), 341–352.

[73]  Maslowski, K.M., Vieira, A.T., Ng, A., Kranich, J., Sierro, F., Yu, D., Schilter, H.C., Rolph, M.S., Mackay, F., Artis, D. and Xavier, R.J. 2009. Regulation of inflammatory responses by gut microbiota and chemoattractant receptor GPR43. Nature, 461(7268), 1282–1286.

[74]  Wu, W., Sun, M., Chen, F., Cao, A.T., Liu, H., Zhao, Y., Huang, X., Xiao, Y., Yao, S., Zhao, Q. and Liu, Z. 2017. Microbiota metabolite short-chain fatty acid acetate promotes intestinal IgA response to microbiota which is mediated by GPR43. Mucosal Immunology, 10(4), 946–956.

[75]  Singh, N., Gurav, A., Sivaprakasam, S., Brady, E., Padia, R., Shi, H., Thangaraju, M., Prasad, P.D., Manicassamy, S., Munn, D.H. and Lee, J.R. 2014. Activation of Gpr109a, receptor for niacin and the commensal metabolite butyrate, suppresses colonic inflammation and carcinogenesis. Immunity, 40(1), 128–139.

[76]  Park, J., Kim, M., Kang, S.G., Jannasch, A.H., Cooper, B., Patterson, J. and Kim, C.H. 2015. Short-chain fatty acids induce both effector and regulatory T cells by suppression of histone deacetylases and regulation of the mTOR–S6K pathway. Mucosal Immunology, 8(1), 80–93.

[77]  Roller, M., Rechkemmer, G. and Watzl, B. 2004. Prebiotic inulin enriched with oligofructose in combination with the probiotics Lactobacillus rhamnosus and Bifidobacterium lactis modulates intestinal immune functions in rats. The Journal of Nutrition, 134(1), 153–156.

[78]  Hoentjen, F., Welling, G.W., Harmsen, H.J., Zhang, X., Snart, J., Tannock, G.W., Lien, K., Churchill, T.A., Lupicki, M. and Dieleman, L.A. 2005. Reduction of colitis by prebiotics in HLA-B27 transgenic rats is associated with microflora changes and immunomodulation. Inflammatory Bowel Diseases, 11(11), 977–985.

[79]  Wang, J., Chen, W.D. and Wang, Y.D. 2020. The relationship between gut microbiota and inflammatory diseases: The role of macrophages. Frontiers in Microbiology, 11, 1065.

[80]  Kechagia, M., Basoulis, D., Konstantopoulou, S., Dimitriadi, D., Gyftopoulou, K., Skarmoutsou, N. and Fakiri, E.M. 2013. Health benefits of probiotics: A review. International Scholarly Research Notices, 2013, 1–7.

[81]  Roy, S. and Dhaneshwar, S. 2023. Role of prebiotics, probiotics, and synbiotics in management of inflammatory bowel disease: Current perspectives. World Journal of Gastroenterology, 29(14), 2078.

[82]  Maneerat, S., Lehtinen, M.J., Childs, C.E., Forssten, S.D., Alhoniemi, E., Tiphaine, M., Yaqoob, P., Ouwehand, A.C. and Rastall, R.A. 2013. Consumption of Bifidobacterium lactis Bi-07 by healthy elderly adults enhances phagocytic activity of monocytes and granulocytes. Journal of Nutritional Science, 2, e44.

[83]  Macfarlane, S., Cleary, S., Bahrami, B., Reynolds, N. and Macfarlane, G.T. 2013. Synbiotic consumption changes the metabolism and composition of the gut microbiota in older people and

modifies inflammatory processes: A randomised, double-blind, placebo-controlled crossover study. Alimentary Pharmacology & Therapeutics, 38(7), 804–816.

[84] Kim, C.S., Cha, L., Sim, M., Jung, S., Chun, W.Y., Baik, H.W. and Shin, D.M. 2021. Probiotic supplementation improves cognitive function and mood with changes in gut microbiota in community-dwelling older adults: A randomized, double-blind, placebo-controlled, multicenter trial. The Journals of Gerontology: Series A, 76(1), 32–40.

[85] Lin, C.C., Huang, W.C., Su, C.H., Lin, W.D., Wu, W.T., Yu, B. and Hsu, Y.M. 2020. Effects of multi-strain probiotics on immune responses and metabolic balance in helicobacter pylori-infected mice. Nutrients, 12(8), 2476.

[86] Skrzydło-Radomańska, B., Prozorow-Król, B., Cichoż-Lach, H., Majsiak, E., Bierła, J.B., Kanarek, E., Sowińska, A. and Cukrowska, B. 2021. The effectiveness and safety of multi-strain probiotic preparation in patients with diarrhea-predominant irritable bowel syndrome: A randomized controlled study. Nutrients, 13(3), 756.

[87] Kanazawa, A., Aida, M., Yoshida, Y., Kaga, H., Katahira, T., Suzuki, L., Tamaki, S., Sato, J., Goto, H., Azuma, K. and Shimizu, T. 2021. Effects of synbiotic supplementation on chronic inflammation and the gut microbiota in obese patients with type 2 diabetes mellitus: A randomized controlled study. Nutrients, 13(2), 558.

[88] Marion-Letellier, R., Savoye, G. and Ghosh, S. 2015. Polyunsaturated fatty acids and inflammation. IUBMB Life, 67(9), 659–667.

[89] Schmitz, G. and Ecker, J. 2008. The opposing effects of n− 3 and n− 6 fatty acids. Progress in Lipid Research, 47(2), 147–155.

[90] Calder, P.C. 2001. Polyunsaturated fatty acids, inflammation, and immunity. Lipids, 36(9), 1007–1024.

[91] Matsuyama, W., Mitsuyama, H., Watanabe, M., Oonakahara, K.I., Higashimoto, I., Osame, M. and Arimura, K. 2005. Effects of omega-3 polyunsaturated fatty acids on inflammatory markers in COPD. Chest, 128(6), 3817–3827.

[92] Astore, C., Nagpal, S. and Gibson, G. 2022. Mendelian Randomization Indicates a Causal Role for Omega-3 Fatty Acids in Inflammatory Bowel Disease. International Journal of Molecular Sciences, 23(22), 14380.

[93] Freund-Levi, Y., Hjorth, E., Lindberg, C., Cederholm, T., Faxen-Irving, G., Vedin, I., Palmblad, J., Wahlund, L.O., Schultzberg, M., Basun, H. and Eriksdotter Jönhagen, M. 2009. Effects of omega-3 fatty acids on inflammatory markers in cerebrospinal fluid and plasma in Alzheimer's disease: The OmegAD study. Dementia and Geriatric Cognitive Disorders, 27(5), 481–490.

[94] Mosele, J.I., Macià, A., Romero, M.P., Motilva, M.J. and Rubió, L. 2015. Application of in vitro gastrointestinal digestion and colonic fermentation models to pomegranate products (juice, pulp and peel extract) to study the stability and catabolism of phenolic compounds. Journal of Functional Foods, 14, 529–540.

[95] Correa-Betanzo, J., Allen-Vercoe, E., McDonald, J., Schroeter, K., Corredig, M. and Paliyath, G. 2014. Stability and biological activity of wild blueberry (Vaccinium angustifolium) polyphenols during simulated in vitro gastrointestinal digestion. Food Chemistry, 165, 522–531.

[96] Mohar, D.S. and Malik, S. 2012. The sirtuin system: The holy grail of resveratrol?. Journal of Clinical & Experimental Cardiology, 3(11), 216.

[97] Speciale, A.N.T.O.N.I.O., Chirafisi, J.O.S.E.L.I.T.A., Saija, A. and Cimino, F. 2011. Nutritional antioxidants and adaptive cell responses: An update. Current Molecular Medicine, 11(9), 770–789.

[98] Biasutto, L., Mattarei, A. and Zoratti, M. 2012. Resveratrol and health: The starting point. ChemBioChem, 13(9), 1256–1259.

[99] Capiralla, H., Vingtdeux, V., Venkatesh, J., Dreses-Werringloer, U., Zhao, H., Davies, P. and Marambaud, P. 2012. Identification of potent small-molecule inhibitors of STAT 3 with anti-inflammatory properties in RAW 264.7 macrophages. The FEBS Journal, 279(20), 3791–3799.

[100] Leiherer, A., Mündlein, A. and Drexel, H. 2013. Phytochemicals and their impact on adipose tissue inflammation and diabetes. Vascular Pharmacology, 58(1–2), 3–20.

[101] Siddiqui, A.M., Cui, X., Wu, R., Dong, W., Zhou, M., Hu, M., Simms, H.H. and Wang, P. 2006. The anti-inflammatory effect of curcumin in an experimental model of sepsis is mediated by up-regulation of peroxisome proliferator-activated receptor-y. Critical Care Medicine, 34(7), 1874–1882.

[102] Marchiani, A., Rozzo, C., Fadda, A., Delogu, G. and Ruzza, P. 2014. Curcumin and curcumin-like molecules: From spice to drugs. Current Medicinal Chemistry, 21(2), 204–222.

[103] Noorafshan, A. and Ashkani-Esfahani, S. 2013. A review of therapeutic effects of curcumin. Current Pharmaceutical Design, 19(11), 2032–2046.

[104] Gupta, S.C., Prasad, S., Kim, J.H., Patchva, S., Webb, L.J., Priyadarsini, I.K. and Aggarwal, B.B. 2011. Multitargeting by curcumin as revealed by molecular interaction studies. Natural Product Reports, 28(12), 1937–1955.

[105] Bae, J.S. 2012. Role of high mobility group box 1 in inflammatory disease: Focus on sepsis. Archives of Pharmacal Research, 35, 1511–1523.

[106] Tsuda, S., Egawa, T., Ma, X., Oshima, R., Kurogi, E. and Hayashi, T. 2012. Coffee polyphenol caffeic acid but not chlorogenic acid increases 5′ AMP-activated protein kinase and insulin-independent glucose transport in rat skeletal muscle. The Journal of Nutritional Biochemistry, 23(11), 1403–1409.

[107] Akyol, S., Ozturk, G., Ginis, Z., Armutcu, F., Yigitoglu, M.R. and Akyol, O. 2013. In vivo and in vitro antineoplastic actions of caffeic acid phenethyl ester (CAPE): Therapeutic perspectives. Nutrition and Cancer, 65(4), 515–526.

[108] Domitrovic, R. 2011. The molecular basis for the pharmacological activity of anthocyans. Current Medicinal Chemistry, 18(29), 4454–4469.

[109] Kanwar, J., Taskeen, M., Mohammad, I., Huo, C., Chan, T.H. and Dou, Q.P. 2012. Recent advances on tea polyphenols. Frontiers in Bioscience, 4, 111.

[110] Singh, B.N., Shankar, S. and Srivastava, R.K. 2011. Green tea catechin, epigallocatechin-3-gallate (EGCG): Mechanisms, perspectives and clinical applications. Biochemical Pharmacology, 82(12), 1807–1821.

[111] Landis-Piwowar, K., Chen, D., Foldes, R., Chan, T.H. and Dou, Q.P. 2013. Novel epigallocatechin gallate analogs as potential anticancer agents: A patent review (2009–present). Expert Opinion on Therapeutic Patents, 23(2), 189–202.

[112] Ullah, H., Sommella, E., Santarcangelo, C., D'Avino, D., Rossi, A., Dacrema, M., Minno, A.D., Di Matteo, G., Mannina, L., Campiglia, P. and Magni, P. 2022. Hydroethanolic extract of Prunus domestica L.: Metabolite profiling and in vitro modulation of molecular mechanisms associated to cardiometabolic diseases. Nutrients, 14(2), 340.

[113] Karasawa, K., Uzuhashi, Y., Hirota, M. and Otani, H. 2011. A matured fruit extract of date palm tree (Phoenix dactylifera L.) stimulates the cellular immune system in mice. Journal of Agricultural and Food Chemistry, 59(20), 11287–11293.

[114] Wong, C.P., Nguyen, L.P., Noh, S.K., Bray, T.M., Bruno, R.S. and Ho, E. 2011. Induction of regulatory T cells by green tea polyphenol EGCG. Immunology Letters, 139(1–2), 7–13.

[115] Yang, J., Yang, X. and Li, M. 2012. Baicalin, a natural compound, promotes regulatory T cell differentiation. BMC Complementary and Alternative Medicine, 12(1), 1–7.

[116] Wang, H.K., Yeh, C.H., Iwamoto, T., Satsu, H., Shimizu, M. and Totsuka, M. 2012. Dietary flavonoid naringenin induces regulatory T cells via an aryl hydrocarbon receptor mediated pathway. Journal of Agricultural and Food Chemistry, 60(9), 2171–2178.

[117] Ijinu, T.P., De Lellis, L.F., Shanmugarama, S., Pérez-Gregorio, R., Sasikumar, P., Ullah, H., Buccato, D.G., Di Minno, A., Baldi, A. and Daglia, M. 2023. Anthocyanins as immunomodulatory dietary supplements: A nutraceutical perspective and micro-/nano-strategies for enhanced bioavailability. Nutrients, 15(19), 4152.

[118] Cianciulli, A., Calvello, R., Porro, C., Trotta, T., Salvatore, R. and Panaro, M.A. 2016. PI3k/Akt signalling pathway plays a crucial role in the anti-inflammatory effects of curcumin in LPS-activated microglia. International Immunopharmacology, 36, 282–290.

[119] Zhang, Q.Y., Mo, Z.N. and Liu, X.D. 2010. Reducing effect of curcumin on expressions of TNF-alpha, IL-6 and IL-8 in rats with chronic nonbacterial prostatitis. Zhonghua nan ke xue= National Journal of Andrology, 16(1), 84–88.

[120] Liu, L., Shang, Y., Li, M., Han, X., Wang, J. and Wang, J. 2015. Curcumin ameliorates asthmatic airway inflammation by activating nuclear factor-E2-related factor 2/haem oxygenase (HO)-1 signalling pathway. Clinical and Experimental Pharmacology and Physiology, 42(5), 520–529.

[121] Allemand, A., Leonardi, B.F., Zimmer, A.R., Moreno, S., Romao, P.R.T. and Gosmann, G. 2016. Red pepper (*Capsicum* baccatum) extracts present anti-inflammatory effects in vivo and inhibit the production of TNF-α and NO in vitro. Journal of Medicinal Food, 19(8), 759–767.

[122] Oyagbemi, A.A., Saba, A.B. and Azeez, O.I. 2010. Capsaicin: A novel chemopreventive molecule and its underlying molecular mechanisms of action. Indian Journal of Cancer, 47(1), 53–58.

[123] Qian, K., Wang, G., Cao, R., Liu, T., Qian, G., Guan, X., Guo, Z., Xiao, Y. and Wang, X. 2016. Capsaicin suppresses cell proliferation, induces cell cycle arrest and ROS production in bladder cancer cells through FOXO3a-mediated pathways. Molecules, 21(10), 1406.

[124] Bachiega, T.F., De Sousa, J.P.B., Bastos, J.K. and Sforcin, J.M. 2012. Clove and eugenol in noncytotoxic concentrations exert immunomodulatory/anti-inflammatory action on cytokine production by murine macrophages. Journal of Pharmacy and Pharmacology, 64(4), 610–616.

[125] Pan, C. and Dong, Z. 2015. Antiasthmatic effects of eugenol in a mouse model of allergic asthma by regulation of vitamin D3 upregulated protein 1/NF-κB pathway. Inflammation, 38, 1385–1393.

[126] Manikandan, P., Vinothini, G., Vidya Priyadarsini, R., Prathiba, D. and Nagini, S. 2011. Eugenol inhibits cell proliferation via NF-κB suppression in a rat model of gastric carcinogenesis induced by MNNG. Investigational New Drugs, 29, 110–117.

[127] Pal, D., Banerjee, S., Mukherjee, S., Roy, A., Panda, C.K. and Das, S. 2010. Eugenol restricts DMBA croton oil induced skin carcinogenesis in mice: Downregulation of c-Myc and H-ras, and activation of p53 dependent apoptotic pathway. Journal of Dermatological Science, 59(1), 31–39.

[128] Muhammad, J.S., Zaidi, S.F., Shaharyar, S., Refaat, A., Usmanghani, K., Saiki, I. and Sugiyama, T. 2015. Anti-inflammatory effect of cinnamaldehyde in Helicobacter pylori induced gastric inflammation. Biological and Pharmaceutical Bulletin, 38(1), 109–115.

[129] Mondal, S. and Pahan, K. 2015. Cinnamon ameliorates experimental allergic encephalomyelitis in mice via regulatory T cells: Implications for multiple sclerosis therapy. PLoS One, 10(1), e0116566.

[130] Rathi, B., Bodhankar, S., Mohan, V. and Thakurdesai, P. 2013. Ameliorative effects of a polyphenolic fraction of Cinnamomum zeylanicum L. bark in animal models of inflammation and arthritis. Scientia Pharmaceutica, 81(2), 567–590.

[131] Umar, S., Hedaya, O., Singh, A.K. and Ahmed, S. 2015. Thymoquinone inhibits TNF-α-induced inflammation and cell adhesion in rheumatoid arthritis synovial fibroblasts by ASK1 regulation. Toxicology and Applied Pharmacology, 287(3), 299–305.

[132] Wang, D., Qiao, J., Zhao, X., Chen, T. and Guan, D. 2015. Thymoquinone inhibits IL-1β-induced inflammation in human osteoarthritis chondrocytes by suppressing NF-κB and MAPKs signaling pathway. Inflammation, 38, 2235–2241.

[133] Su, X., Ren, Y., Yu, N., Kong, L. and Kang, J. 2016. Thymoquinone inhibits inflammation, neoangiogenesis and vascular remodeling in asthma mice. International Immunopharmacology, 38, 70–80.

[134] Iacobellis, N.S., Lo Cantore, P., Capasso, F. and Senatore, F. 2005. Antibacterial activity of Cuminum cyminum L. and Carum carvi L. essential oils. Journal of Agricultural and Food Chemistry, 53(1), 57–61.

[135] Santos, F.A., Silva, R.M., Campos, A.R., De Araujo, R.P., Júnior, R.L. and Rao, V.S.N. 2004. 1, 8-cineole (eucalyptol), a monoterpene oxide attenuates the colonic damage in rats on acute TNBS-colitis. Food and Chemical Toxicology, 42(4), 579–584.

[136] Zhao, C., Sun, J., Fang, C. and Tang, F. 2014. 1, 8-cineol attenuates LPS-induced acute pulmonary inflammation in mice. Inflammation, 37, 566–572.

[137] Lima, P.R., De Melo, T.S., Carvalho, K.M.M.B., De Oliveira, Í.B., Arruda, B.R., De Castro Brito, G.A., Rao, V.S. and Santos, F.A. 2013. 1, 8-cineole (eucalyptol) ameliorates cerulein-induced acute pancreatitis via modulation of cytokines, oxidative stress and NF-κB activity in mice. Life Sciences, 92(24–26), 1195–1201.

[138] Khan, A., Vaibhav, K., Javed, H., Tabassum, R., Ahmed, M.E., Khan, M.M., Khan, M.B., Shrivastava, P., Islam, F., Siddiqui, M.S. and Safhi, M.M. 2014. 1, 8-cineole (eucalyptol) mitigates inflammation in amyloid Beta toxicated PC12 cells: Relevance to Alzheimer's disease. Neurochemical Research, 39, 344–352.

[139] Shin, I.S., Hong, J., Jeon, C.M., Shin, N.R., Kwon, O.K., Kim, H.S., Kim, J.C., Oh, S.R. and Ahn, K.S. 2013. Diallyl-disulfide, an organosulfur compound of garlic, attenuates airway inflammation via activation of the Nrf-2/HO-1 pathway and NF-kappaB suppression. Food and Chemical Toxicology, 62, 506–513.

[140] Lee, H.S., Lee, C.H., Tsai, H.C. and Salter, D.M. 2009. Inhibition of cyclooxygenase 2 expression by diallyl sulfide on joint inflammation induced by urate crystal and IL-1β. Osteoarthritis and Cartilage, 17(1), 91–99.

Imad Ahmad*, Fazle Rabbi, Fiaz Alam

# Chapter 8
# Metabolic disorders

**Abstract:** Metabolic disorders are a group of interrelated disease conditions, which include central adiposity, hyperglycemia, hypertension, hypertriglyceridemia, and hypercholesterolemia. Collectively they still present a challenge to pharmacology and therapeutics. A number of existing medicaments almost from every mode of therapeutics are available; however, the global burden of the syndrome is like a pandemic of noncommunicable diseases. This chapter will give a brief introduction to some of the most common pathologies related to metabolic disorders, available therapeutic options, and their limitations. The main focus of the chapter is then directed towards nutraceuticals that are attracting researchers with its safe and effective but diverse and promising mechanisms. Six types of this diverse class of therapeutic modality are discussed: dietary fibers, prebiotics, probiotics, polyphenols, spices, vitamins, and minerals. It is notable that three of these modify gut microbiome, which ultimately prevents and modifies GIT physiology, even affecting major organs involved in metabolism like liver. The other three classes have their own mechanisms by serving as cofactors, or modifying osmotic/electrolytic pumps to keep/restore a homeostatic balance of the human body.

## 1 Introduction

Metabolic disorders are interrelated pathologies of central adiposity, hyperglycemia, hypertension, hypertriglyceridemia, and hypercholesterolemia. Collectively, the disease cluster of cardiovascular diseases and diabetes is known as "The Deadly Quartet," "The Insulin Resistance Syndrome," or "Syndrome X," [1] Different treatment options are available to manage metabolic disorders. However, a single drug candidate is not able to provide an effective therapeutic effect due to the involvement of multiple metabolic pathways [2]. Allopathic approach provides a much targeted therapeutic option. However, the disease cluster, adverse reactions, and drug resistance are the limiting factors of an effective therapeutic process [3]. In this situation, natural

*Corresponding author: Imad Ahmad**, Department of Pharmacy, Abdul Wali Khan University Mardan, Khyber Pakhtunkhwa, Pakistan; Department of Pharmacy, The Professional Institute of Health Sciences, Mardan, Khyber Pakhtunkhwa, Pakistan, e-mail: imadahmad4574@gmail.com
**Fazle Rabbi,** Department of Pharmacy, Abasyn University Peshawar, Peshawar 25000, Khyber Pakhtunkhwa, Pakistan
**Fiaz Alam,** Department of Pharmacy, COMSATS University Islamabad, Abbottabad Campus, Pakistan

https://doi.org/10.1515/9783111317601-008

plant extracts or their isolated compounds are a possible option. Traditionally and historically, numerous plants and their products have been used in the treatment of metabolic disorders [4]. This chapter provides an overview of common metabolic disorders, currently available therapeutic options and their limitations, as well as various options in the category of nutraceuticals.

## 1.1 Diabetes mellitus

A group of metabolic derangements characterized by carbohydrate, fat, and protein metabolism from defect in insulin action, secretion, or both. The deficient action of insulin in the target tissues creates disturbances in carbohydrate, protein, and fat metabolism. This further worsens the condition by inducing long-term irreversible complications like nephropathy, retinopathy, autonomic neuropathy, peripheral neuropathy, sexual dysfunction, and Charcot joints.

Diabetes mellitus is broadly categorized into two types based on pathophysiology. Type I diabetes mellitus is characterized by complete deficiency of insulin secretion, hence the name *insulin dependent diabetes mellitus*. In contrast, Type II diabetes mellitus involve a combination of insufficient compensatory secretory response of insulin and resistance to the insulin action (insulin resistance) which is known to be *non-insulin-dependent diabetes mellitus* [5, 6].

In response to deficiency in the strictly necessary nutrients, many abnormal biochemical pathways are activated. On the cause of complications, several hypotheses are equally logical, which include the Maillard hypothesis [7, 8], the aldose reductase hypothesis [9], reductive stress (pseudohypoxia) [10, 11], carbonyl stress [12–15], altered lipoprotein metabolism [16, 17], true hypoxia [18], oxidative stress [16, 19, 20], increased protein kinase C activity [7], and altered growth factor [8, 9] or cytokine [9] activities. Since each of these ideas reflects a different underlying pathogenic process, the list of these hypotheses is lengthy. This is most likely caused by the sensitivity of certain tissues to various mechanisms.

Various hypotheses cross over and overlap, for example, the possibility that advanced glycation end products (AGE) production and altered polyol pathway activity leads to oxidative stress. Reductive stress activates protein kinase C and oxidative stress accelerates the production of AGE. Growth factor expression and oxidative stress may be induced by AGEs [21]. Some of the commonly known hypotheses regarding diabetes complications are discussed here.

Increased oxidative stress determines the path towards diabetic complications, i.e., peripheral neuropathy, altered endoneurial metabolism or oxidative damage, defect in neuronal blood flow, or neurotrophic support [7–9, 12–15, 20]. Along with enhanced fructose synthesis and the subsequent precursor development of AGEs, the polyol pathway plays an important role in the pathophysiology of diabetic neuropathy [7].

Enhanced activity in polyol pathway leads to endoneurial myo-inositol depletion as well as a rise in nerve sorbitol levels. Aldo-keto reductase enzyme family is the foundation of the polyol pathway. Comprehensive research has been done on this pathway, which is blocked by aldose reductase inhibitors. Aldose reductase catalyzes the conversion of glucose to sorbitol along with oxidative dehydrogenation of NADPH to $NADP^+$. As a result, NADPH levels drop, which also affects the level of reduced glutathione. The bioavailability of glutathione, an antioxidant crucial for diabetes mellitus, may be impacted by changes to the polyol pathway. Thus, the polyol production route is an additional mechanism that causes enhanced oxidative stress in diabetes mellitus [22]. Previously, it was thought that the enzyme aldose reductase turned glucose into sorbitol. In the presence of the enzyme sorbitol dehydrogenase and the co-factor NAD, sorbitol is converted to fructose.

AGEs are created when glycating substances, such as glucose, react nonenzymatically. These glucose-derived glycating chemicals promote fatty acid oxidation. In the endothelial cells of the heart and arteries, (dicarbonyls such as 3-deoxyglucosone, glyoxal, and methylglyoxal) oxidize fatty acids in the presence of proteins [14, 15]. AGEs are abundant in the extracellular matrix in diabetes mellitus [11, 16–19]. Cells are harmed by intracellular synthesis of AGEs in three different ways:

a.  AGEs can change the function of intracellular proteins.
b.  Extracellular matrix (ECM) elements that have been altered by AGE precursors interact improperly with some other ECM elements and ECM receptors (integrins) that are expressed on the surface of the cell.
c.  AGEs precursor, which altered plasma proteins and connected to receptors found on cells, e.g., vascular endothelial cells, vascular smooth muscle cells and macrophages.

In mammalian tissues, there are many members of the protein kinase C family, which includes roughly 11 isoforms. Diacylglycerol (DAG), which is substantially increased by $Ca^{2+}$ ions and phosphatidyl serine are essential for the activity of isoforms [11]. The third common route, which is mediated by excessive and prolonged activation of various protein kinase C isoforms, mediates tissue damage brought on by diabetes mellitus (ROS-induced). Glucose is converted to diacylglycerol using triose phosphate. The glycolytic enzyme GAPDH's activity is inhibited by elevated ROS. As a result, the level of DAG, a precursor to triose phosphate, increases within the cells [23].

## 1.2 Obesity

Low-grade inflammation is a defining feature of obesity. Many inflammatory markers have higher circulating levels in obese persons than in lean individuals, and it is thought that this may contribute to metabolic disturbances [24]. These higher levels of circulating inflammatory markers are both released by adipocytes and invading macrophages. Following weight loss, inflammatory marker blood levels are decreased. High-fat and high-glucose meals both have the potential to cause postprandial inflammation. AGEs make the latter worse, however specific antioxidants or meals that contain antioxidants can help offset this effect to some extent [25, 26]. In the metabolic syndrome (MS), obesity frequently co-occurs with other cardiovascular risk factors, e.g., dyslipidemia, hypertension, and hyperglycemia [27, 28].

Individuals with at least three of these conditions are classified as having MS:
- *Blood pressure* ≥ 130/85 mmHg
- *Fasting triglycerides* ≥ 150 mg/dL (1.70 mmol/L)
- *Fasting-glucose* ≥ 100 mg/dL (5.55 mmol/L)
- *HDL-cholesterol* < 40 mg/dL (1.03 mmol/L) in males, and < 50 mg/dL (1.29 mmol/L) in females
- *Waist circumference* ≥ 102 cm in males, and ≥ 88 cm in females

However, the IDF (International Diabetes Federation), AHA (American Heart Association), and NHLBI (National Heart, Lung, and Blood Institute) could not agree on the concept of abdominal obesity [29]. According to the IDF, men should have a waist circumference of 94 cm, and women should have a waist circumference of 80 cm to be considered obese. In contrast, the AHA/NHLBI suggested cut points of ≥ 102 and ≥ 88 cm for the two sexes, respectively. Representatives from the IDF and AHA/NHLBI have recently made an effort to iron out any lingering discrepancies between the definitions of MS. Both sides settled that the presence of any three of the five risk factors is sufficient to make the diagnosis of MS; however, abdominal obesity should not be a requirement for diagnosis and should instead be one of the five criteria.

There has been increasing evidence that smoking, eating meals high in fat, and not exercising all contribute to the emergence of obesity and related disorders [30]. Metabolic risk and exercise have shown dose–response connections in both adults and children. In adults with prediabetes, the Diabetes Prevention Program found that exercise and diet decreased chances of developing diabetes mellitus [31].

Even without weight reduction, regular exercise helps reduce total, visceral, and subcutaneous fat. It also lowers glycemia, increases free FA (FFA) oxidation, and reduces insulin resistance. Due to the improved ability of the muscle to absorb circulating glucose as a result of lower intramuscular fat stores, exercise might be seen as "insulin-like" action. In terms of mechanism, exercise raises the expression of the insulin receptor substrate (IRS1) and the glucose transporter (GLUT-4) as well as the posttranscriptional control of PI3-kinase expression [32]. Exercise raises HDL content, enhances the utilization of FFA as an energy source, and activates lipolytic activity, which lowers plasma triglycerides. Improved antioxidant defense and improved tolerance to chronic ROS generation are both strong indicators of oxidative stress adaptation in exercise-trained people. Additionally, exercise appears to lower low-grade chronic inflammation, albeit this effect may be influenced by changes in fasting blood sugar, fat mass, and glycated hemoglobin (HbA1c) [33].

In the short term, nicotine may reduce hunger and rise energy expenditure, which may help explain why smoking cessation is commonly followed by weight gain and why smokers typically have lower body weight than nonsmokers [34]. Contrarily, heavy smokers tend to be heavier than nonsmokers, which may be due to a combination of dangerous habits (e.g., smoking, inactivity, and poor eating) that contribute to weight gain. Additionally, it was discovered that smoking increased the quantity and concentration of phospholipids (oxPAPC) oxidation products in peripheral blood mono-nuclear cells, which led to the production of ROS via NOX activation as a result of depleted glutathione. Additionally, smoking raised hs-CRP readings and activated NF-κB [35].

A core gut microbiome also affects the metabolic patterns. Obese people, however, have a lower variety and changed metabolic pathways in their microbiota. Our microbiota's composition may be fundamentally impacted by our diet [36]. The gut microbiome can be effectively and quickly (within a day) modified by some diets, e.g., high-fat diet, according to early studies. Overall, changes in the consumption of specific food components (micronutrients, carbohydrates, prebiotics, and probiotics) have an effect on the composition of the gut microbiota as well as the expression of genes in host tissues like intestine, liver, and muscle. Due to aberrant metabolic pathways linked to intestinal barrier function and systemic immunology, this may then either promote or hinder the production of fat mass [37]. Gram-negative bacterial cell walls contain a substance called lipopolysaccharide (LPS); when it binds to the CD14-TLR4 complex, it can cause inflammation. This has been demonstrated in recent research to be the mechanism by which gut bacteria can switch the inflammatory state of obesity and insulin resistance (IR). The discovery that the deletion of TLR4 reduced high-fat diet-induced IR served as confirmation of the significance of TLR4 pathways for metabolic disease. Furthermore, Cd14 knockout rats that displayed a diminished inflammatory response to LPS were resistant to weight gain [38].

Tissue fatty acid composition is modified due to a decrease in *Lactobacilli* and *Bifidobacteria*. This ultimately has an impact on the host metabolism and inflammatory state. In fact, free linoleic acid can be converted into bioactive conjugated linoleic acid isomers that have antidiabetic, anti-atherosclerotic, immune-modulating, and anti-obesity activities by mammalian intestinal *Lactobacilli* and *Bifidobacteria*. Supplementing with *Bifidobacterium breve* and linoleic acid increased the levels of cis-9, trans-11-conjugated linoleic acid, docosahexaenoic acid and eicosapentaenoic acid in the intestinal, hepatic, and AT tissues by two to three times, as well as a decrease in the expression of the proinflammatory cytokines Inf-$\gamma$, IL-6, and TNF-$\alpha$ [39].

## 1.3 Thyroid disorders

Patients suffering from metabolic syndrome have been found to have more thyroid abnormalities, most notably subclinical hypothyroidism, than the general population.

Hypothyroidism and the metabolic syndrome both function as separate CVD risk factors. A significant overlap exists in the pathogenic pathways of atherosclerotic cardiovascular disease caused by metabolic syndrome and hypothyroidism, which may enhance the risk for CVD if both disorders are present [40]. Thyroid stimulating hormone (TSH) levels have been found to be greater in metabolic syndrome patients than in healthy controls, and metabolic syndrome is more common in people with elevated TSH levels than in people with normal TSH levels. The relationship between thyroid dysfunction and the components of the metabolic syndrome is still up for dispute [41].

## 1.4 Familial hypercholesterolemia

One of the most prevalent inherited metabolic disorders is familial hypercholesterolemia (FH). This is a cluster of genetic conditions characterized by high circulating LDL cholesterol levels (LDL-C). Such abnormally high LDL-C levels promote the formation of atherosclerotic plaque in the arteries and at a young age with significantly higher risk of coronary heart disease (CHD). Khachadurian, who studied phenotypes in large Lebanese families in the 1960s, was the first to describe this. He pointed out that homozygotes had twice the LDL-C levels of heterozygotes due to autosomal codominant inheritance. Around the same time, Fredrickson demonstrated that FH was brought on by the improper low-density lipoprotein metabolism [42]. Instead of cerebral or peripheral artery disease, patients with FH primarily have an excess of CHD. Heterozygous FH increases the risk of early CHD by roughly 20 times, with young, untreated men having the highest risk [43].

FH often has over 90% penetrance and dominant or codominant inheritance. FH has been associated with mutations in the low density lipoprotein receptor (LDLR), proprotein convertase subtilin/kexin 9 (PCSK 9), low-density lipoprotein receptor adaptor protein 1 (LDLRAP1) and apo-lipoprotein B (Apo B) genes. Over 80% of people with FH have known variations in these genes. An abnormality in one allele that is harmful results in heterozygous FH. On the other hand, true homozygotes (biallelic mutations in one of the known genes) or compound heterozygosity (two separate mutations in the same or different candidate genes known to cause FH) results in homozygous FH [44].

# 2 Pharmacological treatments and limitations

An appropriate pharmacologic therapy for diabetes depends upon whether the person is insulin-resistant, insulin-deficient, or both. Two main types of treatment options are noninsulin therapies and insulin therapies. Noninsulin therapies include insulin sensitizers, pramlintide, secretagogues, sodium-glucose cotransporter-2 (SGLT-2) blockers, incretins, and alpha-glucosidase inhibitors, while insulin therapies include insulin and insulin analogues [45]. By increasing the sensitivity of peripheral tissues to the effects of insulin, insulin sensitizers reduce glycemic load. Biguanides and thiazolidinediones are two families of oral hypoglycemic insulin sensitizers. They may be taken singly in monotherapy, in conjunction with other drugs such as insulin or sulfonylureas, or both together [46]. Regarding biguanides, only metformin is currently in use. The increase in muscle and fat insulin sensitivity is caused by the reduction of hepatic glucose production. Metformin primarily reduces fasting glycemia, although it can also improve postprandial glucose levels [47]. Despite these beneficial effects, some of the most common side effects are GIT complaints, including abdominal discomfort, metallic taste, nausea, and diarrhea. A limiting effect is rise in the blood lactate concentrations, which may be life-threatening (<1 in 100,000) [46].

Rosiglitazone and pioglitazone are thiazolidinediones, which are agonists of peroxisome proliferator-activated receptor gamma and categorized as insulin sensitizers. Besides, its role in fasting and postprandial glycemic control, it has limitations of weight gain, increase in subcutaneous adipose tissues, and fluid retention [48]. Patients with moderate to severe functional heart failure should not use it. Similarly, insulin secretagogues promote the secretion of insulin from pancreas. Insulin released into blood further modifies glucose utilization. This class includes sulfonylureas and glinides. Sulfonylureas are effective in lowering fasting and postprandial glucose levels [49]. Second generation sulfonylureas are very effective and one of the most widely prescribed drugs. However, limiting factors of weight gain and hypoglycemia come across as disrupting compliance. Sometimes hypoglycemic episodes require urgent medical care and may lead to coma [50].

Glinides such as nateglinide and repaglinide are similar in action as sulfonylureas, with shorter duration and faster onset of action. Patients with erratic meal timings have the option of glinides. There is a lower risk of weight gain but a precaution must be exercised in patients with liver dysfunction. Blocking the activity of enzyme α-glucosidase in the brush borders of the small intestine delays carbohydrate absorption [51]. They effectively target postprandial hyperglycemia without triggering hypoglycemia. However, major GI complaints prevail amongst patients, e.g., abdominal cramps, bloating, diarrhea, and flatulence. Another concern is that, patients with severe renal or hepatic impairment cannot avail its benefits [52].

Incretin as therapy requires injectable glucagon like peptide-1 [GLP-1] receptor agonists or orally active formulations of dipeptidyl peptidase-4 [DPP-4] blockers. Incretin-based medicines pose high risk of acute pancreatitis. Discontinuity of these medications is required along with medical assessment if abdominal pain develops [53]. The GLP-1 agonists have a dual mechanism to control glycemic excursions, suppress glucagon secretion after meals, and stimulate insulin in a glucose concentration-dependent manner. Exenatide is a synthetic analogue of a hormone exendin 4 present in the Gila monster saliva [54]. GLP-1 is a normal product of the human small intestine, which stimulates secretion of insulin, along with inhibition of glucagon secretion and liver glucose production in a glucose-dependent manner. However, a delaying effect on gastric emptying leads to nauseating feelings, vomiting, and diarrhea. Another concern is the only parenteral dosage form availability, which requires subcutaneous injection at least 60 min before meal [55]. Dipeptidyl peptidase-4 (DPP-4) is a cell membrane-based enzyme, which rapidly degrades GLP-1 as well as a glucose-dependent insulinotropic polypeptide. An antidiabetic approach is to suppress dipeptidyl peptidase-4 to achieve higher insulin level and lower levels of glucagon in a glucose-dependent manner [56]. This effectively controls postprandial and fasting blood glucose levels. It is one of well-tolerated drugs, with headache and nasopharyngitis as the common side effects. The FDA has approved four DPP-4 inhibitors for diabetic patients, i.e., sitagliptin, saxagliptin, linagliptin, and alogliptin. These drugs can be used alone or in conjunction with metformin, sulfonylureas, thiazolidinediones, or insulin. Pramlintide, a synthetic version of the amylin hormone generated by beta cells, blocks central pathways to reduce glucagon output, delay gastric emptying, and decrease hunger. Pramlintide reduce insulin requirements by up to 50% as an advantage. However, subcutaneous administration is required before each meal [57].

SGLT-2 inhibitors are the newest drugs for type 2 diabetes mellitus. SGLT-2 is a proteinaceous sodium-glucose cotransporter located in the proximal tubules of kidney. Ninety percent (90%) of the filtered glucose is reabsorbed into circulation as part of its function. Patients with type 2 diabetes mellitus benefit from improved glycemic control due to the inhibition of this transporter, which causes glucose to be excreted in the urine. However, urinary tract infections, vaginal yeast infections, and polyuria are very common with its use [58].

The huge burden of diabetes and its complications on the health care system mandates efforts to more optimal treatment approach. An early and intensive intervention is the key to reduce the risk of microvascular and macrovascular complications. Despite availability of a large spectrum of therapeutic options, many challenges prevail among patients. Some of the current challenges in diabetes management include optimization of available therapies for adequate glycemic control, alongside blood pressure and lipid control [59]. The current therapies have limitations and some present a challenge to patient compliance. Sometime a different drug response is noted amongst individuals. Failure to achieve good enough glycemic control and with reduced adverse effects is another limitation. Exploring more effective therapeutic strategies for patients with diabetes is crucial [60].

A number of approaches are available for hyperthyroidism including antithyroid drugs, β-blockers, radioiodine therapy, and thyroidectomy. Antithyroid medications slowly ease symptoms by preventing the thyroid gland from synthesizing hormones. These include propylthiouracil and methimazole. Symptoms may begin to improve in weeks to months of therapy, even 12–18 months. The dosage may be reduced after that, if symptoms improve. A rare but serious side effect is liver damage, which occurs with antithyroid medications especially propylthiouracil. Some people are allergic to these medicines and the result is fever, hives, joint pain, or skin rashes [61, 62]. β-blockers do not affect thyroid hormone levels, but they can diminish hyperthyroidism symptoms. They actually control heart palpitations, rapid heart rate, and tremors. However, their use is associated with asthmatic attacks, fatigue, and sexual problems [63]. Radioiodine therapy causes thyroid gland to shrink because radioiodine is taken up by the gland. This slows down thyroid activity and makes it underactive. After that, hypothyroidism occurs, which then requires thyroxin as treatment [64]. Thyroidectomy is the surgical removal of thyroid gland. It is an option in case of pregnancy or those unable to take antithyroid medicine and radioiodine therapy. Surgery is associated with damage to vocal cords and parathyroid glands along with lifelong treatment with levothyroxine [65].

The main focus of FH is the reduction in high levels of LDL cholesterol in order to reduce heart attack risk [66]. The first line option is the statins therapy, which blocks cholesterol synthesis. Examples include pravastatin, atorvastatin, simvastatin, fluvastatin, rosuvastatin, and lovastatin. Ezetimibe is a drug that limits cholesterol absorption contained in the food. PCSK9 inhibitors are newer drugs (Alirocumab and Evolocumab) that help the liver to absorb more LDL cholesterol, thus reducing circulating cholesterol in the blood. They are expensive and required to be injected under the skin every few weeks. These factors make them cost-ineffective; the injection requires a technique, which may lead to noncompliance [67].

# 3 Nutraceuticals and metabolic system disorders

Nutraceuticals are used in conjunction with nutritional therapy since food not only gives energy and nutrients, but it also has health advantages. Nutraceuticals help us detoxify our bodies and improve our eating patterns and digestion [68]. The scientific community is in agreement that alteration in the gut microbiota is a key factor in the development of metabolic disorders. Alteration in the composition of gut microbiota leads to metabolic endotoxemia, increased adiposity, oxidative stress, β-cell dysfunction, and systemic inflammation. Prebiotics and probiotics have ameliorated CVD and diabetes mellitus through restoration of gut microbiota, which improved insulin signaling stimulation and cholesterol-lowering effects [69].

A nutraceutical can be a food that is naturally nutrient-rich, like spirulina, garlic, or soya, or it can be a specific part of a food, such as omega-3 oil from salmon. The demand for both herbal and non-herbal extracts is rising steadily on a global scale. Both *Ginkgo biloba* and green tea have been used extensively as nutraceuticals to cure cancer and aid in weight loss. Several studies revealed the role of nutraceuticals in metabolic disorders. Probiotics and prebiotics have improved lipid profiles along with reduction of LDL-cholesterol (low-density lipoprotein), triglycerides and serum total cholesterol, or an increment of HDL-cholesterol (high-density lipoprotein) in the context of alleviating CVD complications [70–72]. Similar to this, some probiotics encourage the production of short-chain fatty acids, which alter the secretion of incretin hormones and reduce cholesterol synthesis [73]. Various nutraceutical classes are identified, which are discussed here along with their mechanism and possibility of their therapeutic use.

## 3.1 Dietary fibers

Several studies have commended the advantages of a high-fiber diet throughout the years. Along with the polyphenolic structural component lignin, fibrous molecules also include complex carbohydrates like cellulose, pectin, mucilages, hemicelluloses, and algal polysaccharides. Although fiber is not by definition a dietary necessity, the health advantages of a higher fiber diet have made this group of nutrients quite well known in the quickly growing nutraceutical industry. The United States Food and Drug Administration (FDA) has authorized the use of a number of health claims relating to a food's particular or overall fiber content [74]. Different types of fibers, their characteristics, sources, and extent of microbial fermentation are summarized in Table 1.

**Table 1:** Types of fibers, food sources, their characteristics and bacterial degradation/fermentation.

| Types of fiber | Solubility character | General characteristics | Natural sources | Degradation via bacterial fermentation |
|---|---|---|---|---|
| Pectin | Soluble | 1–4% plant polysaccharides, vegetables, fruits | Apples, whole wheat flour, cabbage, bran, root vegetables, beans | + |
| Gums | Soluble | | Oat meal, legumes, dried beans | +++ |
| Mucilage | Soluble | Synthesized by plant cells | Food additives | +++ |
| Cellulose | Insoluble | A structural framework, fruits, cereals, vegetables, grains, legumes | Root vegetables, whole wheat flour, bran, beans, cabbage family, apples, peas | + |
| Hemicellulose | Insoluble | 15–30% cell wall matrix for cellulose | Whole grain, bran, cereal | + |
| Lignin | Insoluble | Mature cell wall | Wheat, mature vegetables | – |

Based on its qualities of water solubility, fiber is often subclassified. Pectin (pectic compounds), mucilages, and gums are examples of soluble fibers. Insoluble fibers include cellulose, modified cellulose, hemicellulose, and lignin. Geographical variations can be found in the quantity of fiber that humans consume. Fruits, legumes, cereals, grains, and some vegetables have all been identified as sources of soluble and insoluble fiber. Fiber content of some food items is summarized in Table 2.

**Table 2:** Fiber content of different food items.

| Food | Fiber (% weight) |
| --- | --- |
| Almond | 3 |
| Apple | 1 |
| String beans | 1 |
| Lima beans | 2 |
| Carrots | 1 |
| Broccoli | 1 |
| Whole wheat flour | 2 |
| White wheat flour | <1 |
| Pears | 2 |
| Oat flakes | 2 |
| Pecan | 2 |
| Popcorn | 2 |
| Strawberry | 1 |
| Walnut | 2 |
| Wheat germ | 3 |

The consumption of fiber is comparatively lower in more industrially developed nations. For instance, Americans consume an average amount of 12–15 grams of fiber per day. This consumption is much lower than the range recommended by World health Organization (WHO), i.e., 25–40 grams of fiber per day. Less than half of the dietary carbohydrates consumed by Americans come from whole grains, fruits, and vegetables. Contrarily, some African tribes are known to have people consume up to 50 grams of fiber each day [75].

In plant-based nutrients like starch and, to a much lesser extent, in glycogen of meats, repeated monosaccharide units (1–4 linkages) are present. Both salivary and pancreatic secretions contain amylase, which can easily break down these connections. Branch points in the chains of starch and glycogen are connected by β-1-6 linkages, which are digested by the enzyme β-1-6 dextrinase (isomaltase) in pancreatic secretions. Instead of β-1-4 linkages between monosaccharides in fibrous polysaccharides, plants create alpha 1–4 links. Such polysaccharides offer resistance to the digestive action of human amylase, which is not able to hydrolyze alpha 1–4 covalent linkages. Microbial fermentation of products in the human digestive tract liberates different biochemical entities. Short-chain fatty acids (SCFAs) are formed in the intestinal lumen and derived from indigestible dietary fibers. SCFAs are formed by colonic microbial fermentation, which modifies some of the host physiological processes. SCFAs mediate signals of nutritional and gut microbiome to the host metabolic pathways. SCFAs have the ability to inhibit LPS-driven inflammatory responses accompanied with metabolic disorders [76]. Some of these mechanisms are summarized in Figure 1.

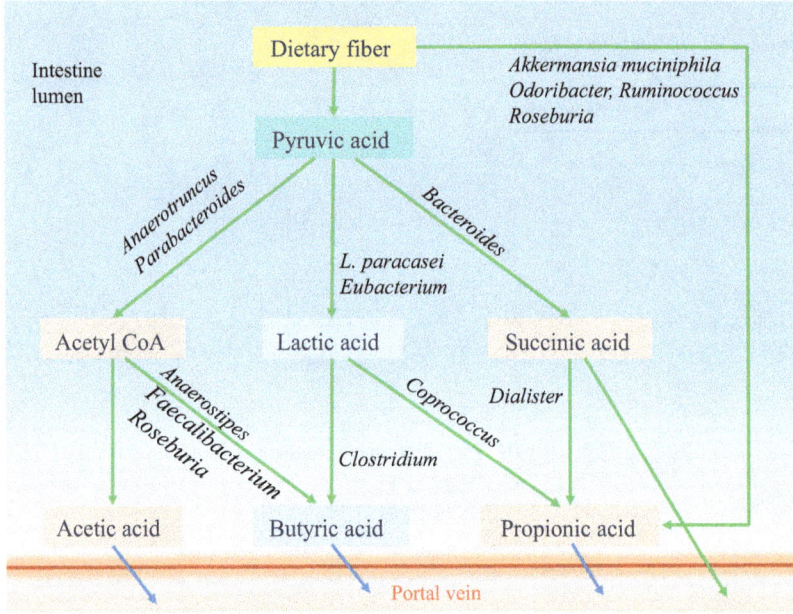

**Figure 1:** Formation of short-chain fatty acids by colonic bacteria via fermentation of indigestible dietary fibers. Various types of bacteria carrying out these biochemical reactions are mentioned with each respective pathway. Finally, the short-chain fatty acids are absorbed by the intestinal cells and then enter portal circulation. Butyric acid may serve as the main source of energy for colonic cells.

## 3.2 Prebiotics

Prebiotics, as defined by Roberfroid, are "selectively fermented ingredients capable of changing the composition or activity of gut microflora that benefits the host health." [77] The prebiotic category includes fiber or specifically fermented nutrients that support changes in gastrointestinal bacteria and their activity, which has positive impacts on the host's health. They act as probiotic bacteria's fertilizers in the colon. They are polysaccharides that can withstand enzyme and acidic digestion in the small intestine and can be used by probiotics and gut microflora for their growth and health-promoting activities in the large intestine, such as boosting mineral absorption and immunity, preventing colon cancer and other GIT diseases, and reducing cholesterol

[78]. Inulin is one example; upon hydrolysis, it yields galacto-oligosaccharides and oligo-fructose.

Prebiotics, such as oligolactose, inulin, oligofructose, and lactulose, are produced by certain enzymes from starch or other carbohydrates, or are found naturally in many plants [79]. A number of southern Thai plants, including jackfruit *(Artocarpus hetero-phyllus Lam.)*, rambutan *(Nephelium lappaceum L.)*, okra *(Abelmoschus esculentus Moench.)*, jampadah *(Artocarpus integer Merr.)*, and palm fruit *(Borrassus flabellifer L.)*, contain a significant amount of polysaccharides that possess prebiotic potential [80]. Prebiotics are occasionally combined with probiotic cultures and used as ingredients in functional foods. They are marketed as sophisticated products known as "nutraceuticals" in any convenient dosage form.

Prebiotics can provide health benefits when they become part of nutraceuticals or functional foods. Here are some examples where a combination of these two is used.

a)  Functional drinks with inulin and oligofructose to strengthen bones, enhance immunity to illnesses, and improve the digestive system's performance.
b)  Breakfast cereals that include more dietary fiber and prebiotics improve the function of the digestive system and promote the development of probiotics.
c)  Prebiotics added to infant meals enhance the effectiveness of the digestive system and boost immunity to sickness. During the first six weeks of life, adding inulin, oligofructose, and galacto-oligosaccharides to infant food upregulated *Lactobacilli* and *Bifidobacteria* in the digestive system of infants from 31–59% of the total gut normal flora, similar to breastfeeding children.
d)  Dairy items, such as yogurt and fermented dairy drinks that contain probiotics and turn into synbiotics when prebiotics are added. In addition, they aid in enhancing $Ca^{+2}$ absorption.
e)  Nutraceuticals, where probiotics and prebiotics are combined to create pills or capsules to boost their presence in the digestive system.
f)  Other products, such as those for weight management, in which prebiotics substitute sugar. Some of these products have a sweetness that is comparable to sugar. When adding prebiotics to food products, special care must be taken. Understanding how these substances interact with other substances in terms of their nutritional (stability, health benefits, dose, possible dangers), physical (solubility, particle size, viscosity), and sensory (flavor, taste, color) qualities is crucial. When used correctly, they can be an extra tool for treating diseases and enhancing customers' health and wellbeing.

Prebiotics are able to withstand harsh digestion in the small intestine because the human intestine lacks the enzymes necessary to hydrolyze their polymer linkages. These prebiotics are subsequently transported by the human body to the large intestine where they are specifically fermented and destroyed by the intestinal flora to create different secondary metabolites. Such metabolites are absorbed into the intestinal epithelium and transported to the liver via portal circulation, where they can have positive effects on the physiological functions of the host, such as regulating immunity, warding off pathogens, enhancing intestinal barrier function, and improving nutrient absorption [81–85]. Acetate, butyrate, and propionate are the most prevalent short-chain fatty acids that are important for sustaining intestinal and systemic health and digested by beneficial intestinal bacteria. Prebiotics' encouragement of the growth of the target microorganisms is also mentioned as a special advantage. Consuming particular prebiotics, for example, can protect or encourage the synthesis of advantageous fermentation products, which can help the proliferation of beneficial flora to compete with other species. Additionally, prebiotics have a protective impact on the body's immune, cardiovascular, and central neurological systems in addition to its gastrointestinal system [81, 86].

## 3.3 Probiotics

Probiotics are living, beneficial microorganisms that are used as dietary components. Their actions include adhering to the digestive tract at particular locations, and their continued existence results in the eradication of pathogens [87]. They provide resistance to acids and bile salts in the colon and aid in maintaining the balance of microorganisms in the gut. *Lactobacilli* and *Bifidobacteria*, which are present in the human digestive tract and are able to ferment lactate make up the majority of probiotics. Some probiotics can create antimicrobial substances like bacteriocins and promote bowel regularity, both of which can benefit both human and animal overall health and wellbeing [88]. Probiotics may have the following characteristics:

a) capable of surviving higher than necessary levels of stomach acidity.
b) able to endure bile salts, which the liver secretes at a concentration of 0.15–0.30% into the small intestine to aid in the digestion of fatty foods.
c) capable of colonizing the intestinal wall to stop pathogens from settling there and resisting the movement of food through the intestines during peristalsis, which promotes more typical food digestion and absorption.
d) capable of preventing the growth of pathogenic organisms through the following mechanisms: compete with pathogens for food; compete with pathogens for colonization of the intestinal wall; produce inhibitor chemicals like organic acids, $H_2O_2$, and bacteriocins.
e) Capable of boosting the host's defenses. It has been discovered that *Lactobacillus* increases the production of immunoglobulins and interferons and improves macrophage activity in the removal of infections from the body.
f) Capable of lowering the levels of cancer-causing enzymes, such as β-glucoronidase, azoreductase, nitrate reductase, and β-glucosidase, to lower the risk of colon cancer.
g) Buke and Gilliland (1990) isolated *Lactobacillus acidophilus* from the stools of nine participants. *L. acidophilus* strains 016 and C14 were discovered to be capable of absorbing 50.9 mg/mL and 47.1 mg/mL of cholesterol, respectively.

Probiotics have a positive impact on the body through four primary methods, including pathogen prevention and the suppression of their growth, enhancement of the gut's barrier function, body immunomodulation, and modulation of the host processes through the release of neurotransmitters [89]. By influencing the immune system of the host, *Oelschlaeger* found that probiotics can directly act or effect other host products, microbial products, or food ingredients [81]. However, a point of worry is that the probiotics' ability to enhance health depends on the probiotic flora that is introduced, and the probiotic's ability to perform a certain function depends on its metabolic properties and surface molecules or secreted components of these microorganisms [90]. A combination of live multiple strains of *Acetobacter, Bifidobacterium, Lactobacillus*, and *Propionibacterium* alleviated adiponectin level, insulin resistance, obesity, and pro-inflammatory cytokines production, as compared with lyophilized single strain or even multi-strains [91]. A mixture of omega-3 fatty acids with live probiotics containing *Lactococcus, Propionibacterium, Bifidobacterium*, and *Lactobacillus* reduced lipid accumulation and hepatic steatosis as compared to probiotics alone [92]. Some other mechanisms can be summarized here from various studies and include, restoring *Akkermansia muciniphila* and *Rikenellaceae*, upregulation of PPARγ and lipoprotein lipase expression, enhancing insulin sensitivity, and TG clearance decreasing *Lactobacillaceae* [93], upregulation of IL-10 and GLP1 expression [94], improvement of intestinal barrier, suppression of LPS production, downregulation of NF-κB-linked TNF-α expression [95], improvement of intestinal barrier, and upregulation of ZO-1, GLP1,and GLUT4 expression [96]. All these mechanisms are summarized in Figure 2.

## 3.4 Polyunsaturated fatty acids (PUFAs)

The carboxylic acid family, which includes polyunsaturated fatty acids (PUFAs), has several carbon–carbon double bonds (C = C). The two important subgroups of PUFAs are, omega-3 (*n*-3) and omega-6 (*n*-6) fatty acids. They are distinguished by the position of the first C = C double bond, counting from the methyl group. Since they cannot be synthesized biochemically inside the human body and are necessary for physiological integrity, they must be received from diet. Linoleic acid, linolenic acid, and arachidonic acid are referred to as essential fatty acids.

Fatty acids make up the majority of fats (lipids) and dietary oils from both animal and plant sources. They play an important structural, physiological, and metabolic role in the body as the main constituents of phospholipids, triglycerides, and cholesterol esters. A triglyceride molecule is created by the esterification of three fatty acid molecules to a glycerol backbone. Fatty acids are nonpolar, organic hydrocarbon chains with at least a single carboxylic acid at their terminal position. While trans-fatty acids do exist in some types of animals and partially hydrogenated fats, naturally occurring fatty acids are often in cis structure. Position of double bonds, chain length and number of double bonds within hydrocarbon chains, and the configuration of the

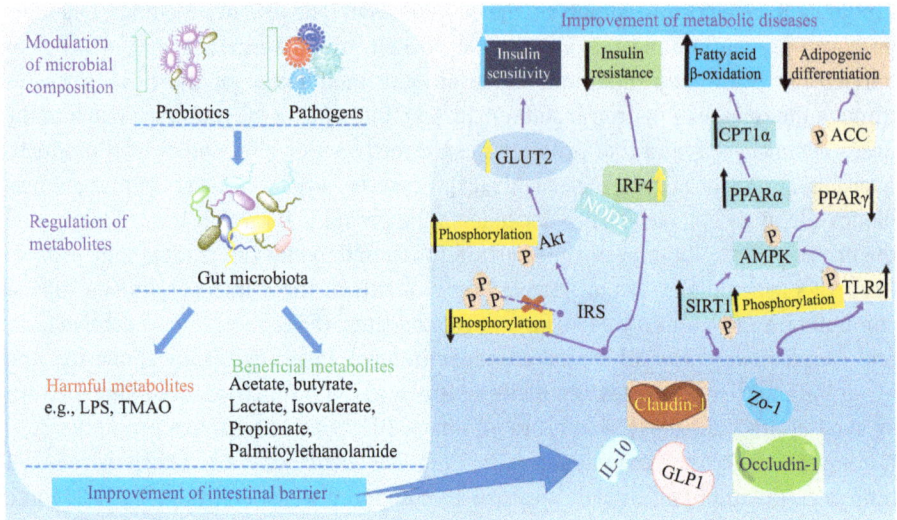

**Figure 2:** Mechanisms of prebiotics and probiotics in metabolic disorders by modifying intestinal microflora. They alleviate metabolic diseases by increasing colonization of beneficial bacteria and decreasing harmful bacteria. The beneficial gut microbiota promotes the production of acetate, butyrate, propionate, isovalerate, lactate, and palmitoylethanolamide, while suppressing LPS, TMAO, and reducing the bile acid pools. This leads to the improvement of intestinal barrier function along with upregulation of IL-10, GLP1, occludin 1, claudin 1, and ZO-1 expressions. The metabolic system is improved via summation of these effects by upregulating IRS1-Akt-GLUT2 pathway leading to increased insulin sensitivity and NOD2-IRF4 pathway, leading to decreased insulin resistance; SIRT1-AMPK-PPARα-CPR1α pathway leading to increased fatty acid beta-oxidation; TLR2-PPARγ-AMPK signaling pathway leading to decreased adipogenic differentiation. GLUT2: glucose transporter 2; IRS1: insulin receptor substrate 1; IRF4: interferon regulatory factor 4; ZO-1: zonula occludens-1; LPS: lipopolysaccharide; SIRT1: sirtuin 1; PPARα/γ: peroxisome proliferators activated receptor α/γ; TLR2: Toll-like receptor 2; ACC: acetyl-CoA carboxylase; AMPK: adenosine monophosphate kinase; Akt: protein kinase B; CPR1α: carnitine palmitoyltransferase 1α; TMAO: trimethylamine-n-oxide; NOD2: oligomerization domain-containing protein 2.

hydrogen atom with respect to the double bond (cis or trans) are just a few of the elements that contribute to the physiological and physical characteristics of fatty acids. In contrast to unsaturated fatty acids, which have at least one double bond in the hydrocarbon chain, saturated fatty acids have none. Two or more double bonds can be found in the hydrocarbon chain of PUFAs, depending on how far away from the methyl end the double bond is.

All of the necessary fatty acids, aside from two, can be produced by humans. Both linolenic acid (ALA, C18: 3, n-3) and linoleic acid (LA, C18: 2, n-6) are precursors of the omega-3 and omega-6 family of fatty acids, respectively. Due to the fact that the body is unable to produce them, they are known as dietary essential fatty acids. PUFAs affect various biological processes, i.e., blood pressure, blood clotting, healthy body growth, and neurological system depending on the position and quantity of double bonds. Additionally, the metabolites of fatty acids known as eicosanoids produced from docosahexaenoic acid (DHA 22:6, n-3) and eicosapentaenoic acid (EPA 20:5, n-3) in the omega-3 series and arachidonic acid (AA 20:4, n-6) in the omega-6 series, play crucial roles in the control of inflammation and the immune system. Unsaturated fatty acids are actively gathered and retained by the human body. The extraordinary care with which these fatty acids are accumulated and conserved in some tissues suggests that they may be a necessary component of some cells. DHA is greatly concentrated in and long-preserved in brain and retinal tissues [97].

Multiple mechanisms have been identified for PUFAs [98]. There are four main ways that n-3 PUFAs might influence cell and tissue activity to produce their physiological effects:

- n-3 PUFAs may affect the levels of hormones and/or metabolites, which in turn may affect how cells and tissues behave.
- The behavior of cells and tissues may be affected by n-3 PUFAs by influencing other variables (such as oxidation of LDL and oxidative stress).
- n-3 PUFAs directly affect cellular mechanisms by altering cell surface or intracellular "sensors" or "receptors."
- Modifications in the phospholipid content of cell membranes as a result of the effects of n-3 PUFAs on cellular activity.

## 3.5 Micronutrients

Vitamins play important roles in preserving healthy metabolism and state of wellness. Any vitamin deficiency can result in distinct clinical signs. Common vitamins including retinol, vitamin B complex, ascorbic acid, cholecalciferol, and tocopherol are present in the majority of nutraceutical and nutritional therapy products. Since a substantial amount of human vitamin intake comes from plant meals, plant biotechnology has been used to increase the vitamin content of crops [99]. Vitamins have a distinct kind of function in the metabolic activities. They are necessary for the functional operation of enzymes that are involved in anabolic and catabolic reactions coupled with energy release and storage. Vitamin B and its subtypes serve as coenzymes in both the process of breaking down food and the production of large molecules, e.g., protein, DNA, and RNA. In contrast to other dietary components referred to as macronutrients (such as fats, carbohydrates, and proteins), vitamins regulate metabolic events [100]. Dietary supplements and natural products lower blood pressure without significant side effects. Among these nutrients, ascorbic acid, magnesium, and potassium supplements improve blood pressure [101]. As compared to healthy populations, higher levels of vitamin D are necessary in diabetes mellitus and obesity. Similarly, in case of hyperparathyroidism, serum vitamin D concentration is lower as compared to healthy individuals, probably due to lower levels of vitamin D-binding protein [102]. In a meta-analysis with 4,896 prediabetes patients, Vitamin D supplementation lowered the risk of T2D and increased the reversion of prediabetes to normoglycemic state [103].

Minerals play an important role in maintaining and restoring human physiology. Minerals are inorganic chemical elements found in rock substances, underground water, plant sources, and fortified mineral water sources. According to their requirements they are further classified as major minerals, trace minerals, ultra-trace minerals. Calcium, chloride, magnesium, phosphorus, potassium, sodium, and sulfur are major minerals required in more than 100 mg per day. Trace and ultra-trace minerals include fluoride, iron, manganese, zinc, chromium, copper, iodine, molybdenum, and selenium [104]. Serious health issues could result from a deficiency in any of these. Both animal and plant meals provide dietary Ca, Zn, Fe, and other minerals. Mineral deficiencies, primarily Zn, Ca, and Fe deficits, are the leading causes of illness in poor nations, especially in newborns and young children. However, a key tactic to improve mineral nutrition is to increase dietary Ca, Zn, and Fe in plant foods [105]. In a randomized, double-blind, controlled, cross-over design having 12 healthy men, moderately hypercholesterolemic (2.20–3 g/L), and aged 20–60 years, the effect of minerals was studied. There was a significant decrease in triglyceridemia (23%), serum very-low-density lipoprotein (31%), and tendency to decrease VLDL cholesterol (p = 0.066) even at fasting state [106]. This study demonstrates that drinking 1 liter of bicarbonate sodium rich mineral water daily in young, moderately hypercholesterolemic people lowers systolic blood pressure, Apo B and CVD risk scores, LDL cholesterol (by 10%), total cholesterol (by 6.3%), and other lipid profiles [107]. Another study in healthy postmenopausal women showed that mineral water rich in sodium, silicon, and bicarbonate reduced cardiovascular risk. The effect was decrease in LDL cholesterol, total cholesterol, as well as the adhesion molecules of early atherosclerosis. High-density lipoproteins (HDL) were increased, while fasting blood glucose and postprandial insulin were decreased [108]. In order for cells to catalyze metabolic processes, minerals are necessary. The body's capacity for metabolism and, ultimately, health status, are heavily influenced by these chemical processes. Mineral deficiencies can affect metabolism and make it difficult to control weight. Hundreds of enzymes involved in metabolism use minerals as cofactors. Blood function depends on copper, iron, and zinc [109].

## 3.6 Polyphenols

In whole plant diets which include vegetables, fruits, whole grain cereal, legumes, coffee, tea, and chocolate, more than 8,000 polyphenolic compounds have been discovered [110]. These compounds are secondary metabolites that plants make to defend themselves against pathogens, oxidants, and ultraviolet radiation. Polyphenols can be categorized into a variety of groups according to the number of phenol rings present and the structural elements that link these rings together. Phenolic acids, which can be divided into two main groups, make up around one-third of the polyphenolic compounds found in food [111]. One of the classes includes compounds that are derived from hydroxybenzoic acid, such as protocatechuic acid, gallic acid, and p-hydroxybenzoic acid. Foods strong in phenolic acids, e.g., coumaric acid, caffeic acid, chlorogenic acid, and coumaric acid fall under the group of hydroxycinnamic acid derivatives, which also includes berry fruits, cherry, coffee, apple, kiwi, pear, and chicory [112]. By inhibiting the synthesis of the active species and precursors of free radicals, polyphenols reduce the rate of oxidation and slow down the growth of free radicals. In the series of lipid peroxidation reactions, they frequently serve as direct radical scavengers [113].

## 3.7 Spices

Spices are flavor-enhancing, rather than nutritional, fragrant vegetable components that can be consumed whole, broken, or pulverized. The primary flavors, aromas, and pungency of these spices give food its distinctive flavor. The pungency is derived by volatile oil. Spices are also responsible for flavor, fragrance, and oleoresin [114]. Spices are commonly used in native remedies, nutraceuticals, pharmaceuticals, beverages, aroma therapy, cosmetics, perfumes, natural colors, preservatives, dental preparations and botanicals as pesticides, in addition to flavoring and seasoning, and they thus play an important role in the economy of a nation. These qualities result from a wide variety of compounds that these spices have created. It has been demonstrated that spices like black pepper, saffron, turmeric, garlic, rosemary, red pepper, clove, ginger, coriander and cinnamon have anti-neurodegenerative properties [115]. The antioxidant, anti-inflammatory, and antidiabetic effects of spices have long been recognized. The bioactive compounds in spices have a wide range of health advantages. The well-known antidiabetic effects of spices may be caused by stimulation of insulin production by the pancreas, inhibition of glucose absorption, or insulin-sparing effects of the bioactive compounds (i.e., phenolic compounds) [116]. Spices contain a lot of antioxidants that can aid with diabetes management [117].

## 4 Summary

We have discussed nutraceutical classes which are promising in alleviating metabolic disorders. Dietary fibers, prebiotics, probiotics, polyphenols, spices, vitamins, and minerals have proven pharmacological mechanisms to modify the pathways involved in, or leading to, one of these metabolic disorders. Being of natural origin, they are available with diversity of selection according to the desired properties. Further progress towards their clinical availability is the target for formulation and translational scientists to bring them to the life of a common man.

# References

[1] Alberti, G. 2005. Introduction to the metabolic syndrome. European Heart Journal Supplements, 7(suppl_D), D3–D5.

[2] Tabassum, N. and Ahmad, F. 2011. Role of natural herbs in the treatment of hypertension. Pharmacognosy Reviews, 5(9), 30.

[3] De la Monte, S.M. and Tong, M. 2014. Brain metabolic dysfunction at the core of Alzheimer's disease. Biochemical Pharmacology, 88(4), 548–559.

[4] Ullah, H., De Filippis, A., Khan, H., Xiao, J. and Daglia, M. 2020. An overview of the health benefits of Prunus species with special reference to metabolic syndrome risk factors. Food and Chemical Toxicology, 144, 111574.

[5] Association, A.D. 2006. Diagnosis and classification of diabetes mellitus. Diabetes Care, 29(1), S43.

[6] Galicia-Garcia, U., Benito-Vicente, A., Jebari, S., Larrea-Sebal, A., Siddiqi, H., Uribe, K.B., et al. 2020. Pathophysiology of type 2 diabetes mellitus. International Journal of Molecular Sciences, 21(17), 6275.

[7] Desrochers, G., Bergeron, S., Khalifé, S., Dupuis, M.-J. and Jodoin, M. 2010. Provoked vestibulodynia: Psychological predictors of topical and cognitive-behavioral treatment outcome. Behaviour Research and Therapy, 48(2), 106–115.

[8] Lotery, H.E., McClure, N. and Galask, R.P. 2004. Vulvodynia. The Lancet, 363(9414), 1058–1060.

[9] Ponte, M., Klemperer, E., Sahay, A. and Chren, -M.-M. 2009. Effects of vulvodynia on quality of life. Journal of the American Academy of Dermatology, 60(1), 70–76.

[10] Nunns, D. and Mandal, D. 1997. Psychological and psychosexual aspects of vulvar vestibulitis. Genitourinary Medicine, 73(6), 541–544.

[11] Arnold, L.D., Bachmann, G.A., Kelly, S., Rosen, R. and Rhoads, G.G. 2006. Vulvodynia: Characteristics and associations with co-morbidities and quality of life. Obstetrics and Gynecology, 107(3), 617.

[12] Bergeron, S., Binik, Y.M., Khalifé, S., Pagidas, K., Glazer, H.I., Meana, M., et al. 2001. A randomized comparison of group cognitive–behavioral therapy, surface electromyographic biofeedback, and vestibulectomy in the treatment of dyspareunia resulting from vulvar vestibulitis. Pain, 91(3), 297–306.

[13] Masheb, R.M., Kerns, R.D., Lozano, C., Minkin, M.J. and Richman, S. 2009. A randomized clinical trial for women with vulvodynia: Cognitive-behavioral therapy vs. supportive psychotherapy. Pain®, 141(1), 31–40.

[14] Masheb, R.M., Nash, J.M., Brondolo, E. and Kerns, R.D. 2000. Vulvodynia: An introduction and critical review of a chronic pain condition. Pain, 86(1), 3–10.

[15] Glazer, H.I. and Ledger, W.J. 2002. Clinical management of vulvodynia. Reviews in Gynaecological Practice, 2(1), 83–90.

[16] Giesecke, J., Reed, B.D., Haefner, H.K., Giesecke, T., Clauw, D.J. and Gracely, R.H. 2004. Quantitative sensory testing in vulvodynia patients and increased peripheral pressure pain sensitivity. Obstetrics & Gynecology, 104(1), 126–133.

[17] Haefner, H.K., Collins, M.E., Davis, G.D., Edwards, L., Foster, D.C., Hartmann, E.D.H., et al. 2005. The vulvodynia guideline. Journal of Lower Genital Tract Disease, 9(1), 40–51.

[18] Lamont, J., Randazzo, J., Farad, M., Wilkins, A. and Daya, D. 2001. Psychosexual and social profiles of women with vulvodynia. Journal of Sex &marital Therapy, 27(5), 551–555.

[19] Graziottin, A., Castoldi, E., Montorsi, F., Salonia, A. and Maga, T. 2001. Vulvodynia: The challenge of "unexplained" genital pain. Journal of Sex &marital Therapy, 27(5), 503–512.

[20] Edwards, L. 2003. New concepts in vulvodynia. American Journal of Obstetrics and Gynecology, 189(3), S24–S30.

[21] Baynes, J.W. and Thorpe, S.R. 1999. Role of oxidative stress in diabetic complications: A new perspective on an old paradigm. Diabetes, 48(1), 1–9.

[22] Sytze van Dam, P. 2002. Oxidative stress and diabetic neuropathy: Pathophysiological mechanisms and treatment perspectives. Diabetes/metabolism Research and Reviews, 18(3), 176–184.

[23] Giacco, F. and Brownlee, M. 2010. Oxidative stress and diabetic complications. Circulation Research, 107(9), 1058–1070.

[24] Barrea, L., Caprio, M., Watanabe, M., Cammarata, G., Feraco, A., Muscogiuri, G., Verde, L., Colao, A. and Savastano, S. 2023. Could very low-calorie ketogenic diets turn off low grade inflammation in obesity? Emerging evidence. Critical Reviews in Food Science and Nutrition, 63(26), 8320–8336.

[25] Khanna, D., Khanna, S., Khanna, P., Kahar, P. and Patel, B.M. 2022. Obesity: A chronic low-grade inflammation and its markers. Cureus, 14(2).

[26] Le, N.-A. 2020. Postprandial triglycerides, oxidative stress, and inflammation. Apolipoproteins, Triglycerides and Cholesterol, 1–12.

[27] Hayden, M.R. 2023. Overview and New insights into the metabolic syndrome: Risk factors and emerging variables in the development of type 2 diabetes and cerebrocardiovascular disease. Medicina, 59(3), 561.

[28] Li, L., Song, Q. and Yang, X. 2019. Lack of associations between elevated serum uric acid and components of metabolic syndrome such as hypertension, dyslipidemia, and T2DM in overweight and obese Chinese adults. Journal of Diabetes Research, 2019, 3175418.

[29] Sigit, F.S., Tahapary, D.L., Trompet, S., Sartono, E., Willems van Dijk, K., Rosendaal, F.R., et al. 2020. The prevalence of metabolic syndrome and its association with body fat distribution in middle-aged individuals from Indonesia and the Netherlands: A cross-sectional analysis of two population-based studies. Diabetology & Metabolic Syndrome, 12, 1–11.

[30] Egbuna, C. and Hassan, S., editors. 2021. Dietary phytochemicals: a source of novel bioactive compounds for the treatment of obesity, cancer and diabetes. Springer Nature: Berlin, Germany.

[31] Gruss, S.M., Nhim, K., Gregg, E., Bell, M., Luman, E. and Albright, A. 2019. Public health approaches to type 2 diabetes prevention: The US national diabetes prevention program and beyond. Current Diabetes Reports, 19, 1–11.

[32] Lan, S. and Albinsson, S. 2020. Regulation of IRS-1, insulin signaling and glucose uptake by miR-143/145 in vascular smooth muscle cells. Biochemical and Biophysical Research Communications, 529(1), 119–125.

[33] Mezzaroba, L., Simão, A.N.C., Oliveira, S.R., Flauzino, T., Alfieri, D.F., De Carvalho Jennings Pereira, W.L., et al. 2020. Antioxidant and anti-inflammatory diagnostic biomarkers in multiple sclerosis: A machine learning study. Molecular Neurobiology, 57, 2167–2178.

[34] Schwartz, A. and Bellissimo, N. 2021. Nicotine and energy balance: A review examining the effect of nicotine on hormonal appetite regulation and energy expenditure. Appetite, 164, 105260.

[35] Yadav, R.S., Kant, S., Tripathi, P.M., Pathak, A.K. and Mahdi, A.A. 2022. Transcription factor NF-κB, interleukin-1β, and interleukin-8 expression and its association with tobacco smoking and severity in chronic obstructive pulmonary disease. Gene Reports, 26, 101453.

[36] Leeming, E.R., Johnson, A.J., Spector, T.D. and Le Roy, C.I. 2019. Effect of diet on the gut microbiota: Rethinking intervention duration. Nutrients, 11(12), 2862.

[37] Khan, S., Luck, H., Winer, S. and Winer, D.A. 2021. Emerging concepts in intestinal immune control of obesity-related metabolic disease. Nature Communications, 12(1), 2598.

[38] Tilg, H., Zmora, N., Adolph, T.E. and Elinav, E. 2020. The intestinal microbiota fuelling metabolic inflammation. Nature Reviews Immunology, 20(1), 40–54.

[39] Hutchinson, A.N., Tingö, L. and Brummer, R.J. 2020. The potential effects of probiotics and ω-3 fatty acids on chronic low-grade inflammation. Nutrients, 12(8), 2402.

[40] Paschou, S.A., Bletsa, E., Stampouloglou, P.K., Tsigkou, V., Valatsou, A., Stefanaki, K., et al. 2022. Thyroid disorders and cardiovascular manifestations: An update. Endocrine, 75(3), 672–683.

[41] Teixeira, P.D.F.D.S., Dos Santos, P.B. and Pazos-Moura, C.C. 2020. The role of thyroid hormone in metabolism and metabolic syndrome. Therapeutic Advances in Endocrinology and Metabolism, 11, 2042018820917869.

[42] Chemello, K., García-Nafría, J., Gallo, A., Martín, C., Lambert, G. and Blom, D. 2021. Lipoprotein metabolism in familial hypercholesterolemia. Journal of Lipid Research, 62, 100062.

[43] Tokgozoglu, L. and Kayikcioglu, M. 2021. Familial hypercholesterolemia: Global burden and approaches. Current Cardiology Reports, 23, 1-13.

[44] Nohara, A., Tada, H., Ogura, M., Okazaki, S., Ono, K., Shimano, H., et al. 2021. Homozygous familial hypercholesterolemia. Journal of Atherosclerosis and Thrombosis, 28(7), 665-678.

[45] Bastaki, S. 2005. Diabetes mellitus and its treatment. Dubai Diabetes And Endocrinology Journal, 13(3), 111-134.

[46] Hundal, R.S., Krssak, M., Dufour, S., Laurent, D., Lebon, V., Chandramouli, V., et al. 2000. Mechanism by which metformin reduces glucose production in type 2 diabetes. Diabetes, 49(12), 2063-2069.

[47] Zhou, G., Myers, R., Li, Y., Chen, Y., Shen, X., Fenyk-Melody, J., et al. 2001. Role of AMP-activated protein kinase in mechanism of metformin action. The Journal of Clinical Investigation, 108(8), 1167-1174.

[48] Wilding, J. 2012. PPAR agonists for the treatment of cardiovascular disease in patients with diabetes. Diabetes, Obesity and Metabolism, 14(11), 973-982.

[49] Sola, D., Rossi, L., Piero Carnevale Schianca, G., Maffioli, P., Bigliocca, M., Mella, R., et al. 2015. Sulfonylureas and their use in clinical practice. Archives of Medical Science, 11(4), 840-848.

[50] Panten, U., Schwanstecher, M. and Schwanstecher, C. 1996. Sulfonylurea receptors and mechanism of sulfonylurea action. Experimental and Clinical Endocrinology & Diabetes, 104(01), 1-9.

[51] Kimura, A. 2000. Molecular anatomy of Alpha-Glucosidase. Trends in Glycoscience and Glycotechnology, 12(68), 373-380.

[52] Dhabi, A.S., Bhatt, N.R. and Shah, M.J. 2013. Voglibose: An alpha glucosidase Inhibitor. Journal of Clinical and Diagnostic Research, 7(12), 3023.

[53] Drucker, D.J. and Nauck, M.A. 2006. The incretin system: Glucagon-like peptide-1 receptor agonists and dipeptidyl peptidase-4 inhibitors in type 2 diabetes. The Lancet, 368(9548), 1696-1705.

[54] Müller, T.D., Finan, B., Bloom, S., D'Alessio, D., Drucker, D.J., Flatt, P., et al. 2019. Glucagon-like peptide 1 (GLP-1). Molecular Metabolism, 30, 72-130.

[55] Shin, S., Le Lay, J., Everett, L.J., Gupta, R., Rafiq, K. and Kaestner, K.H. 2014. CREB mediates the insulinotropic and anti-apoptotic effects of GLP-1 signaling in adult mouse β-cells. Molecular Metabolism, 3(8), 803-812.

[56] Mulvihill, E.E. and Drucker, D.J. 2014. Pharmacology, physiology, and mechanisms of action of dipeptidyl peptidase-4 inhibitors. Endocrine Reviews, 35(6), 992-1019.

[57] Chen, X.W., He, Z.X., Zhou, Z.W., Yang, T., Zhang, X., Yang, Y.X., et al. 2015. Clinical pharmacology of dipeptidyl peptidase 4 inhibitors indicated for the treatment of type 2 diabetes mellitus. Clinical and Experimental Pharmacology and Physiology, 42(10), 999-1024.

[58] Rosenwasser, R.F., Sultan, S., Sutton, D., Choksi, R. and Epstein, B.J. 2013. SGLT-2 inhibitors and their potential in the treatment of diabetes. Diabetes, Metabolic Syndrome and Obesity: Targets and Therapy, 6, 453-467.

[59] Skyler, J.S. 2004. Effects of glycemic control on diabetes complications and on the prevention of diabetes. Clinical Diabetes, 22(4), 162-166.

[60] Valerón, P.F. and De Pablos-velasco, P.L. 2013. Limitations of insulin-dependent drugs in the treatment of type 2 diabetes mellitus. Medicina Clinica, 141, 20-25.

[61] Mansourian, A.R. 2010. A review on hyperthyroidism: Thyrotoxicosis under surveillance. Pakistan Journal of Biological Sciences: PJBS, 13(22), 1066-1076.

[62] Wang, Y., Sun, Y., Yang, B., Wang, Q. and Kuang, H. 2022. The management and metabolic characterization: Hyperthyroidism and hypothyroidism. Neuropeptides, 102308.

[63]  Daroff, R.B., Frishman, W.H., Lederman, R.J. and Stewart, W.C. 1993. Beta-blockers: Beyond cardiology. Patient Care, 27(11), 47–58.

[64]  Lee, S.L. 2012. Radioactive iodine therapy. Current Opinion in Endocrinology, Diabetes and Obesity, 19(5), 420–428.

[65]  Fewins, J., Simpson, C.B. and Miller, F.R. 2003. Complications of thyroid and parathyroid surgery. Otolaryngologic Clinics of North America, 36(1), 189–206.

[66]  Benito-Vicente, A., Uribe, K.B., Jebari, S., Galicia-Garcia, U., Ostolaza, H. and Martin, C. 2018. Familial hypercholesterolemia: The most frequent cholesterol metabolism disorder caused disease. International Journal of Molecular Sciences, 19(11), 3426.

[67]  Raal, F.J., Hovingh, G.K. and Catapano, A.L. 2018. Familial hypercholesterolemia treatments: Guidelines and new therapies. Atherosclerosis, 277, 483–492.

[68]  Premi, M. and Bansal, V. 2021. Nutraceuticals for management of metabolic disorders. In *Treating Endocrine and Metabolic Disorders with Herbal Medicines*. IGI Global: Hershey, PA 17033, USA, pp. 298–320.

[69]  Yoo, J.Y. and Kim, S.S. 2016. Probiotics and prebiotics: Present status and future perspectives on metabolic disorders. Nutrients, 8(3), 173.

[70]  Ejtahed, H., Mohtadi-Nia, J., Homayouni-Rad, A., Niafar, M., Asghari-Jafarabadi, M., Mofid, V., et al. 2011. Effect of probiotic yogurt containing Lactobacillus acidophilus and Bifidobacterium lactis on lipid profile in individuals with type 2 diabetes mellitus. Journal of Dairy Science, 94(7), 3288–3294.

[71]  Amar, J., Chabo, C., Waget, A., Klopp, P., Vachoux, C., Bermúdez-Humarán, L.G., et al. 2011. Intestinal mucosal adherence and translocation of commensal bacteria at the early onset of type 2 diabetes: Molecular mechanisms and probiotic treatment. EMBO Molecular Medicine, 3(9), 559–572.

[72]  Qin, J., Li, Y., Cai, Z., Li, S., Zhu, J., Zhang, F., et al. 2012. A metagenome-wide association study of gut microbiota in type 2 diabetes. Nature, 490(7418), 55–60.

[73]  Ryan, P.M., Ross, R.P., Fitzgerald, G.F., Caplice, N.M. and Stanton, C. 2015. Functional food addressing heart health: Do we have to target the gut microbiota? Current Opinion in Clinical Nutrition & Metabolic Care, 18(6), 566–571.

[74]  Van Soest, P.J. 1978. Dietary fibers: Their definition and nutritional properties. The American Journal of Clinical Nutrition, 31(10), S12–S20.

[75]  Miller, K.B. 2020. Review of whole grain and dietary fiber recommendations and intake levels in different countries. Nutrition Reviews, 78(Supplement_1), 29–36.

[76]  Zhang, S., Zhao, J., Xie, F., He, H., Johnston, L.J., Dai, X., et al. 2021. Dietary fiber-derived short-chain fatty acids: A potential therapeutic target to alleviate obesity-related nonalcoholic fatty liver disease. Obesity Reviews, 22(11), e13316.

[77]  Roberfroid, M. 2007. Prebiotics: The concept revisited. The Journal of Nutrition, 137(3), 830S–7S.

[78]  Cao, Z., Guo, Y., Liu, Z., Zhang, H., Zhou, H. and Shang, H. 2022. Ultrasonic enzyme-assisted extraction of comfrey (Symphytum officinale L.) polysaccharides and their digestion and fermentation behaviors in vitro. Process Biochemistry, 112, 98–111.

[79]  Manzoor, S., Wani, S.M., Mir, S.A. and Rizwan, D. 2022. Role of probiotics and prebiotics in mitigation of different diseases. Nutrition, 96, 111602.

[80]  Thammarutwasik, P., Hongpattarakere, T., Chantachum, S., Kijroongrojana, K., Itharat, A., Reanmongkol, W., Tewtrakul, S. and Ooraikul, B. 2009. Prebiotics-a review. Songklanakarin Journal of Science and Technology, 31(4), 401–408.

[81]  You, S., Ma, Y., Yan, B., Pei, W., Wu, Q., Ding, C., et al. 2022. The promotion mechanism of prebiotics for probiotics: A review. Frontiers in Nutrition, 9, 1000517.

[82]  Frei, R., Akdis, M. and O'Mahony, L. 2015. Prebiotics, probiotics, synbiotics, and the immune system: Experimental data and clinical evidence. Current Opinion in Gastroenterology, 31(2), 153–158.

[83]  Newman, A.M. and Arshad, M. 2020. The role of probiotics, prebiotics and synbiotics in combating multidrug-resistant organisms. Clinical Therapeutics, 42(9), 1637–1648.

[84] Peredo-Lovillo, A., Romero-Luna, H. and Jiménez-Fernández, M. 2020. Health promoting microbial metabolites produced by gut microbiota after prebiotics metabolism. Food Research International, 136, 109473.

[85] Yan, S., Tian, Z., Li, M., Li, B. and Cui, W. 2019. Effects of probiotic supplementation on the regulation of blood lipid levels in overweight or obese subjects: A meta-analysis. Food & Function, 10(3), 1747–1759.

[86] Kaewarsar, E., Chaiyasut, C., Lailerd, N., Makhamrueang, N., Peerajan, S. and Sirilun, S. 2023. Optimization of mixed inulin, fructooligosaccharides, and galactooligosaccharides as prebiotics for stimulation of probiotics growth and function. Foods, 12(8), 1591.

[87] Télessy, I.G. 2019. Nutraceuticals. In *The Role of Functional Food Security in Global Health*. Elsevier, pp. 409–421.

[88] Mustar, S. and Ibrahim, N. 2022. A sweeter pill to swallow: A review of honey bees and honey as a source of probiotic and prebiotic products. Foods, 11(14), 2102.

[89] Olaimat, A.N., Aolymat, I., Al-Holy, M., Ayyash, M., Abu Ghoush, M., Al-Nabulsi, A.A., et al. 2020. The potential application of probiotics and prebiotics for the prevention and treatment of COVID-19. Npj Science of Food, 4(1), 17.

[90] Mazziotta, C., Tognon, M., Martini, F., Torreggiani, E. and Rotondo, J.C. 2023. Probiotics mechanism of action on immune cells and beneficial effects on human health. Cells, 12(1), 184.

[91] Kobyliak, N., Falalyeyeva, T., Tsyryuk, O., Eslami, M., Kyriienko, D., Beregova, T., et al. 2020. New insights on strain-specific impacts of probiotics on insulin resistance: Evidence from animal study. Journal of Diabetes & Metabolic Disorders, 19, 289–296.

[92] Kobyliak, N., Falalyeyeva, T., Bodnar, P. and Beregova, T. 2017. Probiotics supplemented with omega-3 fatty acids are more effective for hepatic steatosis reduction in an animal model of obesity. Probiotics and Antimicrobial Proteins, 9, 123–130.

[93] Alard, J., Lehrter, V., Rhimi, M., Mangin, I., Peucelle, V., Abraham, A.L., et al. 2016. Beneficial metabolic effects of selected probiotics on diet-induced obesity and insulin resistance in mice are associated with improvement of dysbiotic gut microbiota. Environmental Microbiology, 18(5), 1484–1497.

[94] Alard, J., Cudennec, B., Boutillier, D., Peucelle, V., Descat, A., Decoin, R., et al. 2021. Multiple selection criteria for probiotic strains with high potential for obesity management. Nutrients, 13(3), 713.

[95] Kim, D.-E., Kim, J.-K., Han, S.-K., Jang, S.-E., Han, M.J. and Kim, D.-H. 2019. Lactobacillus plantarum NK3 and Bifidobacterium longum NK49 alleviate bacterial vaginosis and osteoporosis in mice by suppressing NF-κ B-Linked TNF-α expression. Journal of Medicinal Food, 22(10), 1022–1031.

[96] Archer, A.C., Muthukumar, S.P. and Halami, P.M. 2021. Lactobacillus fermentum MCC2759 and MCC2760 alleviate inflammation and intestinal function in high-fat diet-fed and streptozotocin-induced diabetic rats. Probiotics and Antimicrobial Proteins, 13(4), 1068–1080.

[97] Rathnakumar, K., Rathnakumar, K., Pandiselvam, R. and Kothakota, A. 2019. Polyunsaturated Fatty Acids as Nutraceuticals. In *Advances in Food Bioproducts and Bioprocessing Technologies*. CRC Press, pp. 475–496.

[98] Abshirini, M., Ilesanmi-Oyelere, B.L. and Kruger, M.C. 2021. Potential modulatory mechanisms of action by long-chain polyunsaturated fatty acids on bone cell and chondrocyte metabolism. Progress in Lipid Research, 83, 101113.

[99] Gupta, M., Aggarwal, R., Raina, N. and Khan, A. 2020. Vitamin-loaded nanocarriers as nutraceuticals in healthcare applications. Nanomedicine for Bioactives: Healthcare Applications, 451–470.

[100] Christakos, S., Dhawan, P., Verstuyf, A., Verlinden, L. and Carmeliet, G. 2016. Vitamin D: Metabolism, molecular mechanism of action, and pleiotropic effects. Physiological Reviews, 96(1), 365–408.

[101] Borghi, C., Tsioufis, K., Agabiti-Rosei, E., Burnier, M., Cicero, A.F., Clement, D., et al. 2020. Nutraceuticals and blood pressure control: A European Society of Hypertension position document. Journal of Hypertension, 38(5), 799–812.

[102] Bilezikian, J.P., Formenti, A.M., Adler, R.A., Binkley, N., Bouillon, R., Lazaretti-Castro, M., et al. 2021. Vitamin D: Dosing, levels, form, and route of administration: Does one approach fit all? Reviews in Endocrine and Metabolic Disorders, 22(4), 1201–1218.

[103] Pramono, A., Jocken, J.W., Blaak, E.E. and Van Baak, M.A. 2020. The effect of vitamin D supplementation on insulin sensitivity: A systematic review and meta-analysis. Diabetes Care, 43(7), 1659–1669.

[104] Gomes, C. and Rautureau, M. 2021. *Minerals Latu Sensu and Human Health: Benefits, Toxicity and Pathologies*. Springer: Switzerland.

[105] Souyoul, S.A., Saussy, K.P. and Lupo, M.P. 2018. Nutraceuticals: A review. Dermatology and Therapy, 8, 5–16.

[106] Zair, Y., Kasbi-Chadli, F., Housez, B., Pichelin, M., Cazaubiel, M., Raoux, F., et al. 2013. Effect of a high bicarbonate mineral water on fasting and postprandial lipemia in moderately hypercholesterolemic subjects: A pilot study. Lipids in Health and Disease, 12, 1–7.

[107] Pérez-Granados, A.M., Navas-Carretero, S., Schoppen, S. and Vaquero, M.P. 2010. Reduction in cardiovascular risk by sodium-bicarbonated mineral water in moderately hypercholesterolemic young adults. The Journal of Nutritional Biochemistry, 21(10), 948–953.

[108] Schoppen, S., Perez-Granados, A.M., Carbajal, A., Oubina, P., Sanchez-Muniz, F.J., Gomez-Gerique, J.A., et al. 2004. A sodium-rich carbonated mineral water reduces cardiovascular risk in postmenopausal women. The Journal of Nutrition, 134(5), 1058–1063.

[109] O'Connell, B.S. 2001. Select vitamins and minerals in the management of diabetes. Diabetes Spectrum, 14(3), 133–148.

[110] Piccolella, S., Crescente, G., Candela, L. and Pacifico, S. 2019. Nutraceutical polyphenols: New analytical challenges and opportunities. Journal of Pharmaceutical and Biomedical Analysis, 175, 112774.

[111] Sun, W. and Shahrajabian, M.H. 2023. Therapeutic potential of phenolic compounds in medicinal plants – Natural health products for human health. Molecules, 28(4), 1845.

[112] Khan, H., Sureda, A., Belwal, T., Çetinkaya, S., Süntar, İ., Tejada, S., et al. 2019. Polyphenols in the treatment of autoimmune diseases. Autoimmunity Reviews, 18(7), 647–657.

[113] Rana, A., Samtiya, M., Dhewa, T., Mishra, V. and Aluko, R.E. 2022. Health benefits of polyphenols: A concise review. Journal of Food Biochemistry, 46(10), e14264.

[114] Procopio, F.R., Ferraz, M.C., Paulino, B.N., Do Amaral Sobral, P.J. and Hubinger, M.D. 2022. Spice oleoresins as value-added ingredient for food industry: Recent advances and perspectives. Trends in Food Science & Technology, 122, 123–139.

[115] Manasa, V., Chaudhari, S.R. and Tumaney, A.W. 2020. Spice fixed oils as a new source of γ-oryzanol: Nutraceutical characterization of fixed oils from selected spices. RSC Advances, 10(72), 43975–43984.

[116] Pradeep, S.R. and Srinivasan, K. 2019. Synergy among dietary spices in exerting antidiabetic influences. Bioactive Food as Dietary Interventions for Diabetes: Elsevier, 407–424.

[117] Okaiyeto, K., Adeoye, R.I. and Oguntibeju, O.O. 2021. Some common West African spices with antidiabetic potential: A review. Journal of King Saud University-Science, 33(6), 101548.

Galvina Pereira*, Saasha Vinoo, Pranali Yadhav

# Chapter 9
# Multiple gastrointestinal tract disorders

**Abstract:** The gastrointestinal (GI) tract plays a vital role in absorption of essential nutrients and hence gut disorders affect the overall physical and mental well-being of an individual. Disturbance in the normal anatomy and/or physiology of GI tract may lead to number of digestive issues including diarrhea, constipation, dyspepsia, ulcers, Irritable Bowel Syndrome, Crohn's disease, celiac disease, and GI cancer, affecting the quality of life. The intestinal barrier function and microflora composition are responsible to maintain the homeostasis and gut health. Nutraceuticals and functional food ingredients are natural compounds that aid in maintaining gut health, enhancing immunity, and restoring the gut microbiome via array of mechanistic targets. They are easily available and cost-effective, with favorable safety profile. Their antioxidant and anti-inflammatory potential may positively influence gut health. In this chapter we have summarized different classes of nutraceuticals, viz., dietary fibers, prebiotics, probiotics, polyunsaturated fatty acids, polyphenols, spices, and vitamins, with a special focus on the prevention, mitigation, and treatment of certain GI disorders.

## 1 Introduction

Gastrointestinal (GI) disorders are diseases affecting our digestive tract beginning from our mouth right up to the anus, and they currently affect around 40% of the global population. The incidence of GI diseases is on the rise globally, owing to increased consumption of unhealthy food products, lifestyle changes, infections, psychosocial reasons, and increased levels of contaminants in food such as metals, synthetic additives, microplastics, pesticides, etc. The GI tract is generally quiet, discreet, and effective, but at least one in three people experience problems with it. When symptoms are mild, conventional medical treatments are sufficient. However, in a few people, persistent and consistent symptoms, severe enough to be incapacitating, and for which there is no known cause/s develop [1].

*Corresponding author: Galvina Pereira,** Institute of Chemical Technology, Matunga, Mumbai 400016, Maharashtra, India; e-mail: galvinaferr@gmail.com
**Saasha Vinoo,** Research Scholar, Bombay College of Pharmacy, Kalina, Mumbai 400098, Maharashtra, India, e-mail: saasha.vinoo@bcp.edu.in
**Pranali Yadhav,** Research Scholar, Bombay College of Pharmacy, Kalina, Mumbai 400098, Maharashtra, India, e-mail: parnali.yadav@bcp.edu.in

https://doi.org/10.1515/9783111317601-009

GI disorders have a strong association with emotional state, as we say, "gut feeling." Our gut is significantly influenced by diet. The GI system is significantly impacted by nutrients, which also controls mucosal growth and turnover [2]. It is crucial for maintaining the normal function, microbial composition, and their metabolites in the gut. Gut immunity, mucosal barrier function, and gut inflammation are all regulated by dietary contents [3]. Vast research has been done on the impact of various food substances on the development of chronic GI disorders. It has been shown that, depending on the geography and the type of prevailing food in a specific environment, the epidemiology of some diseases might differ significantly [4]. Apart from diet, several medicaments that are consumed on a daily basis by patients suffering from chronic illnesses also impact gut health. Some of these drugs have tendency to alter the function of GIT leading to diseases state.

The intestinal barrier plays a vital role in the absorption of nutrients and the maintenance of gut health. The food we consume contains billions of bacteria and pathogens, where the GI barrier acts as a protective layer and prevents them from entering the systemic circulation. The intestinal barrier principally consists of the mucosal layer, gut epithelium, gut vasculature, and lamina propria region. The innermost mucosal layer comprised of mucus is secreted by the goblet cells located in the gut epithelium and is made up of highly glycosylated proteins (Figure 1). The mucus layer is much thicker in the colonic region (800 microns) and can be divided into two layers. The inner layer is firmly adherent and contains antibacterial peptides such as defensins and lysozyme. The outer layer is loosely adherent and is rich in gut microbiota. The gut microbiota plays a vital role in gut immunity and in maintaining intestinal homeostasis. Gut integrity is maintained by a single layer of columnar intestinal epithelium, which covers an area of 400 m$^2$. The gut epithelium consists of Paneth cells, enteric cells, goblet cells, and dendritic ends of the neurons. The Paneth cells in the gut epithelium secrete antimicrobial peptides, lysozymes, and mucin. The intestinal epithelial cells are firmly attached by tight junctions [5]. The tight junctions, which seal the gut epithelium, consist of several transmembrane proteins such as occludin, claudins, and junctional adhesion molecules [6]. The GI barrier function is found to be compromised in a few disease conditions. The tissue side of the intestinal barrier is innervated by enteric neurons. The lamina propria is located below the basement membrane of the intestinal epithelium and is involved in providing innate and acquired immune responses. This layer is rich in cytokines, chemokines, mast cells, and IgA. The GI system is associated with the largest network of lymphoid tissue containing mature lymphocytes. The Gut Associated Lymphoid Tissue (GALT), which lies in the lamina propria region comprises of Peyer's patches, mesenteric lymph nodes, lymphoid follicles, and interepithelial lymphocytes that play a major role in immunity and defend the gut from pathogens [7]. As the food exposes our gut to a large number of antigens, the GIT provides the major site for lymphocyte contact with these antigens, thus providing immunity [8].

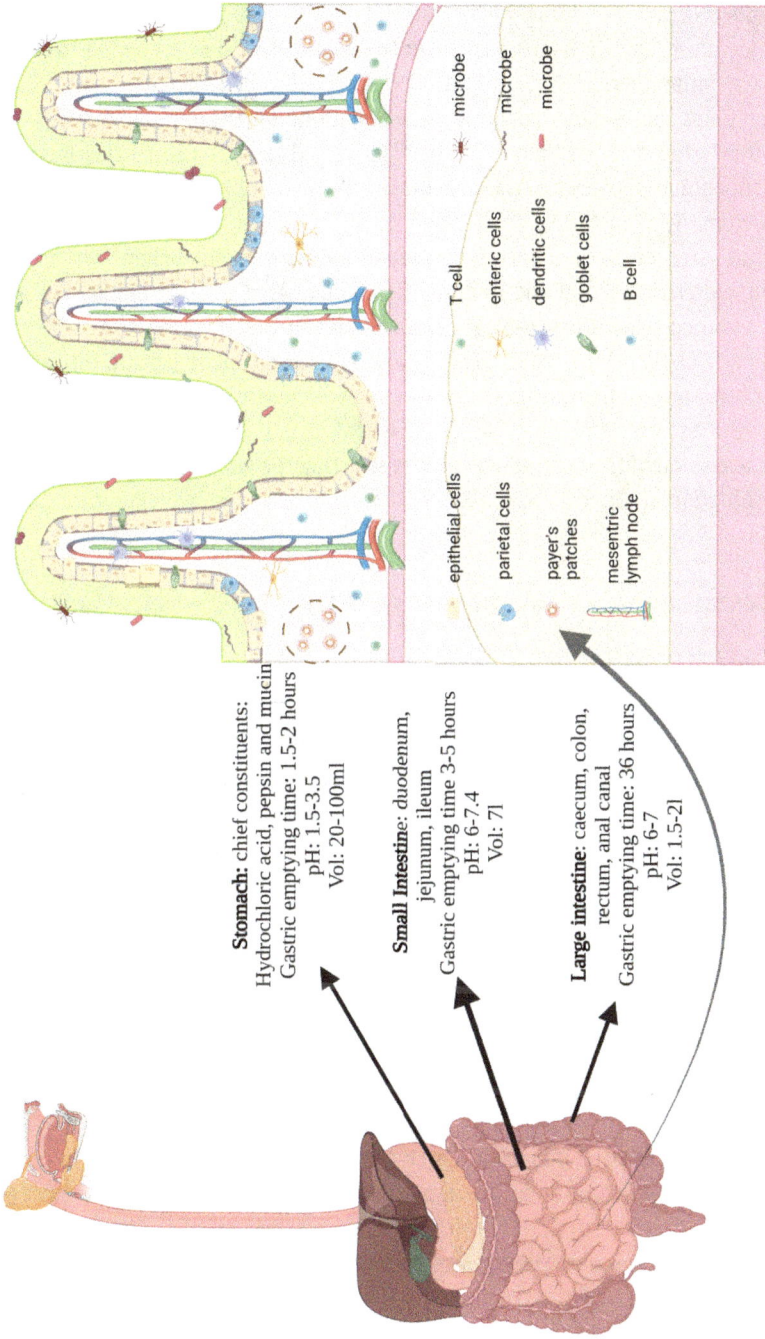

**Stomach:** chief constituents:
Hydrochloric acid, pepsin and mucin
Gastric emptying time: 1.5-2 hours
pH: 1.5-3.5
Vol: 20-100ml

**Small Intestine:** duodenum,
jejunum, ileum
Gastric emptying time 3-5 hours
pH: 6-7.4
Vol: 7l

**Large intestine:** caecum, colon,
rectum, anal canal
Gastric emptying time: 36 hours
pH: 6-7
Vol: 1.5-2l

epithelial cells

parietal cells

payer's
patches

mesentric
lymph node

T cell

enteric cells

dendritic cells

goblet cells

B cell

microbe

microbe

microbe

**Figure 1:** The gastrointestinal tract showing detailed structure of gut barrier.

GI disorders often dictate the absorption of nutrients. A diseased gut may lead to nutritional deficiency disorders. GI disorders disrupt the barrier function of the gut leading to infections and inflammation. Gut health is also reported to be associated with disorders such as anxiety, depression, Alzheimer's disease, multiple sclerosis, and Parkinsonism. Overall, they affect the quality of life. The disorders of the GIT can be classified as structural or functional depending upon the underlying conditions. The functional disorders as per Rome IV convention are defined as "morphologic and physiological abnormalities that often occur in combination including motility disturbance, visceral hypersensitivity, altered mucosal and immune function, altered gut microbiota, and altered central nervous system processing" [9]. Examples of functional GI disorders include irritable bowel syndrome (IBS), constipation, and diarrhea. Structural disorders are those that arise out of deformities in the GI tract which can be physically seen and assessed. This includes Inflammatory Bowel Disease (IBD), colon polyps, and cancer. The following are some of the common disorders affecting the global population, which require a combination of different approaches for treatment and management.

## 1.1 Gastroesophageal reflux disease (GERD)

Gastroesophageal Reflux Disease (GERD) is a diseased condition in which the contents of the stomach often regurgitate back into the esophagus, maybe because of the weakened muscles below the esophageal sphincter. The contents which regurgitate may or may not be acidic, but generally the acidic contents are responsible for the symptoms such as heartburn, irritation, chest pain, etc. These symptoms may increase at night after lying down or after consumption of a heavy meal [10]. The primary treatment is the use of antacids, proton pump inhibitors, and histamine (H2) blockers. The incidence of GERD is high in patients who are overweight, alcoholics, people who smoke, pregnant women, in certain medical conditions, and in people who consume drugs such as corticosteroids, NSAIDs, or methotrexate [11].

## 1.2 Dyspepsia

Dyspepsia, often referred to as indigestion is characterized by symptoms such as pain, discomfort, and burning sensation in the upper part of the upper abdomen. It is prevalent in 20–25% of the western population. Dyspepsia can be classified into two categories, viz., structural and functional. Organic causes include peptic ulcer, GERD, cancers of the GI system, and pancreatic and biliary disorders. Functional disorders are associated with infections, inflammation, hypersensitivity, psychosocial factors, and autonomic central nervous system dysregulation. Symptoms of dyspepsia include delayed gastric emptying, impaired gastric accommodation, and hypersensitivity to gastric distention [12]. Management strategies for this condition depend upon the cause and include consumption of acid-suppressive drugs, eradication of infections especially, *H. pylori*, prokinetic agents, fundus-relaxing drugs, antidepressants, and psychological interventions [13].

## 1.3 Diarrhea

Diarrhea is characterized by a rapid bowel movement and watery stools, and it accounts for 1.3 million deaths annually. This may also be accompanied by other symptoms such as abdominal pain, cramping, nausea, or vomiting. Acute diarrhea is caused by the consumption of certain drugs, food intolerance, certain medical conditions, and infections of the GI system generally by bacteria, viruses, or intestinal parasites. These include bacterial infections caused by *E. coli, Salmonella, Campylobacter, Shigella, Clostridium,* and viruses such as adenovirus, astrovirus, and sapovirus [14]. The symptoms of diarrhea can be managed by many over-the-counter drugs such as loperamide and bismuth subsalicylate. However, they might not treat the underlying cause.

## 1.4 Constipation

Constipation is characterized by altered gastrointestinal motility, which affects around 11–20% of the adult population annually [15]. Constipation is diagnosed by a decrease in the frequency of defecation, dry fecal knot, or need for excessive straining or defecation insufficiency. Constipation might cause abdominal pain, discomfort, bloating, fecal impaction, intestinal obstruction, nausea, vomiting, and blood in stools. Generally, the coordination of several factors is responsible for the smooth functioning of the peristaltic movement. These include gut microbiota and their metabolites, digestive enzymes, mucus secretion, bile acids, enteric nervous system, and immune system. The enteric nervous system maintains gut motility via the transmission of myogenic and neurogenic signals across the gut, which is controlled by the CNS. The imbalance of any of these factors might result in constipation [16].

Constipation can either be organic or functional in nature. Organic constipation generally arises out of a physical cause which might be structural or metabolic, arising out of a deficiency of a particular enzyme, hormone, or neurological condition. Functional constipation is the most common type of constipation, with no physical cause and can be further divided into three categories.

1. Normal transit constipation, when the muscles show incorrect contraction and relaxation during peristalsis.
2. Slow transit constipation, a delay in colonic transit time due to neural colonic motor irregularities.
3. Defecation disorders, when the muscle/s of the colon become dysfunctional.

The current therapy for constipation includes laxatives, suppositories, and enemas. A range of potential surgical and nonsurgical treatment options exist for malignant bowel obstruction, aiming to reduce symptom burden and improve quality of life [17]. Patients suffering from constipation are often dissatisfied with current therapy because of its ineffectiveness, diarrhea, drug dependence, and discomfort. Complementary techniques such as abdominal massage in geriatric patients were found to significantly reduce the severity of constipation. A diet rich in fibers and electrolytes is often recommended to reduce constipation; however, the average dietary intake does not suffice this need and external supplementation with nutraceuticals is recommended [18].

## 1.5 Irritable bowel syndrome

Irritable Bowel Syndrome (IBS) is a disorder of gut–brain interaction affecting around 7–18% of the world population [19]. It is characterized by altered gastric motility, altered mucosal and immune function, altered gut microbiota, and enhanced visceral hypersensitivity. The enhanced GI sensitivity may be attributed to peripheral and/or central sensitization. Rome IV criteria (2016), the gold standard for IBS diagnosis includes "the presence of persistent recurrent bouts of abdominal pain and altered bowel habits for at least three months" to differentiate it from other GI disorders. The incidence of this condition is more prevalent in women and teenagers. IBS is divided into four categories depending upon the prevalence of predominant symptoms, especially the stool pattern, viz., IBS-D (diarrhea), IBS-C (constipation), IBS-M (mixed – diarrhea and constipation), and IBS-U (stool pattern cannot be defined by the above three classes).

Common triggers for IBS include food and stress. Studies have revealed a possible association between IBS and psychological problems, such as anxiety and depression. Up to 90% of patients suffering from IBS tend to exclude certain food to alleviate the symptoms of IBS [20]. Peripheral pathophysiologic mechanisms in IBS include alterations in neuronal function, luminal and tissue mediators, immune response, intestinal permeability, bile acid processing, serotonin signaling, and gut microbiota [21]. IBS is often treated with drugs that alleviate the symptoms of diarrhea, constipation, dietary modification to decrease the triggering food ingredients, and behavioral therapy [22].

## 1.6 Inflammatory bowel disease

Inflammatory Bowel Disease (IBD) is a chronic inflammatory bowel disorder resulting in morbidity. The gut epithelial barrier function is compromised in this condition owing to epithelial damage, dysfunctional immune response, and dysbiosis. Specific segments of the GIT undergo inflammation. Loss of integrity of the intestinal epithelial barrier leads to increased permeability. The inflammatory infiltration leads to changes in the gut microenvironment, the nature of defensive peptides secreted, and the composition of gut microbiota [23]. IBD can be classified into three categories Crohn's disease, ulcerative colitis, and unclassified IBD. Crohn's disease involves granulomatous inflammatory reactions in different parts of the gut right from the buccal mucosa to the stomach. Ulcerative colitis (UC) is a nonspecific inflammatory response associated with the colon. The symptom of this condition includes rectal urgency, pain in the GIT, and the presence of blood in the stool. The immune cells such as B-cells proliferate into the inflamed region in UC leading to an abnormal humoral response to antibodies [24,25]. The current therapies for IBD include anti-inflammatory and immunosuppressive drugs that provide only symptomatic relief and there is a need for drugs that repair the inflamed epithelium and homeostasis. Surgical treatments are also associated with numerous complications and lead to discomfort for the patients [5].

## 1.7 GI cancers

GI cancers are the second most prevalent cause of cancer-related death and the fourth most in terms of incidences of new cases the world over. The incidence is high in subjects aged 30–80 years and is more prevalent in males. The mortality-to-incidence ratio of 0.75 indicates a bad prognosis and poor survival rate. GI cancers may occur in all organs of the digestive system, which includes the stomach, intestine, colon, rectum, liver, and pancreas [26], and could be classified as follows:

a) **Gastrointestinal stromal tumors:** These are rare, generally benign in the stomach, originate in stomach epithelial cells, and have a higher propensity to spread in different regions of the body.

b) **Neuroendocrine tumors (including carcinoids):** This type of tumor grows from neuroendocrine cells, is generally benign, and causes an imbalance of hormones.

c) **Lymphomas:** These involve lymphocytes located principally in the lymphatic tissues. The incidence of this disease is highest in the stomach, followed by the small intestine and ileocecal region. Diffuse large B-cell lymphoma is the most common of all types of GI lymphomas [27].

d) **Adenocarcinomas:** Adenocarcinomas of the GIT start with mucus-producing cells and is the most prevalent form of GI cancer. Depending upon its location it is divided into two categories: cardiac gastric cancer when it occurs within 5 cm of the proximal stomach and noncardiac gastric cancer when it occurs in parts of the intestine and colon [28].

e) **Other cancers:** The stomach can also be the site of other cancers, such as squamous cell carcinomas, small cell carcinomas, and leiomyosarcomas; however, these tumors are not so common.

*H. pylori* infection, smoking, alcohol, genetics, obesity, and red meat are recognized as significant risk factors responsible for the initiation and progression of GI cancers. Diet, geography, age, and sex also have an important association with the incidence of cancer. Patients with GI cancers are generally asymptomatic until there is metastasis. The symptoms include bloating, discomfort, and pain. The gastric obstruction leads to weight loss, ascites, and bleeding [29]. Genetic instability is seen in all subtypes of cancer [30]. The treatment involves surgery to remove the infected part, especially by using laparoscopy, chemotherapy using specific drugs, radiation, and immunotherapies. These treatments may be used individually or one after the other [31]. These therapies are associated with many adverse effects because of their action on healthy cells. The side effects of surgery include pain, heartburn, nausea, etc., whereas those with radiation therapy include fatigue, intestinal bleeding, and loose stools. The adverse effects of chemotherapy include bleeding, hair loss, anorexia, low blood counts, and loss of appetite. The adverse effects of monoclonal antibodies include nausea, diarrhea, fever, and chills. Some of these effects can be reduced by using targeted therapy.

## 1.8 Celiac disease

Celiac disease (CD) also called gluten-sensitive enteropathy of the intestine, is triggered after the consumption of foods containing gluten in certain genetically predisposed populations. It is most prevalent in European and South American populations. It is more common in subjects in whom genes HLA-DQ2 (90% frequency) and HLA-DQ8 (5%–10%) are most closely connected to class II human leukocyte antigens (HLA-II). Certain foods such as wheat, triticale, rye, farina, and barley are rich in gluten, and thus possess a potential risk factor for the pathogenesis of the disease. The gliadin fragment of gluten makes it through the enterocytes by breaking the tight junctions into the lamina propria region where they undergo deamination by transglutaminase. These deaminated gliadin molecules then link specifically to HLA-DQ2 and HLA-D8 molecules. The gliadin-HLA complexes are then detected by specific T-cell receptors initiating the immune response and cytotoxic activity leading to damage to the intestinal mucosa [32]. Duodenal villous atrophy along with antibodies against tissue transglutaminase is diagnosed in people suffering from CD. Symptoms of CD include diarrhea, abdominal pain, anemia, malabsorption, osteoporosis, and neuropathy, though the disease may be asymptomatic in certain cases. Often the patients diagnosed with this condition are advised to avoid foods containing gluten, as intake of food containing gluten could delay gastric healing. The symptoms of CD such as mucosal lesions improved within two weeks after consumption of gluten-free diet, while the GI inflammation may take 3–12 months to improve [33,34].

## 1.9 Peptic ulcer

Peptic ulcer is characterized by lesions observed in the lining of the stomach or the intestinal tract (especially proximal duodenum) either because of gastric acid secretion or the activity of the enzyme pepsin. These lesions penetrate deeply to the muscularis propria layer. Depending upon the location of the ulcer the symptoms include epigastric pain 15–30 min after consumption of a meal in case of gut ulcer or 2–3 h after meal in case of duodenal ulcer. In addition, duodenal ulcers are usually associated with night-time pain. The other symptoms associated with this condition include gastric pain, hematemesis melena, bloating burning sensation, nausea, vomiting, and weight loss. *H. pylori* infection and certain medications such as NSAIDs, chemotherapy, and antidepressants are often reported to cause peptic ulcers. *H. pylori* are found to be associated with almost 90% of peptic ulcer cases and has the ability to strongly adhere to inflamed mucosa. It secretes the enzyme urease, which converts the urea into ammonia thereby raising the pH of the stomach and intestinal lining. The surface antigens bring about the disruption of the gut mucosal lining. The NSAIDs inhibit the enzyme cyclooxygenase-1 (COX-1) and block the synthesis of prostaglandins, which is responsible for protecting the gastric mucosa [35].

Other factors that contribute to peptic ulcers include Zollinger-Ellison 1 Syndrome, Crohn's disease, and cancers of the GI tract. The therapy often prescribed for peptic ulcer disease includes proton pump inhibitors and H2 blockers, which provide antisecretory action thereby providing symptomatic treatment. Antacids are commonly used to neutralize the burning sensation of stomach contents and antibiotics are used to counter *H. pylori* infection. Sucralfate and misoprostol have also been used in the management of peptic ulcers with very limited success. If the patient is unresponsive to the above therapy or if the ulcer size is greater than 5 mm, surgical intervention is required.

# 2 Nutraceuticals in management of GI disorders

Nutraceuticals are looked upon as a safe and economical alternative for people suffering from chronic illness. The market for nutraceutical ingredients is on the rise as they help in the management of disease conditions with minimal adverse effects and a wider margin of safety. It allows people to take control of their health. Many of these nutraceuticals act as preventive agents and delay the onset of certain disorders such as diabetes, cancer, and arthritis. They may also help in dietary deficiencies and in maintaining the homeostasis in diseased conditions such as endocrine disorders. Nutraceuticals and functional food ingredients used in GI disorders often include phytochemicals possessing antioxidant and anti-inflammatory activities as well as beneficial effects on bowel movement [36].

## 2.1 Dietary fibers

Dietary fibers are generally plant-derived polymers mainly belonging to the carbohydrates class and associated molecules, which cannot be digested by humans. They generally occur in the cell walls of plants and tissues. Depending upon their solubility in water they can be classified as soluble dietary fibers (SDFs) and insoluble dietary fibers (IDFs). The proportions of SDF and IDF vary from plant to plant. Although DFs are generally obtained from plant sources, certain fibers such as xanthan and gellan have been sourced from microorganisms.

Soluble dietary fibers are freely dispersible in the gut fluids and form a viscous gel like mush. They imbibe water, swell, and provide a sense of satiety and fullness. The SDFs because of their viscosity are known to decrease the enzymatic diffusion, slow the rate of digestion, and retard the absorption of nutrients. This may be beneficial as they retard the absorption of sugars and lipids in diabetic patients. They also bind to cholesterol bile complex and provide hypercholesteremic effect. The SDFs act as bulking agents, soften the stool, and provide relief in constipation. They are fermented by the gut microflora into short-chain fatty acids (SCFAs) such as butyrate, propionate, and acetate, which reduce the pH of intestines. A few examples of SDFs, which are commercially used as nutraceuticals include pectin, xanthan, hemicellulose, β-glucans, and oligosaccharides.

Insoluble dietary fibers include indigestible carbohydrates such as cellulose, hemicellulose, chitosan, and lignin, which do not form viscous gel when in contact with water. They have a relatively compact structure, their water-holding capacity is poorer compared to soluble fibers, and hence they do not swell to form a mush. They play a vital role in gut motility [37]. They show a slow and incomplete fermentation leading to increased peristalsis and reduced fecal time. Given below is the brief description of some common dietary fibers with nutraceutical potential in GI health and disease.

### 2.1.1 Pectin

Pectin is SDF and is generally found in fruits such as apple, citrus, sugar beet pulp, plums, and numerous fruits, playing an important role in maintaining the shape of fruits. It is biosynthesized in raw fruits as protopectin, an insoluble fiber, which then gets converted to soluble pectin as the fruit ripens. Pectin gets hydrolyzed into component sugars making the fruit softer. It is a high molecular weight polymer (50–150 kDa) made up of few hundreds to thousands of residues of galacturonic acid linked via α-1,4 glycosidic bonds. In pectin molecule the hydroxyl group of galacturonic acid shows methyl esterification [38]; depending upon the nature of esterification the pectin can be divided into high methoxyl pectin (> 50% esterified galacturonic acid residues) and low methoxyl pectin (< 50% esterified galacturonic acid residues).

### 2.1.2 Xanthan

It is a high molecular weight polysaccharide obtained from a gram-negative bacterium *Xanthomonas campestris* by fermentation of sugars. The xanthan polymer is made up of D-glucose units with a trisaccharide side chain branch made up of two mannoses separated by guluronic acid. The polymer is completely soluble in cold water yielding a clear viscous gel [39].

### 2.1.3 Hemicellulose

It is a heteropolysaccharide present in plants as a cementing material for cellulose fibrils. It comprises of almost 30% of the mass of the plant cell wall and consists of about 100–200 sugar units. In comparison to cellulose, it is a short, branched polymer, which possesses substitution at multiple sites. The substituents include acetyl, ferulic, and/or coumaric acid moieties. The substituent sugars present in the branching site vary depending upon its source and nature of the plant. The substituent sugars include hexoses such as glucose, galactose, mannose, and pentoses such as arabinose and xylose. The hardwood hemicellulose is rich in pentoses such as xylose whereas the soft woods predominantly contain hexoses such as glucose, mannose, and galactose [40]. Hemicelluloses can be classified as xylans, xyloglucans, and heteromannans depending upon their constituent sugars.

Xylan is made up of D-xylose sugar backbone, the residues of which are linked by β- 1,4 linkage. The polymer may show branching with α- 1,2-linked D-glucuronic acid or D- glucuronic acid with methyl substitution at O-4 position. D- arbinofuranose may be attached at α- 1,3, or α- 1,2 and the D-xylose residues in backbone may be acetylated. The arabinofuranose may show ferulic acid esters at O-5 [41]. Xyloglucan is a linear polysaccharide made up of β-1,4 -linked D- glucan substituted with xylose. The pattern of branching varies depending upon the plant species. It is abundantly found in tamarind, impatiens, and Annona [42]. Heteromannans are divided into four categories depending upon the constituent sugars: mannans, galactomannans, glucomannans, and galactoglucomannans. Of these mannans contain β- 1,4-linked mannose whereas glucomannans and galactoglucomannans contain both mannose and glucose units attached by β-1,4-linkage in their backbone. In galactomannan and galactoglucomannan, the galactosyl residue is linked by α-1,6 linkage [43]. Galactomannans are abundantly found in guar gum, locust bean gum, fenugreek gum, and tara gum [42].

### 2.1.4 Fructans

Fructans are polysaccharides made up of about four to several hundred fructose units, produced by plants and certain microorganisms. Depending upon their structure the fructans can be classified as inulins, levans, and gramminans. Inulins consist of either one or two β-1,2-linked fructose attached to sucrose chain. Their degree of polymerization lies between 10 and 20. It is partially degraded in the large intestine into short-chain fatty acids and lactate. They are found in yacon (*Smallanthus sonchifolius*), garlic, Chinese garlic (*Allium sativum*), barley (*Hordeum vulgare*), sweet leaf (*Stevia rebaudiana*), Jerusalem artichoke (*Helianthus tuberosus*), chicory (*Cichorium intybus*), asparagus (*Asparagus sp.*), agave (*Agave sp.*) dandelion (*Taraxacum officinale*), burdock (*Arctium sp.*) suma (*Pfalia glomerate*), and onions (*Allium cepa*) [44]. Levans consist of β-2,6-linked fructose attached to the sucrose chain and are generally found in grasses. Gramminans are mixed fructans consisting of β-2,6-linked fructose residues with β-1,2 branching [45]. The β- configuration of these sugars makes them resistant to our digestive enzymes. Fructans are used as emulsifiers in the food industry and show beneficial effect as prebiotic.

### 2.1.5 β-glucans

It is polysaccharide found in cell walls of yeast, algae, mushrooms, barley, and wheat. The solubility of β-glucans in water depends on the size and branching of polymer. Both water-soluble and insoluble fractions can be extracted by manipulating the extraction conditions. The β-glucans obtained from bacteria and algae are unbranched and are linked via β-1- > 3 glycosidic linkage. Those obtained from yeast, mushroom, oats, and barley show branching. The monomers are linked via β-1- > 3 and β-1- > 6 glycosidic linkage in yeast and mushrooms and via β-1- > 3 and β-1- > 4 glycosidic linkage in oats and barley [46]. The polysaccharides containing higher proportions of β-1- > 4 linkages and high molecular weight are insoluble in water whereas those possessing higher β-1- > 3 glycosidic linkage and low molecular weight are classified under SDF [47].

### 2.1.6 Cellulose

Cellulose is a homopolysaccharide most abundantly found in plants. It is made up of thousands of glucose residues joined by β-1- > 4 glycosidic linkage. Cellulose cannot be digested by enzymes secreted in our gut. Cellulose is also produced by certain bacteria such as *Acetobacter xylinum*. As compared to the plant cellulose, which shows water retention values of 60%, bacterial cellulose can show retention values as high as 1,000% by weight [46].

### 2.1.7 Lignin

Unlike other previously discussed fibers, lignin is not a polysaccharide. It is a hetero-polymer derived from phenylpropanoid building blocks such as coniferyl, sinapyl, and p-coumaryl alcohol linked by ester or carbon-carbon linkage. The degree of poly-merization and methoxy substitution vary from plant to plant.

Dietary fibers are commonly found in peels of fruits, bran, and seeds. The chemi-cal nature of the fiber affects its solubility, nutritional value, viscosity, and fermenta-tion rate in the gut. The stages of maturity of plant and ripening of fruits have effects on their fiber content [48]. The SDFs, owing to their high viscosity and gel-forming ability act as excellent emulsifiers and reduce the absorption of cholesterol [49]. High-fiber diets are often advocated and proven beneficial in conditions such as diabetes, cancer, IBD, and constipation. Dietary fibers provide the gut microbiota with source of food and regulate their composition and population. Rice bran dietary fiber con-tains about 90% IDFs composed of cellulose, hemicellulose, and insoluble pectin, while a small percent of SDFs composed of hemicellulose and pectin [50]. *Psyllium* seed husk is rich in SDF made up of highly branched arabinoxylan and shows a slower fermentation rate compared to cereal arabinoxylan due to the special struc-ture of its backbone [47]. Fibers are used as food additives to enhance texture and taste [50]. The recommended dose of dietary fiber in a well-balanced diet is 28–35 g for adults. High fiber diet has shown to increase the levels of SCFAs and microbiome alpha diversity [51]. Numerous techniques have been reported for extraction and processing of DFs. These include acid and enzymatic hydrolysis, microbial fermenta-tion, radiation, steam explosion, and ultrasound treatment. These treatments are known to alter or enhance the water-holding, oil-holding, or swelling capacity, solubil-ity, and H-bond formation within the polymer [23].

## 2.2 Prebiotics

Prebiotics are the components of food responsible for providing nutrition to the gut microbiota, which in turn produce metabolites that show positive effect on gut health. Prebiotics selectively stimulate the growth of certain bacteria found in gut. Many of the abovementioned dietary fibers act as prebiotics since they are resistant to gastric pH and not digestible by enzymes secreted in the gut but can be metabolized by the gut microflora. The carbohydrate prebiotic fibers generally show degree of polymeri-zation ≥ 3.

The prebiotics, on digestion, are converted to bioactive metabolites such as buty-rate, propionate and acetate, which possess beneficial effects for gut health. The pre-biotics act by promoting satiety, creating hostile environment for the pathogenic organisms, and maintaining the gut barrier function. Prebiotics are also reported to enhance bowel movement by acting as bulking and stool softening agents in case of water-soluble fibers and increased peristalsis in case of IDF. Prebiotics also possess immunomodulatory activities, mainly through the regulation of the growth and/or composition of gut microbiota. The commonly used prebiotics are dietary fibers, such as fructo-oligosaccharides, galacto-oligosaccharides, transgalacto-oligosaccharides, lactulose, resistant starch, and poly dextrose. Their regular consumption may alleviate the symptoms of IBS and Crohn's disease and decrease the incidences of colorectal cancer. In addition, they are also reported to improve cognition and decrease sugar, lipid, and cholesterol absorption [52].

## 2.3 Probiotics

The group of microorganisms that thrive in our digestive tract is called "gut microbiota," which is composed of bacteria, fungi, viruses, and intestinal parasites. The human gut may contain about $3.8 \times 10^{13}$ microorganisms at a given time. The gut composition is almost uniform in most individuals but may show a slight variability depending upon the environment, age, health, or use of medicines [53]. They are responsible for maintaining the intestinal homeostasis. They play a vital role in the maintaining the integrity of GIT. Bacteria thriving in the gut are responsible for producing neurotransmitters, enzymes, vitamins, antibiotic compounds, antimicrobial peptides, and certain important metabolites responsible for absorption of vitamins and minerals. Several factors are found to responsible for bringing about imbalance in gut microbiota leading to disease state. These factors include diet, surgery, psychosocial state, consumption of certain medicines like antibiotics, and presence of other disease conditions. In such cases restoration of gut microbiota by external supplementation can help regulate the gut and alleviate the symptoms resulting out of the same.

Probiotics are nonpathogenic living microorganisms, which when administered in sufficient quantity, aid in restoration of gut microflora. The common probiotics include fermented foods such as yogurt, buttermilk, kefir, sauerkraut, pickles, tempeh, natto, kimchi, miso, kombucha, etc. Of these, yogurt holds predominance in the probiotic market due to its high acceptability and availability globally. Once ingested orally the microorganisms must survive the harsh acidic conditions of the stomach and assault by digestive enzymes to make their way into the intestines. To do that, the "therapeutic minimum" recommended dose of probiotics by US FDA is $10^6$ CFU/mL. The strain of the probiotic organism and the vehicle for oral delivery are deciding factors that influence the dose of probiotic organism. Recently, several novel approaches have been devised to protect the probiotic organisms from the effect of digestive enzymes and pH of stomach. These formulation and processing techniques include inducing sporulation, encapsulation, emulsification, lyophilization, and spray-drying with a polymer matrix [54]. Such formulation techniques may protect them from the stressful environmental conditions and release the probiotic organism in favorable conditions. The small fraction of bacteria that survives and reaches the small intestine then colonizes the intestine, bringing beneficial effects. The probiotic organisms have the ability to kill harmful bacteria and digest certain types of food such as dietary fibers to produce useful metabolites and boost immunity. The probiotic bacteria produce SCFAs such as acetic, lactic, and propionic acid, which are responsible for decrease in luminal pH facilitating the absorption of vital nutrients. Probiotics are also responsible for cell-mediated immune response. Para-probiotics are inactivated microorganisms or their fractions, which confer health benefits. The inactivation techniques include heat, gamma irradiation, exposure to extreme pH, ultrasound, and supercritical carbon dioxide [55]. The probiotic strains should meet a few criteria before use in human subjects: (i) they should be a component of gut microbiota; (ii) they should be nonpathogenic and safe for oral consumption; (iii) they should be stable to gastric pH, digestive enzymes, bile, and gases present in gut; and (iv) they should produce beneficial effect on gut health.

*Lactobacilli* are the most abundantly used class of probiotics, most commonly found in the mouth, gut, and genitourinary system and are involved in digestion of sugars into lactose and lactic acid. They also metabolize lactose in our diet and benefit the lactose intolerant population. *Lactobacilli* grow between 2–53 °C with an optimum growth at 30–40 °C at a pH between 5.5 and 6.2. They produce numerous enzymes such as lactase, proteases, peptidases, fructanases, amylases, bile salt hydrolases, phytases, and esterase and aid in digestion [56]. The common sources of *Lactobacilli* include dairy products such as yoghurt, cottage cheese and fermented milk, and non-dairy products such as sourdough bread, kombucha, kimchi, and sauerkraut [57]. Due to their abundance in food and dairy products the *Lactobacilli* are regarded as safe for human consumption (GRAS) by US FDA. Some of the commercially available strains include *L. amylovorous, L. acidophilus, L.casei L. johnsonoo, L.johnsonii L. crispatus, L. paracasei, L. casei, L. delbrueckii, L. gallinarum, L. gasseri, L. rhamnosus L. plantarum,* and *L. reuteri.*

*Bifidobacterium* species are gram-negative, V-type or Y-type, nonspore-forming, anaerobic, and nonmotile organisms found in the gut and genitourinary tract. The genus *Bifidobacterium* comprises of more than 90 species of which the following are been used commercially: *B. adolescentis, B. animalis, B. lactis B. longum, B.breve B. bifidum, B. infantis, B. thermophilum, and B. paseudocatenulatum.* The human strain grows optimally between 36–38 °C and a pH range of 6.5–7. *Bifidobacteria* metabolizes linoleic acid and some other unsaturated fatty acids to produce conjugated linolic acid (CLA), where CLA and its derivatives are reported for its anti-infective, anti-inflammatory, and anticancer activity. *Bifidobacteria* also aids in saccharides metabolism to produce short- and medium-chain fatty acids (SCFAs and MCFAs) and fatty acid ethyl esters. The enzyme fructokinase produced by *Bifidobacteria* strains metabolizes fructose in GI tract, as high levels of fructose are correlated to gut inflammation. *Bifidobacteria* can also metabolize lactose and hence are beneficial in subjects with lactose intolerance.

*Enterococcus* species are gram-positive, catalase-negative cocci, which are normal inhabitants of the human gut. They grow in the temperature range of 10–45 °C making them one of the most thermostable nonspore-forming bacteria. They are also used in starter cultures for preparation of cheese. Enterococcus consists of a number of species; however only a few show probiotic potential, viz., *E. faecalis, E. faecium, E. lactis, E. durans, E. avium,* and *E. hirae.* Of these the most commercially used ones are *E. faecalis* and *E. Faecium* [58]. They show the ability to produce biofilms and adhere to intestinal epithelium. *Enterococcus* produces inhibitory compounds called enterocins, which belong to Class II bacteriocins that help in counteracting harmful pathogens such as *E. coli, Salmonella,* and *Shigella. Enterococcus* also produces folate, an essential vitamin required for cell multiplication.

Two important species of yeast *Saccharomyces,* which have been commercially used as probiotics are *S. cervisiae* and *S. boulardii.* Of these two strains *S. boulardii* is the commercially favored yeast due to its attributes such as optimum growth temperature, i.e., 37 °C and viability even at a lower pH value. It was found that in patients with antibiotic therapy, which have disturbed microbiota, the levels of *S. boulardii* are high [59]. *Escherichia coli* (Nissle, 1917) is a gram-negative nonpathogenic probiotic bacterium, which is widely studied for its beneficial effects in GI Disorders such as diarrhea, diverticular disease, and UC [60].

Table 1 enlisting the most common probiotic strains and their benefits to GI health.

**Table 1:** Probiotic strains with their major role in GI health.

| Probiotics | Sources | Strains | Major role | References |
|---|---|---|---|---|
| *Lactobacillus* | Fermented millets, yoghurt, cottage cheese, fermented milk, sourdough bread, kombucha, kimchi, sauerkraut | *L. amylovorous, L. acidophilus, L.casei L. johnsonoo, L.johnsonii L. crispatus, L. paracasei, L. casei, L. delbrueckii, L. gallinarum, L. gasseri, L. rhamnosus L. plantarum, L. reuteri* | Conversion of sugars to lactic acid and lactic acid derivatives, production of SCFA | [61] |
| *Bifidobacteria* | Fermented dairy products | *B. adolescentis, B. animalis, B. lactis B. longum, B. breve, B. bifidum, B. infantis, B. thermophilum, B. paseudocatenulatum* | Carbohydrate metabolism, restoration of gut microbiome, prevention of pathogenic infections and diarrhea, beneficial in lactose intolerance | [62] |

**Table 1** (continued)

| Probiotics | Sources | Strains | Major role | References |
|---|---|---|---|---|
| *Enterococcus* | Human gut | *Ent. faecalis, Ent. Faecium, Ent. lactis, Ent. durans Ent. avium, Ent. hirae* | Production of Bacteriocin anti-infective, antioxidant, production of folate | [63] |
| *Saccharomyces* | Fermented food, viz., rice | *Saccharomyses cervisiae* and *Saccharomyses boulardii* | Fermentation of sugars, prophylaxis and treatment of diarrhea, anti-infective against pathogenic microorganisms (*C. difficile*), modulation of immune response | [64] |
| *Escherichia coli* (Nissle, 1917) | Human gut | *Escherichia coli* (Nissle, 1917) | Defense against pathogenic bacteria, diarrhea, IBD, Crohn's disease and UC, antagonizes pathogenic bacteria, secretion of immune factors, anti-inflammatory | [65] |

## 2.4 Polyunsaturated fatty acids (PUFAs)

Polyunsaturated fatty acids are fatty acid derivatives with more than one double bond. PUFAs come in multifarious forms, including *n*-3, i.e., alpha-linolenic acid (ALA), eicosapentaenoic acid (EPA), and docosahexaenoic acid (DHA) and *n*-6 fatty acids, i.e., linolenic acid (LA) and arachidonic acid (AA). *n*-3 PUFAs have important roles in the lungs, immune system, heart, blood vessels, as well as energy production. PUFAs can be obtained from plant sources such as linseed, sunflower, walnuts, and soya bean or from marine sources such as cod, shark, salmon, and mackerel. Fatty fishes like herring, mackerel, trout, salmon, and tuna, where the fat is deposited in the muscles, constitute the primary source of *n*-3 PUFAs as compared to white fish (such as cod, pollock, hake, and haddock liver). Additional sources of *n*-3 PUFAs may include lean beef and chicken. Dietary ALA is transformed into *n*-3 PUFA, in the brain, liver, and testes, which is then absorbed into phospholipids of cell membranes, particularly in the retina and brain [66].

Unlike *n*-3 PUFAs, *n*-6 PUFAs (LA or AA) are associated with pro-inflammatory effects, primarily through the production of eicosanoid mediators such as thromboxanes, prostaglandins, and leukotrienes [67]. Long-chain $n$-3 PUFAs appear to compete with AA, to halt the generation of eicosanoids, thus preventing inflammatory processes. $n$-3 PUFAs may also reduce the expression of adhesion molecules, which are involved in inflammatory interactions between leukocytes and endothelial cells and generating the well-known inflammatory cytokines like interleukin-1, interleukin-6 and TNF-α [68]. Studies have revealed that intake of supplements based on $n$-3 PUFAs offers protection against the pathogenesis of numerous chronic degenerative illnesses, such as IBD, cancer, depression, rheumatoid arthritis, and cardiovascular diseases [69]. Leukotriene B4 levels in the intestinal membrane are decreased by consumption of long-chain $n$-3 PUFAs, which also prevent adhesion molecule gene expression. The inhibition of toll-like receptor-4 (TLR4), PPAR-γ, and NF-kB by $n$-3 PUFAs could significantly contribute to the reduction of intestinal inflammation [70]. The concentration of prostaglandins in the stomach increases by several thousand times when PUFAs are given intragastrically, enhancing mucosal protection and hastening the healing of mucosal ulcers [71]. Some research studies also suggest the prevention of the risk factors for gastric caners with $n$-3 PUFAs supplementation [72].

Docosahexaenoic acid (DHA) prevents the growth of *H. pylori in vitro* and the colonization of mice stomach mucosa. *H. pylori* is a significant risk factor for gastric cancer, duodenal ulcers, and chronic active gastritis while PUFAs consumption tends to limit the bacterial growth by rupturing cell membranes and causing them to lyse. Intake of $n$-3 PUFA-rich diet prevented tumor growth in a mouse xenograft experiment; supported *in vitro* experimental model successfully assessed the reduction of cancer metastasis via apoptosis and the reduction of MMP-10 expression through ERK and STAT3 phosphorylation [73]. Combining ω-3 PUFA with 5-fluorouracil showed synergistic effects in lowering gastric cancer cell growth. $n$-3 PUFAs shielded esophageal epithelial cells against acid damage via increasing Nrf2 expression that mediated NLRP3 inflammasome activation [74]. Moreover, $n$-3 PUFAs may regulate the gut microbiota diversity and abundance, as studies showed the inhibitory effects on *Enterobacteria* and enhancing the growth and/or activity of *Bifidobacteria* [67].

## 2.5 Polyphenols

Polyphenols are naturally occurring di- or tri- hydroxy substituents of phenol. Fruits, vegetables, and cereals are rich sources of dietary polyphenols. Fruits including grapes, apples, pears, cherries, and berries contain up to 200–300 mg of polyphenols per 100 g of fresh weight [75]. Polyphenols are one of the potent antioxidants that may reduce the amount of ROS and RNS and hence benefit patients with IBD and gastric ulcers [76]. Lignans are dimeric phenlypropanoids derivatives found in small quantities in sesame, soy, rapeseed, linseed, rye, and barley. The chief lignans found in diet are pinoresinol, lariciresinol, matairesinol, secoisolariciresinol, mediorecinol, and syringaresinol. They are known to possess antioxidant, anti-inflammatory, antiproliferative and anticancer properties. The lignans aglycones are poorly absorbed and reach the colon where they are converted by gut microbiota into enterolignans such as enterolactone and enterodiol [77]. Stilbenes are a class of plant secondary metabolites containing two phenyl rings linked by ethanol or ethylene linkage. They are present as monomers, polymers, or heteromers. They are present in plants such as grapes (*Vitis vinifera*), joint fir (*Gnetum parvifolium*), and Japanese knotweed (*Polygonum cuspidatum*). Resveratrol, a commonly occurring stilbene derivative, is reported to possess antitumor, anti-inflammatory, and antioxidant activities [78]. Resveratrol has been found effective against the majority of cancer types, including lung, skin, breast, prostate, gastric, and colorectal cancer. Additionally, it has been shown to prevent angiogenesis and metastasis. According to considerable data from human cell cultures, resveratrol has the capacity to change a number of pathways involved in cell proliferation, death, and inflammation. The enzymes cyclooxygenase, hydroperoxidase, protein kinase C, Bcl-2 phosphorylation, Akt, focal adhesion kinase, NF-B, matrix metalloproteinase-9, and cell cycle regulators have all been demonstrated to be suppressed by resveratrol. Flavonoids are polyphenolic compounds possessing phenyl-benzo-γ-pyrone nucleus with numerous hydroxyl substitution referred to as flavan nucleus. They act as potent antioxidant, antibacterial, antiviral, and anti-inflammatory. The antioxidant and anti-inflammatory potential of flavonoids is attributed to its protective effects on gut epithelium [76]. Phenolic acid refers to a phenolic compound having at least one carboxylic acid substitution. They are classified into two groups: hydroxybenzoic acid and hydroxy cinnamic acid derivatives. Contrary to hydroxybenzoic acids, hydroxycinnamic acids (i.e., ferulic acid, caffeic acid, coumaric acid, and sinapic acid) are more commonly found in dietary sources. Caffeic acid, chlorogenic acid, ferulic acid, p-coumaric acid, and sinapic acid are hydroxycinnamic acid derivatives that exhibit high antioxidant properties by preventing lipid oxidation and scavenging reactive oxygen species (ROS). Chlorogenic acid and caffeic acid limit the development of mutagenic and carcinogenic substances by inhibiting the *N*-nitrosation mechanism.

The protective effects of polyphenols on human cancer cell lines commonly results in a delay in the development or spread of tumors. These effects have been observed at a range of sites, including the mouth, stomach, duodenum, colon, liver, lung, mammary gland, or skin. Quercetin, catechins, isoflavones, lignans, flavanones, ellagic acid, red wine polyphenols, resveratrol, and curcumin are just a few of the polyphenols that have been investigated. In certain models, all of these polyphenols revealed therapeutic potential despite having different mechanisms of action. Myriad mechanisms of action, as well as anti-proliferation, cell cycle arrest or induction of apoptosis, prevention of oxidation, induction of detoxification enzymes, regulation of the host immune system, anti-inflammatory activity, and improvements in cellular signaling, have been interconnected to the ability of polyphenols in preventing cancer. By regulating the levels of cytochrome P450 enzymes involved in their activation to carcinogens, polyphenols somewhat have a consequence on metabolism of pro-carcinogens. By boosting the expression of phase II conjugating enzymes, they might also simplify the process for their excretion. The cytotoxicity of polyphenols may be the reason for this upregulation of phase II enzymes. Consuming polyphenols may then cause these enzymes to begin their own detoxification processes, thus enhancing overall resistance to hazardous xenobiotics by our bodies.

Numerous polyphenols containing herbs and phytoconstituents have been traditionally used and documented for their anti-inflammatory effects in management of IBS. A few of these include phenols-enriched plants such as *Perilla frutescens, Aloe vera, Mentha piperita, Cynara scolymus, Acacia catechu, Camellia sinensis. A. vera* is known to act as a laxative thereby regulating gut motility. The anthraquinone derivatives in aloe, viz., aloin, aloe-emodin, aloin, aloesin, 2'-O-feruloylaloesin, aloeresin A, isobarbaloin, aloenin, and aloe-emodin show laxative effect. *A. vera* is also reported to reduce gut hypersensitivity and can be useful in management of IBS. It shows anti-inflammatory effect by preventing the formation of prostaglandin E2, downregulating assorted transcription factors, and suppressing the events of lipoxygenase and COX-2 enzymes [79]. The hydrolysable tannins present in triphala are converted by microbiota to urolithin metabolites responsible for health benefits. They are responsible for maintaining intestinal homeostasis in gut bacteria, preventing inflammation and protecting the gut against oxidative damage. Green tea is an abundant source of catechins such as epigallocatechin-3-gallate (EGCG), epicatechin, epigallocatechin, and epicatechin-3-gallate. These have been thought to provide considerable anti-inflammatory and antioxidant benefits, which can aid in the treatment and prevention of pathologies like cancer, heart disease, diabetes, obesity, and inflammatory illnesses.

Polyphenols appear to have a part in decreasing TNF-α, COX-2, and NF-kB transcription. Additionally, they control the pathways controlled by nuclear erythroid 2-related factor 2 (Nrf2), mitogen-activated protein kinases (MAPKs), and signal transducer and transcription activator 1/3 (STAT1/3). Additionally, they enhance beneficial functions as apoptosis inducers and inhibitors of colorectal cancer. Gallic acid effects on a BALB/c mouse colitis model were examined, and it was found that gallic acid dramatically decreased the disease activity index, colon shortening, damage, and inflammatory process via the suppression of IL-21 and IL-23 production.

## 2.6 Spices

Herbs and spices have been used traditionally in food industry for flavor enhancement, digestive aids, and preservation. Traditional and medicinal uses of dietary spices have also been well documented in range of ailments. The application of licorice roots in colic symptoms, ginger tea for indigestion, and turmeric for anti-inflammatory properties are known in traditional medicine. Spices are known for their antibacterial and anti-inflammatory activity and carminative activities. Several spices are known to promote digestion and improve bowel movement. Spices such as ginger, oregano, black pepper, cayenne pepper, and rosemary may promote the growth of *Bifidobacterium* and inhibit *Clostridium* species.

### 2.6.1 Ginger

The polysaccharides present in ginger were reported to alleviate symptoms of UC by inhibiting pro-inflammatory cytokine levels and repairing the intestinal barrier regulating the levels of occluding-1 and ZO-1 [80]. Gingerols are reported to possess anticancer activity by modulating apoptosis, DNA damage, and epigenetic regulation. Shagols, which are formed because of heat and processing conditions also possess antioxidant, anti-inflammatory, and anticancer effects. Zingerone possesses antiemetic and antimicrobial properties [81]. Garlic contains inulin, which acts as a prebiotic to support intestinal flora. Consumption of black garlic is reported to decrease oxidative stress in the esophagus and suppressed proinflammatory cytokines in esophagus of rats [82].

### 2.6.2 Peppermint

Peppermint oil acts as a carminative and is reported to relax gastrointestinal smooth muscles by reducing $Ca^{++}$ influx and attenuate contractile responses in guinea pig taenia coli to different drugs such as acetylcholine, histamine, and 5-hydroxyl tryptamine [83]. The anti-inflammatory properties of *Mentha piperita* L. render them useful in subjects with IBS symptoms. Both xylene- and acetic acid-induced colitis in rats and mice are prevented by peppermint oil. It has been demonstrated that menthol inhibits generation of inflammatory mediators by human monocytes. TRP channels are known to be present in immune cells. Given that the activation of TRPM8 in mouse models reduces chemically induced colitis, it is assumed that the anti-inflammatory actions of peppermint oil may be at least partly mediated through this receptor. It was observed that menthol induced circular smooth muscle relaxation in the human colon, by directly inhibiting the contractility via blocking the $Ca^{++}$ influx through sarcolemma L-type $Ca^{++}$ channels. When given orally or intraperitoneally to IBS patients, peppermint oil lessened the visceral pain by inhibiting the gut TRPM8 and/or TRPA1 receptors.

### 2.6.3 Cumin and black cumin

Cumin has been used traditionally to treat digestive spasms, abdominal cramps, and bloating associated with irritable bowel syndrome. Black cumin or *Nigella sativa* has been reported to possess gastric ulcer healing activity in mouse models. The bioactive component found in black cumin, i.e., rhamnogalacturonan-I-type pectic polysaccharide is documented to increase gastric mucin content, COX-2, prostaglandin E2, and MMP-2 and decreases MMP-9 [84].

Table 2 has summarized the spices and their bioactive components that are beneficial for GI health.

**Table 2:** Spices and their beneficial effects on GI health.

| Spices | Bioactive components | Major role | References |
|---|---|---|---|
| *Zingiber officinale* | Gingerols, shagols, oleo-resin, polysaccharides | Modulation of gut microbiota, maintaining intestinal barrier function, peptic ulcers, and UC | [80] |
| *Nigella sativa* | Rhamnogalacturonan-I | Increase MMP-2 and decrease MMP-9, increase mucin, prostaglandin E2 and Cox-2 in gastric ulcers | [84] |
| *Curcuma longa* | Curcumin Pectic polysaccharide | Muco-protective, antisecretory, anti-inflammatory effects mediated via IL-10, anti-ulcer, chemoprotective | [85] |
| *Cuminum cyminum* | Total DF = 59%, SDF = 48.5%, IDF = 10.5%, Cellulose, hemicellulose and lignin, essential oil | As prebiotic, digestive aid | [86] |
| *Capsicum annuum* | Capsacin, oleoresin, Capsorubin | Gastroprotective effects, GERD, gastric ulcer, dyspepsia | [87] |
| *Crocus sativus* L | Crocin, crocetin, picrocrocin, safranal | Shows protective effect in UC and colon cancer, antioxidant | [88] |
| *Ferula foetida* | Oleo-gum resin | Shows anti-ulcer and anti-inflammatory effects | [89] |
| *Foeniculum vulgare* Mill. | Galactomannan, 4-hydroxyisolucine, diosgenin | Production of SCFA | [90] |
| *Allium sativum* L. | Polysaccharides, Alliin, Sulfur-containing compounds | Regulation of gut microbiota, increase in population of Bifidobacterium, Lachnospiraceae NK4A136 production of SCFA, Chemoprotective effects | [91] |
| *Chicorium intybus* | Inulin | Increase in stool frequency, softening of stool, regulation of gut microbiota | [92] |

## 2.7 Vitamins

Vitamins are micronutrients required in trace amounts, which play a vital role in metabolic functions of our body. They act as cofactors in a number of enzymatic processes. Vitamins and their derivatives especially vitamins A, B2, D, and E are reported to benefit the intestinal flora, in particular, that in the colonic region. Vitamins B2, C, and E modulate the gut microbiome and this enhances SCFA production [93]. Vitamin A derivatives such as retinoids play a vital role in mucin production, cell growth, and differentiation, which is responsible for maintaining gut barrier function. Several studies have found association between dietary intake of Vitamin A and gut microbiota composition [94]. Studies showed that in Sudanese and Brazilian children, the regular and adequate dietary intake of vitamin A maybe associated with the reduction of diarrhea incidence [95]. Vitamin D also modulates gut microflora, and exerts anti-inflammatory effects on the GI tract. Deficiency of vitamin D could be associated with intestinal inflammatory diseases like Crohn's and UC, perhaps through the loss of epithelial junctional proteins and inhibition of inflammatory mediators [96]. Vitamin E is known extensively for its antioxidant and protective effect on the gut epithelial membrane [97]. Vitamin B2 is reported to support the microbial diversity. In a clinical study, in 11 healthy adults who received adequate dosage of vitamin B2 (100 mg) supplementation, the count of butyrate-producing bacteria *Faecalibacterium prausnitzii* and gut microbial species possessing anti-inflammatory properties increased [98]. Vitamin C is critical for host defense mechanism and immunity. It shows antioxidant and antimicrobial effect and hence can modulate the gut microbial communities [99]. Vitamin C shows antioxidant effect and exerts anti-inflammatory effect on the cells of the gut epithelium [100].

# 3 Summary

Nutraceuticals are seen as safe, effective, cost-effective adjuvants and alternatives for chronic ailments that require lifelong medication. Several nutraceuticals and functional food ingredients show beneficial effect on gut health. The relationship between gut health and physical and mental well-being is well established. The dietary fibers help regulate gut homeostasis by regulating gut microflora, peristalsis, and generating beneficial metabolites. Herbs containing polyphenol possess antioxidant, anti-inflammatory, and protective effects. Diet rich in PUFAs inhibits the synthesis of inflammatory mediators. Numerous spices and vitamins have demonstrated to regulate gut health by numerous mechanisms. They may delay the onset and progression of several diseased conditions of the GI tract. A number of nutraceuticals are subjected to clinical trials for the assessment of their safety and efficacy in subjects with GI disorders. With the advancement in formulation and processing technology it is now possible to enhance the absorption and retention of bioactives in the gut, providing effective health benefits.

# References

[1]   Lennard-Jones, J.E. 1983. Functional gastrointestinal disorders. New England Journal of Medicine, 308(8), 431–435.
[2]   O'keefe, S.J.D. 1996. Nutrition and gastrointestinal disease. Scandinavian Journal of Gastroenterology, 31(sup220), 52–59.

[3]     Sahu, P., Kedia, S., Ahuja, V. and Tandon, R.K. 2021. Diet and nutrition in the management of inflammatory bowel disease. Indian Journal of Gastroenterology, 40(3), 253–264.

[4]     Corsello, A., Pugliese, D., Gasbarrini, A. and Armuzzi, A. 2020. Diet and nutrients in gastrointestinal chronic diseases. Nutrients, 12(9), 2693.

[5]     Kotla, N.G. and Rochev, Y. 2023. IBD disease-modifying therapies: Insights from emerging therapeutics. Trends in Molecular Medicine, 29(3), 241–253.

[6]     Casula, E., Pisano, M.B., Serreli, G., Zodio, S., Melis, M.P., Corona, G., Costabile, A., Cosentino, S. and Deiana, M. 2023. Probiotic lactobacilli attenuate oxysterols-induced alteration of intestinal epithelial cell monolayer permeability: Focus on tight junction modulation. Food and Chemical Toxicology, 172, 113558.

[7]     Montgomery, R.K., Mulberg, A.E. and Grand, R.J. 1999. Development of the human gastrointestinal tract: Twenty years of progress. Gastroenterology, 116(3), 702–731.

[8]     Kurian, S.J., Baral, T., Sekhar, M. and Rao, M. 2022. Role of probiotics and prebiotics in digestion, metabolism, and immunity. In Bagchi, D., Ohia, S. (Eds.), *Nutrition and Functional Foods in Boosting Digestion, Metabolism and Immune Health*. Academic Press, pp. 501–522.

[9]     Drossman, D.A. 2016. Functional gastrointestinal disorders: History, pathophysiology, clinical features, and Rome IV. Gastroenterology, 150(6), 1262–1279. e2.

[10]    Ribolsi, M., Guarino, M.P.L., Tullio, A. and Cicala, M. 2020. Post-reflux swallow-induced peristaltic wave index and mean nocturnal baseline impedance predict PPI response in GERD patients with extra esophageal symptoms. Digestive and Liver Disease, 52(2), 173–177.

[11]    Cryer, B. and Spechler, S.J. 2000. Effects of non-steroidal anti-inflammatory drugs (NSAIDs) on acid reflux in patients with gastroesophageal reflux disease (GERD). Gastroenterology, 118(4), A862.

[12]    Lee, K.-J., Kindt, S. and Tack, J. 2004. Pathophysiology of functional dyspepsia. Best Practice & Research Clinical Gastroenterology, 18(4), 707–716.

[13]    Oustamanolakis, P. and Tack, J. 2012. Dyspepsia. Journal of Clinical Gastroenterology, 46(3), 175–190.

[14]    Bhat, A., Rao, S.S., Bhat, S., Vidyalakshmi, K. and Dhanashree, B. 2023. Molecular diagnosis of bacterial and viral diarrhoea using multiplex-PCR assays: An observational prospective study among paediatric patients from India. Indian Journal of Medical Microbiology, 41, 64–70.

[15]    Schiller, L.R. 2019. Chronic constipation: New insights, better outcomes?. The Lancet Gastroenterology & Hepatology, 4(11), 873–882.

[16]    Du, Y., Li, Y., Xu, X., Li, R., Zhang, M., Cui, Y., Zhang, L., Wei, Z., Wang, S. and Tuo, H. 2022. Probiotics for constipation and gut microbiota in Parkinson's disease. Parkinsonism & Related Disorders, 103, 92–97.

[17]    Boland, J.W. and Boland, E.G. 2022. Constipation and malignant bowel obstruction in palliative care. Medicine, 50(12), 775–779.

[18]    Santucci, N.R., Chogle, A., Leiby, A., Mascarenhas, M., Borlack, R.E., Lee, A., Perez, M., Russell, A. and Yeh, A.M. 2021. Non-pharmacologic approach to pediatric constipation. Complementary Therapies in Medicine, 59, 102711.

[19]    Zeeshan, M.H., Vakkalagadda, N.P., Sree, G.S., Kishore Anne, K., Devi, S., Parkash, O., Fawwad, S.B.U., Haider, S.M.W., Mumtaz, H. and Hasan, M. 2022. Irritable bowel syndrome in adults: Prevalence and risk factors. Annals of Medicine & Surgery, 81, 104408.

[20]    McGowan, A. and Harer, K.N. 2021. Irritable bowel syndrome and eating disorders. Gastroenterology Clinics of North America, 50(3), 595–610.

[21]    Videlock, E.J. and Chang, L. 2021. Latest insights on the pathogenesis of irritable bowel syndrome. Gastroenterology Clinics of North America, 50(3), 505–522.

[22]    Kassebaum-Ladewski, A. 2021. Irritable bowel syndrome - strategies for diagnosis and management. Physician Assistant Clinics, 6(4), 637–653.

[23]  Li, X., Wang, B., Hu, W., Chen, H., Sheng, Z., Yang, B. and Yu, L. 2022. Effect of γ-irradiation on structure, physicochemical property and bioactivity of soluble dietary fiber in navel orange peel. Food Chemistry: X, 14, 100274.

[24]  Li, S., Zhang, F. and Zhang, Q. 2022. Pathological features-based targeted delivery strategies in IBD therapy: A mini review. Biomedicine & Pharmacotherapy, 151, 113079.

[25]  Wehkamp, J., Götz, M., Herrlinger, K., Steurer, W. and Stange, E.F. 2016. Inflammatory bowel disease: Crohn's disease and ulcerative colitis. Deutsches Ärzteblatt International, 113(5), 72.

[26]  Smyth, E.C., Nilsson, M., Grabsch, H.I., Van Grieken, N.C. and Lordick, F. 2020. Gastric cancer. The Lancet, 396(10251), 635–648.

[27]  Ghimire, P. 2011. Primary gastrointestinal lymphoma. World Journal of Gastroenterology, 17(6), 697.

[28]  Balakrishnan, M., George, R., Sharma, A. and Graham, D.Y. 2017. Changing trends in stomach cancer throughout the world. Current Gastroenterology Reports, 19(8), 36.

[29]  Zali, H., Rezaei-Tavirani, M. and Azodi, M. 2011. Gastric cancer: Prevention, risk factors and treatment. Gastroenterology and Hepatology from Bed to Bench, 4(4), 175–185.

[30]  Grabsch, H.I. and Tan, P. 2013. Gastric cancer pathology and underlying molecular mechanisms. Digestive Surgery, 30(2), 150–158.

[31]  Nunobe, S., Kumagai, K., Ida, S., Ohashi, M. and Hiki, N. 2016. Minimally invasive surgery for stomach cancer. Japanese Journal of Clinical Oncology, 46(5), 395–398.

[32]  Caio, G., Volta, U., Sapone, A., Leffler, D.A., De Giorgio, R., Catassi, C. and Fasano, A. 2019. Celiac disease: A comprehensive current review. BMC Medicine, 17, 1–20.

[33]  Kivelä, L., Caminero, A., Leffler, D.A., Pinto-Sanchez, M.I., Tye-Din, J.A. and Lindfors, K. 2021. Current and emerging therapies for coeliac disease. Nature Reviews Gastroenterology & Hepatology, 18(3), 181–195.

[34]  Lebwohl, B., Sanders, D.S. and Green, P.H.R. 2018. Coeliac disease. The Lancet, 391(10115), 70–81.

[35]  Malfertheiner, P., Chan, F.K. and McColl, K.E. 2009. Peptic ulcer disease. The Lancet, 374(9699), 1449–1461.

[36]  Kumar, C., Kumar, S., Prabu, S. and Suriyaprakash, T. 2012. Nutraceuticals and their medicinal importance. International Journal of Health & Allied Sciences, 1(2), 47–47.

[37]  Bai, X., He, Y., Quan, B., Xia, T., Zhang, X., Wang, Y., Zheng, Y. and Wang, M. 2022. Physicochemical properties, structure, and ameliorative effects of insoluble dietary fiber from tea on slow transit constipation. Food Chemistry: X, 14, 100340.

[38]  Mudgil, D. 2017. The interaction between insoluble and soluble fiber. In Samaan, R.A. (Ed.), *Dietary Fiber for the Prevention of Cardiovascular Disease*. Academic Press, pp. 35–59.

[39]  Feiner, G. 2006. Additives: Phosphates, salts (sodium chloride and potassium chloride, citrate, lactate) and hydrocolloids. In Feiner, G. (Ed.), *Meat Products Handbook: Practical Science and Technology*. Elsevier, pp. 72–88.

[40]  Saha, B.C. 2003. Hemicellulose bioconversion. Journal of Industrial Microbiology and Biotechnology, 30(5), 279–291.

[41]  Petzold-Welcke, K., Schwikal, K., Daus, S. and Heinze, T. 2014. Xylan derivatives and their application potential-Mini-review of own results. Carbohydrate Polymers, 100, 80–88.

[42]  Nishinari, K., Takemasa, M., Zhang, H. and Takahashi, R. 2007. Storage Plant polysaccharides: Xyloglucans, galactomannans, glucomannans. In Kamerling, H. (Ed.), *Comprehensive Glycoscience*. Elsevier, pp. 613–652.

[43]  Pauly, M., Gille, S., Liu, L., Mansoori, N., De Souza, A., Schultink, A. and Xiong, G. 2013. Hemicellulose biosynthesis. Planta, 238(4), 627–642.

[44]  Nasrollahzadeh, M., Sajjadi, M., Nezafat, Z. and Shafiei, N. 2021. Polysaccharide biopolymer chemistry. In Nasrollahzadeh, M. (Ed.), *Biopolymer-Based Metal Nanoparticle Chemistry for Sustainable Applications*. Elsevier, pp. 45–105.

[45]  Stick, R.V. and Williams, S.J. 2009. Disaccharides, Oligosaccharides and Polysaccharides. In Stick, R.V.
      & Williams, S. (Eds.), *Carbohydrates: The Essential Molecules of Life*. Elsevier, pp. 321–341.

[46]  Mitmesser, S. and Combs, M. 2017. Prebiotics: Inulin and Other Oligosaccharides. In Floch, M.H.,
      Ringel, Y., & Walker, W.A. (Eds.), *The Microbiota in Gastrointestinal Pathophysiology*. Elsevier,
      pp. 201–208.

[47]  Hartikainen, K. and Katina, K. 2012. Improving the quality of high-fibre breads. In Cauvain, S.P. (Ed.),
      *Breadmaking*. Elsevier, pp. 736–753.

[48]  Khorasaniha, R., Olof, H., Voisin, A., Armstrong, K., Wine, E., Vasanthan, T. and Armstrong, H. 2023.
      Diversity of fibers in common foods: Key to advancing dietary research. Food Hydrocolloids, 139,
      108495.

[49]  Dong, R., Liao, W., Xie, J., Chen, Y., Peng, G., Xie, J., Sun, N., Liu, S., Yu, C. and Yu, Q. 2022.
      Enrichment of yogurt with carrot soluble dietary fiber prepared by three physical modified
      treatments: Microstructure, rheology and storage stability. Innovative Food Science & Emerging
      Technologies, 75, 102901.

[50]  Wu, X., Li, F. and Wu, W. 2022. Effect of rice bran rancidity on the structure and antioxidant
      properties of rice bran soluble dietary fiber. Journal of Cereal Science, 105, 103469.

[51]  Zhang, F., Fan, D., Huang, J. and Zuo, T. 2022. The gut microbiome: Linking dietary fiber to
      inflammatory diseases. Medicine in Microecology, 14, 100070.

[52]  De Filippis, A., Ullah, H., Baldi, A., Dacrema, M., Esposito, C., Garzarella, E.U., Santarcangelo, C.,
      Tantipongpiradet, A. and Daglia, M. 2020. Gastrointestinal disorders and metabolic syndrome:
      Dysbiosis as a key link and common bioactive dietary components useful for their treatment.
      International Journal of Molecular Sciences, 21(14), 4929.

[53]  Dahl, S.M., Rolfe, V., Walton, G.E. and Gibson, G.R. 2023. Gut microbial modulation by culinary herbs
      and spices. Food Chemistry, 409, 135286.

[54]  Qin, X.-S., Gao, Q.-Y. and Luo, Z.-G. 2021. Enhancing the storage and gastrointestinal passage
      viability of probiotic powder (Lactobacillus Plantarum) through encapsulation with pickering high
      internal phase emulsions stabilized with WPI-EGCG covalent conjugate nanoparticles. Food
      Hydrocolloids, 116, 106658.

[55]  Almada, C.N., Almada-Érix, C.N., Bonatto, M.S., Pradella, F., Dos Santos, P., Abud, Y.K.D., Farias, A.S.,
      Martínez, J., Sant'Anna Filho, C.B., Lollo, P.C., Costa, W.K.A., Magnani, M. and Sant'Ana, A.S. 2021.
      Obtaining paraprobiotics from Lactobacilus acidophilus, Lacticaseibacillus casei and
      Bifidobacterium animalis using six inactivation methods: Impacts on the cultivability, integrity,
      physiology, and morphology. Journal of Functional Foods, 87, 104826.

[56]  Maske, B.L., De Melo Pereira, G.V., da Vale, A.S., De Carvalho Neto, D.P., Karp, S.G., Viesser, J.A., De
      Dea Lindner, J., Pagnoncelli, M.G., Soccol, V.T. and Soccol, C.R. 2021. A review on enzyme-producing
      lactobacilli associated with the human digestive process: From metabolism to application. Enzyme
      and Microbial Technology, 149, 109836.

[57]  Al-Yami, A.M., Al-Mousa, A.T., Al-Otaibi, S.A. and Khalifa, A.Y. 2022. Lactobacillus species as
      probiotics: Isolation sources and health benefits. Journal of Pure and Applied Microbiology, 16(4),
      2270–2291.

[58]  Flint, S. 2002. Enterococcus faecalis and Enterococcus faecium. In Roginski, H., Fuquay, J.W., & Fox,
      P.F. (Eds.) *Encyclopedia of Dairy Sciences*. Academic Press, pp. 904–907.

[59]  Kelesidis, T. and Pothoulakis, C. 2012. Efficacy and safety of the probiotic *Saccharomyces boulardii* for
      the prevention and therapy of gastrointestinal disorders. Therapeutic Advances in
      Gastroenterology, 5(2), 111–125.

[60]  Behrouzi, A., Mazaheri, H., Falsafi, S., Tavassol, Z.H., Moshiri, A. and Siadat, S.D. 2020. Intestinal
      effect of the probiotic Escherichia coli strain Nissle 1917 and its OMV. Journal of Diabetes &
      Metabolic Disorders, 19(1), 597–604.

[61] Qayyum, N., Shuxuan, W., Yantin, Q., Ruiling, W., Wang, S., Ismael, M. and Lü, X. 2023. Characterization of Short-chain fatty acid-producing and cholesterol assimilation potential probiotic Lactic acid bacteria from Chinese fermented rice. Food Bioscience, 52, 102404.

[62] Li, L.-Q., Chen, X., Zhu, J., Zhang, S., Chen, S.-Q., Liu, X., Li, L. and Yan, J.-K. 2023. Advances and challenges in interaction between heteroglycans and Bifidobacterium: Utilization strategies, intestinal health and future perspectives. Trends in Food Science & Technology, 134, 112–122.

[63] Daba, G.M., El-Dien, A.N., Saleh, S.A.A., Elkhateeb, W.A., Awad, G., Nomiyama, T., Yamashiro, K. and Zendo, T. 2021. Evaluation of Enterococcus strains newly isolated from Egyptian sources for bacteriocin production and probiotic potential. Biocatalysis and Agricultural Biotechnology, 35, 102058.

[64] Graff, S., Chaumeil, J.-C., Boy, P., Lai-Kuen, R. and Charrueau, C. 2008. Influence of pH conditions on the viability of Saccharomyces boulardii yeast. The Journal of General and Applied Microbiology, 54(4), 221–227.

[65] Zhao, Z., Xu, S., Zhang, W., Wu, D. and Yang, G. 2022. Probiotic *Escherichia coli* NISSLE 1917 for inflammatory bowel disease applications. Food & Function, 13(11), 5914–5924.

[66] De Deckere, E.A., Korver, O., Verschuren, P.M. and Katan, M.B. 1998. Health aspects of fish and n-3 polyunsaturated fatty acids from plant and marine origin. European Journal of Clinical Nutrition, 52(10), 749–753.

[67] Fu, Y., Wang, Y., Gao, H., Li, D., Jiang, R., Ge, L., Tong, C. and Xu, K. 2021. Associations among dietary omega-3 polyunsaturated fatty acids, the gut Microbiota, and Intestinal Immunity. Mediators of Inflammation, 2021, 1–11.

[68] Calder, P.C. 2006. Polyunsaturated fatty acids and inflammation. Prostaglandins, Leukotrienes and Essential Fatty Acids, 75(3), 197–202.

[69] Costantini, L., Molinari, R., Farinon, B. and Merendino, N. 2017. Impact of Omega-3 Fatty Acids on the Gut Microbiota. International Journal of Molecular Sciences, 18(12), 2645.

[70] Mozaffari, H., Daneshzad, E., Larijani, B., Bellissimo, N. and Azadbakht, L. 2020. Dietary intake of fish, n-3 polyunsaturated fatty acids, and risk of inflammatory bowel disease: A systematic review and meta-analysis of observational studies. European Journal of Nutrition, 59(1), 1–17.

[71] Hollander, D. and Tarnawski, A. 1986. Dietary essential fatty acids and the decline in peptic ulcer disease–a hypothesis. Gut, 27(3), 239–242.

[72] Dai, J., Shen, J., Pan, W., Shen, S. and Das, U.N. 2013. Effects of polyunsaturated fatty acids on the growth of gastric cancer cells in vitro. Lipids in Health and Disease, 12(1), 71.

[73] Park, J.-M., Kwon, S.-H., Han, Y.-M., Hahm, K.-B. and Kim, E.-H. 2013. Omega-3 polyunsaturated fatty acids as potential chemopreventive agent for gastrointestinal cancer. Journal of Cancer Prevention, 18(3), 201–208.

[74] Chen, Y.-H., Jiang, Y., Wei, -J.-J., Li, X.-D., Zhang, P.-H., Lian, -T.-T. and Zhuang, Z.-H. 2022. N-3 polyunsaturated fatty acids protect esophageal epithelial cells from acid exposure. Food Research International, 162, 111943.

[75] Pandey, K.B. and Rizvi, S.I. 2009. Plant polyphenols as dietary antioxidants in human health and disease. Oxidative Medicine and Cellular Longevity, 2(5), 270–278.

[76] Ronzio, R.A. 2020. Naturally occurring antioxidants. In Pizzorno, E. & Murray, M.T. (Eds.), *Textbook of Natural Medicine*. Elsevier, pp. 731–751.

[77] Yeung, A.W.K., Tzvetkov, N.T., Balacheva, A.A., Georgieva, M.G., Gan, R.-Y., Jozwik, A., Pyzel, B., Horbańczuk, J.O., Novellino, E., Durazzo, A., Lucarini, M., Camilli, E., Souto, E.B., Atanasov, A.G. and Santini, A. 2020. Lignans: Quantitative analysis of the research literature. Frontiers in Pharmacology, 11, 37.

[78] Su, X., Zhou, D. and Li, N. 2022. Bioactive stilbenes from plants. Studies in Natural Products Chemistry, 73, 265–403.

[79]   Roudsari, N.M., Lashgari, N.-A., Momtaz, S., Farzaei, M.H., Marques, A.M. and Abdolghaffari, A.H. 2019. Natural polyphenols for the prevention of irritable bowel syndrome: Molecular mechanisms and targets; a comprehensive review. DARU Journal of Pharmaceutical Sciences, 27(2), 755–780.

[80]   Hao, W., Chen, Z., Yuan, Q., Ma, M., Gao, C., Zhou, Y., Zhou, H., Wu, X., Wu, D., Farag, M.A., Wang, S. and Wang, Y. 2022. Ginger polysaccharides relieve ulcerative colitis via maintaining intestinal barrier integrity and gut microbiota modulation. International Journal of Biological Macromolecules, 219, 730–739.

[81]   Kiyama, R. 2020. Nutritional implications of ginger: Chemistry, biological activities and signaling pathways. The Journal of Nutritional Biochemistry, 86, 108486.

[82]   Filocamo, A., Nueno-Palop, C., Bisignano, C., Mandalari, G. and Narbad, A. 2012. Effect of garlic powder on the growth of commensal bacteria from the gastrointestinal tract. Phytomedicine, 19(8–9), 707–711.

[83]   Hills, J.M. and Aaronson, P.I. 1991. The mechanism of action of peppermint oil on gastrointestinal smooth muscle. Gastroenterology, 101(1), 55–65.

[84]   Manjegowda, S.B., Rajagopal, H.M. and Dharmesh, S.M. 2017. Polysaccharide of Black cumin (*Nigella sativa*) modulates molecular signaling cascade of gastric ulcer pathogenesis. International Journal of Biological Macromolecules, 101, 823–836.

[85]   Rajagopal, H.M., Manjegowda, S.B., Serkad, C. and Dharmesh, S.M. 2018. A modified pectic polysaccharide from turmeric (*Curcuma longa*) with antiulcer effects via anti–secretary, mucoprotective and IL–10 mediated anti–inflammatory mechanisms. International Journal of Biological Macromolecules, 118, 864–880.

[86]   Sowbhagya, H.B., Suma, P.F., Mahadevamma, S. and Tharanathan, R.N. 2007. Spent residue from cumin – A potential source of dietary fiber. Food Chemistry, 104(3), 1220–1225.

[87]   Maji, A.K. and Banerji, P. 2016. Phytochemistry and gastrointestinal benefits of the medicinal spice, Capsicum annuum L. (Chilli): A review. Journal of Complementary and Integrative Medicine, 13(2), 97–122.

[88]   Khorasany, A.R. and Hosseinzadeh, H. 2016. Therapeutic effects of saffron (*Crocus sativus* L.) in digestive disorders: A review. Iranian Journal of Basic Medical Sciences, 19(5), 455–469.

[89]   Vijayasteltar, L., Jismy, I.J., Joseph, A., Maliakel, B., Kuttan, R. and K, I.M. 2017. Beyond the flavor: A green formulation of Ferula asafoetida oleo-gum-resin with fenugreek dietary fibre and its gut health potential. Toxicology Reports, 4, 382–390.

[90]   Shtriker, M.G., Hahn, M., Taieb, E., Nyska, A., Moallem, U., Tirosh, O. and Madar, Z. 2018. Fenugreek galactomannan and citrus pectin improve several parameters associated with glucose metabolism and modulate gut microbiota in mice. Nutrition, 46, 134–142.

[91]   Zhao, R., Qiu, Z., Bai, X., Xiang, L., Qiao, Y. and Lu, X. 2022. Digestive properties and prebiotic activity of garlic saccharides with different-molecular-weight obtained by acidolysis. Current Research in Food Science, 5, 2033–2044.

[92]   Watson, A.W., Houghton, D., Avery, P.J., Stewart, C., Vaughan, E.E., Meyer, P.D., De Bos Kuil, M.J.J., Weijs, P.J.M. and Brandt, K. 2019. Changes in stool frequency following chicory inulin consumption, and effects on stool consistency, quality of life and composition of gut microbiota. Food Hydrocolloids, 96, 688–698.

[93]   Pham, V.T., Dold, S., Rehman, A., Bird, J.K. and Steinert, R.E. 2021. Vitamins, the gut microbiome and gastrointestinal health in humans. Nutrition Research, 95, 35–53.

[94]   Li, L., Krause, L. and Somerset, S. 2017. Associations between micronutrient intakes and gut microbiota in a group of adults with cystic fibrosis. Clinical Nutrition, 36(4), 1097–1104.

[95]   Barreto, M.L., Farenzena, G.G., Fiaccone, R.L., Santos, L.M.P., Assis, A.M.O., Araújo, M.P.N. and Santos, P.A.B. 1994. Effect of vitamin A supplementation on diarrhoea and acute lower-respiratory-tract infections in young children in Brazil. The Lancet, 344(8917), 228–231.

[96] Meckel, K., Li, Y.C., Lim, J., Kocherginsky, M., Weber, C., Almoghrabi, A., Chen, X., Kaboff, A., Sadiq, F., Hanauer, S.B., Cohen, R.D., Kwon, J., Rubin, D.T., Hanan, I., Sakuraba, A., Yen, E., Bissonnette, M. and Pekow, J. 2016. Serum 25-hydroxyvitamin D concentration is inversely associated with mucosal inflammation in patients with ulcerative colitis. The American Journal of Clinical Nutrition, 104(1), 113–120.

[97] Howard, A.C., McNeil, A.K. and McNeil, P.L. 2011. Promotion of plasma membrane repair by vitamin E. Nature Communications, 2(1), 597.

[98] Lopez-Siles, M., Khan, T.M., Duncan, S.H., Harmsen, H.J.M., Garcia-Gil, L.J. and Flint, H.J. 2012. Cultured representatives of two major phylogroups of human colonic faecalibacterium prausnitzii can utilize pectin, uronic acids, and host-derived substrates for growth. Applied and Environmental Microbiology, 78(2), 420–428.

[99] Rumsey, S.C. and Levine, M. 1998. Absorption, transport, and disposition of ascorbic acid in humans. The Journal of Nutritional Biochemistry, 9(3), 116–130.

[100] Mousavi, S., Bereswill, S. and Heimesaat, M.M. 2019. Immunomodulatory and antimicrobial effects of vitamin C. European Journal of Microbiology and Immunology, 9(3), 73–79.

Zubair Ahmad, Musarat Riaz, Abdur Rauf*, Hassan Zeb,
Taghrid S. Alomar

# Chapter 10
# Bone health

**Abstract:** Bone health and arthritis disorders are significant health concerns, especially in the aging population. Traditional treatments for these conditions can be costly, invasive, and often come with adverse side effects. This chapter explores the potential of nutraceuticals as a safe, cost-effective, and noninvasive approach to promoting bone health and alleviating the symptoms of arthritis disorders. This chapter provides a comprehensive review of the classification of nutraceuticals, sources, and their application in bone health. Nutraceuticals such as vitamins, minerals, antioxidants, and plant extracts possess anti-inflammatory, analgesic, and immune-modulating effects, making them a promising alternative to traditional treatments. The mechanisms by which nutraceuticals work and their potential therapeutic applications for bone health and arthritis disorders are examined in detail. It also discusses the challenges and limitations of using nutraceuticals. The safety and efficacy of nutraceuticals in the prevention and management of bone health and arthritis disorders are also examined. The future directions of research in this field are discussed as well.

# 1 Introduction

In recent years, there has been increasing interest in nutraceuticals as a means of promoting bone health and alleviating the symptoms of arthritis disorders. Bone health is a vital aspect of overall well-being, particularly as we age. Poor bone health can lead to conditions such as osteoporosis, which can result in fractures, decreased mobility, and chronic pain. Arthritis, on the other hand, can cause significant discomfort, making it difficult to perform even the simplest tasks. Traditional treatments for

---

*Corresponding author: Abdur Rauf, Department of Chemistry, University of Swabi, Anbar 23561, Khyber Pakhtunkhwa, Pakistan; Department of Chemistry, College of Science, Princess Nourah bint Abdulrahman University, Riyadh 11671, Saudi Arabia, e-mail: abdurrauf@uoswabi.edu.pk
Zubair Ahmad, Department of Chemistry, University of Swabi, Anbar 23561, Khyber Pakhtunkhwa, Pakistan
Musarat Riaz, Department of Chemistry, Sardar Bahadar Khan Women's University, Quetta 87300, Balochistan, Pakistan
Hassan Zeb, Department of Statistics, Islamia College University Peshawar 25120, Khyber Pakhtunkhwa, Pakistan
Taghrid S. Alomar, Department of Chemistry, College of Science, Princess Nourah bint Abdulrahman University, Riyadh 11671, Saudi Arabia

https://doi.org/10.1515/9783111317601-010

these conditions can be costly, invasive, and often come with adverse side effects [1, 2].

Nutraceuticals offer a promising alternative to traditional treatments, offering a safe, cost-effective, and noninvasive approach to promoting bone health and alleviating the symptoms of arthritis. Nutraceuticals such as vitamins, minerals, antioxidants, and various plant extracts have been found to possess a wide range of beneficial properties for bone health and arthritis disorders. These properties include anti-inflammatory, analgesic, and immune-modulating effects [3, 4]. A nutraceutical, sometimes known as a bioceutical, is a pharmacological alternative that purports to provide physiological advantages. Nutraceuticals are chemicals that give physiological advantages or defense against chronic illnesses. Nutraceuticals can be used to promote health, slow the process of aging, prevent chronic illnesses, extend life, or maintain the body's function or structure [5]. The word "nutraceutical" was invented in 1989 by the Organization for Innovation in Medicine (New York, US; an educational foundation formed in the United States to foster medical discoveries) to offer a label for this rapidly expanding area of biomedical research. A nutraceutical is any substance that may be regarded a food or component of a food that delivers medical or health advantages such as illness prevention and treatment [4]. Nutraceuticals can include everything from extracted nutrients, nutritional supplements, and foods to genetically altered "designer" diets, herbal items, and packaged meals including cereals, soups, and drinks [5]. People are profoundly worried about the management, governance, and pricing of their health care. They are dissatisfied with contemporary medicine's pricey, high-tech disease treatment strategy; the customer is looking for supplementary or alternative helpful items; and the red tape of care services makes nutraceuticals more enticing. With the development of various nutraceuticals, obtaining an optimal or maximum state of nutrition and health is becoming less difficult. The health advantages of nutraceuticals are numerous [4, 5]. Various products promise to prevent or treat not just cancer and heart disease, but also hypertension, high cholesterol, obesity, osteoporosis, diabetes, and arthritis. The inflammation or swelling of one or more joints is referred as arthritis [6]. It addresses over 100 disorders that affect the joints, surrounding tissues, and other connective tissues. The signs of arthritis vary depending on the kind, but they commonly comprise stiffness and pain in joints [7]. There are numerous arthritis treatments available, but each has a number of side effects that limit their use [8]. People are looking for alternative ways to cure arthritis and improve bone health, and nutraceuticals are one of them [1, 7, 9].

This chapter will explore the latest research on nutraceuticals for bone health and arthritis disorders. It will examine the mechanisms by which these nutraceuticals work and their potential therapeutic applications. The chapter will also explore the safety and efficacy of nutraceuticals in the prevention and management of bone health and arthritis disorders. Furthermore, it will discuss the challenges and limitations of using nutraceuticals in clinical practice and the future directions of research in this field.

# 2 Nutraceuticals in bone health

Strong and healthy bones are essential for our overall health and well-being. Nutraceuticals have been gaining attention for their potential to improve bone health, and some studies have suggested that they may even play a role in the prevention and treatment of bone-related diseases such as osteoporosis. However, it is important to be cautious when considering the use of nutraceuticals for bone health and to seek guidance from a healthcare professional [10, 11]. Nutrients such as calcium, vitamin D, and magnesium are crucial for maintaining strong bones [12]. While these can be obtained through a balanced diet, many people may not consume enough of these nutrients on a regular basis. Nutraceuticals, such as calcium supplements or vitamin D-fortified foods, can help fill these nutrient gaps and support bone health. In addition to these essential nutrients, other compounds found in nutraceuticals may also benefit bone health. For example, plant-based compounds such as flavonoids and polyphenols have been shown to have anti-inflammatory properties and may help reduce bone loss [13]. Omega-3 fatty acids found in fish oil supplements have also been associated with improved bone density in some studies [14]. While the potential benefits of nutraceuticals for bone health are promising, it is important to remember that not all products are created equal. Quality and purity can vary greatly among nutraceuticals, and some may even contain harmful contaminants. It is crucial to choose reputable brands and to always consult with a healthcare professional before adding any new supplements to one's routine. Overall, incorporating nutraceuticals into a healthy diet and lifestyle may be a beneficial way to support bone health and prevent bone-related diseases (Figure 1).

## 2.1 Polyphenols

Polyphenols are a diverse group of naturally occurring compounds found in various plant-based foods and beverages such as fruits, vegetables, tea, and red wine [15]. These bioactive compounds are renowned for their antioxidant and anti-inflammatory properties. Polyphenols are known to combat oxidative stress by neutralizing harmful free radicals in the body, reducing inflammation, and supporting overall health [16]. In the context of bone health, polyphenols have shown promise in several studies. These compounds may help reduce bone loss by mitigating inflammation and oxidative stress in the skeletal system, ultimately contributing to improved bone health. The reduction of inflammation is particularly crucial in the prevention of osteoporosis and other bone-related diseases, as chronic inflammation can lead to accelerated bone resorption and impaired bone formation [17]. One prominent group of polyphenols is flavonoids, which includes quercetin, epicatechin, and resveratrol. These have been studied for their potential to enhance bone density and quality. Resveratrol, found in red grapes

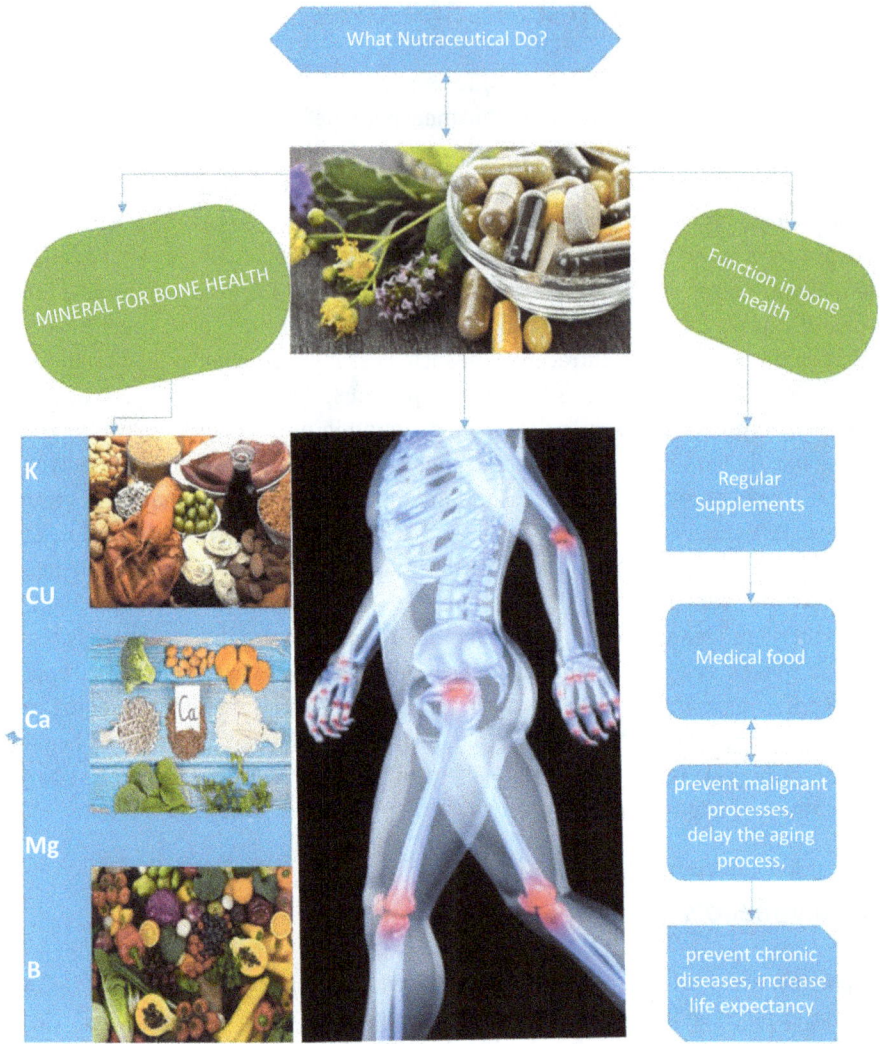

**Figure 1:** Benefits of minerals-based supplements in bone health.

and red wine, for instance, has shown promise in increasing bone mineral density and bone strength in animal studies [18].

Additionally, some polyphenols have been associated with the modulation of bone formation and resorption markers like epicatechin, commonly found in cocoa and dark chocolate, which has demonstrated the ability to stimulate bone formation while suppressing bone resorption. These mechanisms suggest that a diet rich in poly-phenol-containing foods may have a positive impact on bone health [19]. Incorporating a variety of polyphenol-rich foods in diet, such as berries, green tea, and colorful vegetables, can be a step toward supporting overall bone health.

## 2.2 Vitamin K$_2$

Vitamin K$_2$, a lesser-known member of the vitamin K family, plays a pivotal role in bone health by facilitating the proper distribution of calcium in the body. This essential vitamin activates proteins responsible for calcium incorporation into bone, preventing its accumulation in soft tissues and arterial walls. The result is a reduced risk of osteoporosis and fractures, making it a vital nutraceutical for maintaining strong and healthy bones [20]. One of the key proteins influenced by vitamin K$_2$ is osteocalcin, which regulates the deposition of calcium in bone tissue. Without adequate vitamin K$_2$, this vital process may be compromised, potentially leading to decreased bone mineral density and an increased risk of fractures [21]. Research has shown that vitamin K$_2$ supplementation can lead to increased bone mineral density and reduced fracture risk, particularly in postmenopausal women who are at a higher risk of osteoporosis [22, 23]. Vitamin K$_2$ can be obtained through dietary sources such as fermented foods and certain cheeses, as well as through supplements [24]. Its role in maintaining bone health is increasingly recognized, and healthcare professionals may recommend vitamin K$_2$ supplementation to individuals at risk of bone-related conditions. The adequate intake of vitamin K$_2$, combined with other essential nutrients, a balanced diet, and weight-bearing exercises, is a comprehensive approach to promoting strong and resilient bones throughout life.

## 2.3 Curcumin

Curcumin, the active compound found in the spice turmeric, has gained significant attention for its potent anti-inflammatory and antioxidant properties [25]. While it is well-known for its potential benefits in managing inflammatory conditions, emerging research suggests that curcumin may also support bone health. Inflammation plays a crucial role in bone health, as chronic inflammation can lead to accelerated bone resorption and impaired bone formation. Curcumin, by virtue of its anti-inflammatory properties, has the potential to mitigate this process. It may reduce the activity of osteoclasts (cells responsible for bone resorption) and enhance the function of osteoblasts (cells responsible for bone formation) [2]. Furthermore, curcumin's antioxidant properties can protect bone cells from oxidative stress, which can damage bone tissue. This dual action makes curcumin an appealing nutraceutical for those looking to support their bone health. Several animal and cell culture studies have provided promising results regarding curcumin's effects on bone health. For example, some studies have shown that curcumin can enhance bone density and strength in rats with osteoporosis [26]. Curcumin supplements are widely available, but it is advisable to consult with a healthcare professional before incorporating them into routine. Incorporating turmeric, and by extension, curcumin, into diet through curries, teas, or supplements may provide an additional layer of support for bone health, alongside a balanced diet and regular exercise.

## 2.4 Glucosamine

Glucosamine, a naturally occurring compound in the body, is often associated with joint health due to its role in the formation and repair of cartilage. However, the impact of glucosamine impact extends beyond joints, potentially benefiting bone health as well. While it indirectly supports bones through its promotion of joint function and overall musculoskeletal well-being, it plays a significant role in maintaining the integrity of cartilage and connective tissues [27]. The cartilage that lines joint surfaces and acts as a cushion is essential for joint mobility and protection. This cartilage is made up of specialized cells called chondrocytes, which produce and maintain the extracellular matrix, providing a smooth, low-friction surface for joint movement. Glucosamine, as a precursor to molecules involved in the formation and repair of cartilage, can enhance the health and function of chondrocytes. In turn, this supports the maintenance of cartilage and joint structures, which is crucial for overall musculoskeletal health, including the bones that articulate with these joints. In addition to supporting joint integrity, glucosamine may indirectly contribute to bone health by encouraging physical activity. Joint discomfort and pain can limit an individual's mobility and activity levels, potentially leading to muscle weakening and reduced bone density [28]. By promoting joint comfort and flexibility, glucosamine can help maintain an active lifestyle, which is a key component of overall bone health. Glucosamine supplements are widely available and can be a valuable addition to a comprehensive approach to bone and joint wellness [27]. However, it is important to note that the efficacy of glucosamine can vary among individuals, and not everyone may experience the same level of benefit. Incorporating glucosamine into the health regimen, alongside other bone-supportive measures such as a balanced diet and weight-bearing exercises, can be a proactive step toward preserving both joint and bone health.

## 2.5 Hyaluronic acid

Hyaluronic acid, a naturally occurring substance in the body, is primarily known for its role in maintaining skin hydration [29]. However, it has a broader impact on overall musculoskeletal health. By supporting joint lubrication and cushioning, hyaluronic acid contributes to joint comfort and flexibility, indirectly promoting bone health through its beneficial effects on the entire musculoskeletal system. The synovial fluid in joints, which helps reduce friction and cushion joint movement, contains hyaluronic acid. This fluid is essential for maintaining the function and health of joints, and by extension, bones. Hyaluronic acid acts as a lubricant and shock absorber within the joint space, enabling smooth and pain-free movement [30, 31]. Age-related changes or joint injuries can lead to a decrease in synovial fluid and hyaluronic acid content, resulting in joint discomfort and reduced mobility [32]. When joints are less functional, individuals may become less active, leading to potential muscle weakening and bone density loss. By supporting joint lubrication and reducing joint discomfort, hyaluronic acid indirectly encourages physical activity, which is crucial for maintaining overall bone health. Additionally, hyaluronic acid has been explored for its potential to enhance the body's natural healing processes [33]. Incorporating hyaluronic acid into the health regimen, alongside other measures like a balanced diet and weight-bearing exercises, can help maintain joint comfort, mobility, and overall bone health.

# 3 Modern treatments of arthritis through nutraceuticals

Arthritis is a disease that affects joints, causing inflammation and pain. There are over 100 types of arthritis, each with different symptoms, but stiffness and joint pain are common [34]. The causes of arthritis are not fully understood, but some factors that contribute to its development include genetics, immune system problems, metabolic disorders, and environmental factors like obesity and joint damage [35, 36]. Self-management skills, stress reduction techniques, and regular physical activity are important ways to prevent and manage arthritis. Losing weight can also reduce stress on joints and improve overall health [37].

In recent years, there has been growing interest in the use of nutraceuticals for the prevention and treatment of arthritis [11]. Arthritis is a degenerative joint condition that affects millions of people worldwide and can significantly impact their quality of life [38]. Nutraceuticals are compounds found in certain foods and supplements that may have health benefits beyond their basic nutritional value. One of the key benefits of nutraceuticals in arthritis is their ability to combat inflammation [39]. Inflammation is a common symptom of arthritis and can cause pain, swelling, and stiffness in the joints [40]. Nutraceuticals such as antioxidants, bioactive phenolic compounds, phytosterols, and polyunsaturated fatty acids have been shown to reduce inflammation and joint damage associated with osteoarthritis [39]. Traditional medicine has long utilized nutraceuticals and dietary supplements made from herbs for the treatment of arthritis [11]. In the future, genetic engineering methods or cell-based therapies such as autologous stem cell transplantation and CAR-T-cell therapy may provide a more effective cure for rheumatic diseases, although these approaches are currently still risky and expensive [41, 42]. Nutritional supplements also appear to be involved in events that have an impact on articular cartilage. The balance of anabolic and catabolic signals in joints may be affected by nutraceuticals, which can help maintain the structural integrity of cartilage [43]. Some of the nutraceuticals used for osteoarthritis include fish oil, glucosamine sulfate, chondroitin sulfate, and hyaluronic acid, olive oil, methionine, undenatured type II collagen, and various botanical extracts [44]. It is important to note that while nutraceuticals may be beneficial in the treatment of arthritis, nutraceuticals should not be considered a replacement for traditional medical treatments but rather a complementary approach to overall health and wellness.

## 3.1 Osteoarthritis

Osteoarthritis (OA) is the most common type of arthritis and affects over 35.5 million adults in the US [45]. It is caused by the degradation of cartilage in the joints, leading to pain, stiffness, decreased range of motion, and swelling [46] (Figure 2). Risk factors for OA include age, gender, obesity, and family history. There is no known cure for OA, but treatments to address symptoms include exercise, physical therapy, weight loss, medications, and assistive devices [46]. Nutraceutical treatments such as glucosamine/chondroitin [47], curcuminoids [48], collagen, Terminalia chebula extract [49], eggshell membrane [50], and cucumber extract [51] have shown effectiveness in clinical research [52].

Nutraceutical treatment is one of the options available for managing the symptoms of osteoarthritis (OA) [53]. Various nutraceuticals have been studied, and human clinical research supports their effectiveness in treating OA [51]. Examples of such nutraceuticals include glucosamine/chondroitin, curcuminoids, collagen (types 1 and 2), Terminalia chebula extract, and eggshell membrane [50, 54]. These nutraceuticals are considered safe for use as they have no harmful side effects. One particularly interesting nutraceutical for OA treatment is cucumber extract [51]. Cucumber contains a small amount of iminosugar called ido-BR1, which has been found to reduce the inflammatory cytokine tumor necrosis factor alpha (TNF-alpha). TNF-alpha exacerbates numerous inflammatory responses, including those related to OA. Ido-BR1 reduces TNF-alpha through a novel mechanism that involves binding to the hyaluronic acid receptor (CD44) or preventing the release of an enzyme called sialidase, which increases the capacity of microorganisms to invade and damage tissue, and is released in response to this receptor's interaction with it [55]. However, many cucumber types have undergone extensive breeding to completely eliminate ido-BR1. To ensure the presence of ido-BR1, a water extract made from the fruit of the cucumber (*Curcumis sativus*) called Q-actin can be used. A daily dose of only 20 mg of Q-actin has been shown to be effective, which is the lowest of any nutraceutical used for treating OA [55].

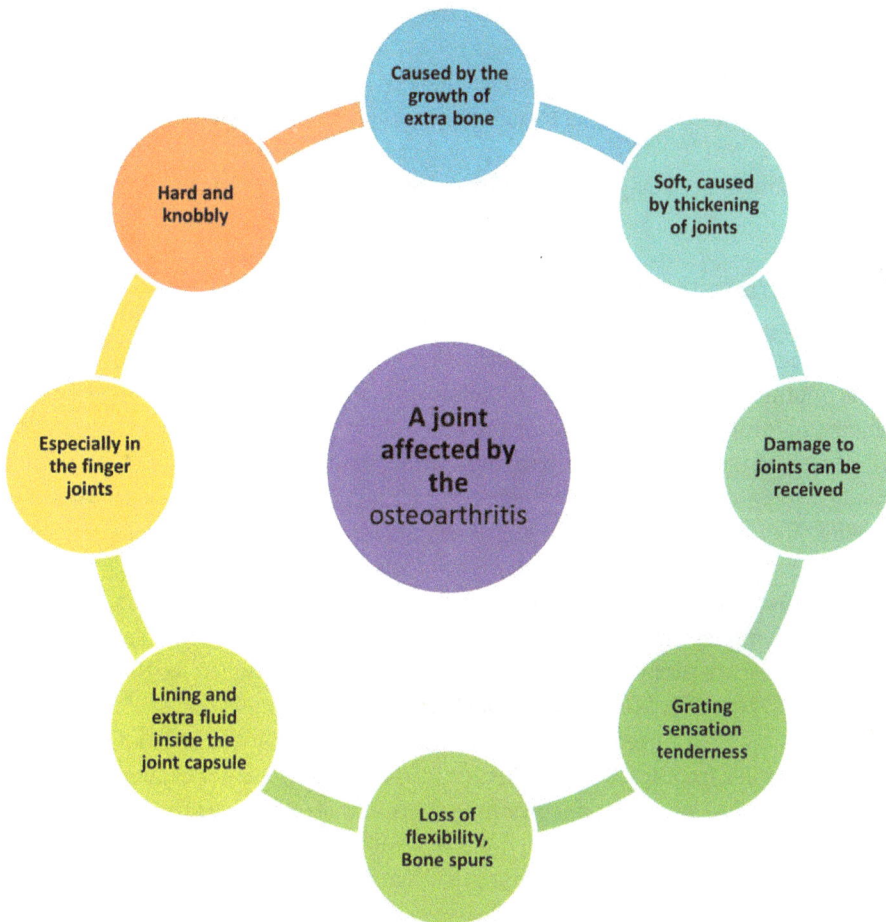

**Figure 2:** Major symptoms of osteoarthritis, which cause problems in bones.

### 3.1.1 Cucumber Ido-BR1: nutraceutical treatment for osteoarthritis

Inflammation plays a crucial role in the development of osteoarthritis (OA) [56]. The inflammatory cytokine TNF-alpha) is generated during acute inflammation, leading to cartilage breakdown in the affected joint. High levels of TNF-α exacerbate the inflammatory response and worsen OA [57]. Fortunately, cucumbers contain a little-known iminosugar called ido-BR1 that can reduce TNF-α levels [58]. Ido-BR1 appears to reduce TNF-α through a novel mechanism involving the hyaluronic acid receptor (CD44) and the enzyme sialidase. However, many cucumber varieties have had ido-BR1 bred out of them. Q-actin, a water extract made from cucumber fruit, has been chosen for its ido-BR1 content [55]. Daily consumption of just 20 mg of Q-actin is sufficient to reduce TNF-α levels and alleviate OA symptoms, making it the lowest-dose nutraceutical used for treating OA [59].

Overall, cucumbers and their extracts hold promise as a natural treatment for OA inflammation, but care must be taken to ensure that the cucumber variety used contains ido-BR1. With further research, cucumber-derived nutraceuticals could become a key component in managing OA and other inflammatory conditions.

## 3.2 Rheumatoid arthritis

Rheumatoid arthritis is a chronic inflammatory disorder that affects joints and can damage other parts of the body. The immune system attacks healthy joint tissues, causing painful swelling, bone erosion, and joint distortion. It is more common in women and middle-aged people, and those with a family history of the disease or who smoke. Symptoms start in smaller joints and then spread to other joints in the body. Patients may also experience fatigue, fever, tender and warm joints, joint stiffness, and eye dryness and pain, among other symptoms [34, 60, 61].

Treatment of rheumatoid arthritis often involves the use of medication, including biological therapies that suppress the immune response causing inflammation [62]. In addition, there are nutraceutical treatments that have been studied for their potential to alleviate symptoms of rheumatoid arthritis [9]. For example, omega-3 fatty acids found in fish oil have been shown to reduce inflammation in the body and may provide some relief from joint pain and stiffness [63]. Curcumin, a compound found in turmeric, also has anti-inflammatory properties and may help reduce joint pain and inflammation [64]. Ginger and green tea have also been studied for their potential anti-inflammatory effects [65]. While more research is needed to fully understand the potential benefits of these nutraceuticals, they may offer a complementary approach to traditional medications for managing rheumatoid arthritis symptoms. It is important to consult with a healthcare professional before starting any new treatments or supplements.

# 4 Summary

Nutraceuticals offer a promising alternative to traditional treatments for bone health and arthritis disorders. The potential benefits of using nutraceuticals include their safety, cost-effectiveness, and noninvasive nature. Nutraceuticals such as vitamins, minerals, antioxidants, and plant extracts possess anti-inflammatory, analgesic, and immune-modulating effects, which can improve bone health and alleviate the symptoms of arthritis disorders. However, the use of nutraceuticals also poses challenges and limitations, such as the lack of standardization and regulation.

Future research should focus on standardizing the production and testing of nutraceuticals and evaluating their long-term safety and efficacy. Additionally, there is a need for more clinical trials to determine the optimal dosage and duration of nutraceuticals for specific bone health and arthritis disorders. Integrating nutraceuticals with other treatments, such as physical therapy and exercise, may also enhance their therapeutic potential. Overall, the potential of nutraceuticals in promoting bone health and managing arthritis disorders warrants further investigation and exploration.

**Acknowledgements:** All of the authors would like to express their gratitude for the joint collaborative work between the Department of Chemistry at the University of Swabi, Khyber Pakhtunkhwa, Pakistan, Department of Statistics Islamia College University Peshawar, Pakistan and the Department of Chemistry at Sardar Bahadur Khan Women's University, Quetta, Pakistan.

# References

[1]   Frestedt, J.L., et al. 2008. A natural mineral supplement provides relief from knee osteoarthritis symptoms: A randomized controlled pilot trial. Nutrition Journal, 7(1), 1–8.

[2]   Pandey, M.K., et al. 2018. Dietary nutraceuticals as backbone for bone health. Biotechnology Advances, 36(6), 1633–1648.

[3]   Espín, J.C., García-Conesa, M.T. and Tomás-Barberán, F.A. 2007. Nutraceuticals: Facts and fiction. Phytochemistry, 68(22–24), 2986–3008.

[4]   Pandey, M., Verma, R.K. and Saraf, S.A. 2010. Nutraceuticals: New era of medicine and health. Asian Journal of Pharmaceutical and Clinical Research, 3(1), 11–15.

[5]   Shinde, N., et al. 2014. Nutraceuticals: A Review on current status. Research Journal of Pharmacy and Technology, 7(1), 110–113.

[6]   Meenan, R.F., Gertman, P.M. and Mason, J.H. 1980. Measuring health status in arthritis. Arthritis & Rheumatism, 23(2), 146–152.

[7]   Lin, Y.-J., Anzaghe, M. and Schülke, S. 2020. Update on the pathomechanism, diagnosis, and treatment options for rheumatoid arthritis. Cells, 9(4), 880.

[8]   Lo, J., Chan, L. and Flynn, S. 2021. A systematic review of the incidence, prevalence, costs, and activity and work limitations of amputation, osteoarthritis, rheumatoid arthritis, back pain, multiple sclerosis, spinal cord injury, stroke, and traumatic brain injury in the United States: A 2019 update. Archives of Physical Medicine and Rehabilitation, 102(1), 115–131.

[9]   Aghamohammadi, D., et al. 2020. Nutraceutical supplements in management of pain and disability in osteoarthritis: A systematic review and meta-analysis of randomized clinical trials. Scientific Reports, 10(1), 20892.

[10]  Keizo, U. 2021. Regulatory aspects of nutraceuticals: Japanese perspective. In Ramesh C. Gupta, Lall, R., Srivastava, A., (Eds.), *Nutraceuticals*, Second Edition, Academic Press, 1299–1307.

[11]  Oppedisano, F., et al. 2021. The role of nutraceuticals in osteoarthritis prevention and treatment: Focus on n-3 PUFAs. Oxidative Medicine and Cellular Longevity, 2021.

[12]  Bruno, J. and Sum, M. 2021. Nutrition and lifestyle approaches to optimize skeletal health. Osteoporosis: A Clinical Casebook, 17–29.

[13]  Gul, K., Singh, A. and Jabeen, R. 2016. Nutraceuticals and functional foods: The foods for the future world. Critical Reviews in Food Science and Nutrition, 56(16), 2617–2627.

[14] Yashodhara, B., et al. 2009. Omega-3 fatty acids: A comprehensive review of their role in health and disease. Postgraduate Medical Journal, 85(1000), 84–90.

[15] Abbas, M., et al. 2017. Natural polyphenols: An overview. International Journal of Food Properties, 20(8), 1689–1699.

[16] Scalbert, A., Johnson, I.T. and Saltmarsh, M. 2005. Polyphenols: Antioxidants and beyond. The American Journal of Clinical Nutrition, 81(1), 215S–217S.

[17] Rao, L., Kang, N. and Rao, A. 2012. Polyphenol antioxidants and bone health: A review. Phytochemicals – a global perspective of their role in nutrition and health.

[18] Inchingolo, A.D., et al. 2022. Effects of resveratrol, curcumin and quercetin supplementation on bone metabolism – A systematic review. Nutrients, 14(17), 3519.

[19] Babosova, R., et al. 2020. The impact of flavonoid epicatechin on compact bone microstructure in rabbits. Biologia, 75, 935–941.

[20] Myneni, V.D. and Mezey, E. 2018. Immunomodulatory effect of vitamin K2: Implications for bone health. Oral Diseases, 24(1–2), 67–71.

[21] Wen, L., et al. 2018. Vitamin K-dependent proteins involved in bone and cardiovascular health. Molecular Medicine Reports, 18(1), 3–15.

[22] Huang, Z.-B., et al. 2015. Does vitamin K2 play a role in the prevention and treatment of osteoporosis for postmenopausal women: A meta-analysis of randomized controlled trials. Osteoporosis International, 26, 1175–1186.

[23] Iwamoto, J. 2014. Vitamin K2 therapy for postmenopausal osteoporosis. Nutrients, 6(5), 1971–1980.

[24] Simes, D.C., et al. 2020. Vitamin K as a diet supplement with impact in human health: Current evidence in age-related diseases. Nutrients, 12(1), 138.

[25] Lestari, M.L. and Indrayanto, G. 2014. Curcumin. Profiles of Drug Substances, Excipients and Related Methodology, 39, 113–204.

[26] Vaiserman, A., et al. 2020. Curcumin: A therapeutic potential in ageing-related disorders. PharmaNutrition, 14, 100226.

[27] Kang, H.E., et al. 2022. Pharmacokinetic comparison of chitosan-derived and biofermentation-derived glucosamine in nutritional supplement for bone health. Nutrients, 14(15), 3213.

[28] Kirkham, S. and Samarasinghe, R. 2009. Glucosamine. Journal of Orthopaedic Surgery, 17(1), 72–76.

[29] Juncan, A.M., et al. 2021. Advantages of hyaluronic acid and its combination with other bioactive ingredients in cosmeceuticals. Molecules, 26(15), 4429.

[30] Zhai, P., et al. 2020. The application of hyaluronic acid in bone regeneration. International Journal of Biological Macromolecules, 151, 1224–1239.

[31] Salwowska, N.M., et al. 2016. Physiochemical properties and application of hyaluronic acid: A systematic review. Journal of Cosmetic Dermatology, 15(4), 520–526.

[32] Temple-Wong, M.M., et al. 2016. Hyaluronan concentration and size distribution in human knee synovial fluid: Variations with age and cartilage degeneration. Arthritis Research & Therapy, 18(1), 1–8.

[33] Chen, L.H., et al. 2018. Hyaluronic acid, an efficient biomacromolecule for treatment of inflammatory skin and joint diseases: A review of recent developments and critical appraisal of preclinical and clinical investigations. International Journal of Biological Macromolecules, 116, 572–584.

[34] Firestein, G.S. 2003. Evolving concepts of rheumatoid arthritis. Nature, 423(6937), 356–361.

[35] Berenbaum, F., et al. 2018. Modern-day environmental factors in the pathogenesis of osteoarthritis. Nature Reviews Rheumatology, 14(11), 674–681.

[36] Reynolds, M.A. 2000, 2014. Modifiable risk factors in periodontitis: At the intersection of aging and disease. Periodontology, 64(1), 7–19.

[37] Keefe, F.J., Somers, T.J. and Martire, L.M. 2008. Psychologic interventions and lifestyle modifications for arthritis pain management. Rheumatic Disease Clinics of North America, 34(2), 351–368.

[38] Pincus, T. and Callahan, L.F. 1993. What is the natural history of rheumatoid arthritis? Rheumatic Disease Clinics of North America, 19(1), 123–151.

[39] Kalra, E.K. 2003. Nutraceutical-definition and introduction. Aaps Pharmscience, 5(3), 27–28.

[40] Spel, L. and Martinon, F. 2020. Inflammasomes contributing to inflammation in arthritis. Immunological Reviews, 294(1), 48–62.

[41] Franceschetti, T. and De Bari, C. 2017. The potential role of adult stem cells in the management of the rheumatic diseases. Therapeutic Advances in Musculoskeletal Disease, 9(7), 165–179.

[42] Orvain, C., et al. 2021. Is there a place for chimeric antigen receptor–T cells in the treatment of chronic autoimmune rheumatic diseases? Arthritis & Rheumatology, 73(11), 1954–1965.

[43] Staff, U. 2022. Nutraceutical treatment in knee osteoarthritis – arthritis-presentation of a dietary plan by means of additional nutraceuticals in lieu of the detrimental side effects of current conventional medicine. Open Journal of Orthopedics, 12(7), 303–326.

[44] Desai, A., Shendge, P.N. and Anand, S.S. 2021. Evidence-based nutraceuticals for osteoarthritis: A. International Journal of Orthopaedics, 7(2), 846–853.

[45] Dillon, C.F., et al. 2006. Prevalence of knee osteoarthritis in the United States: Arthritis data from the Third National Health and Nutrition Examination Survey 1991–94. The Journal of Rheumatology, 33(11), 2271–2279.

[46] Wieland, H.A., et al. 2005. Osteoarthritis – An untreatable disease?. Nature Reviews Drug Discovery, 4(4), 331–344.

[47] Deal, C.L. and Moskowitz, R.W. 1999. Nutraceuticals as therapeutic agents in osteoarthritis: The role of glucosamine, chondroitin sulfate, and collagen hydrolysate. Rheumatic Disease Clinics of North America, 25(2), 379–395.

[48] Henrotin, Y., Priem, F. and Mobasheri, A. 2013. *Curcumin: A New Paradigm and Therapeutic Opportunity for the Treatment of Osteoarthritis: Curcumin for Osteoarthritis Management.* Vol. 2, Springerplus, pp. 1–9.

[49] Lopez, H., et al. 2017. Effects of dietary supplementation with a standardized aqueous extract of Terminalia chebula fruit (AyuFlex®) on joint mobility, comfort, and functional capacity in healthy overweight subjects: A randomized placebo-controlled clinical trial. BMC Complementary and Alternative Medicine, 17(1), 1–18.

[50] Ruff, K.J., et al. 2009. Eggshell membrane in the treatment of pain and stiffness from osteoarthritis of the knee: A randomized, multicenter, double-blind, placebo-controlled clinical study. Clinical Rheumatology, 28, 907–914.

[51] Pérez-Piñero, S., et al. 2022. Effectiveness of a cucumber extract supplement on articular pain in patients with knee osteoarthritis: A randomized double-blind controlled clinical trial. Applied Sciences, 13(1), 485.

[52] Henrotin, Y., Sanchez, C. and Balligand, M. 2005. Pharmaceutical and nutraceutical management of canine osteoarthritis: Present and future perspectives. The Veterinary Journal, 170(1), 113–123.

[53] Ragle, R.L. and Sawitzke, A.D. 2012. Nutraceuticals in the management of osteoarthritis: A critical review. Drugs & Aging, 29, 717–731.

[54] Clegg, D.O., et al. 2006. Glucosamine, chondroitin sulfate, and the two in combination for painful knee osteoarthritis. New England Journal of Medicine, 354(8), 795–808.

[55] Nash, R.J., et al. 2020. Iminosugar idoBR1 isolated from Cucumber Cucumis sativus reduces inflammatory activity. ACS Omega, 5(26), 16263–16271.

[56] Abusudah, W. 2020. *The Interrelationships among Vitamin D, Inflammation and Obesity.* Howard University.

[57] Jang, D.-I., et al. 2021. The role of tumor necrosis factor alpha (TNF-α) in autoimmune disease and current TNF-α inhibitors in therapeutics. International Journal of Molecular Sciences, 22(5), 2719.

[58] Nash, R.J., Mafongang, A., Singh, H., Singwe-Ngandeu, M., Penkova, Y.B., Kaur, T. and Akbar, J. 2023. Standardised ido-BR1 Cucumber extract improved parameters linked to moderate osteoarthritis in a placebo-controlled study. Current Rheumatology Reviews, 19(3), 345–351.

[59]  Grover, A.K. and Samson, S.E. 2015. Benefits of antioxidant supplements for knee osteoarthritis: Rationale and reality. Nutrition Journal, 15(1), 1–13.

[60]  Guidelines, A.C.o.R.S.o.R.A. 2002. Guidelines for the management of rheumatoid arthritis: 2002 update. Arthritis & Rheumatism, 46(2), 328–346.

[61]  McInnes, I.B. and Schett, G. 2011. The pathogenesis of rheumatoid arthritis. New England Journal of Medicine, 365(23), 2205–2219.

[62]  Van Schouwenburg, P.A., Rispens, T. and Wolbink, G.J. 2013. Immunogenicity of anti-TNF biologic therapies for rheumatoid arthritis. Nature Reviews Rheumatology, 9(3), 164–172.

[63]  Cleland, L.G., James, M.J. and Proudman, S.M. 2003. The role of fish oils in the treatment of rheumatoid arthritis. Drugs, 63, 845–853.

[64]  Razavi, B.M., Ghasemzadeh Rahbardar, M. and Hosseinzadeh, H. 2021. A review of therapeutic potentials of turmeric (Curcuma longa) and its active constituent, curcumin, on inflammatory disorders, pain, and their related patents. Phytotherapy Research, 35(12), 6489–6513.

[65]  Fechtner, S., et al. 2017. Molecular insights into the differences in anti-inflammatory activities of green tea catechins on IL-1β signaling in rheumatoid arthritis synovial fibroblasts. Toxicology and Applied Pharmacology, 329, 112–120.

Hemanth Kumar Manikyam*

# Chapter 11
# Medicinal plants and alternative therapies for reproductive system health

**Abstract:** Reproductive healthcare is a critical component of the overall well being of individuals, encompassing various dimensions related to the functioning of the reproductive system and the mental, physical, and social states of individuals. Global organizations like the WHO and the UN have been actively engaged in addressing reproductive health issues, aiming to achieve sustainable development and poverty eradication. This overview explores the multifaceted nature of reproductive healthcare, with a specific focus on the significance of nutritional and pharmacological interventions. The importance of reproductive health care services for both men and women is discussed, including gynecological exams, contraception, infertility treatments, pregnancy care, and STI testing and treatment. Despite its critical importance, reproductive healthcare faces challenges related to access, stigma, cost, and cultural barriers. The abstract also delves into women's reproductive health, emphasizing the need for protection and care, through initiatives targeting birth control, depression management, infertility treatments, and addressing gynecologic cancers and STIs. Men's reproductive health is highlighted as a vital component of societal development, with common male reproductive issues examined. Additionally, the abstract underscores the role of pharmacological therapies in managing reproductive health pathologies while also emphasizing the importance of preventive care medicines. Furthermore, the significance of micronutrients in the reproductive health of both men and women is discussed, along with the role of specific vitamins and minerals. Overall, this overview emphasizes the complexities of reproductive healthcare and the potential benefits of various interventions to improve global reproductive health outcomes.

# 1 Introduction

All issues related to the reproductive system and its functions and processes, including the state of mental, physical and social well being, is termed as reproductive health of an individual. Sexual and reproductive health is fundamental to all individuals, irrespective of the gender, and to the socio-economic development of the country

---
*Corresponding author: Hemanth Kumar Manikyam, Department of Chemistry, Natural Products Chemistry Division, Faculty of Science, North East Frontier Technical University, Aalo Post Office, West Siang Dist., Aalo 791001, Arunachal Pradesh, India, e-mail: phytochem2@gmail.com

https://doi.org/10.1515/9783111317601-011

and its communities [1]. The WHO and the United Nations collectively work in advising the global policy makers, along with NGOs, government and private agencies, and other research institutions[1]. As part of the Millennium goals, many countries agreed to take measures towards sustainable development and eradicating poverty. Improving maternal health, reducing early birth defects, reducing puberty related problems, and reduction in STDs (Sexually Transmitted diseases) through sex education etc. [1] are key agenda taken up by WHO and the United Nations to improve the reproductive and sexual health of individuals.

Apart from socio-economical uplifting, nutritional and dietary habits of an individual directly play important roles in the reproductive and sexual health of an individual. Management of metabolic and mental stress is a key criterion that plays a crucial role in the reproductive health status of an individual [2]. Reproductive healthcare refers to the medical and healthcare services that are designed to support and promote reproductive health in individuals, particularly in the areas of sexual health, fertility, and family planning [3, 4]. Reproductive healthcare is important for both men and women, as reproductive health issues can have significant impact on both physical and emotional well-being [3, 4].

In this chapter, we will discuss the different aspects of reproductive healthcare, including the importance of reproductive health, the different types of reproductive healthcare services available, and the key issues and challenges in the field of reproductive healthcare.

## 1.1 Importance of reproductive health

Reproductive health is a crucial aspect of the overall health and well-being of an individual. It encompasses a wide range of issues, including sexual health, fertility, pregnancy, childbirth, and menopause. Poor reproductive health can have significant physical and emotional impact, including infertility, sexually transmitted infections, pregnancy complications, and mental health issues [1, 3–5]. Reproductive healthcare services play a critical role in promoting and maintaining reproductive health. These services include preventative care such as regular gynecological exams and screenings for sexually transmitted infections, as well as treatments for reproductive health issues, such as infertility treatment and contraception [6, 7].

There are many different types of reproductive healthcare services available, including:

**Gynecological exams:** Regular gynecological exams are an important aspect of reproductive healthcare for women. These exams typically involve a pelvic exam, a Pap test to screen for cervical cancer, and other tests and screenings as needed [8, 9].

**Contraception**: Contraception is an important aspect of reproductive healthcare for both men and women. There are many different types of contraception available, including hormonal and nonhormonal methods [9].

**Infertility treatments:** Infertility treatments are available for couples who are struggling to conceive. These treatments may include fertility medications, intrauterine insemination (IUI), or *in vitro* fertilization (IVF) [8, 9].

**Pregnancy care:** Pregnancy care involves monitoring the health of the mother and fetus throughout pregnancy, and may include regular checkups, ultrasound exams, and prenatal testing [8, 9].

**Sexually transmitted infection (STI) testing and treatment:** STI testing and treatment are important aspects of reproductive healthcare, as untreated STIs can lead to serious health complications [8, 9].

## 1.2 Challenges in reproductive healthcare

Despite the importance of reproductive healthcare, there are still many challenges in the field. These challenges include:
–   Access to reproductive healthcare can be a challenge, particularly for individuals who live in rural or remote areas or who do not have access to transportation [8, 10].
–   Reproductive health issues, such as infertility or STIs, can be stigmatized, which can make it difficult for individuals to seek care or discuss these issues with their healthcare providers [8, 10].
–   Reproductive healthcare services can be expensive, particularly infertility treatments and some types of contraception. This can make it difficult for individuals to access the care they need [8, 10].
–   Cultural and linguistic barriers can make it difficult for individuals to access reproductive healthcare services or to communicate effectively with their healthcare providers [8, 10].

## 1.3 Women's reproductive health and related issues

As per the Center for Disease Control and Prevention (CDC), woman's reproductive health is sensitive and a complex system that needs to be protected from infections, injuries, and long-term health issues. Protection of women's reproductive health means control over health and preventing birth defects and also menopause problems [11]. Millennium goals of the WHO and the UN give priority to improve women's reproductive health, particularly in providing healthcare services during birth control through contraceptives, depression management, infertility, menopause, and other health issues like Endometriosis, Menorrhagia, Polycystic ovary syndrome (PCOs), Interstitial cystitis (IC) uterine fibroids, gynecologic cancers like cervical cancer (caused by Human papillomavirus (HPV)), ovarian cancer, uterine cancer, vaginal and vulvar cancers, and sexually transmitted diseases like HIV/AIDS [11].

## 1.4 Men's reproductive health

Men's reproductive health also has equal importance in communities for the socioeconomic development, which is overlooked in discussions of sexual health and reproductive health issues – particularly, contraception and infertility are perceived as female-related issues [12, 13]. Men and their partners needed to protect themselves from STDs and maintain their reproductive health by preserving fertility. Men are also prone to cancers associated with the reproductive system, which need to be addressed in the healthcare policies [13]. Most common problems that cause infertility in males include problem of ejection of semen, low levels of sperms, low motility sperms, and low morphology of sperms [12, 13].

# 2 Pharmacological therapies and their limitations

In developing countries, growing poor reproductive health has become a major concern; however, reproductive healthcare medicines are not available to a majority of the population in these countries. The wellbeing of families and communities are threatened by lack of reproductive health medicines [13, 14]. To ensure accessibility to essential reproductive health medicines like contraceptives, medicines to prevent or to treat STDs like HIV/AIDS, and other medicines for healthy pregnancy and birth defects, there is a need for strong long-term commitments from governments for a wide range of policies and activities to ensure finance and good supply chain systems [10, 11, 14]. Because of the inadequate availability of reproductive health medicines, many developing countries have poor reproductive healthcare services and they are unreliable. Good quality and affordable reproductive health medicines are needed, which can significantly reduce reproductive health issues [14]. The national drug policy should ensure:

- **Access:** affordability and availability of essential medicines [15]
- **Quality:** efficacy, quality and safety of the medicines [15]
- **Rational use:** availability of cost-effective and therapeutically valuable medicines for consumers and healthcare professionals [15]

The national drug policy of developing countries already list the essential medicines for reproductive health management in general, such as anesthetics, analgesics, antipyretics, nonsteroidal anti-inflammatory medicines, antidepressants, antiallergics, anaphylaxis medicines, anti-infective medicines, contraceptives, antimalarial, antiviral mostly for HIV/AIDS and micronutrients (vitamins and minerals) [15].

Recreational drugs are another problem faced in reproductive healthcare, where unauthorized drug usage leads to impotency, affecting the reproductive system. Every medication has its own benefits and adverse effects. Introduction of preventive care medicines is always the best option in the reproductive healthcare system, apart from the availability of contraceptives [14].

Nutritional aspects such as poor intake of micro and macro minerals, vitamins, and proteins are mostly associated with poor reproductive health and performance because of imbalance in moods, hormones, and the development of quality sperms in men and poor ovulatory maturation in women [17]. Preventive care is most often associated with nutritional aspects. Traditional medical systems like Ayurveda, Traditional Chinese Medicine (TCM), Tibetan medicine, and Unani address the reproductive well-being issues and provide preventive care [4, 16].

# 3 Role of micronutrients in reproductive health

Vitamins and minerals play an important role in the reproductive health of men and women. Common functions like hormone balancing, mood enhancement, energy, immunity, and regulating the reproductive cycle are key functions of micronutrients [6, 7]. In women, for a healthy reproductive health, micronutrients play a role in ovulation, healthy egg production, thyroid function, menstruation cycle and healthy pregnancy. In men, micronutrients play an important role in stamina building, prostate health, and healthy sperm production. Most birth defects are associated with deficiency of micronutrients, which is a serious health issue that is alarming in developing countries [11–13]. There are several key nutrients that are important for reproductive health in both men and women.

## 3.1 Minerals

Minerals play a vital role in maintaining overall health and well-being, including reproductive health [18]. A balanced diet that includes a variety of nutrient-dense foods can help to ensure that the body is getting all the necessary minerals to support reproductive health. If you have concerns about your reproductive health, it is important to speak with a healthcare professional.

- Zinc is an essential mineral for male and female reproductive health. In males, it is important for the development of healthy sperm and in females, it plays a crucial role in the maturation of eggs. Zinc is also necessary for the production of hormones such as testosterone, which is vital for male reproductive health [18].
- Iron is important for the production of hemoglobin, which is necessary for carrying oxygen in the blood. In women, iron deficiency can lead to anemia, which can cause menstrual irregularities and infertility [18].
- Calcium is important for bone health, and it is also necessary for muscle contractions, including the contractions of the uterus during labor and delivery. It is important for pregnant women to get enough calcium to support the development of their baby's bones and teeth [18].
- Magnesium is important for maintaining a healthy reproductive function in both men and women. It helps to regulate hormones and supports healthy sperm and egg production [18].
- Selenium is important for male reproductive health as it is necessary for the production of healthy sperm. It also has antioxidant properties that help to protect the reproductive system from damage caused by free radicals [18].
- Iodine is important for thyroid function, which plays a crucial role in the regulation of reproductive hormones. It is important for pregnant women to get enough iodine to support the healthy development of their baby's brain and nervous system [18].

## 3.2 Vitamins

Vitamins are essential nutrients that play a crucial role in maintaining the overall health, including reproductive health. A balanced diet that includes a variety of nutrient-dense foods can help to ensure that the body is getting all the necessary vitamins to support reproductive health [19]. If you have concerns about your reproductive health, it is important to speak with a healthcare professional.

- Vitamin A is important for the development and maintenance of healthy reproductive tissues in both men and women. It also plays a role in the development of a healthy fetus during pregnancy [19].
- Vitamin B6 is important for the regulation of reproductive hormones, including progesterone and estrogen. It also helps to reduce symptoms of premenstrual syndrome (PMS) in women [19].
- Vitamin B12 is important for the production of healthy sperm and eggs. It also helps to maintain the health of the nervous system, which is important for sexual function [19].
- Vitamin C is important for the development and maintenance of healthy reproductive tissues in both men and women. It also helps to protect against damage caused by free radicals [19].

- Vitamin D is important for the absorption of calcium, which is necessary for the development and maintenance of healthy bones [19]. It is important for pregnant women to get enough vitamin D to support the development of their baby's bones and teeth. It also plays a role in the regulation of reproductive hormones. It also plays an important role in regulating the menstrual cycle in women [19].
- Vitamin E is important for the development and maintenance of healthy reproductive tissues in both men and women. It also has antioxidant properties that help to protect against damage caused by free radicals [19].
- Folate is important for the healthy development of a baby's neural tube during the early stages of pregnancy. It also plays a role in sperm and egg production [19].
- Vitamin K is important for blood clotting, which is necessary during childbirth [19].

# 4 Herbal medicines in reproductive healthcare

There are a number of herbal medicines that have traditionally been used to support reproductive health in both men and women. Ayurvedic medicine, a traditional Indian system of medicine, offers various herbal remedies to treat reproductive health problems [8–10].

## 4.1 Withania somnifera

*Withania somnifera* (Ashwagandha) is a popular Ayurvedic herb used to treat male infertility, low sperm count, and low libido. It is believed to improve sperm quality and motility, and enhance overall sexual function [20]. It has been used for centuries to treat a wide range of health conditions, including reproductive health problems. In recent years, several clinical studies have investigated the potential benefits of Ashwagandha for reproductive health issues [20].

Ashwagandha has been shown to have a positive effect on male fertility. A study showed that taking Ashwagandha root extract for 90 days resulted in an increase in sperm count, semen volume, and sperm motility in infertile men [20]. Another study showed improved semen quality in men with oligospermia with the intake of Ashwagandha [20].

Ashwagandha may also have benefits for female fertility. Daily intake of Ashwagandha root powder for 20 days resulted in an increase in serum progesterone levels in women with luteal phase defect, a condition that can cause infertility [20]. Ashwagandha may also be helpful in managing PCOS, a common reproductive health condition that can cause infertility. Intake of Ashwagandha root extract for 12 weeks resulted in decreased serum testosterone levels and an improvement in insulin sensitivity in women with PCOS [20]. Ashwagandha may be an effective herbal therapy in managing menopausal symptoms [20], such as hot flashes and sleep disturbances, as shown in a study – consumption of Ashwagandha root extract for 8 weeks resulted in a significant reduction in the frequency and severity of hot flashes in postmenopausal women [20].

Ashwagandha has been shown to have a positive effect on stress and anxiety levels, which can impact reproductive health [20]. Treatment of adults with a history of chronic stress with Ashwagandha root extract for 6 weeks resulted in a significant reduction in stress and anxiety levels, the ultimate result of which could be improvement of overall health and well-being, including reproductive health [20]. Overall, Ashwagandha shows promise as an effective and safe herbal remedy for a variety of reproductive health problems [20]. However, more research is needed to fully understand its mechanisms of action and to determine optimal dosages and durations of treatment.

## 4.2 Asparagus racemosus

*Asparagus racemosus* (Shatavari) is an Ayurvedic herb used to treat female reproductive problems such as menstrual irregularities, infertility, and menopausal symptoms [21]. It is believed to regulate hormones, improve ovulation, and support healthy reproductive tissues.

## 4.3 Tribulus terrestris

*Tribulus terrestris* (Gokshura) is an Ayurvedic herb used to treat male infertility and low libido [22]. It is believed to improve sperm count, motility, and quality, and enhance sexual function.

## 4.4 Symplocos racemosa

*Symplocos racemosa* (Lodhra) is an Ayurvedic herb used to treat female reproductive problems such as menstrual irregularities, heavy menstrual bleeding, and uterine fibroids [23]. It is believed to regulate hormones, reduce inflammation, and support healthy reproductive tissues.

## 4.5 Mucuna pruriens

*Mucuna pruriens* (Kapikacchu) is an Ayurvedic herb used to treat male infertility and low libido [24]. It is believed to improve sperm count, motility, and quality, and enhance sexual function.

## 4.6 Boerhavia diffusa

*Boerhavia diffusa* (Punarnava) is an Ayurvedic herb used to treat male and female reproductive problems such as infertility and urinary tract infections [25]. It is believed to improve kidney function, reduce inflammation, and support healthy reproductive tissues.

## 4.7 Triphala

Triphala is an Ayurvedic combination of three fruits – Amla (*Emblica officinalis*), Haritaki (*Terminalia chebula*), and Bibhitaki (*Terminalia bellerica*) [26]. It is believed to have a range of health benefits, including improving digestive function, reducing inflammation, and supporting healthy reproductive tissues.

# 5 Vajikarna: An ancient branch of Ayurvedic medicine

Vajikarna is a branch of Ayurvedic medicine dated back to 1,500 BCE that deals with the diagnosis and treatment of male and female reproductive health problems [27]. Vajikarna treatment aims to enhance reproductive health by improving the quality and quantity of semen and ova, as well as promoting healthy sexual function [27]. The treatment approach includes a combination of herbal remedies, dietary changes, lifestyle modifications, and yoga and meditation practices [27]. Some commonly used Ayurvedic herbs in Vajikarna treatment include Ashwagandha, Shatavari, Gokshura, Kapikacchu, and Triphala [21–27]. Dietary changes may involve the consumption of nutrient-dense foods such as fruits, vegetables, and whole grains, while avoiding processed and junk food [27]. Lifestyle modifications may include the adoption of a regular exercise routine, practicing stress-management techniques, and getting adequate sleep [27]. Yoga and meditation practices can help to reduce stress and anxiety, which can have a negative impact on reproductive health [27]. Vajikarna treatment can be effective for a range of male and female reproductive health problems such as low libido, erectile dysfunction, premature ejaculation, infertility, and menstrual irregularities [27] (Table 1).

**Table 1:** Aphrodisiac herbs reported in Ayurveda as Vajikaran Rasayan. [28, 29]

| Hindi name | Botanical name | Family | Parts used | Uses |
|---|---|---|---|---|
| Akarkara | *Anacyclus pyrethrum* DC | Asteraceae | Dried roots | Strengthening and aphrodisiac |
| Akharot | *Juglans regia* Linn. | Juglandaceae | Dried cotylcdous | Strengthening, semen enhancer, and aphrodisiac |
| Adarakha | *Zingiber officinalis* Rosc. | Zingiberaceae | Fresh rhizomes | For male reproductive organs |
| Bhrngaraja | *Elcipta alba* nassle | Asteraceae | Whole plant | Regenerative and strengthening |
| Mandukaparni | *Bacopa monnieri* Linn. | Scrophularaceae | Dried whole plant | Longevity |
| Anar | *Punica granatum* Linn. | Punicaceae | Dried seed | For male reproductive organs and strength |
| Gambhari | *Gmeline arborea* Roxb. | Verbenaceae | Dried fruit | Longevity and good for testicles |
| Ganna | *Saccharum officinarum* Linn. | Podceae | Dried stem | Strengthen testicles and semen production |
| Jayata | *Sesbana sesban* Linn. | Fabaceae | Fresh & dried root | Longevity |
| Talmakhana | *Asteracantha longifolia* Nees | Acanthaceae | Whole plant seed | Aphrodisiac |
| Makoya | *Solanum nigrum* Linn. | Solanaceae | Dried whole plant | For male reproductive organs and strength |
| Kaitha | *Feronia limonia* Linn. | Rutaceae | Dried pulp of mature fruit | For male reproductive organs and strength |
| Mahuwa | *Madhuca indica* | Saptoceae | Flower | Good for male reproductive organs and strength |
| Tesu | *Buteamono sperma* Lam. | Fabaceae | Dried stem bark | Good for male reproductive organs and strength |
| Gandhaprasarini | *Paederia foetida* Linn. | Rubiaceae | Whole plant | Good for male reproductive organs and strength |
| Piyal | *Buchanania lanzan* Spreng | Anacardiaceae | Seed | Good for male reproductive organs and strength |
| Chaval | *Oryza sativa* Linn. | Poaceae | Dried root | Good for male reproductive organs, strength, and longevity |

**Table 1** (continued)

| Hindi name | Botanical name | Family | Parts used | Uses |
|---|---|---|---|---|
| Shankhapusphi | *Convolvulus pluricaulis* Chois | Convolulacea | Whole plant | Good for male reproductive organs, strength, and longevity |
| Vidarikanda | *Pueraria tuberosa* DC | Leguminosae | Sliced & dried pieces of tuberous root | Good for male reproductive organs, strength, and longevity |
| Basanaay | *Aconitum chasmanthum* | Ranunculaceae | Dried roots | Good for male reproductive organs, strength, and longevity |
| Jav | *Hordeum vulgare* Linn. | Poaceceae | Dried fruit | Good for male reproductive organs, strength, and longevity |
| Amla | *Emblica officinalis* | Euphorbiaceae | Fresh fruit pulp | Good for male reproductive organs, strength, and longevity |
| Vijayasara | *Pterocarpus marsupium* Roxb. | Leguminosae | Heart wood | Longevity |
| Aswagandha | *Withania somnifera* Dunal | Solanaceae | Dried mature roots | Good for male reproductive organs, strength, and longevity, and aphrodisiac |
| Kunghi | *Abutilon indicum* Linn. | Malvaceae | Roots | Good for male reproductive organs and strength |
| Bela | *Aegle marmelos* | Rutaceaeae | Ripe fruit | Strength |
| Gokhru | *Tribulus terrestris* Linn. | Zygophyllaceae | Root fruit | Good for male reproductive organs, strength, and longevity |
| Giloy | *Tinospora cordifolia* | Menispermaceae | Stem | Good for male reproductive organs, strength, and longevity |
| Gugal | *Commiphora wightii* | Burseraceae | Exudate | Strength |
| Harad | *Terminallia chebula* Retz | Combretaceae | Mature fruit | Longevity |
| Jaiphal | *Myristica fragraus* | Myristicaceae | Dried seeds | Good for male reproductive organs |

**Table 1** (continued)

| Hindi name | Botanical name | Family | Parts used | Uses |
|---|---|---|---|---|
| Kapasa | *Gossypium herbaceum* Linn. | Malvaceae | Seed | Good for male reproductive organs |
| Kasesu | *Scirpus kysoor* Roxb. | Cyperaceae | Rhizome | Semen |
| Kerada | *Pandanus tectorius* sokmel | Pandanaceae | Root | Good for reproductive organs, strength, and longevity |
| Saunt | *Foeniculum vulgare* Mill | Umbelliferae | Ripe fruit | Good for reproductive organs and strength |
| Bhaang | *Cannabis sativa* Linn. | Cannabaceae | Dried leaves | Aphrodisiac |
| Mulethi | *Glycyrrhiza glabra* Linn. | Leguminosae | Root | Good for reproductive organs and strength |
| Hadjod | *Cissus quadrangularis* Linn. | Vitaceae | Dried stem | Testicle health |
| Kewandr | *Mucuna prurita* Hook. | Fabaceae | Mature seed | Good for reproductive organs and testicle health |
| Munkka | *Vitis vinifera* Linn. | Vitaceae | Dried mature fruit | Testicle health |
| Evana | *Ricinus communis* Linn. | Euphorbiaceae | Fresh leaf | Testicle health |
| Bichuhathjori | *Martynia annua* Linn. | Martyniaceae | Dried seed | Longevity |
| Kakoli | *Lillum polyphyllum* D.Don | Liliaceae | Tuberous root | Semen enhancer |
| Kamal kand | *Nelumbo nucifera* Gaertn | Nymphaeaceae | Rhizome | Testicle health |
| Kasa | *Saccharum spontaneum* Linn. | Poaceae | Root stock | Testicle health and strength |
| Kui | *Nymphaea alba* Linn. | Nymphaeaceae | Dried flowers | Strength |
| Lahasun | *Allium sativum* Linn. | Liliaceae | Bulb | Good for reproductive organs and testicle health |
| Pitabala | *Sida rhomifolia* Linn. | Malvaceae | Dried root | Semen enhancer, reproductive organs and testicle health |

**Table 1** (continued)

| Hindi name | Botanical name | Family | Parts used | Uses |
| --- | --- | --- | --- | --- |
| Manjitha | *Rubia cordifolia* Linn. | Rubiaceae | Stem | Testicle health and longevity |
| Mashvan | *Teramnus labialis* Spreny | Fabaceae | Whole plant | Semen enhancer, goo for reproductive organs and testicle health |
| Masur | *Lens culinaris* medic | Fabaceae | Dried seeds | Semen enhancer, good for reproductive organs and testicle health |
| Pan | *Piper betle* Linn. | Piperaceae | Leaf | Semen enhancer, good for reproductive organs and testicle health |
| Nariyal | *Cocos nucifera* Linn. | Arecaceae | Dried endosperm | Semen enhancer, good for reproductive organs and testicle health |
| Raktachandana | *Petrocarpus santalinus* Linn. | Fabaceae | Heard wood | Semen enhancer, good for reproductive organs and testicle health |
| Sarivan | *Desmodium gangetium* DC | Fabaceae | Dried root | Semen enhancer, good for reproductive organs and testicle health |
| Chaval | *Oryza sativa* Linn. | Poaceae | Dried fruit | Semen enhancer, good for reproductive organs and testicle health |
| Sarkand | *Saccharum bengalense* Retz. | Poaceae | Root | Semen enhancer, good for reproductive organs and testicle health |
| Gulab | *Rosa centifolia* Linn. | Rosaceae | Dried flower | Semen enhancer, good for reproductive organs and testicle health |
| Seesam | *Dalbergia sissoo* Roxb. | Fabaceae | Stem bark | Semen enhancer, good for reproductive organs and testicle health |
| Jhuner | *Taxus baccata* Linn. | Taxaceae | Dried leaf | Semen enhancer |
| Safedchandan | *Santalum album* Linn. | Santalaceae | Heart wood | Semen enhancer, good for reproductive organs and testicle health |

**Table 1** (continued)

| Hindi name | Botanical name | Family | Parts used | Uses |
|---|---|---|---|---|
| Tal | *Borassus flabellifer* Linn. | Araceae | Male inflorescence | Semen enhancer, good for reproductive organs and testicle health |
| Louki | *Lagenaria siceraria* | Cucurbitaceae | Fresh fruit | Semen enhancer, good for reproductive organs and testicle health |
| Neel kanal | *Nymphaea stellata* Willd | Nymphaeaceae | Dried flower | Longevity |

# 6 Ayurvedic Rasayana for better reproductive health

Rasayana Shastra is a branch of Ayurveda, the ancient Indian system of medicine, which focuses on the use of medicinal herbs and minerals to promote physical and mental health, and increase longevity [30]. Rasayana treatments are used to rejuvenate the body and enhance its natural healing powers. In the context of reproductive health, Rasayana preparations are used to improve fertility, enhance sexual function, and promote healthy pregnancy and childbirth [30]. Some of the Rasayana preparations used in reproductive health include:

## 6.1 Amalaki Rasayana

Amalaki Rasayana is a popular Rasayana preparation in Ayurveda that is believed to promote longevity and overall health [30]. It is made from the Indian gooseberry, also known as Amalaki or Amla, which is a rich source of antioxidants, vitamin C, and other beneficial phytochemicals [30]. In Ayurveda, Amalaki Rasayana is traditionally used to strengthen the immune system, improve digestion, enhance memory and cognitive function, and promote healthy aging. It is believed to have a rejuvenating effect on the body, helping to repair and regenerate tissues, and protect against the effects of oxidative stress and inflammation [30].

Research has also suggested that Amalaki Rasayana may have antiaging effects on the body [30]. A study demonstrated that Amalaki Rasayana had a significant antiaging effect on laboratory animals, reducing age-related changes in various physiological parameters such as body weight, lipid peroxidation, and antioxidant enzyme levels [30, 32]. Another study showed that Amalaki Rasayana had a positive effect on the aging process in humans [30]. In the study, 108 healthy volunteers were given Amalaki Rasayana for 45 days, and their levels of oxidative stress and inflammation were measured [31]. The results showed that Amalaki Rasayana significantly reduced oxidative stress and inflammation – indicating its potential as an antiaging supplement [31]. Overall, while further research is needed to fully understand the effects of Amalaki Rasayana on longevity, its traditional use in Ayurveda and preliminary scientific evidence suggest that it may have beneficial effects on the aging process and overall health.

## 6.2 Medhya Rasayana

Medhya Rasayana is a branch of Ayurvedic medicine that focuses on promoting mental health and cognitive function [33]. It includes various herbs and formulations that are believed to support memory, concentration, and overall brain function. While Medhya Rasayana is primarily focused on mental health, there are some herbs and formulations within this category that may also have benefits for reproductive health [33]. Below are some of the Medhya Rasayana herbs that may possess benefits for reproductive health.

### 6.2.1 Bacopa monnieri

*Bacopa monnieri* (Brahmi) is a herb that is traditionally used in Ayurveda to enhance cognitive function and support mental health [34]. It is believed to promote healthy brain function and improve memory, concentration, and learning ability. Some studies suggest that Brahmi may also have benefits for male reproductive health, including improving sperm quality and reducing oxidative stress [34].

### 6.2.2 Convolvulus pluricaulis

*Convolvulus pluricaulis* (Shankhapushpi) is a herb that is used in Ayurveda to promote mental clarity and enhance cognitive function [34]. It is believed to have a calming effect on the mind and improve memory and concentration. Some studies suggest that Shankhapushpi may also have benefits for female reproductive health, including reducing menstrual pain and improving hormonal balance [34].

### 6.2.3 Nardostachys jatamansi

*Nardostachys jatamansi* (Jatamansi) is a herb that is traditionally used in Ayurveda to support mental health and promote relaxation [34]. It is believed to have a calming effect on the mind and may also have benefits for female reproductive health, including reducing menstrual pain and improving hormonal balance [34].

### 6.2.4 Withania somnifera

*Withania somnifera* (Ashwagandha) is a herb that is commonly used in Ayurveda to support mental health and promote physical vitality [20, 34]. It is believed to have a calming effect on the mind and reduce stress levels. Some studies suggest that Ashwagandha may also have benefits for male reproductive health, including improving sperm quality and reducing oxidative stress [20]. While these herbs may have benefits for reproductive health, it is important to consult a qualified Ayurvedic practitioner before using any Medhya Rasayana formulation or supplement, as they may interact with other medications or have contraindications in certain medical conditions [34].

## 6.3 Brahma Rasayana

Brahma Rasayana is an Ayurvedic formulation that is believed to promote overall health and longevity [35]. It is considered a Medhya Rasayana, which means that it is focused on promoting mental health and cognitive function. While Brahma Rasayana is not specifically targeted at reproductive health, it contains several herbs that may have benefits for this area. Given below are some of the herbs contained in Brahma Rasayana that may have benefits for reproductive health.

### 6.3.1 Asparagus racemosus

*Asparagus racemosus* (Shatavari) is a herb that is commonly used in Ayurveda to support female reproductive health [35]. It is believed to help balance hormones, promote fertility, and reduce menstrual pain.

### 6.3.2 Withania somnifera

*Withania somnifera* (Ashwagandha) is a herb that is commonly used in Ayurveda to promote physical and mental vitality [20, 34, 35]. It is believed to have a positive effect on male reproductive health, including improving sperm quality and reducing oxidative stress.

### 6.3.3 Sida cordifolia

*Sida cordifolia* (Bala) is a herb that is traditionally used in Ayurveda to promote overall health and vitality [35]. It is believed to have a positive effect on male reproductive health, including improving sperm count and motility.

### 6.3.4 Tinospora cordifolia

*Tinospora cordifolia* (Guduchi) is a herb that is commonly used in Ayurveda to promote overall health and vitality [35]. It is believed to have a positive effect on both male and female reproductive health, including improving fertility and reducing menstrual pain.

Overall, while Brahma Rasayana is not specifically targeted at reproductive health, the herbs contained in this formulation may have benefits for this area.

## 6.4 Chyawanprash

Chyawanprash is an Ayurvedic formulation that has been used for centuries to promote overall health and longevity [51]. It is considered a Rasayana, which means that it is focused on rejuvenating and revitalizing the body. While Chyawanprash is not specifically targeted at reproductive health, it contains several herbs that may have benefits in this area. Given below are some of the herbs contained in Chyawanprash that may have benefits for reproductive health.

### 6.4.1 Asparagus racemosus

*Asparagus racemosus* (Shatavari) is a herb that is commonly used in Ayurveda to support female reproductive health [36]. It is believed to help balance hormones, promote fertility, and reduce menstrual pain.

### 6.4.2 Tribulus terrestris

*Tribulus terrestris* (Gokshura) is a herb that is commonly used in Ayurveda to support male reproductive health [36]. It is believed to help improve sperm quality and quantity, promote testosterone production, and improve erectile dysfunction.

### 6.4.3 Withania somnifera

*Withania somnifera* (Ashwagandha) is a herb that is commonly used in Ayurveda to promote physical and mental vitality [20, 34–36]. It is believed to have a positive effect on male reproductive health, including improving sperm quality and reducing oxidative stress.

### 6.4.4 Emblica officinalis

*Emblica officinalis* (Amla) is a herb that is commonly used in Ayurveda to support overall health and immunity [36]. It is believed to have a positive effect on male reproductive health, including improving sperm quality and reducing oxidative stress.

### 6.4.5 Pueraria tuberosa

*Pueraria tuberosa* (Vidari) is a herb that is used in Ayurveda to support female reproductive health [36]. It is believed to improve fertility, balance female hormones, and promote healthy pregnancy and childbirth.

These Rasayana preparations are typically prepared as herbal formulations, which can be taken orally in the form of powders, capsules, or tablets [36]. They are also used in topical applications, such as herbal oils and creams, for external use [36].

In conclusion, Ayurvedic herbal medicine offers various remedies to treat reproductive health problems. However, it is important to seek advice from a qualified practitioner before using any herbal remedy, to ensure safety and efficacy.

# 7 TCM in treating reproductive health problems

TCM has a long history of treating reproductive health problems, dating back thousands of years. TCM views the body as a complex system of interconnected organs and energy pathways, and seeks to restore balance and harmony among these systems through natural remedies and lifestyle modifications [37]. TCM involves acupuncture, dietary changes, Qi Gong and Tai Chi, cupping therapy, mind-body techniques, and herbal remedies [37]. Acupuncture involves the insertion of thin needles into specific points on the body to stimulate the flow of energy (Qi) and promote healing. It has been used for centuries to treat a variety of reproductive health problems, including infertility, menstrual irregularities, and menopausal symptoms [38]. TCM places a strong emphasis on dietary modifications to support overall health and balance [39]. For example, TCM practitioners may recommend avoiding cold and raw foods, increasing the intake of warm and cooked foods, and avoiding certain foods that are believed to disrupt reproductive function. Qi Gong and Tai Chi are ancient Chinese practices that involve gentle movements and breathing techniques to promote physical, mental, and emotional well-being [40]. These practices can be particularly helpful in reducing stress and anxiety, which are common contributors to reproductive health problems. Cupping therapy involves the use of suction cups to promote blood flow and healing. It can be used to treat a variety of reproductive health problems, including menstrual pain and infertility [41]. TCM also emphasizes the importance of mind-body techniques such as meditation and visualization to support overall health and balance [41]. These techniques can be particularly helpful in reducing stress and anxiety, which are common contributors to reproductive health problems.

TCM uses a variety of herbal remedies to treat reproductive health problems. Some commonly used herbs include dong quai, ginseng, and rehmannia, which are believed to improve blood flow and hormone balance [40, 41].

## 7.1 Angelica sinensis

*Angelica sinensis* (dong quai) is often called the "female ginseng" and is commonly used to treat menstrual cramps, irregular periods, and other menstrual disorders [42]. It is believed to improve blood circulation and hormone balance.

## 7.2 Panax ginseng

*Panax ginseng* (ginseng) is a popular TCM herb that is believed to improve energy and vitality [42]. It is often used to treat fatigue, low libido, and other reproductive health problems.

## 7.3 Rehmannia glutinosa

*Rehmannia glutinosa* (rehmannia) is commonly used to treat infertility and menopausal symptoms [42]. It is believed to nourish the blood and improve hormonal balance.

## 7.4 Lycium barbarum

*Lycium barbarum* (goji berries) is rich in antioxidants and is believed to improve energy, fertility, and sexual function [42].

## 7.5 Paeonia lactiflora

*Paeonia lactiflora* (Chinese peony) is often used to treat menstrual pain and irregularities [42]. It is believed to improve blood flow and reduce inflammation.

## 7.6 Astragalus membranaceus

*Astragalus membranaceus* (astragalus) is commonly used in TCM to improve immune function and energy [42]. It is believed to strengthen the body's vital energy and reduce fatigue.

## 7.7 Eucommia ulmoides

*Eucommia ulmoides* (Eucommia bark) is believed to strengthen the reproductive system and improve fertility [42]. It is also believed to improve joint health and reduce inflammation.

## 7.8 Dioscorea opposita

*Dioscorea opposita* (Chinese yam) is believed to nourish the blood and improve reproductive health [42]. It is often used to treat menstrual irregularities, infertility, and other reproductive health problems.

# 8 Potential use of probiotics and prebiotics in reproductive health care

Probiotics are beneficial bacteria that live in the gut and help support a healthy immune system and digestive function. They may also play a role in reproductive health by improving vaginal and urinary tract health [47–50]. The vagina has a delicate balance of bacteria that helps in keeping it healthy. When this balance is disrupted, it can lead to conditions such as bacterial vaginosis or yeast infections. Probiotics may help restore this balance by increasing the number of beneficial bacteria in the vagina [48, 49]. Several studies have shown that probiotics can be effective in treating and preventing vaginal infections [49]. A meta-analysis reported that probiotics were effective in treating bacterial vaginosis and reducing the risk of recurrent infections [49]. Another study found that women who took probiotics during pregnancy were less likely to develop group B *Streptococcus* infection, which can be harmful to newborns. Probiotics may also play a role in improving male reproductive health. It is also demonstrated that men with infertility had lower levels of beneficial bacteria in their semen compared to fertile men. Supplementing with probiotics could improve sperm count and motility in men with infertility [48, 49].

Prebiotics are supposed to support the growth and/or activity of beneficial bacteria in the gut. In addition to other health benefits, prebiotics may also play a role in reproductive health by improving gut health and reducing inflammation [50]. Research has shown that prebiotics may be beneficial in treating conditions such as PCOS and endometriosis, which are common causes of female infertility. PCOS is characterized by insulin resistance and hormonal imbalances, while endometriosis involves inflammation and immune dysfunction [51]. Prebiotics may help improve insulin sensitivity and reduce inflammation, which can improve these conditions. It was reported that women with PCOS who took a prebiotic supplement for 12 weeks had significant improvements in insulin resistance, as well as reduction in testosterone levels and body weight [52, 53]. A study also showed that women with endometriosis had lower levels of certain beneficial gut bacteria compared to healthy women [54]. Regarding male reproductive health, prebiotic supplementation improved semen quality and reduced oxidative stress in men with infertility.

# 9 Adaptogenic herbs

Adaptogens may help the body adapt to stressors and promote balance and overall well-being [55]. Adaptogens may play a role in reproductive health by helping to reduce stress and inflammation, which can have a negative impact on fertility and overall reproductive health (Table 2). One of the most well-known adaptogens is Ashwagandha, which has been shown to have anti-inflammatory and antioxidant properties, and may help improve testosterone levels and sperm quality in men with infertility [56]. In women, Ashwagandha may help regulate menstrual cycles and improve fertility [57]. Rhodiola has been shown to reduce stress and improve mood, and may also have anti-inflammatory effects. Rhodiola supplementation has been shown to improve sexual function and reduce fatigue in women with hypoactive sexual desire disorder [58]. Ginseng, eleuthero, and maca root have been shown to help reduce stress and improve energy levels, and may also have positive effects on sexual function and hormone balance [42, 46].

**Table 2:** Common herbal adaptogens [55].

| Adaptogen | Possible benefits | Possible side effects |
|---|---|---|
| American ginseng (Panax quinquefolius) | May boost trusted source memory, reaction time, calmness, and immune system | May interact trusted source with blood thinners |
| Ashwagandha (Withania somnifera) | May reduce trusted source stress and anxiety | May cause stomach upset; not safe during pregnancy |
| Astragalus (Astragalus membranaceus) | May combat trusted source fatigue | May interact trusted source with drugs that affect the immune system |
| Cordyceps (Cordyceps militaris) | May boost stamina | May cause trusted source dry mouth, nausea, abdominal distension, throat discomfort, headache, diarrhea, and allergic reactions; may cause lead poisoning; not safe for people with RA, multiple sclerosis, or systemic lupus erythematosus (lupus) |
| Goji berry (Lycium barbarum) | May boost trusted source energy, physical and mental performance, calmness, sense of well-being, can improve sleep | May cause allergic reaction |
| Eleuthero root (Eleutherococcus senticosus) | May improve focus and stave off mental fatigue | May cause stomach upset and headache |
| Jiaogulan (Gynostemma pentaphyllum) | May reduce trusted source stress and boost endurance | No side effects from trusted source recorded as yet |
| Licorice root (Glycyrrhiza glabra) | May reduce trusted source oxidative stress | May cause trusted source high blood pressure, reduced potassium; possibly unsafe for people with kidney disease or cardiovascular problems; not suitable during pregnancy |
| Roseroot (Rhodiola rosea) | May stave off trusted source physical and mental fatigue | May cause trusted source dizziness, dry mouth or excess salivation |
| Schisandra berry / magnolia berry (Schisandra chinensis) | May boost endurance, mental performance, and working capacity | May cause restlessness, sleep problems, breathing difficulty |

**Table 2** (continued)

| Adaptogen | Possible benefits | Possible side effects |
|---|---|---|
| Tulsi / holy basil *(Ocimum sanctum)* | May reduce trusted source physical and mental stress, stress-related anxiety, depression, and improve memory and thinking | Likely safe Trusted Source for most people, but more research is needed |
| Turmeric *(Curcuma longa)* | May reduce trusted source depression | Likely safe trusted source in small amounts |

# 10 Other nutraceutical ingredients

Nutraceutical ingredients are bioactive compounds found in foods or supplements that have potential health benefits beyond basic nutrition. Here are some examples of nutraceutical ingredients that have been studied for their potential to treat reproductive health problems.

## 10.1 Omega-3 fatty acids

Omega-3 fatty acids are essential fats found in fish, nuts, and seeds. They have anti-inflammatory properties and have been shown to improve fertility in both men and women by improving sperm quality and ovarian function [59].

## 10.2 Coenzyme Q10

Coenzyme Q10 is an antioxidant found in fish, meats, and nuts. It is important for energy production in cells and has been shown to improve sperm quality and motility [60].

## 10.3 L-arginine

L-arginine is an amino acid found in meat, dairy, and nuts. It plays a role in nitric oxide production, which improves blood flow to reproductive organs and has been shown to improve erectile dysfunction and fertility in men [61].

## 10.4 Maca root

Maca root is a plant that is native to Peru and is used as a natural remedy for infertility and sexual dysfunction. It has been shown to improve sperm count and motility in men and regulate menstrual cycles in women [58].

# 11 Integrated approach toward reproductive healthcare system

An integrated approach to treating reproductive health problems involves combining different systems and therapies to provide a comprehensive and personalized approach to care. This approach recognizes that there are many factors that can impact reproductive health, including genetics, environment, lifestyle, and stress [62]. An integrated approach to reproductive health care may involve combining conventional allopathic medicine with alternative and complementary therapies, such as herbal medicine, acupuncture, and mindfulness-based stress reduction techniques. This approach may also incorporate nutritional and lifestyle interventions, such as dietary changes and exercise programs [63].

The goal of an integrated approach to reproductive health care is to provide personalized, patient-centered care that addresses the underlying causes of health problems and support overall health and well-being [64]. By combining different systems and therapies, healthcare practitioners can offer a more comprehensive approach to care that addresses the unique needs and concerns of each patient. An integrated approach to reproductive healthcare may be particularly beneficial for individuals with complex or chronic health problems, as well as those who have not found relief from traditional allopathic treatments [65]. This approach also recognizes the importance of treating the whole person, rather than just focusing on individual symptoms or conditions.

Overall, an integrated approach to reproductive healthcare offers a promising path forward for addressing the complex and multifaceted issues related to reproductive health. By combining different systems and therapies, healthcare practitioners can provide more comprehensive care that addresses the unique needs and concerns of each patient, and support optimal reproductive health and overall wellness [66]. The following points should be considered while designing an integrated approach:

- **Assessment:** Begin by conducting a comprehensive assessment of the patient's reproductive health, including medical history, symptoms, lifestyle factors, and any previous treatments or interventions [62].
- **Diagnosis:** Use the information gathered in the assessment to make a diagnosis or identify any underlying health conditions or imbalances that may be contributing to reproductive health problems [63].

- **Conventional treatment:** Consider conventional treatments, such as medications or surgeries, as appropriate for the diagnosis [64].
- **Alternative and complementary therapies:** Explore alternative and complementary therapies that may support reproductive health, such as acupuncture, herbal medicine, or mind-body techniques like meditation or yoga [64].
- **Nutritional and lifestyle interventions:** Consider the role of nutrition and lifestyle factors in reproductive health, and recommend dietary changes, exercise programs, and stress reduction techniques, as appropriate [65].
- **Monitoring and follow-up:** Monitor the patient's progress and adjust the treatment plan, as needed.

By combining conventional treatments with alternative and complementary therapies, nutritional and lifestyle interventions, and ongoing monitoring and follow-up, an integrated approach to reproductive health-care can offer a more comprehensive and personalized approach to care that addresses the unique needs and concerns of each patient.

# 12 Summary

Reproductive healthcare is a crucial aspect of overall health and well-being, and encompasses a wide range of conditions and treatments. Allopathic medicine, or Western medicine, has made significant advances in the diagnosis and treatment of reproductive health problems, including hormonal imbalances, infertility, and sexually transmitted infections. However, allopathic medicine has its limitations, and may not always address the underlying causes of these problems. Alternative and complementary therapies, such as herbal medicine, traditional Chinese medicine, and Ayurveda, may offer additional options for treating reproductive health problems. These therapies often take a more holistic approach to health, and may focus on improving the overall wellness and addressing the root causes of health problems. In recent years, there has also been growing interest in the role of nutrition and lifestyle factors in reproductive health. Nutraceuticals, dietary supplements, probiotics, and prebiotics may all play a role in improving reproductive health by promoting gut health, reducing inflammation, and improving hormonal balance. Despite these advances, reproductive healthcare remains a complex and multifaceted field, and there is still much to be learned about the causes and treatments of reproductive health problems. A comprehensive approach that combines allopathic medicine with alternative therapies and lifestyle interventions may offer the best chance of achieving optimal reproductive health.

In summary, reproductive healthcare is a critical aspect of overall health and well-being, and a variety of approaches are available for addressing reproductive health problems. While allopathic medicine has made significant advances in the field, alternative therapies and lifestyle interventions may also offer benefits. Ultimately, a comprehensive approach that addresses the root causes of reproductive health problems may be the most effective strategy for promoting optimal reproductive health.

# References

[1]    Mason-Jones, A.J., Crisp, C., Momberg, M., Koech, J., De Koker, P. and Mathews, C. 2012. A systematic review of the role of school-based healthcare in adolescent sexual, reproductive, and mental health. Systematic Reviews, 1, 49.

[2]    Surkan, P.J., Kennedy, C.E., Hurley, K.M. and Black, M.M. 2011. Maternal depression and early childhood growth in developing countries: Systematic review and meta-analysis. Bulletin of the World Health Organization, 89, 607.

[3]    Surkan, P.J., Kawachi, I., Ryan, L.M., Berkman, L.F., Carvalho Vieira, L.M. and Peterson, K.E. 2008. Maternal depressive symptoms, parenting self-efficacy, and child growth. American Journal of Public Health, 98, 125_132.

[4]    Singh, A.R. 2010. Modern medicine: Towards prevention, cure, well-being and longevity. Mens Sana Monographs, 8(1), 17–29.

[5]    ShaonLahiri, J.B., Sedlander, E., Munar, W. and Rimal, R. 2023. The role of social norms on adolescent family planning in rural Kilifi county, Kenya. Plos One, 18(2), e0275824.

[6]    RohiniGanjoo, R.N.R., Talegawkar, S.A., Sedlander, E., Pant, I., Bingenheimer, J.B., Chandarana, S., AikaAluc, Y.J. and HagereYilma, B.P. 2022. Improving iron folic acid consumption through interpersonal communication: Findings from the reduction in Anemia through normative innovations (RANI) project. Patient Education and Counseling, 105(1), 81–87.

[7]    Hurley, W.L. and Doane, R.M. 1989. Recent developments in the roles of vitamins and minerals in reproduction. Journal of Dairy Science, 72(3), 784–804.

[8]    Fathalla, M.F. 1991. Reproductive health: A global overview. Annals of the New York Academy of Sciences, 626, 1–10.

[9]    Stephenson, R. and Tsui, A.O. 2002. Contextual influences on reproductive health service use in Uttar Pradesh, India. Studies in Family Planning, 33(4), 309–320.

[10]   Akazili, J., Kanmiki, E.W., Anaseba, D., Govender, V., Danhoundo, G. and Koduah, A. 2020. Challenges and facilitators to the provision of sexual, reproductive health and rights services in Ghana. Sexual and Reproductive Health Matters, 28(2), 1846247.

[11]   Gronowski, A.M. and Schindler, E.I. 2014. Women's health. Scandinavian Journal of Clinical and Laboratory Investigation, Supplement, 244, 2–7.

[12]   Barazani, Y., Katz, B.F., Nagler, H.M. and Stember, D.S. 2014. Lifestyle, environment, and male reproductive health. Urologic Clinics of North America, 41(1), 55–66.

[13]   Roudsari, R.L., Sharifi, F. and Goudarzi, F. 2023. Barriers to the participation of men in reproductive health care: A systematic review and meta-synthesis. BMC Public Health, 23(1), 818.

[14]   Stover, J., Hardee, K., Ganatra, B., et al. 2016. Interventions to improve reproductive health. In Black, R.E., Laxminarayan, R., & Temmerman, M. et al, (Eds.), Reproductive, Maternal, Newborn, and Child Health: Disease Control Priorities. 3rd edition, Vol. 2, Washington (DC): The International Bank for Reconstruction and Development / The World Bank.

[15]   Nikfar, S., Kebriaeezadeh, A., Majdzadeh, R. and Abdollahi, M. 2005. Monitoring of National Drug Policy (NDP) and its standardized indicators; conformity to decisions of the national drug selecting committee in Iran. BMC International Health and Human Rights, 5(1), 5.

[16]   Pandey, M.M., Rastogi, S. and Rawat, A.K. 2013. Indian traditional ayurvedic system of medicine and nutritional supplementation. Evidence-Based Complementary and Alternative Medicine, 2013, 376327.

[17]   Silvestris, E., Lovero, D. and Palmirotta, R. 2019. Nutrition and female fertility: An interdependent correlation. Front Endocrinol (Lausanne), 10, 346.

[18]   Mirnamniha, M., Faroughi, F., Tahmasbpour, E., Ebrahimi, P. and Beigi Harchegani, A. 2019. An overview on role of some trace elements in human reproductive health, sperm function and fertilization process. Reviews on Environmental Health, 34(4), 339–348.

[19]   Kontic-Vucinic, O., Sulovic, N. and Radunovic, N. 2006. Micronutrients in women's reproductive health: I. Vitamins. International journal of fertility and women's medicine, 51(3), 106–115.

[20]   Nasimi Doost Azgomi, R., Zomorrodi, A., Nazemyieh, H., Fazljou, S.M.B., Sadeghi Bazargani, H., Nejatbakhsh, F., Moini Jazani, A. and Ahmadi Asrbadr, Y. 2018. Effects of withania somnifera on reproductive system: A systematic review of the available evidence. BioMed Research International, 2018, 4076430.

[21]   Alok, S., Jain, S.K., Verma, A., Kumar, M., Mahor, A. and Sabharwal, M. 2013. Plant profile, phytochemistry and pharmacology of Asparagus racemosus (Shatavari): A review. Asian Pacific Journal of Tropical Disease, 3(3), 242–251.

[22]   Akhtari, E., Raisi, F., Keshavarz, M., Hosseini, H., Sohrabvand, F., Bioos, S., Kamalinejad, M. and Ghobadi, A. 2014. Tribulus terrestris for treatment of sexual dysfunction in women: Randomized double-blind placebo – Controlled study. Daru, 22(1), 40.

[23]   Acharya, N., Acharya, S., Shah, U., Shah, R. and Hingorani, L. 2016. A comprehensive analysis on Symplocos racemosa Roxb.: Traditional uses, botany, phytochemistry and pharmacological activities. Journal of Ethnopharmacology, 181, 236–251.

[24]   Lampariello, L.R., Cortelazzo, A., Guerranti, R., Sticozzi, C. and Valacchi, G. 2012. The magic velvet bean of Mucuna pruriens. Journal of Traditional and Complementary Medicine, 2(4), 331–339.

[25]   Mishra, S., Aeri, V., Gaur, P.K. and Jachak, S.M. 2014. Phytochemical, therapeutic, and ethnopharmacological overview for a traditionally important herb: Boerhavia diffusa Linn. BioMed Research International, 2014, 808302.

[26]   Peterson, C.T., Denniston, K. and Chopra, D. 2017. Therapeutic uses of Triphala in Ayurvedic Medicine. Journal of Alternative and Complementary Medicine, 23(8), 607–614.

[27]   Dalal, P.K., Tripathi, A. and Gupta, S.K. 2013. Vajikarana: Treatment of sexual dysfunctions based on Indian concepts. Indian Journal of Psychiatry, 55(Suppl 2), S273–6.

[28]   Sharma, V., Thakur, M., Chauhan, N.S. and Dixit, V.K. 2010. Effects of petroleum ether extract of Anacyclus pyrethrum DC. on sexual behavior in male rats. Zhong Xi Yi Jie He Xue Bao, 8(8), 767–773.

[29]   Chauhan, N.S., Sharma, V., Dixit, V.K. and Thakur, M. 2014. A review on plants used for improvement of sexual performance and virility. BioMed Research International, 2014, 868062.

[30]   Nishteswar, K. 2013. Pharmacological expression of Rasayanakarma. Ayu, 34(4), 337–338.

[31]   Kumar, V., Aneesh, K.A., Kshemada, K., Ajith, K.G.S., Binil, R.S.S., Deora, N., Sanjay, G., Jaleel, A., Muraleedharan, T.S., Anandan, E.M., Mony, R.S., Valiathan, M.S., Santhosh, K.T.R. and Kartha, C.C. 2017. Amalaki rasayana, a traditional Indian drug enhances cardiac mitochondrial and contractile functions and improves cardiac function in rats with hypertrophy. Scientific Reports, 7(1), 8588.

[32]   Guruprasad, K.P., Dash, S., Shivakumar, M.B., Shetty, P.R., Raghu, K.S., Shamprasad, B.R., Udupi, V., Acharya, R.V., Vidya, P.B., Nayak, J., Mana, A.E., Moni, R., Sankaran, M.T. and Satyamoorthy, K. 2017. Influence of Amalaki Rasayana on telomerase activity and telomere length in human blood mononuclear cells. Journal of Ayurveda and Integrative Medicine, 8(2), 105–112.

[33]   Bhargavi, S. and Madhan Shankar, S.R. 2021. Dual herbal combination of Withania somnifera and five Rasayana herbs: A phytochemical, antioxidant, and chemometric profiling. Journal of Ayurveda and Integrative Medicine, 12(2), 283–293.

[34]   Kulkarni, R., Girish, K.J. and Kumar, A. 2012. Nootropic herbs (Medhya Rasayana) in Ayurveda: An update. Pharmacognosy Reviews, 6(12), 147–153.

[35]   Guruprasad, K.P., Mascarenhas, R., Gopinath, P.M. and Satyamoorthy, K. 2010. Studies on Brahma rasayana in male Swiss albino mice: Chromosomal aberrations and sperm abnormalities. Journal of Ayurveda and Integrative Medicine, 1(1), 40–44.

[36]   Sharma, R., Martins, N., Kuca, K., Chaudhary, A., Kabra, A., Rao, M.M. and Prajapati, P.K. 2019. Chyawanprash: A traditional Indian bioactive health supplement. Biomolecules, 9(5), 161.

[37]   Xu, X., Yin, H., Tang, D., Zhang, L. and Gosden, R.G. 2003. Application of traditional Chinese medicine in the treatment of infertility. Human fertility (Cambridge), 6(4), 161–168.

[38]   Jiang, D., Li, L., Wan, S. and Meng, F. 2019. Acupuncture and Chinese herbal medicine effects on assisted reproductive technology: Six cases and their clinical significance. Medical Acupuncture, 31(6), 395–406.

[39]   Marshall, A.C. 2020. Traditional Chinese medicine and clinical pharmacology. Drug Discovery and Evaluation: Methods in Clinical Pharmacology, 2, 455–482.

[40]   Van Dam, K. 2020. Individual stress prevention through Qigong. International Journal of Environmental Research and Public Health, 17(19), 7342.

[41]   Wang, J.X., Yang, Y., Song, Y. and Ma, L.X. 2018. Positive effect of acupuncture and cupping in infertility treatment. Medical Acupuncture, 30(2), 96–99.

[42]   Teng, B., Peng, J., Ong, M. and Qu, X. 2017. Successful pregnancy after treatment with Chinese herbal medicine in a 43-year-old woman with diminished ovarian reserve and multiple uterus fibrosis: A case report. Medicines (Basel), 4(1), 7.

[43]   Safarinejad, M.R. and Safarinejad, S. 2012. The roles of omega-3 and omega-6 fatty acids in idiopathic male infertility. Asian Journal of Andrology, 14(4), 514–515.

[44]   Florou, P., Anagnostis, P., Theocharis, P., Chourdakis, M. and Goulis, D.G. 2020. Does coenzyme Q10 supplementation improve fertility outcomes in women undergoing assisted reproductive technology procedures? A systematic review and meta-analysis of randomized-controlled trials. Journal of Assisted Reproduction and Genetics, 37(10), 2377–2387.

[45]   Bodis, J., Farkas, B., Nagy, B., Kovacs, K. and Sulyok, E. 2022. The Role of L-Arginine-NO system in female reproduction: A narrative review. International Journal of Molecular Sciences, 23(23), 14908.

[46]   Beharry, S. and Heinrich, M. 2018. Is the hype around the reproductive health claims of maca (Lepidium meyenii Walp.) justified? Journal of Ethnopharmacology, 211, 126–170.

[47]   Sharma, H. 2016. Ayurveda: Science of life, genetics, and epigenetics. Ayu, 37(2), 87–91.

[48]   Dalal, P.K., Tripathi, A. and Gupta, S.K. 2013. Vajikarana: Treatment of sexual dysfunctions based on Indian concepts. Indian Journal of Psychiatry, 55(Suppl 2), S273–6.

[49]   Hashem, N.M. and Gonzalez-Bulnes, A. 2022. The use of probiotics for management and improvement of reproductive Eubiosis and function. Nutrients, 14(4), 902.

[50]   You, S., Ma, Y., Yan, B., Pei, W., Wu, Q., Ding, C. and Huang, C. 2022. The promotion mechanism of prebiotics for probiotics: A review. Frontiers in Nutrition, 9, 1000517.

[51]   Azizi-Kutenaee, M., Heidari, S., Taghavi, S.A. and Bazarganipour, F. 2022. Probiotic effects on sexual function in women with polycystic ovary syndrome: A double blinded randomized controlled trial. BMC Womens Health, 22(1), 373.

[52]   Singh, S., Pal, N., Shubham, S., Sarma, D.K., Verma, V., Marotta, F. and Kumar, M. 2023. Polycystic ovary syndrome: Etiology, current management, and future therapeutics. Journal of Clinical Medicine, 12(4), 1454.

[53]   Miao, C., Guo, Q., Fang, X., Chen, Y., Zhao, Y. and Zhang, Q. 2021. Effects of probiotic and synbiotic supplementation on insulin resistance in women with polycystic ovary syndrome: A meta-analysis. Journal of International Medical Research, 49(7), 3000605211031758.

[54]  Helli, B., Kavianpour, M., Ghaedi, E., Dadfar, M. and Haghighian, H.K. 2022. Probiotic effects on sperm parameters, oxidative stress index, inflammatory factors and sex hormones in infertile men. Human fertility (Cambridge), 25(3), 499–507.

[55]  Panossian, A.G., Efferth, T., Shikov, A.N., Pozharitskaya, O.N., Kuchta, K., Mukherjee, P.K., Banerjee, S., Heinrich, M., Wu, W., Guo, D.A. and Wagner, H. 2021. Evolution of the adaptogenic concept from traditional use to medical systems: Pharmacology of stress- and aging-related diseases. Medicinal Research Reviews, 41(1), 630–703.

[56]  Todorova, V., Ivanov, K., Delattre, C., Nalbantova, V., Karcheva-Bahchevanska, D. and Ivanova, S. 2021. Plant adaptogens-history and future perspectives. Nutrients, 13(8), 2861.

[57]  Lopresti, A.L., Drummond, P.D. and Smith, S.J. 2019. A randomized, double-blind, placebo-controlled, crossover study examining the hormonal and vitality effects of Ashwagandha (*Withania somnifera*) in aging, overweight males. American Journal of Men's Health, 13(2), 1557988319835985.

[58]  Ajgaonkar, A., Jain, M. and Debnath, K. 2022. Efficacy and safety of Ashwagandha (*Withania somnifera*) root extract for improvement of sexual health in healthy women: A prospective, randomized, placebo-controlled study. Cureus, 14(10), e30787.

[59]  Ivanova Stojcheva, E. and Quintela, J.C. 2022. The effectiveness of Rhodiola rosea L. preparations in alleviating various aspects of life-stress symptoms and stress-induced conditions-encouraging clinical evidence. Molecules, 27(12), 3902.

[60]  Harris, K.M. 2010. An integrative approach to health. Demography, 47(1), 1–22.

[61]  Maharaj, P. 2004. Integrated reproductive health services: The perspectives of providers. Curationis, 27(1), 23–30.

[62]  Zhou, J., Hu, W. and Qu, F. 2022. Integrative medicine research for improving female reproductive health during COVID-19. Integrative Medicine Research, 11(4), 100894.

[63]  Sharma, R., Biedenharn, K.R., Fedor, J.M. and Agarwal, A. 2013. Lifestyle factors and reproductive health: Taking control of your fertility. Reproductive Biology and Endocrinology, 11, 66.

[64]  Walker, M.H. and Tobler, K.J. 2023. Female Infertility. In *StatPearls [Internet]*. Treasure Island (FL): StatPearls Publishing.

[65]  Van Balen, F. and Visser, A.P. 1997. Perspectives of reproductive health. Patient Education and Counseling, 31(1), 1–5.

[66]  Sharma, R., Biedenharn, K.R., Fedor, J.M. and Agarwal, A. 2013. Lifestyle factors and reproductive health: Taking control of your fertility. Reproductive Biology and Endocrinology, 11, 66.

Besma Sehili*

# Chapter 12
# Infectious disorders

**Abstract:** In the early 1900s and even before that, high mortality rates worldwide were caused by infectious diseases. Indeed, viruses, bacteria, and fungi have all been involved in causing infectious disorders, and the human body produces antibodies against the infection agent or develops resistance to it, in response to all infectious agents. Antimicrobials are used to prevent and treat infections and have an important role in healthcare, but increasing antimicrobial resistance poses a serious threat to infection management. However, the significant decrease in the number of approved antimicrobials as well as the continued overuse of antimicrobial drugs are causing severe antimicrobial resistance problems, requiring the use of novel approaches that reduce the risks of developing resistance and prevent and treat infectious diseases. Particularly in the last few decades, nutraceuticals have joined the healthcare system as an attractive and simple approach to disease prevention, including infectious disorders. This chapter summarizes the beneficial effects of nutraceuticals on infectious disorders, including dietary fibers, prebiotics, probiotics, polyunsaturated fatty acids, vitamins, minerals, polyphenols, and spices, as well as the mechanisms of action that are associated with their therapeutic effect.

# 1 Introduction

In general, humans live peacefully with the microorganisms that surround them. An infection may occur if pathogen concentrations reach an unusually high density or when the defensive system is compromised. Most infections remain unrecognized, but the pathogenic microorganisms occasionally elicit a reaction from the human body, leading to clinical manifestations, which are defined as an infectious disease [1]. In the early 1900s and even before that, high mortality rates worldwide were caused by infectious diseases [2]. Bacteria, viruses, and fungi have all been involved in causing infectious diseases, and the body produces antibodies against the infection agent or develops resistance to it in response to all pathogenic microbes [1, 3]. Indeed, fever is a common symptom of most bacterial, fungal, and viral infections. Some examples of infectious disorders are summarized in Table 1.

*Corresponding author: Besma Sehili, Faculty of Medicine, Badji Mokhtar University, BP 205., Annaba 23000, Algeria, e-mail: besma.sehili@univ-annaba.dz, besma.sehili@gmail.com

https://doi.org/10.1515/9783111317601-012

**Table 1:** Some common examples of viral, bacterial, and fungal infections.

| Microbial pathogens | Example | Infectious disease |
|---|---|---|
| **Viruses** | Influenza virus | Influenza |
| | Lentivirus | HIV/AIDS |
| | Hepatitis B virus | Hepatitis |
| | Herpes simplex | Encephalitis |
| | Flavivirus | |
| | Enterovirus | Meningitis |
| | Rhinoviruses | Pneumonia |
| | Coronaviruses | |
| **Bacteria** | *Neisseria gonorrhoeae* | Gonorrhea |
| | *Helicobacter pylori* | Gastric ulcer |
| | *Treponema pallidum* | Syphilis |
| | Meningococci | Meningitis |
| | Streptococci | |
| | Pneumococci | Pneumonia |
| | Streptococci | |
| **Fungi** | *Tinea versicolor* | Mycosis |
| | Bacterial vaginosis | Vaginal infections |
| | *Cryptococcus neoformans* | Cryptococcal meningitis |

The human body is exposed to bacterial infection, and the main modes of bacterial transmission are vectors, contact, vehicular, droplet, and airborne. However, each bacterial species preferentially infects certain organs over others, which may cause gastric ulcers, syphilis, urinary tract infections, Lyme disease, and cholera. *Neisseria meningitidis*, for example, can not only cause meningitis by infecting the meninges but can also infect lungs, to cause pneumonia [3, 4]. Viruses are responsible for common infectious illnesses like the flu and the common cold. They also cause serious diseases, including encephalitis, HIV/AIDS, and hepatitis [3]. In order to cause infection, all viruses must acquire access to a receptive host cell. Many viruses penetrate the body via mucous membranes, following ingestion, contact, or inhalation. Viruses, unlike bacteria, cannot survive without a host. They have been found to infect not only all types of human cells, but the range and nature of cells that viruses might infect vary. Poliovirus, for example, is a neurotropic virus that may infect neurons, although human papillomaviruses (HPVs) show a preference for epithelial tissues [5]. An estimated 1.5 million fungal species exist on our planet. Furthermore, only about 100 fungus species have been linked to human illnesses. Indeed, allergies, extensive mycoses, and superficial infections are some examples of the illnesses caused by pathogenic fungi [6].

This chapter summarizes the beneficial effects of nutraceuticals, including dietary fiber, prebiotics, probiotics, polyunsaturated fatty acids, vitamins, minerals, polyphenols, and spices, on infectious disorders as well as the mechanisms of action that are associated with their therapeutic effect.

# 2 Pharmacological therapy and microbial resistance

Antimicrobials, which include antibiotics, antivirals, and antifungals, are medications used to prevent and treat infections [7]. They have an important role in healthcare, and increasing antimicrobial resistance poses a serious threat to infection management [8]. However, antibiotic resistance has been identified by the World Health Organization (WHO) as among the three most severe problems for public health in the twenty-first century [2]. Generally, the capacity of germs to survive and grow even in the presence of antimicrobial drugs that have previously been proven to be effective against these microorganisms has been defined as "antimicrobial resistance". Although bacterial resistance is the most common, resistance has also been reported in fungi and viruses [9, 10]. The main factors driving the emergence of antimicrobial resistance include antibiotic misuse, overuse of antibiotics, and extensive agricultural antibiotic use.

Bacteria are able to resist drugs through a wide range of mechanisms. Generally, drug enzymatic inactivation, drug removal by active efflux, drug absorption inhibition, and drug target alteration are the main mechanisms of antibiotic resistance [11]. So, antimicrobial resistance can be developed against gram-negative bacteria as well as gram-positive bacteria and can cause severe illnesses. Tuberculosis, for example, caused by *Mycobacterium tuberculosis*, remains one of the world's most deadly infectious illnesses. The most common antibiotic-resistant bacteria are cotrimoxazole-resistant *Escherichia coli*, cephalosporin-resistant *E. coli*, methicillin-resistant *Staphylococcus aureus*, vancomycin-resistant *Staphylococcus aureus*, macrolide-resistant *Streptococcus pneumoniae*, and carbapenem-resistant *Pseudomonas aeruginosa* [12].

Moreover, resistance is only caused by mutations in the antiviral target site or antiviral drug activator genes. Drug resistance can develop as a result of the presence of viruses that are less susceptible to antiviral agents or from the drugs themselves. Under natural conditions, the virus will mutate, and the drug selection pressure causes mutations that result in resistance and, as a result, virus survival [13, 14]. Thus, human viral infections such as AIDS, herpes, and influenza have all been linked to antiviral drug resistance [13]. Because of the mutations in the protease and reverse transcriptase genes, and even in the presence of antiretrovirals, HIV will spread rapidly [14]. For example, herpes simplex virus (HSV) acyclovir resistance is related to DNA polymerase mutations or viral thymidine kinase. Moreover, amantadine-resistant influenza is frequent and can be spread to others without losing virulence.

Apart from antibiotic and antiviral resistance, antifungal resistance has become more common, exacerbating an already difficult therapeutic situation [13]. However,

*Candida auris*, which is ranked as one of the most frequent fungal infections, has been reported to be resistant to amphotericin B, voriconazole, and fluconazole [7]. Combating antimicrobial resistance necessarily requires a global coordinated effort that includes the development of new conventional antibiotics as well as the development of alternative antimicrobial drugs, the development of novel combination therapies, and the quick detection of antimicrobial resistance pathogens [15].

# 3 Nutraceuticals: A possible preventive approach

A considerable reduction in the number of approved antimicrobials and the continued overuse of antimicrobial drugs are causing severe antimicrobial resistance problems, requiring the use of novel approaches that reduce the risks of developing resistance, and prevent and treat infectious diseases [16, 17]. It is crucial for the organism's survival to successfully combat infections caused by microbial pathogens [18]. Particularly in the last few decades, nutraceuticals have joined the healthcare system as an attractive and simple approach to disease prevention, including infectious disorders [19].

## 3.1 Dietary fibers

The native microbiota is now widely recognized as an important barrier to pathogenic microorganisms [20]. As a result, infectious diseases are frequently associated with an unbalanced gut microbiota [21]. A fiber-rich diet aids in the maintenance of a varied and functional gut microbiota by generating short-chain fatty acids (SCFAs). Indeed, high fiber consumption and the production of SCFAs by gut flora, whose fermenting fibers provide energy to the host as well as an immunomodulatory impact, suggest that SCFAs are useful in the treatment of infectious disorders. The production of SCFAs by microflora stimulates the generation of antimicrobial peptides and mucus, as well as tight junction protein expression [22], which contributes to the preservation of epithelial physiology [23].

The vast majority of bacterial pathogens in the gastrointestinal tract [24], like *Enterobacteriaceae*, have to gain access to the intestinal epithelial and infiltrate the mucosal barrier in order to enhance persistence or colonization [20]. The mucus layer that coats the gastrointestinal tract prevents bacterial adhesion and transformation across the epithelial membrane. Therefore, including fermentable fibers in the diet can reduce mucin production, which may decrease the incidence of bacterial translocation through the intestinal barrier [25].

Dietary fibers of plant origin have been reported to reduce the adhesion of pathogenic *Escherichia coli* to intestinal epithelial cells. Soluble fiber extract from plantains, for example, decreased the adhesion of adherent invasive *E. coli* (AIEC), enterotoxigenic *E. coli* (ETEC), and *Shigella* to intestinal epithelial cells. Several dietary fibers, such as chitosan, have demonstrated a direct bactericidal impact by inhibiting pathogen multiplication, specifically enterohemorrhagic *E. coli* (EHEC). Chitosan's antibacterial action is most likely due to intracellular leakage, caused by the binding of positively charged chitosan to a bacterial surface that is negatively charged, resulting in membrane permeability alterations and the death of a cell [20].

Diarrhea is one of the first indications of infections caused by viruses [26]. Research was conducted on 120 patients hospitalized for gastric cancer, comparing the efficacy of fiber and probiotic-containing enteral feeding to a diet with no fiber on diarrhea. The authors reported that after 7 days, patients who consumed fiber and probiotics had better bowel movements and fewer diarrheal episodes. Therefore, the research finds that adding fiber to enteral nutrition formulas prevents diarrhea, compared to enteral nutrition formulations without fiber [27]. Rushdi et al. conducted a randomized controlled trial to evaluate the addition of guar gum, a prebiotic and soluble fiber, in enteral feeding, to reduce diarrheal episodes in intensive care unit patients. The authors reported that using a guar gum-enriched enteral diet reduced diarrheal episodes in intensive care unit patients who already had diarrhea [28].

## 3.2 Prebiotics

Recently, in both human and animal investigations, prebiotics have been researched for their significant health effects. They have especially been assessed for their beneficial effects in the treatment or prevention of gastrointestinal diseases [29, 30]. Prebiotics may promote the growth of healthy gut bacteria, specifically *Bifidobacteria* and *Lactobacilli*, which have health-promoting effects for the host [31, 32]. Therefore, prebiotics, such as nondigestible oligosaccharides, have been suggested as another approach for restoring and improving colonization resistance and reducing infections [33]. If prebiotics boost the host's immune defenses, it is reasonable to expect that they will reduce the severity of the disease [30]. Prebiotics generally provide food for different microorganisms, and they only act on the natural flora [34]. The mechanisms of action by which prebiotics might confer health benefits include increasing the number of gut-beneficial bacteria – the fermentation of prebiotics by the *Bifidobacterium* species produces SCFAs, which have a variety of consequences, such as colonic acidification, which is detrimental to many pathogenic bacterial species; decreasing the intestine pH, which has an impact on the growth of several microorganisms; increasing calcium absorption; and possibly having some immunomodulation effect [29, 30]. Further, prebiotics may potentially influence host's immune function via mechanisms other than the regulation of beneficial bacteria in the gut. It is suggested that carbohydrate moieties on the prebiotic may react with immune cell receptors. Some oligosaccharides, such as oligofructose, can be bound to the receptors of pathogenic microorganisms and prevent them from adhering to this same sugar on the epithelial membrane, inhibiting adhesion [30].

Prebiotics can especially help reduce the recurrence of diarrhea due to *Clostridium difficile*. A randomized controlled trial was conducted for 30 days in which patients suffering from *C. difficile*-induced diarrhea received a placebo or oligofructose, in addition to antibiotic therapy. The authors reported that the occurrence rate significantly decreased, from 34.3% in the control group to 8.3% in the oligofructose recipients [32]. Fructo-oligosaccharides (FOS), gluco-oligosaccharides (GOS), and inulin have been proven *in vitro* to enhance colonic bifidobacterial population growth and bacterial colonization resistance to *C. difficile* [33]. Other *in vitro* fermentation research with fecal bacteria-mixed cultures grown on galacto-oligosaccharide and inulin found that the oligosaccharides, particularly galacto-oligosaccharide, inhibited *C. difficile* growth and toxin production [31].

Viruses cause almost all infectious diarrheas, with rotavirus being the most common pathogen. However, breastfeeding has been documented to decrease the risk of infection in newborns due to the abundance of immunogenic substances such as prebiotics and probiotics. Consequently, prebiotics and probiotics have been added to cow's milk formulas to mimic the immunological capacity of breastfeeding. Research on synbiotics has found that a combination of zinc, fructo-oligosaccharides, and numerous probiotics (*Bifidobacterium lactis*, *Lactobacillus acidophilus*) can reduce the duration of diarrhea [35]. Antibiotic use is common in children, which can lead to antibiotic-associated diarrhea. A randomized, double-blind study was conducted to evaluate the efficacy of a prebiotic-enriched formula provided to infants aged one to two years who were being treated for bronchitis with amoxicillin. Antibiotic use decreased the overall fecal bacteria while increasing clostridia, according to the authors [32]. Another study was conducted in Indonesia on children aged 1–14 years. It reported that the duration of diarrhea was decreased with a fructo-oligosaccharide supplement, and the prevalence of severe diarrhea was also reduced in infants receiving a formula combining both gluco-oligosaccharides and fructo-oligosaccharides [36].

Prebiotics can regulate immune function both directly and indirectly through affecting the balance of the gut microbial communities or by releasing microbial bioactive substances like SCFAs. They can also encourage the growth of beneficial microorganisms such as *Lactobacilli* and *Bifidobacteria* [37]. An in *vivo* study on mice infected with *Candida albicans*, systemic *Listeria monocytogenes*, and *Salmonella typhimurium* was conducted. It was reported that mice that were given oligofructose or inulin for 6 weeks demonstrated improved resistance to microbial infections, decreased mortality, and enhanced T-cell activity [38]. A double-blind, placebo-controlled trial showed that four sources of xylo-oligosaccharides, which may be found in fruit, vegetables, honey, and bamboo shoots, increased the number of fecal *Bifidobacteria* in the elderly subjects with mean age of 78.6 years [31]. In HIV-infected adolescents and children, three months of supplementation with fructo-oligosaccharides (3.75 g/day) led to an increase in lymphocyte proliferation, which is specific to the BCG vaccination that is used to assess cellular immunological response [39].

A randomized, double-blind, placebo-controlled trial was conducted in babies who received a prebiotic formula containing fructo-oligosaccharides and gluco-oligosaccharides or a placebo (maltodextrin) throughout the first six months of their life. The authors reported a considerable reduction in all types of infections as well as upper respiratory tract infection occurrences in the group that received probiotics compared to the placebo group. Further, they noted that infants who received prebiotics had a considerably lower overall incidence of respiratory infections compared to the placebo group [40].

## 3.3 Probiotics

The use of probiotics for illness treatment and prevention is attracting scientific and commercial attention [41]. Probiotics have been shown to have the potential to treat and prevent some infectious disorders, with the majority of probiotic clinical application research focusing on the gastrointestinal tract [40, 42]. According to current research, probiotics are more important in pathogen prevention, gut barrier maintenance, and immune modulation than in microbial colonization. *Lactobacillus casei*, for example, has anti-inflammatory properties in *Shigella*-infected human intestines [22].

Several mechanisms have been suggested, including the production of antimicrobial agents, the digestion of lactose, and competition for nutrition or space [41]. Probiotics provide antimicrobial effects by releasing antimicrobial molecules, such as bacteriocins, hydrogen peroxide, and biosurfactants, into the gut environment and by limiting the proliferation of other microorganisms. The significant advantages of probiotics result from their capacity to metabolize complex carbohydrates, produce lactic acid, and to produce SCFAs like butyrate, which reduce bacterial translocation, increase tight junction structure, and stimulate the generation of mucin-type glycoproteins that maintain the intestinal epithelium's integrity [29, 32].

Some types of diarrheas can be avoided by partial lactose digestion and the enhancement of lactase activity in the intestinal mucosa. *Lactobacilli* in fermenting milk, for example, contain active β-galactosidase, which lowers lactose levels and may influence the severity of rotavirus-caused osmotic diarrhea. Probiotics can also employ enzymatic processes to prevent disease caused by toxins and alter toxin receptors. *Saccharomyces boulardii*, for example, destroys the receptors of *Clostridium difficile* toxin in the rabbit intestinal lumen [41].

Antimicrobial-induced diarrhea is the main adverse effect of antimicrobial treatment [16]. Indeed, antibiotic-induced diarrhea, such as that induced by cephalosporin, clindamycin, and penicillin, is well documented as being caused by *Clostridium difficile* growth in the gut [43]. The efficacy of probiotics such as *Lactobacillus spp.* in both antibiotic-induced diarrhea and diarrhea caused by *C. difficile* has been investigated. For example, *Saccharomyces boulardii* was found to be effective in preventing illness recurrence, but only in people who had more than one *C. difficile* infection in a row [16, 40].

Another study demonstrated that, compared to the placebo group, the administration of 1 g/day of *Saccharomyces boulardii* with vancomycin for eradication considerably reduced recurrence. The mechanism of action of *S. boulardii* is considered to be the ingestion of *C. difficile* toxin B or A, which is crucial for *C. difficile* pathogenicity, as well as the digestion of toxin receptors by a proteolytic enzyme of *S. boulardii* [43].

Combination antibacterial therapy has become the standard for eradicating *Helicobacter pylori*, a major pathogenic agent worldwide due to its carcinogenicity and link to peptic ulcer illness [40]. Urease, a main colonization agent provided by *Helicobacter pylori* [44], hydrolyzes urea to ammonium, raising the pH of the stomach and favoring microorganism colonization [45]. So, both in humans and in experimental animal models, it was reported that strains of *Bifidobacterium*, various and numerous lactic acid bacteria, inhibited proliferation, prevented infection, and increased *H. pylori* urease activity. A study of 120 *H. pylori*-positive patients showed that *H. pylori* was eliminated in 72% of cases in the control group, and rates of eradication rose to 88% and 87%, respectively, indicating a significant eradication-promoting effect [43]. *In vitro* research on *Lactobacilli*, *Bifidobacteria*, and *Bacillus subtilis* strains revealed that the probiotic inhibits *H. pylori* proliferation and adhesion. Further, *Lactobacillus salivarius* has been shown to inhibit *H. pylori* colonization in mice [45].

Bacteria, like bacterial vaginosis, and fungi, like vulvovaginal candidiasis, induce the most prevalent vaginal infections [46]. Indeed, several studies have documented that probiotics are beneficial in the treatment of vaginal infections, including vulvovaginal candidiasis and bacterial vaginosis. *Lactobacillus spp.* constitute a normal, healthy vaginal microbiota, serving in the vagina as a microbial barrier. Therefore, *Lactobacillus spp.* maintain an acidic intravaginal microflora and degrade carbohydrates by producing $CO_2$ and lactic acid and inhibiting the proliferation of pathogenic microorganisms like *Escherichia coli*, *Enterobacteria*, *Candida*, and *Gardnerella vaginalis* [47]. Probiotic treatment and preventative strategies for vaginal infections are based on a diverse range of mechanisms of action, including direct nutrient competition with pathogens, the production of bacteriocins, biosurfactants, and hydrogen peroxide, as well as organic acids that maintain the vaginal pH low enough just to prevent pathogen proliferation or by modulating regional immune responses [46].

Several studies have found that probiotics can help treat or prevent respiratory tract infections caused by respiratory syncytial virus and influenza A virus by reducing infectious signs and duration, producing antiviral components, lowering virus concentrations in nasal washes or the lungs, and boosting immunological activity. Moreover, probiotic supplements combining *Lactobacillus fermentum*, *Lactobacillus casei*, and *Lactobacillus paracasei* substantially decreased the occurrence of upper respiratory infections and influenza-like symptoms in adults [48]. During the winter, Layer et al. evaluate the efficacy of probiotic consumption on influenza-like symptoms and the incidence and duration of colds in healthy children. The authors reported that daily probiotic supplements for 6 months was an effective strategy to reduce symptoms such as cough, fever, and rhinorrhea, as well as the antibiotic prescription and duration of the cold [49]. Another study found that probiotics like *Lacto-bacillus* and *Bifidobacterium*, in combination or alone, as well as a nonpathogenic *Enterococcus faecalis* strain, may have a positive impact on the duration and intensity of the signs and symptoms but may not be effective in lowering the occurrence of illness [40].

## 3.4 Polyunsaturated fatty acids

Data from research has demonstrated that dietary nutrients such as polyunsaturated fatty acids (PUFAs) derived from animals or plants are essential components with a large potential for enhancing metabolic homeostasis and the functioning of immune cells [50]. Eicosapentaenoic (EPA) and docosahexaenoic (DHA) acids are the principal *n*-3 PUFAs of interest that influence immune responses [51]. Although omega-3 fatty acids are commonly used for the treatment of chronic inflammatory diseases, in clinical practice, their use as antimicrobial agents has received limited attention [52, 53].

Generally, long-chain PUFAs can have an active influence on immune cell function and maintenance [54]. α-linolenic, γ-linolenic, docosahexaenoic, arachidonic, and eicosapentaenoic acids are PUFAs that have been shown to serve as endogenous antiviral, antibacterial, and antifungal immunomodulatory agents [55]. Both omega-6 and omega-3 metabolites are involved in the production of various mediators like leukotrienes, prostaglandins, thromboxanes, and protectins that can interact with numerous viruses and influence inflammation [54]. Several mechanisms are implicated in the immunomodulatory effects and the anti-inflammatory response of *n*-3 PUFAs during infection [53]. Antimicrobial mechanisms of PUFAs, especially omega-3, may thus include modifications in membrane hydrophobicity, interruption of cell-to-cell communication, generating cellular leakages by increased membrane poles, disruption of ATP production, and interruption of the electron transport system [52].

Chronic hepatitis C virus (HCV) infection is frequent, and it frequently leads to cirrhosis, which can result in ascites, encephalopathy, variceal hemorrhage, and hepatocellular cancer [56]. A HCV replicon RNA system was used in research conducted by Leu et al., and they demonstrated that many PUFAs may exert anti-HCV activity. Further explanation of the precise anti-HCV mechanism caused by PUFAs may result in the development of agents with powerful anti-hepatitis C viral activity [57]. Moreover, omega-3 fatty acids (8.8% w/w) block HCV-induced lipogenesis gene expression and enhance infection results. Epidemiological investigations have demonstrated that a higher intake of omega-3-rich seafood lowers the possibility of hepatocellular carcinoma development, and omega-3 supplementation for 7 days may help to decrease infection rates and enhance the restoration of liver function in humans, following hepatectomy [53].

During the first three years of life, more than 95% of children suffer from at least one acute respiratory illness [22]. Breast milk contains long-chain PUFAs, which play important roles in the immune system; breastfeeding thus helps to prevent childhood respiratory diseases [58]. As a result, numerous expert groups suggest that particular quantities of long-chain PUFAs be included in the meals of newborns, children, and their mothers. Some research, including randomized clinical trials of docosahexaenoic and arachidonic acid or fish oil supplementation in pregnant women or children, reported decreases in respiratory infections. In a double-blind study of children aged 18–36 months, the high-docosahexaenoic acid group reduced the occurrence of respiratory diseases such as pneumonia, pharyngitis, and bronchitis. Additionally, in the control group, 46% of children suffered more than one respiratory episode, compared to 17% in the high-docosahexaenoic acid group [59].

Generally, infection with *Helicobacter pylori* is linked to numerous gastrointestinal disorders like gastric cancer, chronic gastritis, and peptic ulcers. Antibiotic regimens are used to treat *H. pylori*-related gastric disorders, and omega-3 fatty acids have been found to be efficient antibacterial, antimutagenic, and rejuvenating agents in infections caused by the *H. pylori* bacteria. PUFAs may assist in preventing gastric inflammation caused by *H. pylori*, and the major pathogenic mechanism of this inflammation is the infiltration of neutrophils into the epithelial cell. Several studies have found that anti-*H. pylori* actions are linked to changes in bacterial structure and outer membrane protein composition, all of which increase bacterial adherence and the rate of inflammation caused by *H. pylori* [9, 60].

## 3.5 Micronutrients

Micronutrient deficiencies affect 2 billion people worldwide; poor growth, higher mortality, and susceptibility to infection are the consequences. For example, pregnancy-related nutritional deficits are linked to a poor immune response to infection [61]. Micronutrients, such as minerals and vitamins, have been proven to play important roles in enhancing the ability of the immune system to fight against certain infections and disorders [62]. Therefore, a well-balanced diet including several minerals and vitamins may be beneficial for many diseases [63].

### 3.5.1 Vitamin A

Vitamin A, widely known as the "anti-infection vitamin," functions as an anti-inflammatory component that improves mucosal integrity and immune system function, protecting the body against many infections. Several mechanisms have been proposed for vitamin A's anti-infective activity, including macrophage phagocytic and oxidative activity, the pro-inflammatory TNF-α, and IL-2 modulation, which stimulates macrophage microbial activity [64]. However, a lack of vitamin A increases the risk of respiratory illness [61]. So, vitamin A supplements have been proven to reduce morbidity and mortality caused by a variety of infectious disorders, including measles, diarrhea, and influenza, as well as infections caused by cytomegalovirus, human immunodeficiency virus, hepatitis B virus, and norovirus [63, 65]. When vitamin A supplementation is used in both the treatment and prevention of viral gastrointestinal disorders, the severity and number of days spent ill are reduced [66]. Active retinol derivatives stimulate the production of type-I IFN (α and β), the most potent antiviral mediator, during an acute infection. Consequently, a persistent immune response to the virus will be able to develop [65].

### 3.5.2 Vitamin D

Because it has anti-oxidative and antimicrobial properties and promotes the immune system's defense against infection, vitamin D is important for preventing acute respiratory illnesses [64, 65]. The infections caused by the hepatitis B virus, human papillomavirus, and human immunodeficiency virus are some of the illnesses known to be linked to a deficiency in vitamin D [65]. Indeed, the general and specialized immune systems can be modulated by vitamin D. It promotes the synthesis of antimicrobial proteins and peptides like human lysozyme and cathelicidin, the phagocytosis of macrophages, and the conversion of monocytes into macrophages. Additionally, vitamin D influences not only the function of T cells and B cells but also helps in the development of regulatory T cells. Therefore, a prior expert panel advised individuals to consume 2,000 IU daily to boost their immune systems' defenses, especially against viral infections [67, 68]. According to a meta-analysis on vitamin D, a vitamin D supplement can help prevent respiratory illnesses. Chronic hepatitis B patients had lower vitamin D concentrations than healthy controls, and the relationship between hepatitis B viral loads and vitamin D levels was inverse [64]. Epithelial tight junctions have been destroyed by influenza viruses, increasing the risk of infection. So, vitamin D maintains their integrity [69].

### 3.5.3 Vitamin E

Vitamin E is another immune booster that impacts numerous components of the immune system, such as antibody production, modulation of inflammatory responses and phagocytosis, and T-cell proliferation and differentiation [62, 67]. However, vitamin E deficiency has been linked to an increase in infections. Additionally, it is essential for the immune system's proper performance, especially in elderly individuals [70]. According to recent studies, vitamin E reduces the duration of influenza virus infection. For example, vitamin E research revealed that participants who received 200 IU daily of the vitamin reported fewer cold days per person per year [64]. The University of Helsinki conducted cancer prevention research and showed that the supplementation of vitamin E decreased the rate of pneumonia in smokers by 69% in human trials [71].

### 3.5.4 Vitamin C

It has been demonstrated that vitamin C significantly improves both general and specific immune systems [65] and can prevent infectious diseases [69]. Vitamin C influences a number of immune system functions, including promoting the function of the epithelial barrier, the migration of white blood cells to infection sites, antibody production, microbial killing, and phagocytosis [64]. A vitamin C supplement helps the mechanisms of respiratory defense, protects against viral infections, and lowers their intensity and duration. Vitamin C supplementation also reduces the intensity and duration of pneumonia, even in older adults. Therefore, patients with serious respiratory illnesses, such as tuberculosis or pneumonia, had lower plasma vitamin C levels [69]. Moreover, 200 mg of vitamin C per day has been demonstrated to reduce mortality and relieve respiratory symptoms in elderly people suffering from severe illnesses [64]. Vitamin C is recognized as a preventive strategy for coronavirus disease because of its antiviral properties, antioxidant effects, and immunomodulatory effects [65].

### 3.5.5 Selenium

Selenium has an immunological and preventative function in infections disorders, particularly respiratory viral illnesses, due to its antioxidant and anti-inflammatory effects. Indeed, selenium decreases viral penetration into respiratory cells by preserving the respiratory epithelial barrier's integrity [65]. It is without a doubt the most essential vitamin for AIDS patients. It also protects against the cytomegalovirus's heart-damaging effects [62]. In a South Korean investigation, patients diagnosed with COVID-19 were shown to have a high prevalence of selenium insufficiency. Some of the most important roles of selenium in COVID-19 include decreasing virus pathogenicity, preventing viral infection, reducing oxidative stress, boosting immunity, and preventing inflammation [65].

### 3.5.6 Zinc

Zinc is essential for antiviral immunity due to its numerous cellular functions, such as the maintenance of an effective immune system [65] and its direct interaction with viral replication [66]. Accordingly, a study reported that supplementing with zinc for children decreases the severity and duration of pneumonia and diarrheal disease. Also, a weekly dosage of 70 mg decreased pneumonia occurrence and mortality and reduced the incidence of diarrhea [61, 65]. Zinc improves the preservation of mucosal and skin membrane integrity, and rhinovirus replication is directly inhibited by zinc ions in their unbound state, which also possess antiviral properties [62]. For patients with human immunodeficiency virus, hepatitis C virus, and SARS-CoV-2 infections, zinc may be used in combination with conventional antiviral therapy. It may also influence the viral proteases of poliovirus and picornavirus by modifying the tertiary structure or directly, as in the case of the encephalomyocarditis virus, to prevent viral polyprotein proteolytic activity [65].

### 3.5.7 Copper

Antiviral, antifungal, antibacterial, and anti-inflammatory properties of copper make it a crucial meta in boosting and enhancing the immune system's defenses against infections [65]. The antimicrobial effect of copper mainly affects the cell membrane of bacteria. The bacterial cell's membrane is ripped by copper ion electrostatic interactions with electronegative groups on the membrane cell of bacteria, like thiol or carboxyl. Nosocomial infections have been controlled with copper. In a detailed investigation, 25 different nosocomial bacteria, including *Staphylococci, Enterobacteriaceae,* and *Pseudomonas spp* that have been isolated from health centers in Algeria, were tested for the antimicrobial efficacy of metallic copper. The authors reported that copper reduced the growth of 43.75% of *Enterobacteriaceae*, 25% of *Pseudomonas spp*, and 60% of the isolated *Staphylococci* strains at 400 g/mL [72, 73].

### 3.5.8 Iron

Iron, another vital mineral, is essential for both infection prevention and health. Therefore, iron is required for neutrophils' generation of reactive oxygen species to eradicate germs and for the growth and differentiation of epithelial tissue. Furthermore, 60 mg/day of element Fe is recommended to reduce respiratory infections in children [64]. An iron-binding glycoprotein known as lactoferrin is found in milk and other exogenous secretions. In both *in vivo* and *in vitro* tests, it has been confirmed that partial degradation of lactoferrin by pepsin can result in peptides known as lactoferricin, which have stronger antimicrobial activity. These peptides have been shown to inhibit the growth of a wide range of pathogenic microorganisms, including fungi, viruses, and antibiotic-resistant *E. coli* strains [74]. In an iron overload mouse model, exogenous lactoferrin was shown to reduce the pulmonary *Mycobacterium tuberculosis* burden. The ability of cystic fibrosis patients to prevent opportunistic *Pseudomonas aeruginosa* infections and biofilm development has also been related to the ability of lactoferrin to sequester iron [75].

## 3.6 Polyphenols

Thus far, over 50,000 different phenolic compounds have been discovered in plants. Famous representatives of polyphenols in foods are flavonoids, stilbenes, phenolic acids, lignans, and other polyphenols. Several phenolic compounds are found in a number of plant-based beverages and foods, such as fruits, vegetables, and seeds [76, 77]. Polyphenols, which have been found in the common diet, have major health benefits for the human body as well as antibacterial, antiviral, and antifungal properties [78]. The antimicrobial activity of polyphenols can help prevent infectious diseases [79]. This property can be attributed not only to direct action against microorganisms but also to the inhibition of germ virulence factors [80]. Several polyphenols have antiviral properties, including engaging directly with viral particles; however, the type of virus, including whether it is an RNA or DNA virus, will influence just how binding occurs; in the inhibition of enzymes, like integrase, reverse transcriptase, and proteinase; inhibition of the replication of DNA viruses; and in epidermal growth factor receptor inhibition [78, 79]. Another property of phenolic compounds is their ability to exert action until intracellular replication, which may be linked to their antioxidant properties. This property prevents certain viruses from oxidizing cells by replicating inside them [78].

### 3.6.1 Flavonoids

Flavonoids are widely distributed in plants and account for 60% of dietary polyphenols; there are an estimated 9,000 different flavonoids [76, 78]. Several of them have important benefits for human health, such as antiviral activities. Numerous studies on flavonoids have focused on their ability to inhibit HIV-1 and HIV-2 as well as herpes simplex virus types 1 and 2 [78]. For example, kaempferol is not only known for its antibacterial activities but also has antiviral properties against the influenza virus, herpes simplex virus, and HIV [81]. Quercetin is the flavonoid most frequently investigated as a clinical nutraceutical [82]. Many studies have promoted quercetin as an excellent anti-inflammatory agent with antioxidant properties that can help prevent infections with the influenza A virus [81]. An *in vitro* study revealed quercetin 3-rhamnoside's inhibitory effects on the influenza A virus by inhibiting viral replication until the early stages of infection [83].

Epigallocatechin gallate (EGCG), a catechin present in tea, has attracted the most interest and has been widely researched for its antifungal, antibacterial, and antiviral properties. It has been demonstrated that EGCG protects against infection caused by influenza virus by blocking viral particles from adhering to the target receptor cells and attaching to the viral hemagglutinin [80]. In research involving adults, it was shown that taking green tea supplements twice daily for three months reduced the occurrence of flu and cold symptoms by 32% and reduced the duration of illnesses by almost 23%

[84]. *In vitro* research demonstrated the efficacy of two flavonols, quercetin and kaemp-ferol, in association with rifampicin against clinically significant methicillin-resistant and rifampicin-resistant *Staphylococcus aureus* isolates. As a result, the complex of quercetin, kaempferol, and rifampicin inhibited β-lactamase significantly [80].

### 3.6.2 Tannins

Tannins have several biologically important effects, such as antibacterial activities, which are because of their astringent effect and depend on the tannin molecule's structure [85]. A diet rich in tannins may help prevent *H. pylori* infection since hydro-lysable tannins have potential antibacterial action against *H. pylori* [86]. Proanthocya-nidins, which are present in berries, are known to prevent the proliferation of numerous pathogenic bacteria, such as oxacillin-resistant *S. aureus* and uropatho-genic *E. coli*. Multiple mechanisms are suggested to describe the effect of the proan-thocyanidin A-type on inhibition of bacterial growth, including the permeabilization of the cell membrane, the inhibition of extracellular microbial enzymes, the destabili-zation of the cytoplasmic membrane, and direct actions on microbial metabolism. Proanthocyanidins have also shown antiviral action against the influenza A virus and type 1 herpes simplex virus. The mechanism that was suggested consisted of avoiding the entrance of the virus inside the host cell, which is the first critical step in primary type 1 herpes simplex virus infection [80].

### 3.6.3 Stilbenes

Due to their antiviral effect, stilbenes have been the focus of significant research. Res-veratrol, which belongs to the stilbene group, is a dietary polyphenol that is found in natural sources such as fruit. The effectiveness of stilbenes and their derivates has been investigated against viruses, including HIV, hepatitis C virus, influenza, herpes simplex virus types 1 and 2, and human papillomavirus [81, 87]. Most research on res-veratrol has been focused on its capacity to increase the antiretroviral efficacy of nu-cleoside reverse transcriptase inhibitors against drug-resistant HIV-1 strains and drug-sensitive HIV-1 strains [88]. The antibacterial properties of resveratrol have been shown in both gram-negative and gram-positive bacteria like *Pseudomonas aerugi-nosa, E. coli, Salmonella Typhimurium, Helicobacter pylori*, methicillin-resistant *Staph-ylococcus aureus, Chlamydia pneumoniae, Klebsiella pneumoniae*, and vancomycin-resistant *Enterococcus*. Several mechanisms, including motility, biofilm formation, and quorum sensing, are proposed to explain their antibacterial activities. In addition to the antibacterial and antiviral properties of resveratrol, numerous studies have ex-amined its antifungal effect [89]. Resveratrol, at low concentrations, induces antifun-gal activity on *Candida albicans* by penetrating inside the cell without causing fatal damage to the cell membrane and promoting cytochrome c release, which can induce apoptosis via caspase protease activation [89, 90].

### 3.6.4 Phenolic acids

Phenolic acids might have a positive effect in the treatment of many infections caused by resistant bacteria because they have been reported to exert antimicrobial properties [91, 92]. Chlorogenic acids, the main phenolic acid compounds of coffee and tea, present many biological activities, including antiviral and antibacterial effects [77]. Phenolic acid also showed a strong antimicrobial action on the gram-positive bacteria such as *Listeria monocytogenes* and *Staphylococcus aureus*, as well as on the gram-negative bacteria including *Pseudomonas aeruginosa*, *E. coli*, and *Salmonella enteritidis* [92].

Therefore, phenolic acids from natural sources and their derivatives have inhibitory effects on a variety of human viruses, including the hepatitis C virus, the hepatitis B virus, the herpes simplex virus, the human immunodeficiency virus, the respiratory syncytial virus, and the influenza virus [93]. The antibacterial activities of simple phenols are most likely due to nonspecific interactions with proteins or mediated by interactions with sulfhydryl groups in microbial enzymes [91].

## 3.7 Spices

For centuries, scientists have investigated the effects of spices and their potential health advantages [94]. In addition to flavoring food and beverages, spices are used for therapeutic purposes. They are recognized especially for their antimicrobial, antioxidant, and medicinal properties [95]. Therefore, spices provide significant promise for the development of novel and safe antimicrobial agents [96].

### 3.7.1 Clove

Clove (*Eugenia caryohyllata*), a Myrtaceae family member, is widely used in therapeutics as an antiseptic against infectious disorders, especially periodontal disease, due to its antibacterial activities. Indeed, the antimicrobial properties of clove have been shown in several studies. Clove water extract, for example, was effective against pathogenic bacteria like *E. coli* and *Staphylococcus aureus* in an experimental model of pyelonephritis. Clove may thus disrupt the cell walls and membranes of bacteria and penetrate cytoplasmic membranes, where it will subsequently prevent the normal production of DNA and proteins [96].

According to Tragoolpua and Jatisatienr, clove essential oil has a direct inactivating effect on the particles of the typical herpes simplex virus strains [97]. Eugenol, a major constituent of cloves, has been shown to inhibit influenza virus replication [98]. However, a rise in antibiotic resistance may make it difficult to successfully treat a *Streptococcus suis* infection [99], which can cause serious infections in humans, including meningitis [100]. Wongsawan et al. reported that clove essential oil had inhibitory action against all tested multidrug-resistant *S. suis* isolates [99]. Several studies have reported that eugenol and clove essential oils have antifungal properties against human pathogenic fungi. Therefore, clove essential oils have a very broad spectrum of antimicrobial action against dermatophytes, *Aspergillus*, and *Candida*. The antimicrobial action of clove essential oils involves numerous mechanisms of action, such as hydrophobicity, which is responsible for dividing inside the lipid bilayer of the cell membrane, causing a change in permeability and, as a result, a leakage of the contents of the cell [101].

### 3.7.2 Cinnamon

Cinnamon (*Cinnamomum zeylanicum*), a Lauraceae family member, is commonly used in soups and savory dishes [96]. It has been the focus of so much research due to its many pharmacological activities, like antimicrobial and anti-inflammatory effects [102]. Cinnamyl acetate, cinnamaldehyde, and cinnamyl alcohol are the four principal compounds of cinnamon that are also used as a health-promoting element to treat illnesses such as gastrointestinal disorders, inflammation, and urinary infections [96]. Therefore, fluconazole-resistant *Candida* stains are a current issue. *C. zeylanicum* has been demonstrated *in vitro* to be effective against fluconazole-resistant and susceptible *Candida* isolates [103]. The antimicrobial effect of cinnamaldehyde involves numerous mechanisms of action, including inhibition of specific enzyme activities, cell wall biosynthesis, and membrane function [96]. To assess the efficacy of cinnamon extract, 98 eligible healthy people and *Helicobacter pylori*-infected patients participated in a randomized clinical trial. The authors reported that clinical manifestations like diarrhea, vomiting, and nausea were significantly decreased in the cinnamon group. Because of this, cinnamon, as an adjunctive treatment, is able to reduce the complications of *H. pylori* infection [104]. Also, cinnamon essential oils and cinnamon extracts have been shown to have antibacterial activities by altering the lipid profile, damaging the cell membrane, inhibiting cell division, ATPases, membrane porins, biofilm formation, and motility, as well as through quorum sensing inhibition [105].

### 3.7.3 Ginger

Ginger (*Zingiber officinale*), a Zingiberaceae family member, is frequently used as an ingredient in pharmaceuticals, food, and other sectors [96]. Because of its various benefits, research has also investigated its antimicrobial properties. However, the efficiency of ginger extracts, oleoresins, and essential oils against viruses, fungi, and bacteria has been evaluated in several *in vitro* assays [106]. According to Singh et al., ginger essential oil inhibited *Staphylococcus aureus*, *Klebsiella pneumoniae*, *Escherichia coli*, and *Pseudomonas aeruginosa* better than oleoresins. Therefore, the affinity and higher hydrophobicity of ginger essential oils for the lipids found in mitochondria and cell walls may affect their permeability and structure [107]. At doses below 1 mg/mL, compounds from *Z. officinale*, such as gingerols and gingerdiols, have demonstrated antifungal action against *Trichophyton mentagrophytes*, *Microsporum gypseum*, *Aspergillus fumigatus*, *Pseudallescheria boydii*, and *Wangiella dermatitidis* [108]. *Z. officinale* rhizomes have been shown to have antiviral effects against human respiratory syncytial virus (HRSV) infection by lowering the development of HRSV-induced plaques in respiratory mucosal cell lines. Moreover, replication of the hepatitis C virus was found to be inhibited by the lyophilized juice extract of *Z. officinale* [109]. Several authors attribute ginger's antimicrobial activity to phenolic compounds (like gingerols, eugenol, gingerdiols) and their synergistic interaction with other bioactive compounds like zingiberene, α- farnesene, cis-caryophillene, β-sesquiphellandrene [106].

### 3.7.4 Thyme

Thyme (*Thymus vulgaris*), a Lamiaceae family member, is frequently used as a herbal medicine and classic culinary herb [110]. Thyme is rich in flavonoids, minerals, vitamins, and phytonutrients. Moreover, the positive effects of both the plant extracts and the essential oil, especially carvacrol and thymol, against various illnesses were demonstrated in several studies, with the antimicrobial properties being the most important ones [110, 111].

Fani and Kohanteb evaluated the antimicrobial activities of thyme essential oil against some oral bacterial and fungal pathogens, including *Streptococcus mutans*, *Candida albicans*, and *Streptococcus pyogenes*. The authors reported that thyme essential oil has effective bactericidal and antifungal activities against tested microorganisms [112]. Furthermore, the antifungal activity of thyme oils is primarily due to thymol and carvacrol, which induce fungal hyphae degeneration and appear to empty their cytoplasmic content [110]. The main active compound of thyme essential oil is thymol, which has an antimicrobial effect by adhering to membrane proteins through hydrophobic and hydrogen bonding interactions, as well as by modifying the permeability of the membranes [96].

### 3.7.5 Garlic

Garlic (*Allium sativum*), a Liliaceae family member, has long been recognized to have antimicrobial properties. Garlic may inhibit many bacterial strains, and allicin, a major constituent of garlic produced through the enzymatic conversion of alliin, inhibits several strains considerably more effectively than antibiotics [94, 113]. Allicin has been proven to exhibit antifungal, antibacterial, and antiviral proprieties against a wide variety of microorganisms like methicillin-resistant *Staphylococcus aureus* and multidrug-resistant enterotoxicogenic *E. coli* strains [113, 114]. Moreover, the antibacterial activities of garlic essential oil and its naturally occurring diallylsulfide compounds have been tested *in vitro* against several pathogenic microorganisms, such as human enteric bacteria, *Pseudomonas aeruginosa*, *Candida spp.*, methicillin-resistant *S. aureus*, *Aspergillus spp.*, and *E. coli*. Allicin, for example, improved skin recovery in the case of methicillin-resistant *Staphylococcus aureus* skin disorders. It was reported that some bacteria are absolutely incapable of developing resistance to crushed garlic and allicin [113, 114]. Consequently, the antibacterial effect of garlic appears to be due to the interaction between thiol groups of microbial enzymes (e.g., trypsin or other proteases) and sulfur compounds of garlic (e.g., allicin), which leads to microbial growth inhibition [94].

### 3.7.6 Turmeric

Turmeric (*Curcuma longa*), a Zingiberaceae family member, has been shown in several studies to have broad-spectrum antimicrobial action, including antifungal, antiviral, and antibacterial properties. However, an ethanolic extract of the rhizomes of turmeric has demonstrated significant antimicrobial effects against several microorganisms like *Staphylococcus aureus*, *Aspergillus oryzae*, and *Escherichia coli* [114]. In order to treat osteomyelitis, generally caused by *Staphylococcus aureus*, Zhou et al. reported that the combination of curcumin and erythromycin can suppress the development of *S. aureus*. Furthermore, researchers discovered that erythromycin monotherapy was ineffective at inhibiting bacterial growth, while lowering IL-6 and TNF-α levels. However, curcumin inhibited bacterial growth significantly [115]. Martins et al. tested the antifungal efficacy of curcumin against 23 strains of fungi and reported that curcumin was much more efficient than fluconazole, particularly reducing *Candida spp.* adherence to human buccal epithelial cells of AIDS patients [116].

Generally, the antimicrobial property of turmeric appears to be due to the hydrophobic interaction and hydrogen bonding of different phenolic compounds of turmeric with membrane proteins, which causes electron transport chain disruption, cell wall collapse, and cell membrane distraction. Curcumin, for example, can reestablish bacterial susceptibility by reducing biofilm formation and making germs more sensitive to antibiotics [114].

# 4 Summary

Nutraceuticals have entered the healthcare system as a simple and attractive approach to disease prevention, including infectious disorders. With numerous studies and research, nutraceuticals may be beneficial therapeutic agents, in combination with conventional therapy, for infectious diseases and their complications. There has been a considerable reduction in the number of approved antimicrobials as well as the rise in antimicrobial resistance, requiring the use of novel approaches that reduce the risks of developing resistance and prevent and treat infectious diseases. Nutraceuticals have also been proven to play important roles in enhancing the immune system and protecting against certain infections. Therefore, prebiotics and probiotics have been shown to have the potential to prevent and treat some infectious diseases, with the majority of studies on the clinical application of probiotics focusing on the gastrointestinal tract. Research data have demonstrated that diet components such as vitamins, minerals, and polyunsaturated fatty acids are essential components with a large potential to improve metabolic homeostasis and the functioning of immune cells. Furthermore, spices and phenolic compounds provide significant promise for the development of novel and safe antimicrobial agents.

# References

[1]     De Pauw, B.E. 2011. What are fungal infections? Mediterranean Journal of Hematology and Infectious Diseases, 3(1), e2011001.

[2]     Lima, P.G., Oliveira, J.T.A., Amaral, J.L., Freitas, C.D.T. and Souza, P.F.N. 2021. Synthetic antimicrobial peptides: Characteristics, design, and potential as alternative molecules to overcome microbial resistance. Life Sciences, 278, 119647.

[3]     Cole, L. and Kramer, P.R. 2016. Bacteria, virus, fungi, and infectious diseases. Human Physiology, Biochemistry and Basic Medicine, 2016, 193–196.

[4]     Werth. 2022. *Overview of Bacteria – Infections*. MSD Manual Consumer Version. https://www.msdma nuals.com/home/infections/bacterial-infections-overview/overview-of-bacteria. (accessed on January 14, 2023).

[5]     Sterling, J.C. 2016. Viral Infections. In *Rook's Textbook of Dermatology*, Vol. 1–4. http://www.rooksder matology.com/manual/c25-sec-0001. (accessed on December 30, 2022).

[6]     Horn, F., Heinekamp, T., Kniemeyer, O., Pollmächer, J., Valiante, V. and Brakhage, A. 2012. Systems biology of fungal infection. Frontiers in Microbiology, 3, 108.

[7]     *Antimicrobial resistance*. 2021. https://www.who.int/news-room/fact-sheets/detail/antimicrobial-resistance. (accessed on December 24, 2022).

[8]     Cui, D., Liu, X., Hawkey, P., Li, H., Wang, Q., Mao, Z. and Sun, J. 2017. Use of and microbial resistance to antibiotics in China: A path to reducing antimicrobial resistance. Journal of International Medical Research, 45(6), 1768–1778.

[9]     Cohen, L.F. and Tartasky, D. 1997. Microbial resistance to drug therapy: A review. American Journal of Infection Control, 25(1), 51–64.

[10]  Nainu, F., Permana, A.D., Djide, N.J.N., Anjani, Q.K., Utami, R.N., Rumata, N.R., Zhang, J., Emran, T.B. and Simal-Gandara, J. 2021. Pharmaceutical approaches on antimicrobial resistance: Prospects and challenges. Antibiotics, 10(8), 8.

[11]  Drexler, M., Institute of Medicine (US). 2011. *What You Need to Know About Infectious Disease.* National Academies Press, pp. 13006.

[12]  Jampilek, J. 2022. Drug repurposing to overcome microbial resistance. Drug Discovery Today, 27(7), 2028–2041.

[13]  Ma, Y., Frutos-Beltrán, E., Kang, D., Pannecouque, C., De Clercq, E., Menéndez-Arias, L., Liu, X. and Zhan, P. 2021. Medicinal chemistry strategies for discovering antivirals effective against drug-resistant viruses. Chemical Society Reviews, 50(7), 4514–4540.

[14]  McKeegan, K.S., Borges-Walmsley, M.I. and Walmsley, A.R. 2002. Microbial and viral drug resistance mechanisms. Trends in Microbiology, 10(10), s8–s14.

[15]  Enioutina, E.Y., Teng, L., Fateeva, T.V., Brown, J.C.S., Job, K.M., Bortnikova, V.V., Krepkova, L.V., Gubarev, M.I. and Sherwin, C.M.T. 2017. Phytotherapy as an alternative to conventional antimicrobials: Combating microbial resistance. Expert Review of Clinical Pharmacology, 10(11), 1203–1214.

[16]  Elmer, G.W. 2001. Probiotics: "Living drugs. American Journal of Health-System Pharmacy, 58(12), 1101–1109.

[17]  Foletto, V.S., da Rosa, T.F., Serafin, M.B., Bottega, A. and Hörner, R. 2021. Repositioning of non-antibiotic drugs as an alternative to microbial resistance: A systematic review. International Journal of Antimicrobial Agents, 58(3), 106380.

[18]  Das, U.N. 2011. Infection, inflammation, and polyunsaturated fatty acids. Nutrition, 27(10), 1080–1084.

[19]  González-Sarrías, A., Larrosa, M., García-Conesa, M.T., Tomás-Barberán, F.A. and Espín, J.C. 2013. Nutraceuticals for older people: Facts, fictions and gaps in knowledge. Maturitas, 75(4), 313–334.

[20]  Sauvaitre, T., Etienne-Mesmin, L., Sivignon, A., Mosoni, P., Courtin, C.M., Van de Wiele, T. and Blanquet-Diot, S. 2021. Tripartite relationship between gut microbiota, intestinal mucus and dietary fibers: Towards preventive strategies against enteric infections. FEMS Microbiology Reviews, 45(2), fuaa052.

[21]  Gong, J. and Yang, C. 2012. Advances in the methods for studying gut microbiota and their relevance to the research of dietary fiber functions. Food Research International, 48(2), 916–929.

[22]  Yang, H., Sun, Y., Cai, R., Chen, Y. and Gu, B. 2020. The impact of dietary fiber and probiotics in infectious diseases. Microbial Pathogenesis, 140, 103931.

[23]  Lee, J.Y., Kim, N., Choi, Y.J., Park, J.H., Ashktorab, H., Smoot, D.T. and Lee, D.H. 2020. Expression of tight junction proteins according to functional dyspepsia subtype and sex. Journal of Neurogastroenterology and Motility, 26(2), 248–258.

[24]  Kim, S., Covington, A. and Pamer, E.G. 2017. The intestinal microbiota: Antibiotics, colonization resistance, and enteric pathogens. Immunological Reviews, 279(1), 90–105.

[25]  Schley, P.D. and Field, C.J. 2002. The immune-enhancing effects of dietary fibres and prebiotics. British Journal of Nutrition, 87(S2), S221–S230.

[26]  Goodgame, R.W. 2001. Viral causes of diarrhea. Gastroenterology Clinics, 30(3), 779–795.

[27]  Hajipour, A., Afsharfar, M., Jonoush, M., Ahmadzadeh, M., Gholamalizadeh, M., Hassanpour Ardekanizadeh, N., Doaei, S. and Mohammadi-Nasrabadi, F. 2022. The effects of dietary fiber on common complications in critically ill patients; with a special focus on viral infections; a systematic review. Immunity, Inflammation and Disease, 10(5), e613.

[28]  Rushdi, T.A., Pichard, C. and Khater, Y.H. 2004. Control of diarrhea by fiber-enriched diet in ICU patients on enteral nutrition: A prospective randomized controlled trial. Clinical Nutrition, 23(6), 1344–1352.

[29]   Cho, S. and Finocchiaro, E.T. (Eds.), 2009. *Handbook of Prebiotics and Probiotics Ingredients: Health Benefits and Food Applications.* CRC Press: Boca Raton.

[30]   Lomax, A.R. and Calder, P.C. 2008. Prebiotics, immune function, infection and inflammation: A review of the evidence. British Journal of Nutrition, 101(5), 633–658.

[31]   Macfarlane, S. 2010. Prebiotics in the gastrointestinal tract. In Watson, R.R., Preedy, V.R. (Eds.), *Bioactive Foods in Promoting Health: Probiotics and Prebiotics.* Academic Press: Cambridge, Massachusetts, pp. 145–156.

[32]   Patel, R. and DuPont, H.L. 2015. New approaches for bacteriotherapy: Prebiotics, new-generation probiotics, and synbiotics. Clinical Infectious Diseases, 60(suppl_2), S108–S121.

[33]   Novak, J. and Katz, J.A. 2006. Probiotics and prebiotics for gastrointestinal infections. Current Infectious Disease Reports, 8(2), 103–109.

[34]   Gibson, G.R., Rastall, R.A. and Fuller, R. 2008. The health benefits of probiotics and prebiotics. In Fuller, R., Perdigón, G. (Eds.), *Gut Flora, Nutrition, Immunity and Health.* Blackwell Publishing Ltd: Oxford, United Kingdom, pp. 52–76.

[35]   Vandenplas, Y., De Greef, E., Devreker, T., Veereman-Wauters, G. and Hauser, B. 2013. Probiotics and prebiotics in infants and children. Current Infectious Disease Reports, 15(3), 251–262.

[36]   Thomas, D.W. and Greer, F.R. 2010. Committee on nutrition; section on gastroenterology, h., and nutrition. Probiotics and Prebiotics in Pediatrics. Pediatrics, 126(6), 1217–1231.

[37]   Yahfoufi, N., Mallet, J., Graham, E. and Matar, C. 2018. Role of probiotics and prebiotics in immunomodulation. Current Opinion in Food Science, 20, 82–91.

[38]   De Vrese, M. and Schrezenmeir, J. 2008. Probiotics, prebiotics, and synbiotics. In Stahl, U., Donalies, U.E.B., Nevoigt, E., (Eds.), *Food Biotechnology.* Berlin Heidelberg: Springer, pp. 1–66.

[39]   Delgado, G.T.C., Tamashiro, W.M.D.S.C., Junior, M.R.M., Moreno, Y.M.F. and Pastore, G.M. 2011. The putative effects of prebiotics as immunomodulatory agents. Food Research International, 44(10), 3167–3173.

[40]   Martinez, R.C.R., Bedani, R. and Saad, S.M.I. 2015. Scientific evidence for health effects attributed to the consumption of probiotics and prebiotics: An update for current perspectives and future challenges. British Journal of Nutrition, 114(12), 1993–2015.

[41]   Alvarez-Olmos, M.I. and Oberhelman, R.A. 2001. Probiotic agents and infectious diseases: A modern perspective on a traditional therapy. Clinical Infectious Diseases, 32(11), 1567–1576.

[42]   Penner, R., Fedorak, R. and Madsen, K. 2005. Probiotics and nutraceuticals: Non-medicinal treatments of gastrointestinal diseases. Current Opinion in Pharmacology, 5(6), 596–603.

[43]   Nomoto, K. 2005. Prevention of infections by probiotics. Journal of Bioscience and Bioengineering, 100(6), 583–592.

[44]   Cover, T.L., Berg, D.E., Blaser, M.J. and Mobley, H.L. 2001. H. pylori Pathogenesis. In Groisman, E.A. (Ed.), *Principles of Bacterial Pathogenesis.* Academic Press: Cambridge, Massachusetts, pp. 509–558.

[45]   Sullivan, Å. and Nord, C.E. 2002. The place of probiotics in human intestinal infections. International Journal of Antimicrobial Agents, 20(5), 313–319.

[46]   Palmeira-de-Oliveira, R., Palmeira-de-oliveira, A. and Martinez-de-oliveira, J. 2015. New strategies for local treatment of vaginal infections. Advanced Drug Delivery Reviews, 92, 105–122.

[47]   Kim, J.-M. and Park, Y.J. 2017. Probiotics in the prevention and treatment of postmenopausal vaginal infections: Review article. Journal of Menopausal Medicine, 23(3), 139–145.

[48]   Olaimat, A.N., Aolymat, I., Al-Holy, M., Ayyash, M., Abu Ghoush, M., Al-Nabulsi, A.A., Osaili, T., Apostolopoulos, V., Liu, S.-Q. and Shah, N.P. 2020. The potential application of probiotics and prebiotics for the prevention and treatment of COVID-19. Npj Science of Food, 4(1), 17.

[49]   Leyer, G.J., Li, S., Mubasher, M.E., Reifer, C. and Ouwehand, A.C. 2009. Probiotic effects on cold and influenza-like symptom incidence and duration in children. Pediatrics, 124(2), e172–179.

[50]   Margină, D., Ungurianu, A., Purdel, C., Nițulescu, G.M., Tsoukalas, D., Sarandi, E., Thanasoula, M., Burykina, T.I., Tekos, F., Buha, A., Nikitovic, D., Kouretas, D. and Tsatsakis, A.M. 2020. Analysis of the

intricate effects of polyunsaturated fatty acids and polyphenols on inflammatory pathways in health and disease. Food and Chemical Toxicology, 143, 111558.

[51] Whelan, J., Gowdy, K.M. and Shaikh, S.R. 2016. N-3 polyunsaturated fatty acids modulate B cell activity in pre-clinical models: Implications for the immune response to infections. European Journal of Pharmacology, 785, 10–17.

[52] Chanda, W., Joseph, T.P., Guo, X., Wang, W., Liu, M., Vuai, M.S., Padhiar, A.A. and Zhong, M. 2018. Effectiveness of omega-3 polyunsaturated fatty acids against microbial pathogens. Journal of Zhejiang University-SCIENCE B, 19(4), 253–262.

[53] Husson, M.-O., Ley, D., Portal, C., Gottrand, M., Hueso, T., Desseyn, J.-L. and Gottrand, F. 2016. Modulation of host defence against bacterial and viral infections by omega-3 polyunsaturated fatty acids. Journal of Infection, 73(6), 523–535.

[54] De Cosmi, V., Mazzocchi, A., Turolo, S., Syren, M.L., Milani, G.P. and Agostoni, C. 2022. Long-chain polyunsaturated fatty acids supplementation and respiratory infections. Annals of Nutrition and Metabolism, 78(1), 8–15.

[55] Das, U.N. 2018. Arachidonic acid and other unsaturated fatty acids and some of their metabolites function as endogenous antimicrobial molecules: A review. Journal of Advanced Research, 11, 57–66.

[56] Flamm, S.L. 2003. Chronic hepatitis C virus infection. Jama, 289(18), 2413–2417.

[57] Leu, G.-Z., Lin, T.-Y. and Hsu, J.T.A. 2004. Anti-HCV activities of selective polyunsaturated fatty acids. Biochemical and Biophysical Research Communications, 318(1), 275–280.

[58] Miles, E.A., Childs, C.E. and Calder, P.C. 2021. Long-chain polyunsaturated fatty acids (LCPUFAs) and the developing immune system: A narrative review. Nutrients, 13(1), 247.

[59] Hageman, J.H.J., Hooyenga, P., Diersen-Schade, D.A., Scalabrin, D.M.F., Wichers, H.J. and Birch, E.E. 2012. The impact of dietary long-chain polyunsaturated fatty acids on respiratory illness in infants and children. Current Allergy and Asthma Reports, 12(6), 564–573.

[60] Park, J.-M., Jeong, M., Kim, E.-H., Han, Y.-M., Kwon, S.H. and Hahm, K.-B. 2015. Omega-3 polyunsaturated fatty acids intake to regulate *Helicobacter pylori*-associated gastric diseases as nonantimicrobial dietary approach. BioMed Research International, 2015, e712363.

[61] Katona, P. and Katona-Apte, J. 2008. The Interaction between Nutrition and Infection. Clinical Infectious Diseases, 46(10), 1582–1588.

[62] Alpert, P.T. 2017. The role of vitamins and minerals on the immune system. Home Health Care Management & Practice, 29(3), 199–202.

[63] Kumar, P., Kumar, M., Bedi, O., Gupta, M., Kumar, S., Jaiswal, G., Rahi, V., Yedke, N.G., Bijalwan, A., Sharma, S. and Jamwal, S. 2021. Role of vitamins and minerals as immunity boosters in COVID-19. Inflammopharmacology, 29(4), 1001–1016.

[64] BourBour, F., Mirzaei Dahka, S., Gholamalizadeh, M., Akbari, M.E., Shadnoush, M., Haghighi, M., Taghvaye-Masoumi, H., Ashoori, N. and Doaei, S. 2020. Nutrients in prevention, treatment, and management of viral infections; special focus on Coronavirus. Archives of Physiology and Biochemistry, 129(1), 16–25.

[65] Fath, M.K., Naderi, M., Hamzavi, H., Ganji, M., Shabani, S., Ghahroodi, F.N., Khalesi, B., Pourzardosht, N., Hashemi, Z.S. and Khalili, S. 2022. Molecular mechanisms and therapeutic effects of different vitamins and minerals in COVID-19 patients. Journal of Trace Elements in Medicine and Biology, 73, 127044.

[66] Costagliola, G., Nuzzi, G., Spada, E., Comberiati, P., Verduci, E. and Peroni, D.G. 2021. Nutraceuticals in viral infections: An overview of the immunomodulating properties. Nutrients, 13(7), 2410.

[67] Eggersdorfer, M., Berger, M.M., Calder, P.C., Gombart, A.F., Ho, E., Laviano, A. and Meydani, S.N. 2022. Perspective: Role of micronutrients and omega-3 long-chain polyunsaturated fatty acids for immune outcomes of relevance to infections in older adults – A narrative review and call for action. Advances in Nutrition, 13(5), 1415–1430.

[68] Wang, G. 2014. Human antimicrobial peptides and proteins. Pharmaceuticals, 7(5), 545–594.

[69] Shakoor, H., Feehan, J., Al Dhaheri, A.S., Ali, H.I., Platat, C., Ismail, L.C., Apostolopoulos, V. and Stojanovska, L. 2021. Immune-boosting role of vitamins D, C, E, zinc, selenium and omega-3 fatty acids: Could they help against COVID-19? Maturitas, 143, 1–9.

[70] Moriguchi, S. and Muraga, M. 2000. Vitamin E and immunity. Vitamins and Hormones, 59, 305–336.

[71] Srivastava, A., Gupta, R.C., Doss, R.B. and Lall, R. 2022. Trace minerals, vitamins and nutraceuticals in prevention and treatment of COVID-19. Journal of Dietary Supplements, 19(3), 395–429.

[72] Bisht, N., Dwivedi, N., Kumar, P., Venkatesh, M., Yadav, A.K., Mishra, D., Solanki, P., Verma, N.K., Lakshminarayanan, R., Ramakrishna, S., Mondal, D.P., Srivastava, A.K. and Dhand, C. 2022. Recent advances in copper and copper-derived materials for antimicrobial resistance and infection control. Current Opinion in Biomedical Engineering, 24, 100408.

[73] Zerbib, S., Vallet, L., Muggeo, A., de Champs, C., Lefebvre, A., Jolly, D. and Kanagaratnam, L. 2020. Copper for the prevention of outbreaks of health care–associated infections in a long-term care facility for older adults. Journal of the American Medical Directors Association, 21(1), 68–71.

[74] Yen, -C.-C., Shen, C.-J., Hsu, W.-H., Chang, Y.-H., Lin, H.-T., Chen, H.-L. and Chen, C.-M. 2011. Lactoferrin: An iron-binding antimicrobial protein against Escherichia coli infection. BioMetals, 24(4), 585–594.

[75] Johnson, E.E. and Wessling-Resnick, M. 2012. Iron metabolism and the innate immune response to infection. Microbes and Infection, 14(3), 207–216.

[76] Gutiérrez-del-río, I., Fernández, J. and Lombó, F. 2018. Plant nutraceuticals as antimicrobial agents in food preservation: Terpenoids, polyphenols and thiols. International Journal of Antimicrobial Agents, 52(3), 309–315.

[77] Zhang, L., Han, Z. and Granato, D. 2021. Polyphenols in foods: Classification, methods of identification, and nutritional aspects in human health. Advances in Food and Nutrition Research, 98, 1–33.

[78] Montenegro-Landívar, M.F., Tapia-Quirós, P., Vecino, X., Reig, M., Valderrama, C., Granados, M., Cortina, J.L. and Saurina, J. 2021. Polyphenols and their potential role to fight viral diseases: An overview. Science of the Total Environment, 801, 149719.

[79] Petti, S. and Scully, C. 2009. Polyphenols, oral health and disease: A review. Journal of Dentistry, 37(6), 413–423.

[80] Daglia, M. 2012. Polyphenols as antimicrobial agents. Current Opinion in Biotechnology, 23(2), 174–181.

[81] Rasouli, H., Farzaei, M.H. and Khodarahmi, R. 2017. Polyphenols and their benefits: A review. International Journal of Food Properties, 20, 1700–1741

[82] Formica, J.V. and Regelson, W. 1995. Review of the biology of quercetin and related bioflavonoids. Food and Chemical Toxicology, 33(12), 1061–1080.

[83] Bahramsoltani, R., Sodagari, H.R., Farzaei, M.H., Abdolghaffari, A.H., Gooshe, M. and Rezaei, N. 2016. The preventive and therapeutic potential of natural polyphenols on influenza. Expert Review of Anti-Infective Therapy, 14(1), 57–80.

[84] Reygaert, W.C. 2018. Green tea catechins: Their use in treating and preventing infectious diseases. BioMed Research International, 2018, 1–9.

[85] Kováč, J., Slobodníková, L., Trajčíková, E., Rendeková, K., Mučaji, P., Sychrová, A. and Bittner Fialová, S. 2023. Therapeutic potential of flavonoids and tannins in management of oral infectious diseases – A review. Molecules, 28(1), 158.

[86] Watson, R.R. and Preedy, V.R. 2008. *Botanical Medicine in Clinical Practice*. CABI: Wallingford.

[87] Levy, E., Delvin, E., Marcil, V. and Spahis, S. 2020. Can phytotherapy with polyphenols serve as a powerful approach for the prevention and therapy tool of novel coronavirus disease 2019 (COVID-19)? American Journal of Physiology-Endocrinology and Metabolism, 319(4), E689–E708.

[88] Date, A.A. and Destache, C.J. 2016. Natural polyphenols: Potential in the prevention of sexually transmitted viral infections. Drug Discovery Today, 21(2), 333–341.

[89] Bostanghadiri, N., Pormohammad, A., Chirani, A.S., Pouriran, R., Erfanimanesh, S. and Hashemi, A. 2017. Comprehensive review on the antimicrobial potency of the plant polyphenol Resveratrol. Biomedicine & Pharmacotherapy, 95, 1588–1595.

[90] Saxena, M., Sharma, R.K., Ramirez-Paz, J., Tinoco, A.D. and Griebenow, K. 2015. Purification and characterization of a cytochrome c with novel caspase-3 activation activity from the pathogenic fungus Rhizopus arrhizus. BMC Biochemistry, 16(1), 21.

[91] Aldulaimi, O.A. 2017. General overview of phenolics from plant to laboratory, good antibacterials or not. Pharmacognosy Reviews, 11(22), 123–127.

[92] Kiokias, S. and Oreopoulou, V. 2021. A review of the health protective effects of phenolic acids against a range of severe pathologic conditions (including coronavirus-based infections). Molecules, 26(17), 5405.

[93] Wu, Y.-H., Zhang, B.-Y., Qiu, L.-P., Guan, R.-F., Ye, Z.-H. and Yu, X.-P. 2017. Structure properties and mechanisms of action of naturally originated phenolic acids and their derivatives against human viral infections. Current Medicinal Chemistry, 24(38), 4279–4302.

[94] Wilson, E.A. and Demmig-Adams, B. 2007. Antioxidant, anti-inflammatory, and antimicrobial properties of garlic and onions. Nutrition & Food Science, 37(3), 178–183.

[95] Peter, K.V. and Babu, K.N. 2012. Introduction to herbs and spices: Medicinal uses and sustainable production. In Peter, K.V. (Ed.), *Handbook of Herbs and Spices*. Woodhead Publishing: Sawston, Cambridge, pp. 1–16.

[96] Liu, Q., Meng, X., Li, Y., Zhao, C.-N., Tang, G.-Y. and Li, H.-B. 2017. Antibacterial and antifungal activities of spices. International Journal of Molecular Sciences, 18(6), 6.

[97] Tragoolpua, Y. and Jatisatienr, A. 2007. Anti-herpes simplex virus activities of Eugenia caryophyllus (Spreng.) Bullock & S. G. Harrison and essential oil, eugenol. Phytotherapy Research, 21(12), 1153–1158.

[98] Vicidomini, C., Roviello, V. and Roviello, G.N. 2021. Molecular basis of the therapeutic potential of clove (Syzygium aromaticum L.) and clues to its anti-COVID-19 utility. Molecules, 26(7), 1880.

[99] Wongsawan, K., Chaisri, W., Tangtrongsup, S. and Mektrirat, R. 2020. Bactericidal effect of clove oil against multidrug-resistant streptococcus suis isolated from human patients and slaughtered pigs. Pathogens, 9(1), 1.

[100] Wertheim, H.F.L., Nguyen, H.N., Taylor, W., Lien, T.T.M., Ngo, H.T., Nguyen, T.Q., Nguyen, B.N.T., Nguyen, H.H., Nguyen, H.M., Nguyen, C.T., Dao, T.T., Nguyen, T.V., Fox, A., Farrar, J., Schultsz, C., Nguyen, H.D., Nguyen, K.V. and Horby, P. 2009. Streptococcus suis, an important cause of adult bacterial meningitis in Northern Vietnam. PLoS One, 4(6), e5973.

[101] Pinto, E., Vale-Silva, L., Cavaleiro, C. and Salgueiro, L. 2009. Antifungal activity of the clove essential oil from Syzygium aromaticum on Candida, Aspergillus and dermatophyte species. Journal of Medical Microbiology, 58(11), 1454–1462.

[102] Husain, I., Ahmad, R., Siddiqui, S., Chandra, A., Misra, A., Srivastava, A., Ahamad, T., Khan, M., Mohd, F., Siddiqi, F., Trivedi, Z., Upadhyay, A., Gupta, S., Srivastava, A., Ahmad, A.N., Mehrotra, B., Kant, S., Mahdi, A.A. and Mahdi, F. 2022. Structural interactions of phytoconstituent(s) from cinnamon, bay leaf, oregano, and parsley with SARS-CoV-2 nucleocapsid protein: A comparative assessment for development of potential antiviral nutraceuticals. Journal of Food Biochemistry, 46(10), e14262.

[103] Gruenwald, J., Freder, J. and Armbruester, N. 2010. Cinnamon and health. Critical Reviews in Food Science and Nutrition, 50(9), 822–834.

[104] Imani, G., Khalilian, A., Dastan, D., Imani, B. and Mehrpooya, M. 2019. Effects of cinnamon extract on complications of treatment and eradication of Helicobacter pylori in infected people. Journal of Herbmed Pharmacology, 9(1), 1.

[105] Vasconcelos, N.G., Croda, J. and Simionatto, S. 2018. Antibacterial mechanisms of cinnamon and its constituents: A review. Microbial Pathogenesis, 120, 198–203.

[106] Beristain-Bauza, S.D.C., Hernández-Carranza, P., Cid-Pérez, T.S., Ávila-Sosa, R., Ruiz-López, I.I. and Ochoa-Velasco, C.E. 2019. Antimicrobial activity of ginger (Zingiber officinale) and its application in food products. Food Reviews International, 35(5), 407–426.

[107] Singh, G., Kapoor, I.P.S., Singh, P., de Heluani, C.S., de Lampasona, M.P. and Catalan, C.A.N. 2008. Chemistry, antioxidant and antimicrobial investigations on essential oil and oleoresins of Zingiber officinale. Food and Chemical Toxicology, 46(10), 3295–3302.

[108] Jayathilake, P.A.L., Jayasinghe, M.A., Walpita, J. and Dilani, K.P.R.I. 2021. Turmeric and ginger as health protective food sources – An integrative review. Vidyodaya Journal of Science, 24(02), 7–26.

[109] Dissanayake, K.G.C., Waliwita, W.A.L.C. and Liyanage, R.P. 2020. A review on medicinal uses of Zingiber officinale (Ginger). International Journal of Health Sciences and Research, 6, 142–148.

[110] Stahl-Biskup, E. and Venskutonis, R.P. 2012. Thyme. In Peter, K.V. (Ed.), *Handbook of Herbs and Spices*. Woodhead Publishing, pp. Sawston, Cambridge. 499–525.

[111] Hammoudi Halat, D., Krayem, M., Khaled, S. and Younes, S. 2022. A focused insight into thyme: Biological, chemical, and therapeutic properties of an indigenous Mediterranean herb. Nutrients, 14(10), 2104.

[112] Fani, M. and Kohanteb, J. 2017. In vitro antimicrobial activity of *Thymus vulgaris* essential oil against major oral pathogens. Journal of Evidence-Based Complementary & Alternative Medicine, 22(4), 660–666.

[113] Lai, P. and Roy, J. 2004. Antimicrobial and chemopreventive properties of herbs and spices. Current Medicinal Chemistry, 11(11), 1451–1460.

[114] Prajapati, S.K., Mishra, G., Malaiya, A., Jain, A., Mody, N. and Raichur, A.M. 2021. Antimicrobial application potential of phytoconstituents from turmeric and garlic. In Pal, D., Nayak, A.K. (Eds.), *Bioactive Natural Products for Pharmaceutical Applications*. Springer International Publishing: Cham, Switzerland, pp. 409–435.

[115] Zhou, Z., Pan, C., Lu, Y., Gao, Y., Liu, W., Yin, P. and Yu, X. 2017. Combination of erythromycin and curcumin alleviates Staphylococcus aureus induced osteomyelitis in rats. Frontiers in Cellular and Infection Microbiology, 7, 379.

[116] Martins, C.V.B., da Silva, D.L., Neres, A.T.M., Magalhães, T.F.F., Watanabe, G.A., Modolo, L.V., Sabino, A.A., de Fátima, A. and de Resende, M.A. 2009. Curcumin as a promising antifungal of clinical interest. The Journal of Antimicrobial Chemotherapy, 63(2), 337–339.

Mostafa Pirali Hamedani, Saied Goodarzi*

# Chapter 13
# Medicinal plants for urinary tract problems

**Abstract:** Kidney and urinary tract diseases can occur in all people and at any age, including women, men, and even children. These diseases affect only a certain part of the body. In women, the urinary system is involved, and in men, the urinary system and reproductive organs such as the prostate will be affected. Benign prostate hyperplasia (BPH), urinary tract infections, urinary incontinence, overactive bladder, cystocele, and kidney stone diseases are among the most common problems in this field. BPH is a process in which the normal cells of the prostate are enlarged, causing urinary retention. Dried fruits of Saw palmetto (*Serenoa repens*), by the inhibition of 5-alpha-reductase mechanism, has been used for the treatment of diseases since ancient times and has enough scientific documents. The active compounds of the barks of African palm (*Prunus africana*) inhibit the proliferation of fibroblasts and induce apoptosis in the prostate. The roots of stinging nettle (*Urtica dioica*), flowers of Star grass (*Hypoxis hemerocallidea*) and seeds of Pumpkin (*Cucurbita pepo*) are other drugs that are used in the natural treatment of this problem and have a lot of scientific and historical support. Kidney stone diseases are the most common urinary system problems. Shatavari roots (*Asparagus racemosus*), aerial parts of Stonebreaker (*Phyllanthus niruri*) and oregano (*Origanum vulgare*), the fruits of Barberry (*Berberis vulgaris*) and Black cumin (*Nigella sativa*), calyx of Roselle (*Hibiscus sabdariffa*), and stems of Horsetail (*Equisetum arvense*) are specific examples for treatment. Each of them treats this disease with its own ingredients and by different mechanisms related to these ingredients.

# 1 Prostate problems

One of the prevalent diseases in older men is benign prostate hyperplasia (BPH). The prevalence of the disease increases with age, as about 90% of men show histological evidence of this problem by the age of 80 and 50% by age 60 [1]. BPH is a process in which the normal cells (stromal and epithelial) of the prostate are enlarged, and urination disorders may also occur. BPH is often confused with "benign prostatic hypertrophy," which is an old term that refers to an increase in cell size rather than cell

---

*Corresponding author: Saied Goodarzi**, Medicinal Plants Research Center, Faculty of Pharmacy, Tehran University of Medical Sciences, Tehran, Iran, e-mail: goodarzi_s@sina.tums.ac.ir
**Mostafa Pirali Hamedani**, Medicinal Plants Research Center, Faculty of Pharmacy, Tehran University of Medical Sciences, Tehran, Iran

https://doi.org/10.1515/9783111317601-013

number. Actually, the nonmalignant prostate enlargement causes some difficulties in urination, such as resistance, interruption, weak stream, hesitancy in urinary flow, urinary urgency, and nocturia. These problems begin to decrease urine outflow, followed by urinary retention (Table 1) [2]. Normally, about 20% of the seminal fluid is secreted by the prostate.

**Table 1:** Causes of BPH-associated urinary retention [3–8].

| Cause | Description |
| --- | --- |
| Urinary tract obstruction | Any obstruction such as muscle tone increase around the urethra, urethra channel narrowing, etc. |
| Infections | Several bacterial and viral infections may lead to discomfort in urination. |
| Neurological impairment | Interruption of sympathetic, parasympathetic, and somatic nerves cause lack of coordination between the internal sphincter, bladder neck, and the urethral sphincter, leading to voluntary urination. This neurological interruption may occur following brain stroke, multiple sclerosis (MS), Guillain–Barre syndrome, diabetic neuropathy, spinal cord injury, and other neurological disorders. |
| Muscle dysfunction | Insufficient muscle tonicity to complete emptying of urine cause fluid flow challenge, leading to retention of urine in the bladder. |
| Drugs | Some drug classes such as α-adrenergic agonists, beta-adrenergic agonists, tricyclic antidepressants, antiarrhythmic drugs, anticholinergics agents, antiparkinsonian drugs, hormonal agents, antipsychotics, 1st generation of antihistamines, muscle relaxants, etc. can affect smooth muscle tonicity of organs in the urinary system. |

Although the etiology of BPH has not been fully understood, dihydrotestosterone hypothesis is most dominant than others. In addition to male factor, other risk factors include BPH family history, obesity, smoking, and sexual activity. On the other side, physical activity can protect men from BPH. Despite not causing serious problems in patients, BPH can interfere with daily activities, decreasing the subject's quality of life [9, 10].

One of the important examinations in the evaluation of BPH is digital rectal examination (DRE), to assess the size, shape, symmetry, nodularity, and firmness of the prostate [11]. Clinical assessment of prostate specific antigen (PSA) is also recommended. PSA is a protein that is produced by the prostate cells, the level of which may rise in prostatitis and BPH. World Health Organization (WHO) recommended a symptom index for BPH evaluation, which is based on the International Prostate Symptom Score (IPPS) questionnaire, referring to urine frequency, intermittency, urgency, incomplete emptying, weak stream, straining to initiate, and nocturia. An IPPS score of 0–10 indicates mild, 10–20 indicates moderate, and 20–30 indicates severe form of BPH.

The conventional treatment options include watchful waiting, alpha-1 adrenergic receptor blocking agents, and 5-alpha reductase inhibitors [12]. Watchful waiting is recommended in patients with mild-to-moderate symptoms. It may aid in controlling hyperglycemia and hyperlipidemia, one of the BPH risk factors. Alpha-1 adrenergic receptor blocking agents decrease the tonicity of the smooth muscle in the bladder, prostate, and urethra. The important point is that alpha-blockers do not reduce the long-term risk of urinary retention. Orthostatic hypotension and dizziness are the prominent adverse effects with these medications, and they require gradual dose titration to reduce the risk of adverse reactions [13, 14]. 5-Alpha reductase inhibitors inhibit the conversion of testosterone to dihydrotestosterone, and may prevent the progression of BPH and urinary retention. Decreased dihydrotestosterone levels are known to inhibit prostate growth. Despite the beneficial effective effects of 5-alpha-reductase inhibitors in reducing PSA and urinary retention, they are also associated with erectile dysfunction, loss of libido, and ejaculation issues [15].

## 1.1 Medicinal plants as an alternative treatment approach

Since centuries, medicinal plants have remained part of traditional medicine systems, and currently continue to be in practice in the treatment of various chronic and age-related diseases, where conventional therapies are limited or do not provide a complete cure and cannot be used for long term. Moreover, conventional therapies may also cause risk adverse effects in the aged, and thus may increase the risk-to-benefit ratio. The medicinal plants reported for use in the prevention and treatment of BPH and urinary retention include *Serenoa repens*, *Hypoxis hemerocallidea*, *Pygeum africanum*, *Urtica dioica*, and *Cucurbita pepo* [16].

### 1.1.1 Serenoa repens

*Serenoa repens* (saw palmetto) belongs to the Arecaceae family, and is endemic in Southeastern United States. *Serenoa* genus refers to Serenoa Watson (American botanist), to

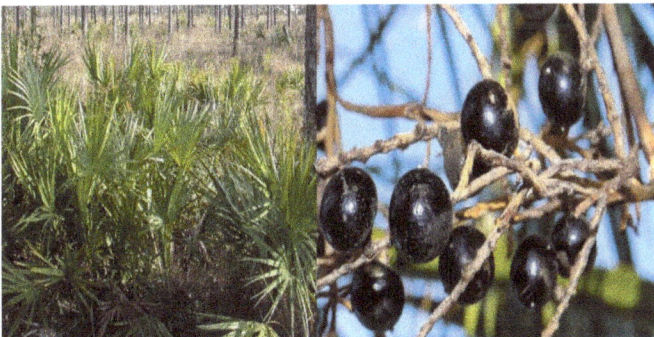

**Figure 1:** *Serenoa repens* tree (left) and fruits (right).

honor him, and repens refers to the creeping habit of this plant. Saw palmetto (Figure 1) grows to 3 meters and its leaves are like the fan palm. Each leaf contains 20 leaflets, each leaflet is 50–100 cm long, and each leaf is 1–2 m in length. The fruits of *S. repens* (Figure 1) are large and reddish black berries. Partially dried or dried fruit (berries) of this medicinal plant are used for prostate problems [17]. Usually, standard fruit extracts that are obtained with relatively lipophilic solvents and containing 70–95% free fatty acids are used to make the product. As shown in Figure 2, the active components of saw palmetto berries are a mixture of fatty acids (such as capric, caprylic, lauric, oleic, myristoleic, palmitic, linoleic, and linolenic acid) and phytosterols (such as β-sitosterol, campesterol, stigmasterol, lupeol, and cycloartenol).

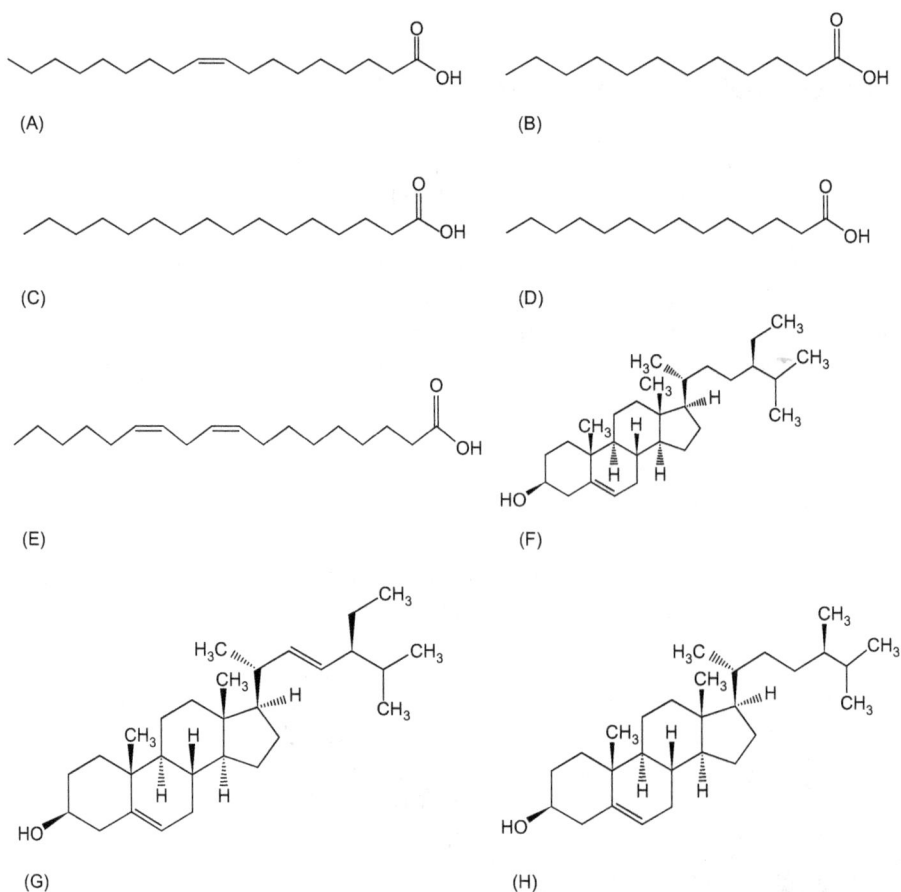

**Figure 2:** Chemical structure of the main constituents of Saw palmetto fruit. (A) Oleic acid, (B) lauric acid, (C) palmitic acid, (D) myristic acid, (E) linoleic acid, (F) β-sitosterol, (G) stigmasterol, (H) campesterol.

The mechanism of action of these active metabolites is inhibition of 5-alpha-reductase, resulting in anti-inflammatory effects. Post-marketing surveillance studies demonstrated that the extract of saw palmetto is well tolerated, have comparative effects, and are safer than finasteride [18]. A study by Carraro et al. demonstrated similar efficacy for finasteride and saw palmetto in terms of quality of life and urine outflow, while saw palmetto was more effective on sexual function [19]. One of the serious concerns with this plant is its use in patients who us anticoagulant drugs, i.e., warfarin, due to its potential to inhibit CYP450 isotype 2C9 [20].

### 1.1.2 Pygeum africanum

*Prunus africana* (Syn. *Pygeum africanum*) or African palm (Figure 3) is a canopy tree from the Rosaceae family, growing to a height of 30–40 m. The leaves of the plant alternate, its petioles are red, and its fruits are drupe shaped with red to brown color. The name, *Prunus*, is derived from the plum shape of fruit and *africana* refers to the montane forests of Africa (the origin of the plant). The extract of *P. africanum* barks (rich in atranorin, atraric acid, *N*-butylbenzene-sulfonamide, ferulic acid, and phytosterols) is being used in southern and central African traditional medicine for urinary tract complications [21, 22]. The bioactive compounds (Figure 4) include atranorin, atraric acid, *N*-butylbenzene-sulfonamide, ferulic acid, and phytosterols that may inhibit 5-alpha reductase and aromatase enzymes, proliferation of fibroblasts, and induce apoptosis in prostate [23, 24]. Also, the isolated *N*-butylbenzene-sulfonamide is reported to inhibit PSA expression [25]. The plant is well tolerated with no serious safety issues being reported across the literature.

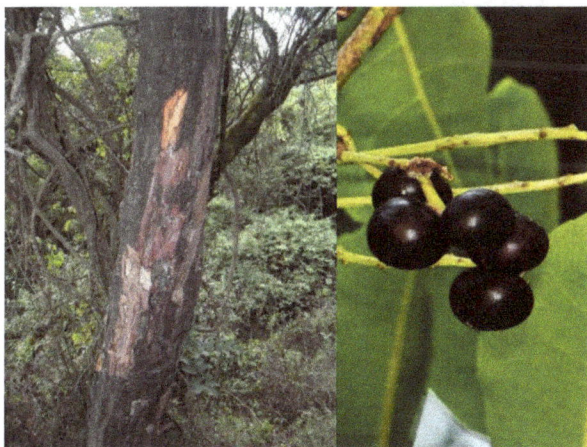

**Figure 3:** *Prunus africana* bark (left) and fruit (right).

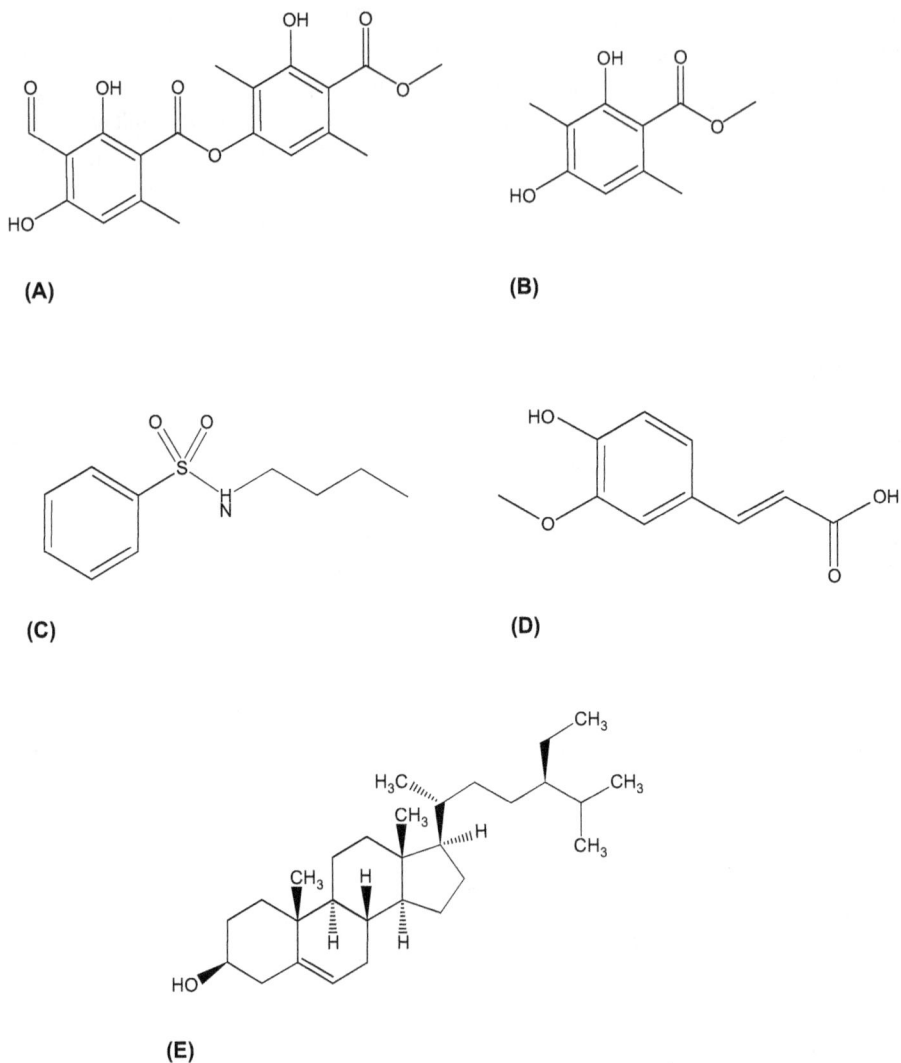

**Figure 4:** Chemical structure of the BPH-active components of *Prunus africana* barks. (A) Atranorin, (B) atraric acid, (C) *N*-butylbenzenesulfunamide, (D) ferulic acid, (E) β-sitosterol.

### 1.1.3 Urtica dioica L

*Urtica dioica* L. (stinging nettle) is a herbaceous perennial plant of the Urticaceae family, found all over the world except in South America, Australia, and Antarctica. *Urtica* means sting and *dioica* means dioecious. Its nettle grows up to 2 m and its leaves have serrated margin (Figure 5). The stems and leaves of *U. dioica* have stinging hair (trichomes) and when touched, these needles injure the skin and release histamine, causing itching and redness. The roots of *U. dioica* are used for BPH while the nettle radix is being used traditionally to treat uterine hemorrhage, cutaneous eruptions, and infantile eczema [26, 27]. The main constituents isolated from nettle (Figure 6) are lignans (pinoresinol, secoisolariciresinol, dehydrodiconiferyl alcohol, and neoolivil), phytosterols (β-sitosterol, stigmasterol, and campesterol), triterpenes (oleanolic and ursolic acid), coumarin (scopoletin), and flavonoids (isorhamnetin, kaempferol, and quercetin). Lignans are documented for 5-alpha reductase inhibitory effects of stinging nettle. Furthermore, some isolated components from the nettle root are known as aromatase inhibitors, thus reliving the burden of BPH symptoms [28].

**Figure 5:** *Urtica dioica* herb (left), roots (right).

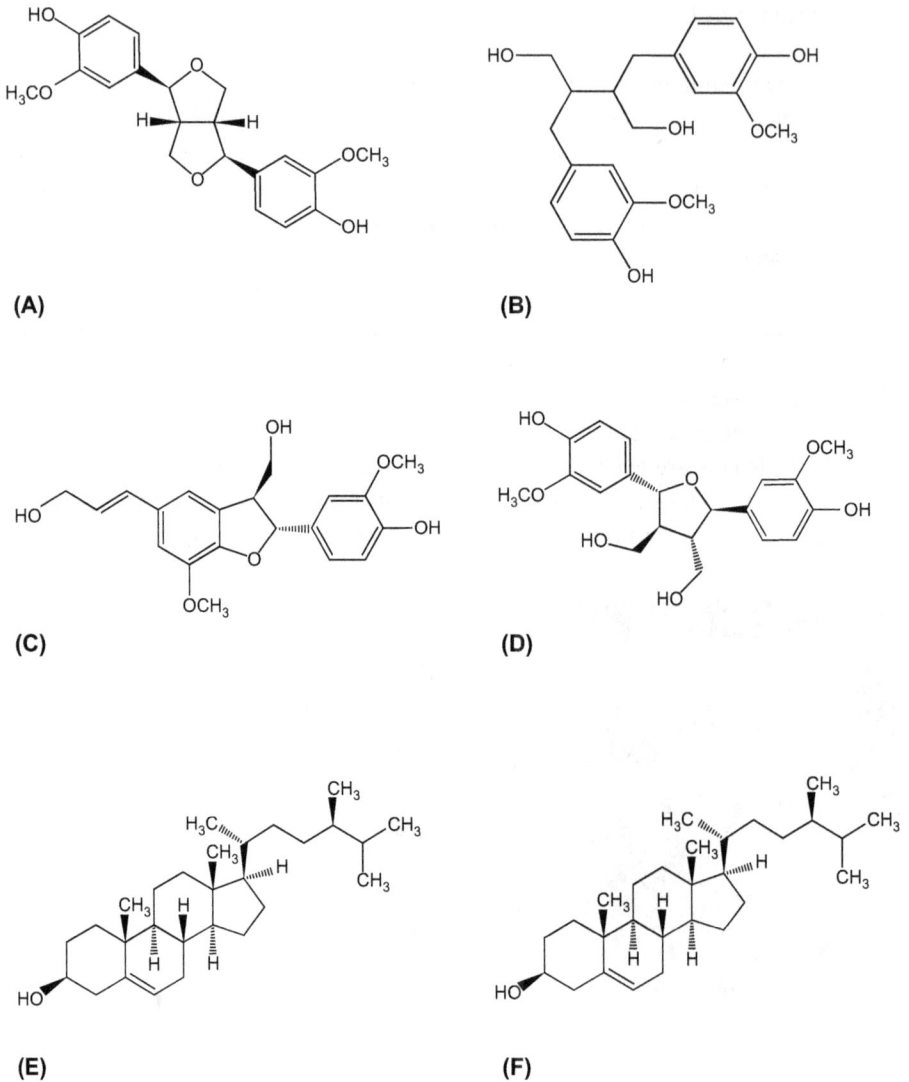

**Figure 6:** Chemical structure of the main constituents of *Urtica dioica* roots. **(A)** Pinoresinol, **(B)** secoisolariciresinol, **(C)** dehydrodiconiferyl alcohol, **(D)** neoolivil, **(E)** campesterol, **(F)** β-sitosterol.

### 1.1.4 Hypoxis hemerocallidea

*Hypoxis hemerocallidea* (Syn. *H. rooperi*) or the South African star grass is a member of Hypoxidaceae, distributed in Southern Africa. Inflorescences of star grass are used for the treatment of bladder and prostate ailments. β-sitosterol is the main isolated compound that may bind prostatic tissue receptors to inhibit BPH progression and improve urine flow and volume [29].

## 1.1.5 Cucurbita pepo

*Cucurbita pepo* (pumpkin) is an annual plant from the Cucurbitaceae family, reported to grow in Mexico about 10,000 years ago. *C. pepo* fruit has a round shape and orange color – it is the nutritional and medicinal part of the plant (Figure 7). The pumpkin seeds are a biological source of polyunsaturated fatty acids such as palmitic acid, stearic acid, oleic acid, and linoleic acid (Table 2) [30]. Pumpkin seeds oil has several biological activities, including their potential use in patients with mild-to-moderate BPH – studies showed a reduction in the risk of prostate cancer after 3 months consumption of pumpkin seeds [31]. One of the possible mechanisms in BPH treatment is the prevention of dihydrotestosterone from binding to androgen receptors [32].

**Figure 7:** *Cucurbita pepo* fruit (left), seeds (right).

**Table 2:** Fatty acid composition of the pumpkin seed oil [33].

| Fatty acid | Content (%) |
|---|---|
| Palmitic (C16:0) | 10.68 ± 0.42 |
| Palmitoleic (C16:1) | 0.58 ± 0.14 |
| Stearic (C18:0) | 8.67 ± 0.27 |
| Oleic (C18:1) | 38.42 ± 0.37 |
| Linoleic (C18:2) | 39.84 ± 0.08 |
| Linolenic (C18:3) | 0.68 ± 0.14 |
| Gadoleic (C20:1) | 1.14 ± 0.00 |
| Total saturated fatty acids | 19.35 ± 0.16 |
| Total unsaturated fatty acids | 80.65 ± 0.16 |

# 2 Kidney stone diseases

Kidney stones (nephrolithiasis) disease or urinary stones (urolithiasis) disease is one of the most common urinary system problems affecting men three times more common than women. Crystals of calcium oxalate, calcium phosphate, uric acid, struvite, and cysteine are the causes of kidney stones, with calcium oxalate being the most common one. The potential risk factors could include positive family history, dehydration, obesity, dietary consumption of excessive salts and sugars, renal tubular acidosis, cystinuria, hyperparathyroidism, and urinary tract infections. Hyperuricosuria and acidic pH of urine may result in uric acid stones. Patients with kidney stones may present several symptoms including back and abdominal pain, painful urination, hematuria, and nausea and vomiting [34, 35].

Low water consumption is the crucial risk factor of stone formation, though inadequate fluids such as hard tap water, mineral or drinking waters with high divalent cations concentration, including calcium carbonate, can also increase the risk for kidney stones. Whereas, mineral waters with high amount of bicarbonate may decrease the risk of calcium oxalate stones formation [36]. In addition, dietary or supplemental proteins can also increase the concentration of calcium in urine, with a decrease in the urinary pH, thus providing a road for stones formation [37]. Data about high carbohydrate and fat consumption in relation to kidney and bladder stone formation are inconsistent. But consumption of foods, vegetables, or fruits that contain high amount of oxalate such as spinach, okra, soybeans, blackberry, kiwi, fig, sesame, almond, hazelnut, black pepper, parsley, etc. and calcium and sodium chloride (salt) can increase the risk of stone formation. A kidney stone may be treated with shockwave lithotripsy, percutaneous nephrolithomy, nephrolithotripsy, or uteroscopy. However, due to the unavailability of successful medical and noninvasive treatment options, most of the population relies on the use of alternative therapies that involve using medicinal plants, especially in the initial course of the disease.

## 2.1 Asparagus racemosus

*Asparagus racemosus* (shatavari) belongs to the Asparagaceae family. It is a perennial plant that grows up to 2 m. The root of shatavari is yellow colored and has a delicious taste. It is used for medicinal purposes [38]. One of the applications of shatavari roots in Ayurvedic medicine is as a cooling agent and as a smooth muscles tonic. The mucilage of this medicinal plant assuages urinary pain as a relaxing agent and treats urinary tract infections. In addition to mucilage, this plant contains steroids (shatavarins and racemosides), flavonoids (rutin, quercetin, and kaempferol), alkaloids (asparagamine), minerals, and vitamins (Figure 8) [39]. *A. racemosus* also increases the outflow of urine in urolithiasis condition and subsequently reduces the urea, creatinine, and uric acid levels in the blood [40].

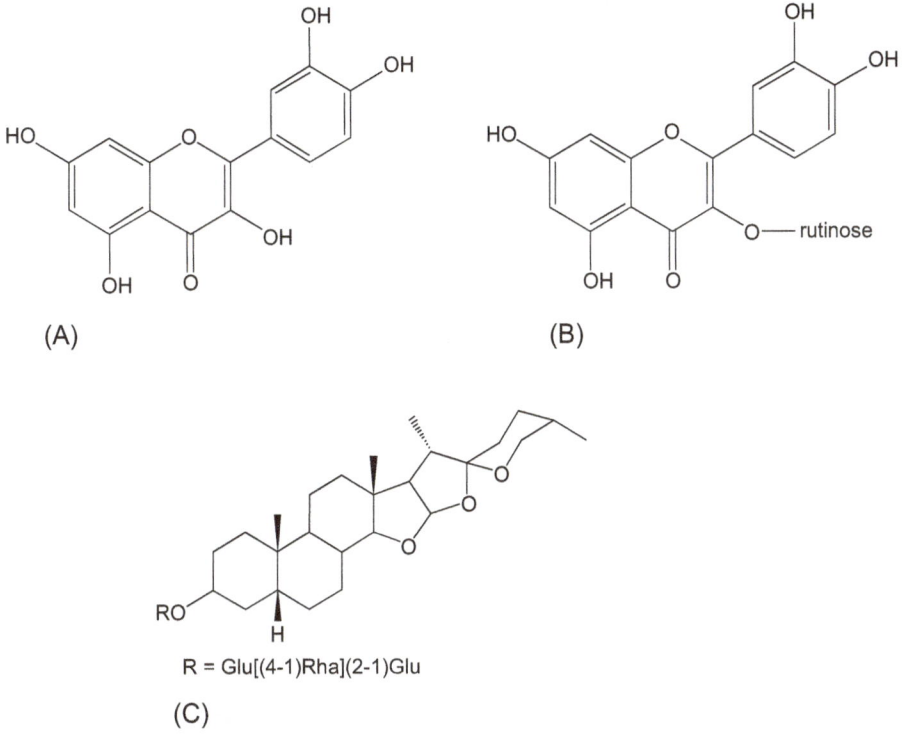

(A)

(B)

R = Glu[(4-1)Rha](2-1)Glu

(C)

**Figure 8:** Chemical structure of the main constituents of *Asparagus racemosus* roots. **(A)** Quercetin, **(B)** rutin, **(C)** shatavarin IV.

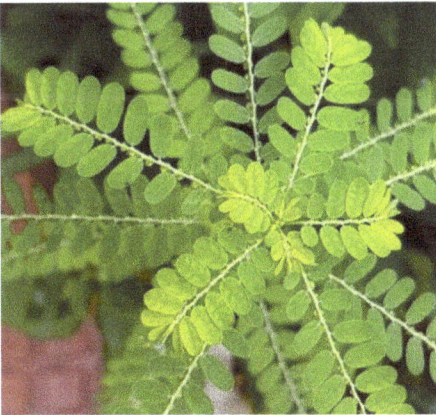

**Figure 9:** *Phyllanthus niruri* L.

## 2.2 Phyllanthus niruri

*Phyllanthus niruri* (stone breaker) belongs to the Euphorbiaceae family and is spread throughout the subtropical and tropical regions. *P. niruri* is a perennial shrub and grows to a height of 50–70 cm **(Figure 9)**. The whole plant is used for medicinal purpose, in-

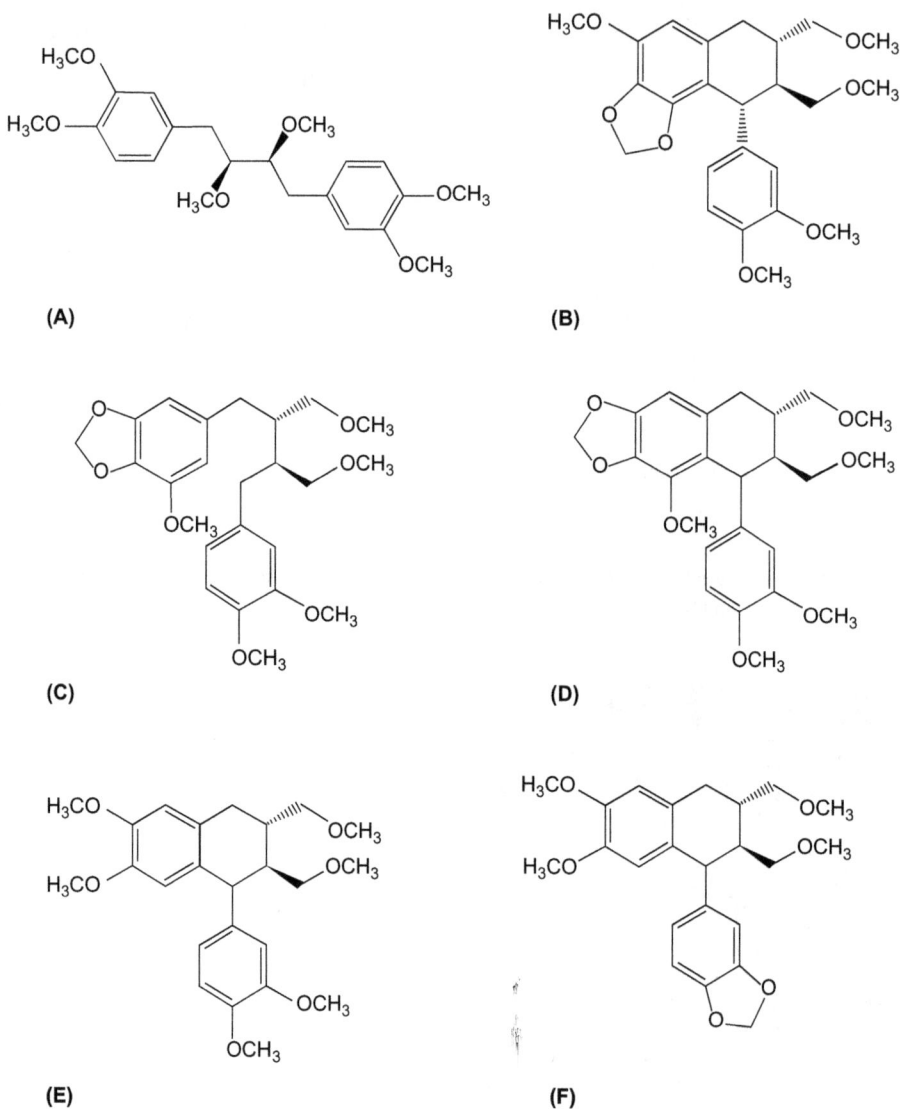

**Figure 10:** Chemical structure of the main constituents of *Phyllanthus niruri* L. **(A)** Phyllantin, **(B)** hypophyllantin, **(C)** lintetralin, **(D)** phyltetralin, **(E)** nirtetralin, **(F)** niranthin.

cluding for kidney problems [41]. Lignans, alkaloids, steroids, flavonoids, and terpenoids are the major classes of secondary metabolites isolated from this plant (Figure 10). One of the main uses of *P. niruri* roots is in the treatment of urinary injuries. *In vitro* studies showed the inhibition of growth and aggregation of calcium oxalate crystals with the treatment of *P. niruri* extract. Other desirable effects include the prevention of nuclear formation, changing of stones shape and texture, and increasing uric acid excretion by its uricosuric effects [42].

## 2.3 Origanum vulgare

*Origanum vulgare* (oregano) belongs to the Lamiaceae family and is native to Mediterranean Sea and Western Eurasia. It is widely used for its aroma and flavor. The plant is perennial with white or purple flowers (Figure 11). Aerial parts of *O. vulgare* demonstrated several pharmacological effects such as anti-inflammatory, antioxidant, antibacterial, anticancer, and antinephrolithiasis behavior [43]. As shown in Figure 12, oregano is rich in essential oil, containing carvacrol, thymol, and sabinene derivatives, and less amounts of other terpenoids. It is responsible for the medicinal actions of the plant such as a diuretic, antiseptic and anti-inflammatory agent [44, 45]. The *O. vulgare* extract demonstrated inhibition of calcium oxalate crystals formation along with its diuretic, hypocalciuric, and hypercitrauric effects [46].

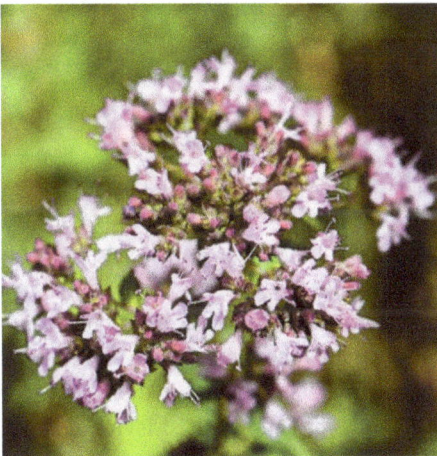

**Figure 11:** *Origanum vulgare* flowers.

(A)          (B)          (C)          (D)

(E)          (F)          (G)          (H)

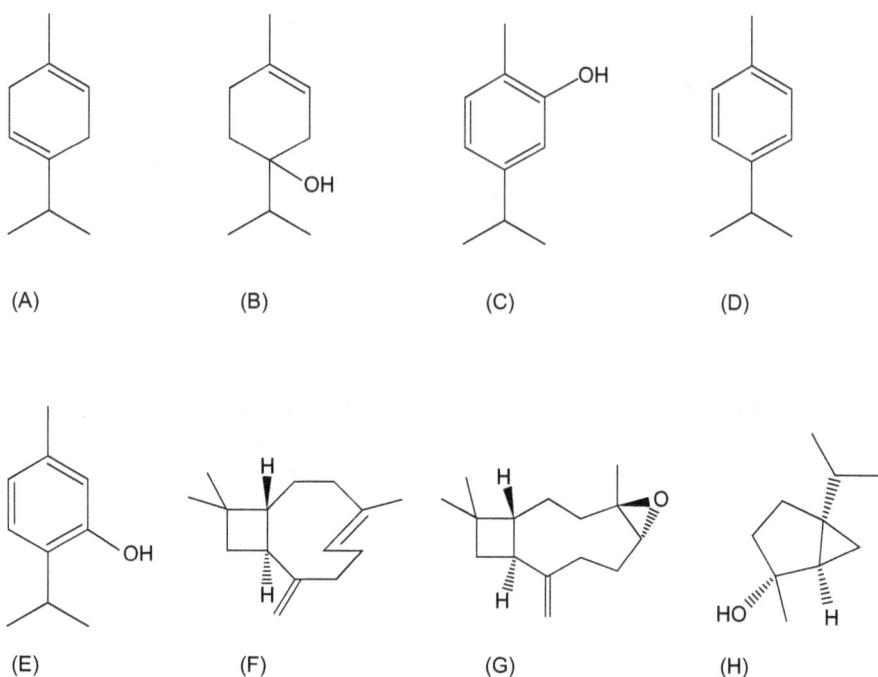

**Figure 12:** Chemical structure of the main constituents of *Origanum vulgare* flowers. (A) *γ*-terpinene, (B) 4-terpineol, (C) carvacrol, (D) p-cymene, (E) thymol, (F) *β*-caryophyllene, (G) *β*-caryophyllene oxide, (H) *trans*-sabinen hydrate.

## 2.4 Barberry (Berberis vulgaris)

*Berberis vulgaris* (barberry) belongs to the Berberidaceae family. It is a bush with yellow flowers and red fruits. Various parts of *B. vulgaris* have been used in traditional medicines for a range of medicinal purposes such as antibacterial, antipyretic, antipruritic, hypotensive, antiarrhythmic, and diuretic activities [47, 48]. The main compound of *B. vulgaris* is alkaloid berberine (Figure 13), which may prevent cardiometabolic disorders [49]. Berberine increases urine outflow, decreases the urinary concentration of calcium, and reduces the size and count of calcium oxalate crystals. Other biological effects attributed to the use of *B. vulgaris* are the prevention of raised levels of blood urea nitrogen and serum creatinine concentrations [50].

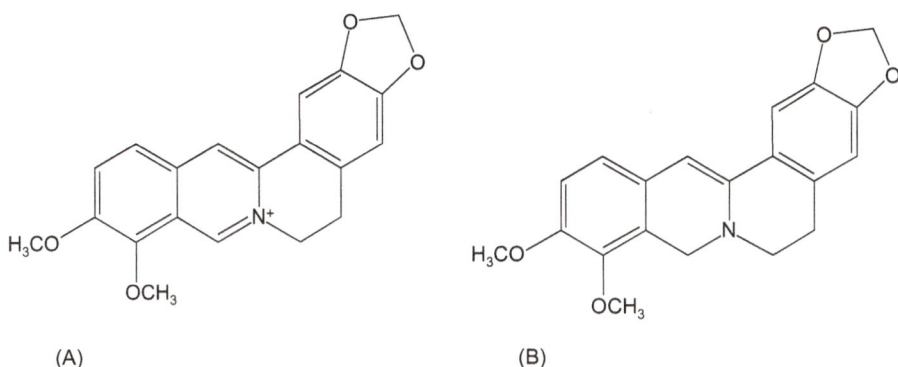

(A)                                                    (B)

**Figure 13:** Chemical structure of berberine alkaloids. **(A)** Berberine; **(B)** dihydroberberine.

## 2.5 Nigella sativa

*Nigella sativa* (black cumin) belongs to the family Ranunculaceae and is widely distributed in the region between the Mediterranean Sea and the Middle East. It is a perennial plant that grows up to 20–30 cm, with blue and white colored flowers and capsuled fruits with numerous seeds **(Figure 14)**. The seeds are mainly used for nutritional and medicinal purposes [51]. This plant is being extensively used in traditional medicine of Western Asia and China for the management of respiratory tract, skin, liver, and urinary tract disorders [52]. Chemically, the seeds are rich in flavonoids (mainly thymoquinone; Figure 15), essential oils, proteins, saponins, and isoquinoline alkaloids [53]. Due to the diuretic effects, *N. sativa* can used as supportive therapy in conditions like hypertension, heart failure, nephritic syndrome, and cirrhosis. A study demonstrated significantly increased urine volume and output, and urea excretion with black cumin treatment [54].

**Figure 14:** *Nigella sativa* flowers (left), seeds (right).

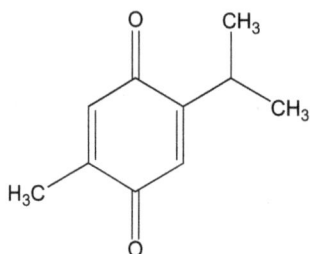

**Figure 15:** Chemical structure of thymoquinone.

## 2.6 Hibiscus sabdariffa

*Hibiscus sabdariffa* (roselle) is a delicacy plant from the Malvaceae family, native to the subcontinent region and Saudi Arabia. It is an annual herbaceous plant that grows 2–3

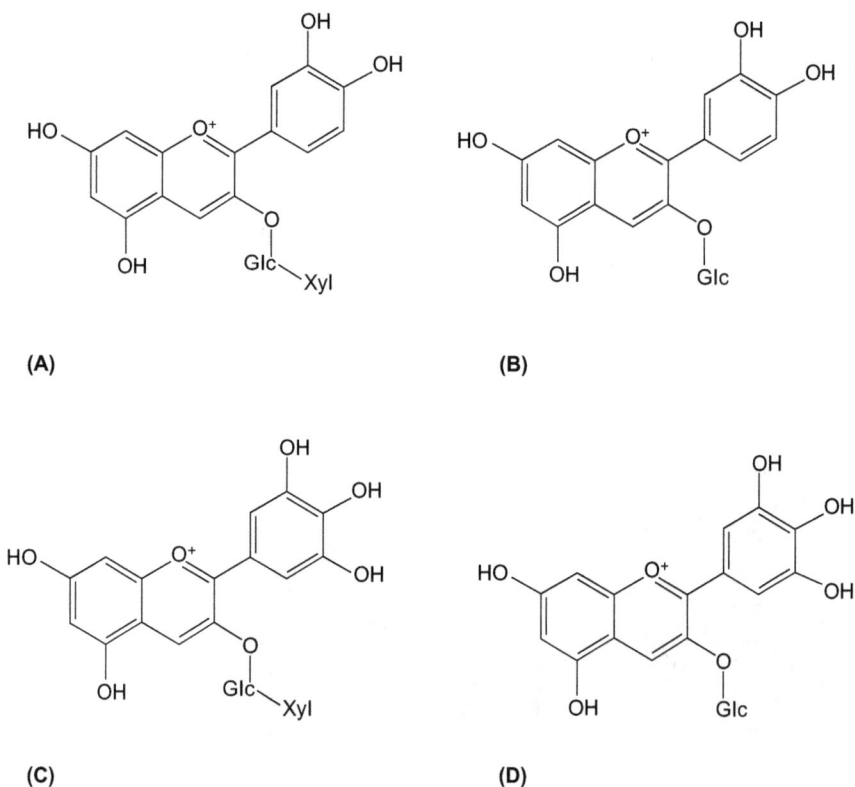

(A)

(B)

(C)

(D)

**Figure 16:** Chemical structure of the main constituents of *Hibiscus sabdariffa* calyx. (A) Cyanidin 3-sambubioside, (B) cyanidin 3-glucoside, (C) delphinidin 3-sambubioside, (D) delphinidin 3-glucoside.

m, with lobed leaves and white flowers with red calyx [55]. Calyces of *H. sabdariffa* are used widely in herbal drinks, beverages, wines, ice creams, and chocolates [56]. The plant is used in traditional medicine for diuretic and hypotensive properties. In modern day practice, roselle calyces are used for decreasing the blood viscosity, reducing the blood pressure, and for diuretic actions. The main constituents of *H. sabdariffa* are organic acids (citric, malic, oxalic, ascorbic acids), anthocyanins (delphinidin, cyaniding, and their derivatives), and flavonoids (hibiscitrin, sabdaritin, gossyptrin, quercetin, and luteolin) (Figure 16) [55]. The potential use of the plant in patients with kidney stones is attributed to the antispasmodic actions by the activation of smooth muscle potassium channels [57] and anti-urolithiatic effects by decreasing calcium and magnesium-ATPase activity and the deposition of calcium crystals [58].

## 2.7 Equisetum arvense

*Equisetum arvense* (horsetail) belongs to the family Equisetaceaem, native to the northern hemisphere regions. *E. arvense* is dimorphic, with sterile (green, branched, and hollow center) and fertile stems (brown, unbranched, and fleshy) (Figure 17). Usually, sterile stems of *E. arvense* are used for medicinal purposes [59]. Horsetail is rich in minerals, especially silicon (present in the form of $SiO_2$). Other compounds that are isolated forms of this genus include flavonoids, phenolic acids, and steroids [60]. In Ayurvedic medicine, the use of this plant for the treatment BPH and urinary incontinence is reported [61]. Literature also supported the medical uses of horsetail for post-traumatic

**Figure 17:** *Equisetum arvense* fertile stem (left), sterile stem (right).

or stasis swelling in cases of bacterial and inflammatory lower urinary tract diseases. Clinical studies demonstrated the effects of the plant on the IPSS score, urinary flow rate, and residual urine [62]. No toxicity has been observed by consuming horsetail up to 2 g/kg/day but due to the presence of nicotine, the plant is not recommended for use in pregnancy, lactation, and children under 12 years of age. Although horsetail is safe, its diuretic effects might cause potassium loss, affecting kidney and heart functioning. In addition, the presence of thiaminase enzyme in horsetail may cause thiamine deficiency, thus co-supplementation of thiamine is recommended [60].

# 3 Summary

As no successful treatment is currently available for the treatment of urinary tract problems like BPH, urinary retention, and kidney stones, the use of medicinal plants as a preventive approach as well as support to conventional medical therapy could be one of the fruitful strategies. Plants are rich in bioactive components that may act through a number of mechanistic targets, resulting in the improvement of disease conditions. The most important thing is that evidence-based studies on these plants are still very limited; so, patients should be monitored closely for any unwanted effects. The potential of phytochemicals to interact with other drugs may result in serious and life-threatening issues; thus, it they are recommended for use under the supervision of qualified healthcare practitioners. Moreover, it is essential to assess the efficacy and safety of these plants in robust randomized clinical trials, as most of the data about these plants in urinary tract problems is based on traditional knowledge.

# References

[1]    Roehrborn, C.G. 2005. Benign prostatic hyperplasia: An overview. Reviews in Urology, 7(Suppl 9), S3.
[2]    Emberton, M. and Anson, K. 1999. Acute urinary retention in men: An age old problem. British Medical Journal, 318(7188), 921–925.
[3]    Nunes, R.L., Antunes, A.A., Silvinato, A. and Bernardo, W.M. 2018. Benign prostatic hyperplasia. Revista da Associação Médica Brasileira, 64, 876–881.
[4]    Verzotti, G., Fenner, V., Wirth, G. and Iselin, C.E. 2016. Acute urinary retention: A mechanical or functional emergency. Revue Medicale Suisse, 12(541), 2060–2063.
[5]    Fong, Y.K., Milani, S. and Djavan, B. 2005. Natural history and clinical predictors of clinical progression in benign prostatic hyperplasia. Current Opinion in Urology, 15(1), 35–38.
[6]    Mancino, P., Dalessandro, M., Falasca, K., Ucciferri, C., Pizzigallo, E. and Vecchiet, J. 2009. Acute urinary retention due to HSV-1: A case report. Le Infezioni in Medicina, 17(1), 38–40.
[7]    Kaplan, S.A., Wein, A.J., Staskin, D.R., Roehrborn, C.G. and Steers, W.D. 2008. Urinary retention and post-void residual urine in men: Separating truth from tradition. The Journal of Urology, 180(1), 47–54.

[8]    Verhamme, K., Sturkenboom, M.C., Stricker, B.H.C. and Bosch, R. 2008. Drug-induced urinary retention. Drug Safety, 31(5), 373–388.

[9]    Girman, C., Jacobsen, S., Rhodes, T., Guess, H., Roberts, R. and Lieber, M. 1999. Association of health-related quality of life and benign prostatic enlargement. European Urology, 35(4), 277–284.

[10]   Girman, C.J., Jacobsen, S.J., Tsukamoto, T., Richard, F., Garraway, W.M., Sagnier, P.P., Guess, H.A., Rhodes, T., Boyle, P. and Lieber, M.M. 1998. Health-related quality of life associated with lower urinary tract symptoms in four countries. Urology, 51(3), 428–436.

[11]   Roehrborn, C.G., Girman, C.J., Rhodes, T., Hanson, K.A., Collins, G.N., Sech, S.M., Jacobsen, S.J., Garraway, W.M. and Lieber, M.M. 1997. Correlation between prostate size estimated by digital rectal examination and measured by transrectal ultrasound. Urology, 49(4), 548–557.

[12]   Parsons, J.K., Dahm, P., Köhler, T.S., Lerner, L.B. and Wilt, T.J. 2020. Surgical management of lower urinary tract symptoms attributed to benign prostatic hyperplasia: AUA guideline amendment 2020. The Journal of Urology, 204(4), 799–804.

[13]   Emberton, M., Cornel, E., Bassi, P., Fourcade, R., Gomez, J. and Castro, R. 2008. Benign prostatic hyperplasia as a progressive disease: A guide to the risk factors and options for medical management. International Journal of Clinical Practice, 62(7), 1076–1086.

[14]   Hellstrom, W.J. and Sikka, S.C. 2006. Effects of acute treatment with tamsulosin versus alfuzosin on ejaculatory function in normal volunteers. The Journal of Urology, 176(4), 1529–1533.

[15]   McConnell, J.D., Bruskewitz, R., Walsh, P., Andriole, G., Lieber, M., Holtgrewe, H.L., Albertsen, P., Roehrborn, C.G., Nickel, J.C., Wang, D.Z. and Taylor, A.M. 1998. The effect of finasteride on the risk of acute urinary retention and the need for surgical treatment among men with benign prostatic hyperplasia. New England Journal of Medicine, 338(9), 557–563.

[16]   Sharma, M., Chadha, R. and Dhingra, N. 2017. Phytotherapeutic agents for benign prostatic hyperplasia: An overview. Mini Reviews in Medicinal Chemistry, 17(14), 1346–1363.

[17]   Bennett, B.C. and Hicklin, J.R. 1998. Uses of saw palmetto (*Serenoa repens*, Arecaceae) in Florida. Economic Botany, 52(4), 381–393.

[18]   Heinrich, M., Barnes, J., Prieto-Garcia, J., Gibbons, S. and Williamson, E.M. 2017. *Fundamentals of Pharmacognosy and Phytotherapy*. 4th edition. Elsevier.

[19]   Carraro, J.C., Raynaud, J.P., Koch, G., Chisholm, G.D., Di Silverio, F., Teillac, P., Da Silva, F.C., Cauquil, J., Chopin, D.K., Hamdy, F.C. and Hanus, M. 1996. Comparison of phytotherapy (Permixon®) with finasteride in the treatment of benign prostate hyperplasia: A randomized international study of 1,098 patients. The Prostate, 29(4), 231–240.

[20]   Baxter, K., Driver, S. and Williamson, E. 2013. *Stockley's Herbal Medicines Interactions*. London, UK: Pharmaceutical Press.

[21]   Rubegeta, E., Makolo, F., Kamatou, G., Enslin, G., Chaudhary, S., Sandasi, M., Cunningham, A.B. and Viljoen, A. 2022. The African cherry: A review of the botany, traditional uses, phytochemistry, and biological activities of Prunus africana (Hook. f.) Kalkman. Journal of Ethnopharmacology, 116004.

[22]   Thompson, R.Q., Katz, D. and Sheehan, B. 2019. The African cherry: Chemical comparison of Prunus africana bark and pygeum products marketed for prostate health. Journal of Pharmaceutical and Biomedical Analysis, 163, 162–169.

[23]   Quiles, M.T., Arbós, M.A., Fraga, A., De Torres, I.M., Reventós, J. and Morote, J. 2010. Antiproliferative and apoptotic effects of the herbal agent Pygeum africanum on cultured prostate stromal cells from patients with benign prostatic hyperplasia (BPH). The Prostate, 70(10), 1044–1053.

[24]   Hartmann, R., Mark, M. and Soldati, F. 1996. Inhibition of 5 α-reductase and aromatase by PHL-00801 (Prostatonin®), a combination of PY102 (*Pygeum africanum*) and UR102 (*Urtica dioica*) extracts. Phytomedicine, 3(2), 121–128.

[25]   Papaioannou, M., Schleich, S., Roell, D., Schubert, U., Tanner, T., Claessens, F., Matusch, R. and Baniahmad, A. 2010. NBBS isolated from Pygeum africanum bark exhibits androgen antagonistic

activity, inhibits AR nuclear translocation and prostate cancer cell growth. Investigational New Drugs, 28(6), 729–743.

[26] Grauso, L., De Falco, B., Lanzotti, V. and Motti, R. 2020. Stinging nettle, Urtica dioica L.: Botanical, phytochemical and pharmacological overview. Phytochemistry Reviews, 19(6), 1341–1377.

[27] Gledhill, D. 2008. *The Names of Plants.* Cambridge University Press, ISBN 978-0-521-86645-3.

[28] Chrubasik, J.E., Roufogalis, B.D., Wagner, H. and Chrubasik, S. 2007. A comprehensive review on the stinging nettle effect and efficacy profiles. Part II: Urticae radix. Phytomedicine, 14(7–8), 568–579.

[29] Berges, R., Windeler, J., Trampisch, H. and Senge, T. 1995. Group β-SS. Randomised, placebo-controlled, double-blind clinical trial of β-sitosterol in patients with benign prostatic hyperplasia. The Lancet, 345(8964), 1529–1532.

[30] Caili, F., Huan, S. and Quanhong, L. 2006. A review on pharmacological activities and utilization technologies of pumpkin. Plant Foods for Human Nutrition, 61(2), 70–77.

[31] Hong, H., Kim, C.-S. and Maeng, S. 2009. Effects of pumpkin seed oil and saw palmetto oil in Korean men with symptomatic benign prostatic hyperplasia. Nutrition Research and Practice, 3(4), 323–327.

[32] Schilcher, H. 1987. Pflanzliche Diuretika. Urologe Ausgabe B, 27(4), 215–222.

[33] Gohari Ardabili, A., Farhoosh, R. and Haddad Khodaparast, M.H. 2011. Chemical composition and physicochemical properties of pumpkin seeds (*Cucurbita pepo* Subsp. pepo Var. Styriaka) grown in Iran. Journal of Analytical Science and Technology, 13(7), 1053–1063.

[34] Siener, R. and Hesse, A. 2003. Fluid intake and epidemiology of urolithiasis. European Journal of Clinical Nutrition, 57(2), S47–S51.

[35] Willis, S., Goldfarb, D.S., Thomas, K. and Bultitude, M. 2019. Water to prevent kidney stones: Tap vs bottled; soft vs hard–does it matter? BJU International, 124(6), 905–906.

[36] Keßler, T. and Hesse, A. 2000. Cross-over study of the influence of bicarbonate-rich mineral water on urinary composition in comparison with sodium potassium citrate in healthy male subjects. British Journal of Nutrition, 84(6), 865–871.

[37] Reddy, S.T., Wang, C.-Y., Sakhaee, K., Brinkley, L. and Pak, C.Y. 2002. Effect of low-carbohydrate high-protein diets on acid-base balance, stone-forming propensity, and calcium metabolism. American Journal of Kidney Diseases, 40(2), 265–274.

[38] Parihar, S. and Sharma, D. 2021. A brief overview on Asparagus racemous. International Journal of Research and Analytical Reviews, 8(4), 96–108.

[39] Singh, A.K., Srivastava, A., Kumar, V. and Singh, K. 2018. Phytochemicals, medicinal and food applications of Shatavari (*Asparagus racemosus*): An updated review. The Natural Products Journal, 8(1), 32–44.

[40] Jagannath, N., Chikkannasetty, S.S., Govindadas, D. and Devasankaraiah, G. 2012. Study of antiurolithiatic activity of Asparagus racemosus on albino rats. Indian Journal of Pharmacology, 44(5), 576.

[41] Lee, N.Y., Khoo, W.K., Adnan, M.A., Mahalingam, T.P., Fernandez, A.R. and Jeevaratnam, K. 2016. The pharmacological potential of Phyllanthus niruri. Journal of Pharmacy and Pharmacology, 68(8), 953–969.

[42] Narendra, K., Swathi, J., Sowjanya, K. and Satya, A.K. 2012. Phyllanthus niruri: A review on its ethno botanical, phytochemical and pharmacological profile. Journal of Pharmacy Research, 5(9), 4681–4691.

[43] Pezzani, R., Vitalini, S. and Iriti, M. 2017. Bioactivities of Origanum vulgare L.: An update. Phytochemistry Reviews, 16(6), 1253–1268.

[44] Kintzios, S. 2012. Oregano. In Peter, K.V. (Ed.), *Handbook of Herbs and Spices.* Sawston, UK: Elsevier, pp. 417–436.

[45] Khokhlenkova, N.V., Buryak, M.V., Povrozina, O.V. and Kamina, T.V. 2019. Principles of the Urolithiasis Phytotherapy. Research Journal of Pharmacy and Technology, 12(9), 4559–4564.

[46]   Khan, A., Bashir, S., Khan, S.R. and Gilani, A.H. 2011. Antiurolithic activity of Origanum vulgare is mediated through multiple pathways. BMC Complementary and Alternative Medicine, 11(1), 1–16.

[47]   Fatehi, M., Saleh, T.M., Fatehi-Hassanabad, Z., Farrokhfal, K., Jafarzadeh, M. and Davodi, S. 2005. A pharmacological study on Berberis vulgaris fruit extract. Journal of Ethnopharmacology, 102(1), 46–52.

[48]   Mokhber-Dezfuli, N., Saeidnia, S., Gohari, A.R. and Kurepaz-Mahmoodabadi, M. 2014. Phytochemistry and pharmacology of berberis species. Pharmacognosy Reviews, 8(15), 8.

[49]   Kong, W., Wei, J., Abidi, P., Lin, M., Inaba, S., Li, C., Wang, Y., Wang, Z., Si, S., Pan, H. and Wang, S. 2004. Berberine is a novel cholesterol-lowering drug working through a unique mechanism distinct from statins. Nature Medicine, 10(12), 1344–1351.

[50]   Bashir, S. and Gilani, A.H. 2011. Antiurolithic effect of berberine is mediated through multiple pathways. European Journal of Pharmacology, 651(1–3), 168–175.

[51]   Orhan, N. 2022. *Adulteration of Nigella (Nigella Sativa) Seed and Seed Oil. Botanical Adulterants Prevention Bulletin*. Austin, TX: ABC-AHP-NCNPR Botanical Adulterants Prevention Program.

[52]   Khattak, K.F. and Simpson, T.J. 2008. Effect of gamma irradiation on the extraction yield, total phenolic content and free radical-scavenging activity of Nigella staiva seed. Food Chemistry, 110(4), 967–972.

[53]   Farag, M.A., Gad, H.A., Heiss, A.G. and Wessjohann, L.A. 2014. Metabolomics driven analysis of six Nigella species seeds via UPLC-qTOF-MS and GC–MS coupled to chemometrics. Food Chemistry, 151, 333–342.

[54]   Amuthan, A., Chogtu, B., Bairy, K. and Prakash, M. 2012. Evaluation of diuretic activity of Amaranthus spinosus Linn. aqueous extract in Wistar rats. Journal of Ethnopharmacology, 140(2), 424–427.

[55]   Da-Costa-Rocha, I., Bonnlaender, B., Sievers, H., Pischel, I. and Heinrich, M. 2014. Hibiscus sabdariffa L.–A phytochemical and pharmacological review. Food Chemistry, 165, 424–443.

[56]   Udayasekhara Rao, P. 1996. Nutrient composition and biological evaluation of mesta (*Hibiscus sabdariffa*) seeds. Plant Foods for Human Nutrition, 49(1), 27–34.

[57]   Sarr, M., Ngom, S., Kane, M.O., Wele, A., Diop, D., Sarr, B., Gueye, L., Andriantsitohaina, R. and Diallo, A.S. 2009. In vitro vasorelaxation mechanisms of bioactive compounds extracted from Hibiscus sabdariffa on rat thoracic aorta. Nutrition & Metabolism, 6(1), 1–12.

[58]   Olatunji, L.A., Usman, T.O., Adebayo, J.O. and Olatunji, V.A. 2012. Effects of aqueous extract of Hibiscus sabdariffa on renal Na (+)-K (+)-ATPase and Ca (2+)-Mg (2+)-ATPase activities in Wistar rats. Zhong Xi Yi Jie He Xue Bao= Journal of Chinese Integrative Medicine, 10(9), 1049–1055.

[59]   Al-Snafi, A.E. 2017. The pharmacology of Equisetum arvense-A review. IOSR Journal of Pharmacy, 7(2), 31–42.

[60]   Carneiro, D.M., Jardim, T.V., Araújo, Y.C.L., Arantes, A.C., De Sousa, A.C., Barroso, W.K.S., Sousa, A.L.L., Da Cunha, L.C., Cirilo, H.N.C., Bara, M.T.F. and Jardim, P.C.B.V. 2019. Equisetum arvense: New evidences support medical use in daily clinic. Pharmacognosy Reviews, 13(26), 51.

[61]   Jain, R., Kosta, S. and Tiwari, A. 2010. Ayurveda and urinary tract infections. Journal of Young Pharmacists, 2(3), 337.

[62]   Song, Y., Li, N.C., Wang, X.F., Ma, L.L., Wan, B., Hong, B.F. and Na, Y.Q. 2005. Clinical study of Eviprostat for the treatment of benign prostatic hyperplasia. Zhonghua Nan Ke Xue= National Journal of Andrology, 11(9), 674–676.

Nusrat K. Shaikh*

# Chapter 14
# Ocular health

**Abstract:** This chapter recapitulates nutraceuticals and food supplements having positive effects in most common ocular ailments like cataract, macular degeneration, glaucoma, dry eye disease, and various inflammatory ophthalmic diseases. The overview of pharmacological treatment and their limitations are also highlighted. The main aim of this chapter is to outline nutraceuticals that have potential biological effects and health benefits, with special focus on eye disease, including polyphenols, xanthophylls, carotenoids, polyunsaturated fatty acids, crocetin, and micronutrients. In a nutshell, many nutritional substances appear to provide advantageous impacts on the overall well-being of the ocular disease.

# 1 Introduction

Your vision is crucial to your overall well-being. The majority of individuals use their eyes to view and comprehend everything that surrounds them. However, certain eye ailments can result in loss of vision, making prompt identification and management of eye conditions crucial. If your doctor demonstrates it or if you notice any fresh difficulties with vision, you must get your eyes examined. Additionally, maintaining good eye health is equally as crucial as maintaining good overall well-being.

Good perception is necessary for efficient brain functioning. Our most important body part, our minds, enables us to lead lives that are complicated. Given that the optic nerve links the brain and the eyes, an optimum interconnected link is essential. By maintaining the health of our eyes and brains, we enhance our entire standard of life. Stronger athletic prowess, stronger motor abilities, greater learning and understanding, and an enhanced standard lifestyle are all influenced by satisfactory vision. Among our more vital sensors is sight, which accounts for 80% of all information we take in. By taking care of our eyes, we may lower our risk of becoming blind and losing our eyesight while also keeping an eye out for any eye condition, such as cataract and glaucoma, that may be growing.

Recent gains in lifespan longevity are certain to have a significant impact on the prevalence of vision impairment and disability in advanced nations, as many eye ill-

*Corresponding author: Nusrat K. Shaikh, Department of Quality Assurance, Smt. N. M. Padalia
Pharmacy College, Gujarat Technological University, Ahmedabad 382210, Gujarat, India,
e-mail: nusratshaikh.pharmacist@gmail.com

https://doi.org/10.1515/9783111317601-014

nesses are age-related. There are approximately 45 million blind people worldwide, the prevalence of which is increasing by 1–2 million annually [1].

## 1.1 Cataract

The eye's lens becomes clouded with a cataract. One leading global contributor to eye impairment is that this clouded lens can appear in one or even in both eyes. Although cataracts can develop at any age, including at birth, they are more common in persons over 50. The patient might not first be aware that they have a cataract. But as cataracts progress, eyesight may become blurred, clouded, or less colorful. The patient can find it challenging to read, see at night-time, or perform other daily tasks. Despite various methods being available for removing cataracts in the past, numerous individuals are reluctant to afford the proper therapy because of various barriers, including financial constraints, expenditure, personal choice, or ignorance [2].

Reports of more than 90% of patients seeing better after having their hazy lens surgically removed and replaced with an artificial lens demonstrates the high success rate of the procedure. The surgery to remove cataracts is secure, and improves vision. The surgeon replaces the impaired lens by means of a fresh, mock lens during cataract surgery, known as intraocular lens (IOL) [3].

Most cataracts are age-related and result from the natural changes that occur in the eyes as people age. However, cataract can develop by different variables, such as after therapy for related optical illness (such glaucoma) or shortly after a corneal assault. Regardless of the cataract's nature, surgery is always the recommended course of action [4]. Patients can prevent cataracts and safeguard their eyes by:

- Protecting eyes from the sun by wearing brimmed glasses and a hat
- Giving up smoking and eating well – such as, plenty of veggies (particularly those with bloomy green foliage like collards, spinach, and kale) and fruits
- Scheduling a comprehensive optical assessment: If the patient is sixty years of age or grownup, they should at least get one checkup every two years.
- At-home care: In the beginning, you might be able to create little adjustments with the objective of treating severe degeneration. Some may utilize stronger lighting at home or at the work place, Reading spectacles with magnification and anti-reflective lenses are also some approaches.

## 1.2 Macular degeneration

Precise and center eyesight are impacted by macular degeneration, often known as age-related macular degeneration or AMD. People require excellent central sight to see items easily and carry out regular responsibilities, including reading and driving a vehicle. The eyeball's macula, the central area of the optic nerve that is used to see minute details, is harmed by it. It is the main contributor to vision loss in adults over 60 [5].

Retinal degeneration comes in two forms: dry and moist [6]. Moist AMD is known as advanced neovascular AMD. It is a condition where peculiar gorepitchers begin to form below the optic nerve and underneath the macular area, eventually triggering leakage of the plasma, which is unsolidified. The arteries and veins can rupture, drip, or wound, endangering the cornea, rapidly reducing vision in the center. As an initial indication of wet AMD, linear patterns seem wave-like.

Dry AMD is also known as atrophic AMD, while the cerebellum's macula gradually ages as a result of which, the center perception gradually blurs. The moist form of AMD grows rapidly, while its dry form is more common, accounting for 70–90% of occurrences. As the macula's effectiveness decreases as time passes, the main point of vision of the affected eye continuously deteriorates. Dry AMD frequently affects both eyes. Perhaps the greatest frequent initial indication of dry AMD is drusen. Drusen syndromes are minute yellow or white deposits inside the skin of the retina. It is common knowledge that individuals aged 60 and above possess them [6]. The presence of limited spots is typical; usually, they do not hinder visibility. However these can become bigger and emerge often in progressive dry AMD or wet AMD, causing danger.

AMD symptoms consist of distorted views, black or dark areas in the core of your sphere of vision, with resemblance of clean lines as curvy or twisted. These symptoms are typically not recognized until the disease has advanced. Diagnosis: Dilated eye exam; the doctor might also advise getting an optical coherence tomography test (OCT). During an OCT examination, an ophthalmologist can use a special device to take photos of the interior of your eye. Vitamin and mineral dietary supplements, injections, and photodynamic therapy (laser and injection) are all forms of treatment.

Depending on the stage and type, AMD can be treated. Early AMD cannot be treated currently; therefore, eye doctors will likely just monitor patients' eyes through routine eye exams. Quitting cigarette smoking, maintaining a nutritious diet, and exercising regularly might be helpful. Assuming you are experiencing moderate AMD in either or both of your eyes, you could possibly avoid late AMD by taking special vitamins and mineral supplements. If you possess only late AMD in a single eye, these dietary supplements may slow the growth of AMD in that eye [7].

Treatment for wet AMD has recently made strides, that includes VEGF (vascular endothelial growth factor) inhibitor intraocular injections and Photodynamic therapy (PDT) – an injectable and laser treatment combination.

At the moment, there is no known treatment for late dry AMD, but researchers are making great efforts to identify one. You can also look for support that can assist you deal with the blurred vision brought on by AMD. Stem cell transplants are being tested in clinical settings for dry AMD. According to research, a person's risk of AMD may be reduced by making the following healthy decisions: give up smoking or refrain from starting, engage in enough exercise to keep lipid and arterial pressure at satisfactory levels, and consume seafood and green vegetables with leaves as part of a balanced diet [8].

## 1.3 Glaucoma

Glaucoma is an eye ailment brought on by an abnormally high fluid pressure inside the eye. Your optic nerve suffers damage from the pressure, which changes how visual information reaches your brain. Blindness in one or both eyes and vision loss are possible outcomes of undiagnosed and untreated glaucoma. Often, glaucoma runs in families. But as recent studies have shown, glaucoma can manifest itself even in the presence of normal pressure within the eyes. Proper management regularly protects the eye against major sight damage [9].

Glaucoma can be "closed angle" or "open angle." Open angle is known as the sneak thief of sight. It develops gradually over an extended stretch of time despite the individual noticing diminished vision, unless the illness is severely advanced. Closed angle is uncomfortable and might appear suddenly. Visual loss can worsen quickly, but the discomfort and pain prompt individuals to seek medical care before serious harm takes place [10].

Migraines, red eyes, rainbow-colored circular rims around illumination, double vision, blind patches, ocular irritation or threat, feeling sick and throwing up are some of the symptoms [9]. Diagnosis: Visual field testing and dilated eye examination. It is an easy and painless exam. The doctor will administer eye medications to expand the pupil before checking for glaucoma and other eye conditions. The assessment includes an inspection of the visual field to gauge side eyesight [11].

The goal of treatments is to lower the ocular pressure, and they often involve medicine (commonly, eye drops), laser therapy, and surgery [12]:
– Medicines: The most popular form of treatment is pharmaceutical sprays for the eyes. They relieve eye strain and protect the optic nervous system from damage.
– Optical laser therapy: To assist the fluid flow out of your eye, doctors can use lasers to relieve eye pressure. The doctor can perform it easily in the clinic.
– Surgery: If medications and laser therapy are ineffective, a doctor may recommend surgery. There are numerous surgical procedures that can aid in the fluid draining from the eye.

# 2 General overview of pharmacological treatments

The complicated sensory organ that is responsible for sight is the eye. Loss of vision can be caused by disease or injury to the ocular system. For both therapeutic and diagnostic purposes, a number of several pertinent, parenteral, and ingested ocular preparations are available. Optic decline, persistent eye irritation, retinopathy caused by diabetes, cataracts, sickness, and redness constitute conditions that are significantly managed by medications [13]. Pharmacological treatment presents a special challenge because of its limits and the ocular fundamental histology. Multiple hydrophobic and hydrophilic layers make up the cornea, and these layers can affect how well topical medications are absorbed. Ophthalmic drugs can sometimes be absorbed into the bloodstream, but this happens infrequently because barriers between the plasma and the eye, bloodstream and water, and serum and translucent separate the eye against systemic vascular access.

While essential for preserving sterility, stabilizers in ocular retinal multidose combination drugs have the potential to be harmful to the ocular surface. Approximately, 70% of the ophthalmic products contain the well-known chemical benzalkonium chloride (BAK), which adversely harms the epithelium of the cornea and conjunctive epithelial cells that reside in the eye that causes cytotoxicity. As a result, ocular surface disease (OSD) symptoms such as ocular surface staining and longer tear evaporation times increase. These negative consequences can appear as soon as seven days after exposure, but they are supplementary significant, with lingering acquaintance, similar to chronic glaucoma treatment. Preservative-free compositions extended-release medication delivery systems and alternative therapies for common eye conditions are just a few of the many methods that can be used to reduce or completely remove BAK exposure [13]. BAK has actions concerning the eye's surface that are carcinogenic and scientific, including nutraceuticals that are both established and developing options for managing ocular diseases – they can reduce or completely eradicate BAK exposure.

For patients with wet AMD, a molecule called VEGF contributes to the development of fresh arteries within the macula. It helps stop the illness from getting worse and might even partially reverse it by preventing the activity of this molecule, but there are very stringent criteria to identify which patients are qualified for this therapy. Ranibizumab, pegaptanib, and aflibercept are anti-VEGF drugs. Although bevacizumab is not approved for the treatment of AMD, it is less expensive and seems to be just as effective. About one in three patients who receive anti-VEGF injections will have eyesight improvement. Treatment will, however, typically preserve vision and stop the problem from getting worse. Approximately 1/10 patients do not respond to the treatment. Previous interventions for wet AMD, such as photodynamic therapy and laser treatment (which is challenging to the macula since it results in scarring and visual loss), were less effective. Verteporfin, a medication used in photodynamic therapy, is injected into a vein, and binds to the newly developed aberrant blood vessels in the macula, making it possible to target those blood vessels with a specific laser. Although it has a good success rate at stopping progression, it would not always succeed to bring back vision [14].

More than 2/3rd of the participants in tests have shown a noticeable improvement in their vision, thanks to intraocular lens systems. However, this approach may be associated with severe side effects, such as increased eye pressure, corneal or new lens fogging, or even injury to the eye itself. As the lenses must be customized for each patient, the cost of this procedure is currently expensive.

Ocular hypertension and glaucoma are usually treated with prostaglandin analogs (i.e., latanoprost) and β-adrenoreceptor topical antagonists (i.e., timolol). Prostaglandin analogs may reduce intraocular pressure (IOP) by enhancing aqueous fluid outflow. Uveoscleral outflow is encouraged by prostaglandin analogues' agonistic action on FP prostanoid receptors. Conspicuous cosmetic alterations brought on by prostaglandin analogues include browning of the upper and lower pupils, thickening the development of eyelash growth and dilating the pupil. Blurred vision, itchy eyes, headaches, and eye discomfort are some other side effects. β-adrenoreceptor topical antagonists can lower IOP by reducing the formation of aqueous humour. By blocking β-receptors, these agents counteract the effects of sympathetic neurotransmitters; nevertheless, it is unclear how they work to treat glaucoma. Even though systemic side effects are uncommon, they may trigger breathing difficulties, cardiac arrest, low pressure, headaches, and reduced tolerance to physical activity when systemically absorbed.

The use of Carbonic anhydrase inhibitors (CAIs) i.e., dorzolamide and brinzolamide may reduce aqueous humour generation. Vision blurring and altered taste are some of the negative consequences. Oral formulation such as acetazolamide can also be considered in the management of IOP reduction. Subjects with sulfonamide allergy should avoid taking CAIs.

# 3 Nutraceuticals: Preventive approach

Plenty of individuals are unlikely to receive adequate nutrition from consuming regular meals. Additionally, the setting in which individuals reside is extremely toxic, with pesticides and pollution interfering with our body's ability to manage it. We are currently plagued by several new diseases that are widespread in our society. Instead of relying on antibiotics that have lost their efficacy, it makes greater economic sense to upgrade the network or the environment. Because medicines are not organic towards the human being, they frequently have detrimental effect, although vitamins of superior quality that the human system is capable of absorbing and exploiting can strengthen our bodies and provide energy.

## 3.1 Polyphenols

Polyphenols are one of the most abundant secondary metabolites and potent antioxidants of the plant origin. Among the most prominent polyphenols are flavan-3-ols, flavonols, flavones, flavanones, and anthocyanins, which make up roughly eight thousand separate categories. Phenylpropanoid metabolism, an extremely branched system, is responsible for producing the majority of polyphenols. Two types of polyphenols that are tend to be found in food products are flavonoids and phenolic acids, among others. Over 10,000 distinct phytonutrients make up the large category of polyphenols, many of which have been shown to have positive impact on health in the form of antimicrobial, antiviral, antibacterial, hypersensitive, anti-inflammatory, and oxidative properties. Figure 1 illustrates some of the potential benefits of adequate consumption of dietary polyphenols in ocular health.

Resveratrol can be found abundantly in red wines and grapes. It has been shown to enhance the activity of mitochondria through SIRT1 and PGC-1 upregulation. Due to its antioxidant and anti-inflammatory qualities, it is also beneficial against eye disorders associated with aging [15]. Resveratrol has been shown to increase the lifespan of retinal ganglion cell (RGC) in glaucoma experimental models and to protect against the loss of dendritic complexity [16, 17]. According to reports, resveratrol lowers IOP both in ocular normotensive and hypertensive animals [18, 19]. In mouse instances of light-induced retinal tissue ageing, resveratrol decreases the apoptosis-related molecular structure, the activator protein 1 (AP-1), and boosts neurologically protective variables such as brain-derived neurotrophic factor (BDNF), leukemia inhibitory factor (LIF), cardiotrophin 1 (CT-1), oncostatin M (OSM), alongside cardiotrophin-like cytokine (CLC) [20, 21]. Resveratrol-based nutritional supplements have been found in case report observations to aid AMD patients by improving retinal pigment epithelium functioning [22].

Quercetin, found in fruits, vegetables, and beverages, possesses a substantial antioxidant potential for removing ROS, and greatly lowers the function of mitochondria through an AMP-activated protein kinase (AMPK)/SIRT1 regulating system [23–26]. Extensive research studies showed quercetin's ability to lower ROS, oxidative cell membrane potential, and prevention of RGC apoptosis [27]. As studied by Zhou and colleagues, quercetin may reduce RGC excitability by increasing GABAergic neural communication and reducing dwarf neurotransmitter glut brain communication [28]. Some research has also shown the inhibition of the activity of RGCs' heat shock protein 72 (HSP 72) with quercetin treatment [29–31]. Quercetin may exert neuroprotective effects on ocular region via downregulation of the activator protein 1 (AP-1 pathway) [32], thus protecting against retinal damage. *In vivo* studies on diabetic experimental models demonstrated the protective effects of quercetin against cell death, neurological inflammation, and ocular oxidative damage [33]. A study showed significant inhibition of choroidal neovascularization *in vitro* and *in vivo*, and improved choroidal blood flow in experimental model of age-related macular degeneration [34].

Epigallocatechin gallate (EGCG) decreased invasion of lymphocytes throughout the submandibular ducts in a mouse model of Sjogren disorder and protected acinar cells against apoptosis [35]. In a dose-dependent fashion, EGCG inhibited the production of numerous IL-1 associated cytokines [36]. In rodents' model, EGCG therapy was

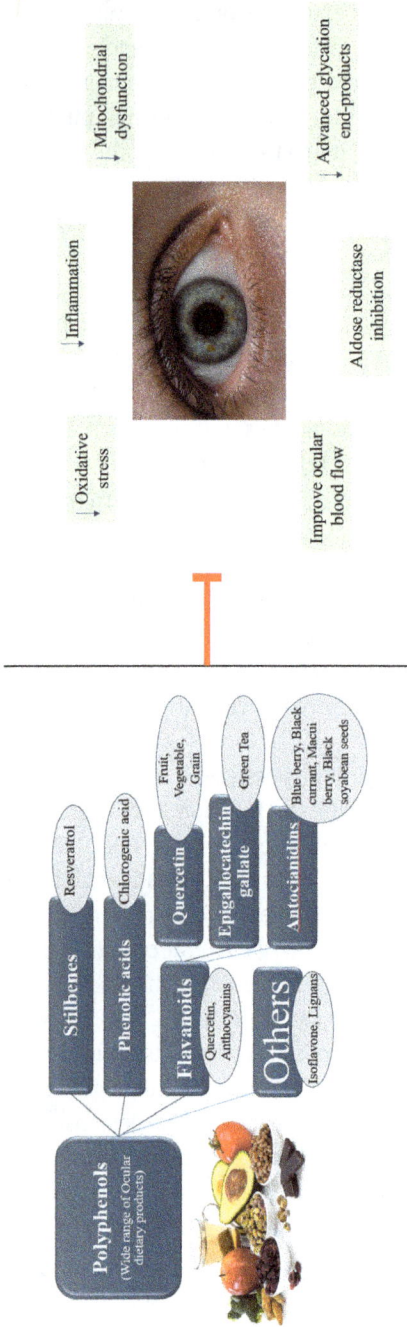

**Figure 1:** Benefits of dietary polyphenols in ocular health.

also linked to decreased corneal epithelial destruction, CD11b+ cell number, and IL-1 expression [37].

Curcumin has potent anti-free radical activity and through nuclear factor Nrf2, also known as the erythroid 2-related factor 2, gradually enhances the mitochondrial functioning [38]. However, curcumin's clinical applications have been constrained by its poor dissolution and low absorption [39]. Delivering curcumin to the lesions using novel formulation methods, such as enhanced products, liposomes, and nanoparticles recipients, might be the best course of action. For instance, with the use of Pluronic F127, a difunctional to prevent polymerization surfactant, and D-tocopherol polyethene glycol 1,000 succinate (TPGS), a non-ionic surfactant, Davis et al. developed curcumin nanocarriers that improved curcumin movement across obstacles in the eyes and boosted curcumin solubility. The topically applied curcumin nanocarrier supported the retinal cells neuro-protection both *in vitro* and *in vivo* [40].

## 3.2 Carotenoids

The natural pigments called carotenoids are produced by numerous plants, where they also impart color to some foods, like pumpkins and carrots. The first biochemical identification of carotenoids was made in 1907 by Richard Willstatter, who categorized them into xanthophyll and carotenoid families. The lutein and zeaxanthin stereoisomer, oxygenated carotenoids formerly referred to as xanthophylls, are located in healthy macula as macular pigments [41]. They are not synthesized endogenously; thus, they should be obtained from dietary sources such as spinach, egg yolks, and wolfberries. Xanthophylls have ROS scavenging, anti-inflammatory, and neuroprotective properties [42]. Lutein offers numerous benefits to ocular health via reduction of endoplasmic reticulum stress, scavenging of free radicals, and anti-apoptotic actions [43, 44]. The *in vivo* studies showed the enhanced mitochondrial biogenesis and upregulated carotenoid metabolic genes with xanthophylls supplementation [45]. Lutein is recommended for use in the prevention of AMD, though it is (with or without zeaxanthin) less likely to offer benefit against the progression to late AMD [46].

The carotenoids, -carotene, -carotene, and –cryptoxanthin, are referred to as provitamin A, as they convert into retinol or vitamin A in the body. Carotenoids are the precursors of retinal, while retinal is a precursor of other forms of vitamin A, which all are well-known for their significant role in vision. Rhodopsin, the primary pigment found in rod and cone cells in the retina of the eye, is made up of retinal photoreceptor protein of bipartite structure. Rhodopsin is stimulated when light strikes the rod and cone photoreceptors in the eye, which then sends an electrical signal to the brain to interpret what is visible. The retinoid cycle, which is a series of additional processes, is also triggered by rhodopsin. All the chemicals required for light to stimulate eye receptors are replenished through this cycle [47].

AMD is the main reason for vision impairment and blindness at advanced ages. The macula has been reported to be affected by chronic exposure to blue light. Such disorders can be avoided by adequate consumption of carotenoids i.e., lutein, zeaxanthin, and meso-zeaxanthin, either through diet or supplements. Cataract is known to cause vision impairment and eye cloudiness, where studies supported the potential role of carotenoids intake in decreasing the risk factors of cataract development [48].

## 3.3 Polyunsaturated fatty acids

Considering that PUFAs (Figure 2) must be obtained from food in order to preserve physiologic authenticity, they are also referred to as "essential fatty acids", as they are not produced by the body. In response to a study examining scientific research on the medical management of dry eye condition with *n*-3 and *n*-6 essential fatty acids, *n*-3 intake demonstrated a beneficial effect on inflammation, restricting the synthesis of prostaglandin precursor molecules, combating cell death of glandular epithelium cells in the crimson gland, and eliminating meibomitis. This allowed a less thick, more flexible cholesterol layer-by-layer to safeguard the tear sac and corneal surface.

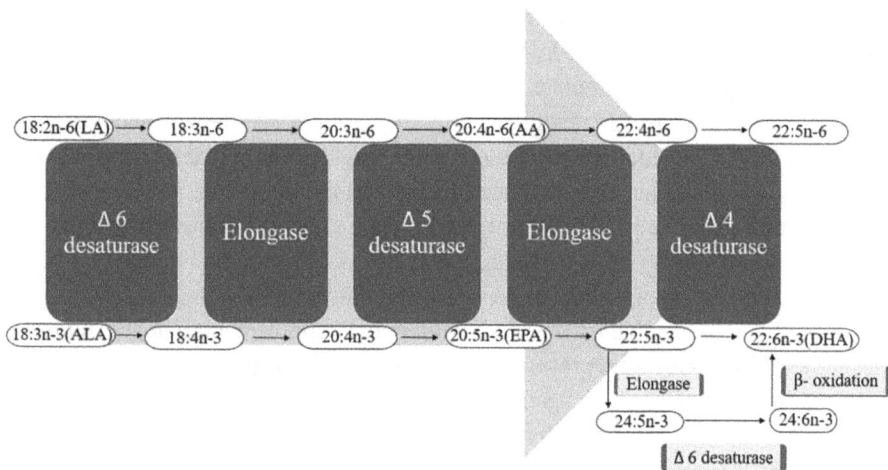

**Figure 2:** The processes through which *n*-3 and *n*-6 polyunsaturated fatty acids remain metabolized. Competition for the same elongation and desaturation enzymes is necessary for the two families of fatty acids. In contrast to faster elongation phases, it should be noted that desaturation steps typically tend to be slow and rate-limiting. Eicosapentaenoic acid (EPA) and arachidonic acid (AA) are sources of eicosanoids i.e., prostaglandins, leukotrienes, and thromboxanes and can mediate a variety of physiological activities.

By increasing ATP generation, mitochondrial biogenesis, and gene expression of proton differential uncoupling proteins throughout living things, $n$-3 PUFAs may alleviate mitochondrial dysfunction and oxidative stress [49]. When exposed to oxidative damage, docosahexanoic acid is transformed into neuroprotectin D1 by the enzyme lipoxygenase, exerting anatomical and neuroprotective functions in the optic nerve [50]. Neuroprotectin D1 (NPD1) also suppresses the proliferation of genes related to necrosis and inflammation, which helps photoreceptors survive. Dietary supplementation of older rats with $n$-3 PUFAs diminished intraocular pressure and showed marked enhancements in permeability capability and decreased ocular rigidity [51]. A diet rich in $n$-3 PUFAs showed significant effects in preserving retina in diabetic rats with AMD-like retinal lesions [52], where recent studies demonstrated the prevention of the production of lipofuscin granules, thus protecting the photoreceptor layer [53, 54]. As part of this pathway, levels of myelin basic protein (MBP), myelin proteolipid peptide (MPP), neuronal fibrillar acidic protein, and brain regulatory factor-like peptide may rise, producing beneficial effects. Moreover, consuming $n$-3 PUFAs is also linked to a reduction in cerebral cartographic shrinkage and neovascular AMD development [55].

## 3.4 Crocetin

The gardenia fruit (*Gardenia jasminoides* Ellis) and saffron crocus (*Crocus starus* L.) are rich in apocarotenoid, known as crocetin [56, 57]. Crocetin exhibits protective properties against retinal degeneration *in vivo* and restores disruption caused by hydrogen peroxide ($H_2O_2$) or tunicamycin through inhibition of caspase activity [58]. In a mouse model of ischemia/reperfusion-induced retinal damage, it has been shown to suppress oxidative stress via p38, the extrinsic signal-regulated protein kinases (ERK), mitogen-induced protein kinases (MAPK), c-Jun N-terminal kinases (JNK), as well as the redox-sensitive NF-B and c-Jun circuit [59]. Increased IOP levels and stimulated microglial neurons are alleviated by a standardized extract of saffron comprising 3% crocin [60]. Clinical investigations also demonstrated the benefits of oral supplementation of saffron or crocin in adults with ocular ailments, including AMD, primary open angle glaucoma, and diabetic macular edema [61]. However, subjects with ATP coupling sequence subgroup variant 4 (ABCA4) mutation in the genome and with Stargardt disease/fundus flavimaculatus (STG/FF) did not benefit (in terms of visual perception) with short-term saffron intake.

## 3.5 Micronutrients

There is currently a growing body of research pointing to a potential function for several micronutrients in the treatment therapy of ophthalmic superficial ailment. In this regard, the impact of vitamins A, B12, C, and selenium on the ocular surface was thoroughly examined, especially in reducing signs and symptoms of dry eye disease. Vitamin A is particularly important for the ocular surface epithelium's metabolism, development, and differentiation. In the developing world, vitamin A deficiency is one of the leading causes of preventable myopia [62]. Vitamin A insufficiency is rare in Western nations, and it has been linked to malabsorption disorders like alcoholism, cystic fibrosis, and weight loss surgery. Nyctalopia (night blindness) is the initial clinical sign of vitamin A insufficiency.

Vitamin B12 plays a critical role in the metabolism of fatty acids and amino acids, as well as in maintaining the nerve's myelin sheath. A deficiency of vitamin B12 has been linked to myelopathy, peripheral neuropathy, neuropsychiatric disorders, and retinal degeneration. Supplementation of vitamin B12 (in eye drops or injectables) may improve severe dry eye disease in subjects with or without neuralgia ocular pain [63, 64].

Vitamin C, in addition to aldehyde dehydrogenase enzymes (ALDH1A1 and ALDH3A1), supports the antioxidant functions of corneal layer, thus protecting the eye from UV-damage [65]. Supplements containing vitamins C and E are reported to significantly reduce the nitric oxide concentrations, improving antioxidant responses in the body as well as increasing the goblet cell density in diabetic individuals, resulting in improving hydration and lubrication of the mucosal surfaces [66].

Selenium is an essential trace element with the capability to fight against oxidative and nitrosative stresses. Meat, fish, seafood, and grains are the primary food sources. The lacrimal gland generates selenium-transport protein called selenoprotein P, which possesses potent antioxidant properties and may protect against peroxynitrite ($ONOO^-$) toxicity. The high selenium content of selenoprotein P suggests that it may be involved in the intracellular storage and transport of selenium. The supplementation of rodents with selenoprotein P suppressed the oxidative stress markers and improved the dry eye index in rat dry eye model. However, clinical evidence for the protective role of selenium in age-related eye ailments (i.e., cataract, macular degeneration, and retinitis pigmentosa) is still lacking, possibly due to unestablished link of selenium deficiencies and eye diseases [67]. Moreover, supplementing selenium beyond the dietary reference intake (i.e., 55 µg/day) is discouraged in order to improve eye health.

# 4 Summary

The fundamental processes of metabolic balance and high-energy expenditure that are crucial to the retina are outlined in this chapter. Oxidative stress is the main factor associated with mitochondrial dysfunction, where nutraceuticals and food bioactive ingredients target mitochondria to reestablish molecular adaptability. Nutrient-rich diets may aid in preventing vision difficulties, including AMD. The antioxidant actions of carotenoids, antioxidant vitamins, flavonoids, lutein, green tea and *n*-3 PUFAs may prevent the risk factors of presbyopia and cataract. Zeaxanthin and lutein may help to improve eyesight and lower the risk factors for cataract development. Rice bran is rich in folic acid and essential fatty acids, promoting eye health. However, assessment of these nutraceuticals and food bioactives in robust clinical trials on larger scale is essential for their efficacy and tolerability, before their recommendation for regular use in clinical settings.

# References

[1]   Delmas, D., Clarisse, C.C., Flavie, C., Jianbo, X. and Virginie, A. 2021. New highlights of resveratrol: A review of properties against ocular diseases. International Journal of Molecular Sciences, 22(3), 1295.

[2]   Dennis, L., Srinivas, K.R., Vineet, R., Yizhi, L., Paul, M., Jonathan, K., Tassignon, M., Jost, J., Chi, P.P. and David, F.C. 2015. Cataract. Nature Reviews Disease Primers, 1, 1–14.

[3]   Livingston, P.M., Carson, C.A. and Taylor, H.R. 1995. The epidemiology of cataract: A review of the literature. Ophthalmic Epidemiol, 2(3), 151–164.

[4]   Khayam, N., Jack, G. and David, B. 2020. Cataract surgery and dry eye disease: A review. European Journal of Ophthalmology, 30(5), 840–855.

[5]   Monika, F., Tiarnan, D.L., Robyn, H.G., Usha, C., Valckenberg, S.S., Caroline, C.K., Wai, T.W. and Emily, Y.C. 2021. Age-related macular degeneration. Nature Reviews Disease Primers, 7, 1–12.

[6]   Andreea, G., Labib, M. and Ovidiu, M. 2015. Age-related macular degeneration. Romanian Journal of Ophthalmology, 59(2), 74–77.

[7]   Yolanda, J., David, A.A., Mario, B.B., Lorena, P., Felisa, R., Laura, O., Marta, V., Ignacio, F., Francisco, P., Indira, S. and Miguel, G. 2022. Novel treatments for age-related macular degeneration: A review of clinical advances in sustained drug delivery systems. Pharmaceutics, 14(7), 1473–65.

[8]   Dtsch, A., Andreas, S. and Med, D. 2020. The diagnosis and treatment of age-related macular degeneration. International, 117(29–30), 513–520.

[9]   Cook, C. and Foster, P. 2012. Epidemiology of glaucoma: What's new? Canadian Journal of Ophthalmology. Journal Canadien D'ophtalmologie, 47(3), q223–226.

[10]  Jonas, J.B., Aung, T., Bourne, R.R., Bron, A.M., Ritch, R. and Panda-Jonas, S. 2017. Glaucoma. Lancet (London, England), 11 390(10108), 2183–2193.

[11]  Huang, C.P., Lin, Y.W., Huang, Y.C. and Tsa, F.J. 2020. Mitochondrial dysfunction as a novel target for neuroprotective nutraceuticals in ocular diseases. Nutrients, 12, 1–23.

[12]  Robert, N.W., Tin, A. and Felipe, A.M. 2015. The pathophysiology and treatment of glaucoma a review. The Journal of the American Medical Association, 14 311(18), 1901–1911.

[13]  Jost, B.J., Tin, A., Rupert, R.B., Alain, M.B., Robert, R. and Songhomitra, P. 2017. Glaucoma, 390(10108), P2183–2193.

[14]  Hebatallah, B.M., Basma, N.A., Dina, F. and Ehab, A.F. 2022. Current trends in pharmaceutical treatment of dry eye disease: A review. European Journal of Pharmaceutical Sciences, 175, 106206.

[15]  Lagouge, M., Argmann, C., Gerhart-Hines, Z., Meziane, H., Lerin, C., Daussin, F., Messadeq, N., Milne, J., Lambert, P., Elliott, P., et al. 2006. Resveratrol improves mitochondrial function and protects against metabolic disease by activating sirt1 and pgc-1alpha. Cell, 127, 1109–1122.

[16]  Abu-Amero, K.K., Kondkar, A.A. and Chalam, K.V. 2016. Resveratrol and ophthalmic diseases. Nutrients, 8, 200.

[17]  Pirhan, D., Yuksel, N., Emre, E., Cengiz, A. and Kursat Yildiz, D. 2016. Riluzole-and resveratrol-induced delay of retinal ganglion cell death in an experimental model of glaucoma. Current Eye Research, 41, 59–69.

[18]  Razali, N., Agarwal, R., Agarwal, P., Kumar, S., Tripathy, M., Vasudevan, S., Crowston, J.G. and Ismail, N.M. 2015. Role of adenosine receptors in resveratrol-induced intraocular pressure lowering in rats with steroid-induced ocular hypertension. Clinical & Experimental Ophthalmology, 43, 54–66.

[19]  Cao, K., Ishida, T., Fang, Y., Shinohara, K., Li, X., Nagaoka, N., Ohno-Matsui, K. and Yoshida, T. 2020. Protection of the retinal ganglion cells: Intravitreal injection of resveratrol in mouse model of ocular hypertension. Investigative Ophthalmology & Visual Science, 61(3), 13–13.

[20]  Natesan, S., Pandian, S., Ponnusamy, C., Palanichamy, R., Muthusamy, S. and Kandasamy, R. 2017. Co-encapsulated resveratrol and quercetin in chitosan and peg modified chitosan nanoparticles: For efficient intra ocular pressure reduction. International Journal of Biological Macromolecules, 104, 1837–1845.

[21]  Kubota, S., Kurihara, T., Ebinuma, M., Kubota, M., Yuki, K., Sasaki, M., Noda, K., Ozawa, Y., Oike, Y., Ishida, S., et al. 2010. Resveratrol prevents light-induced retinal degeneration via suppressing activator protein-1 activation. The American Journal of Pathology, 177, 1725–1731.

[22]  Bola, C., Bartlett, H. and Eperjesi, F. 2014. Resveratrol and the eye: Activity and molecular mechanisms. Graefe's Archive for Clinical and Experimental Ophthalmology, 252, 699–713.

[23]  Boots, A.W., Haenen, G.R. and Bast, A. 2008. Health effects of quercetin: From antioxidant to nutraceutical. European Journal Pharmacology, 585, 325–337.

[24]  Qiu, L., Luo, Y. and Chen, X. 2018. Quercetin attenuates mitochondrial dysfunction and biogenesis via upregulated ampk/sirt1 signaling pathway in oa rats. Biomedicine & Pharmacotherapy, 103, 1585–1591.

[25]  Panche, A.N., Diwan, A.D. and Chandra, S.R. 2016. Flavonoids: An overview. Journal of Nutritional Science, 5, 47.

[26]  Wang, D.M., Li, S.Q., Wu, W.L., Zhu, X.Y., Wang, Y. and Yuan, H.Y. 2014. Effects of long-term treatment with quercetin on cognition and mitochondrial function in a mouse model of alzheimer's disease. Neurochemical Research, 39, 1533–1543.

[27]  Gao, F.J., Zhang, S.H., Xu, P., Yang, B.Q., Zhang, R., Cheng, Y., Zhou, X.J., Huang, W.J., Wang, M., Chen, J.Y., et al. 2017. Quercetin declines apoptosis, ameliorates mitochondrial function and improves retinal ganglion cell survival and function in in vivo model of glaucoma in rat and retinal ganglion cell culture in vitro. Frontiers in Molecular Neuroscience, 10, 285.

[28]  Zhou, X., Li, G., Yang, B. and Wu, J. 2019. Quercetin enhances inhibitory synaptic inputs and reduces excitatory synaptic inputs to off- and on-type retinal ganglion cells in a chronic glaucoma rat model. Frontiers in Neuroscience, 13, 672.

[29]  Park, K.H., Cozier, F., Ong, O.C. and Caprioli, J. 2001. Induction of heat shock protein 72 protects retinal ganglion cells in a rat glaucoma model. Investigative Ophthalmology & Visual Science, 42, 1522–1530.

[30]  Caprioli, J., Ishii, Y. and Kwong, J.M. 2003. Retinal ganglion cell protection with geranylgeranylacetone, a heat shock protein inducer, in a rat glaucoma model. Transactions of the American Ophthalmological Society, 101, 39–50.

[31]  Li, N., Li, Y. and Duan, X. 2014. Heat shock protein 72 confers protection in retinal ganglion cells and lateral geniculate nucleus neurons via blockade of the sapk/jnk pathway in a chronic ocular-hypertensive rat model. Neural Regeneration Research, 9, 1395–1401.

[32]  Koyama, Y., Kaidzu, S., Kim, Y.C., Matsuoka, Y., Ishihara, T., Ohira, A. and Tanito, M. 2019. Suppression of light-induced retinal degeneration by quercetin via the ap-1 pathway in rats. Antioxidants, 8, 79.

[33] Kumar, B., Gupta, S.K., Nag, T.C., Srivastava, S., Saxena, R., Jha, K.A. and Srinivasan, B.P. 2014. Retinal neuroprotective effects of quercetin in streptozotocin-induced diabetic rats. Experimental Eye Research, 125, 193–202.

[34] Zhuang, P., Shen, Y., Lin, B.Q., Zhang, W.Y. and Chiou, G.C. 2011. Effect of quercetin on formation of choroidal neovascularization (CNV) in age-related macular degeneration (AMD). Eye Science, 26(1), 23–29.

[35] Hsu, S.D., Dickinson, D.P., Qin, H., Borke, J., Ogbureke, K.U.E., Winger, J.N., Camba, A.M., Bollag, W.B., Stoppler, H.J., Sharawy, M.M., et al. 2007. Green tea polyphenols reduce autoimmune symptoms in a murine model for human Sjogren's syndrome and protect human salivary acinar cells from TNF-a-induced cytotoxicity. Autoimmunity, 40, 138–147.

[36] Cavet, M.E., Harrington, K.L., Vollmer, T.R., Ward, K.W. and Zhang, J.-Z. 2011. Anti-inflammatory, and anti-oxidative effects of the green tea polyphenol epigallocatechin gallate in human corneal epithelial cells. Molecular Vision, 17, 533–542.

[37] Lee, H.S., Chauhan, S.K., Okanobo, A., Nallasamy, N. and Dana, R. 2011. Therapeutic efficacy of topical epigallocatechin gallate in murine dry eye. Cornea, 30, 1465–1472.

[38] Trujillo, J., Granados-Castro, L.F., Zazueta, C., Anderica-Romero, A.C., Chirino, Y.I. and Pedraza-Chaverri, J. 2014. Mitochondria as a target in the therapeutic properties of curcumin. Archiv der Pharmazie, 347, 873–884.

[39] Lopez-Malo, D., Villaron-Casares, C.A., Alarcon-Jimenez, J., Miranda, M., Diaz-Llopis, M., Romero, F.J. and Villar, V.M. 2020. Curcumin as a therapeutic option in retinal diseases. Antioxidants, 9, 48.

[40] Davis, B.M., Pahlitzsch, M., Guo, L., Balendra, S., Shah, P., Ravindran, N., Malaguarnera, G., Sisa, C., Shamsher, E., Hamze, H., et al. 2018. Topical curcumin nanocarriers are neuroprotective in eye disease. Scientific Reports, 8, 11066.

[41] Obana, A., Gohto, Y., Asaoka, R., Gellermann, W. and Bernstein, P.S. 2021. Lutein and zeaxanthin distribution in the healthy macula and its association with various demographic factors examined in pseudophakic eyes. Antioxidants, 10(12), 1857.

[42] Aziz, E., Batool, R., Akhtar, W., Rehman, S., Shahzad, T., Malik, A., Shariati, M.A., Laishevtcev, A., Plygun, S., Heydari, M. and Rauf, A. 2020. Xanthophyll: Health benefits and therapeutic insights. Life Science, 240, 117104.

[43] Chae, S.Y., Park, S.Y. and Park, G. 2018. Lutein protects human retinal pigment epithelial cells from oxidative stress-induced cellular senescence. Molecular Medicine Reports, 18(6), 5182–5190.

[44] Zhang, C., Wang, Z., Zhao, J., Li, Q., Huang, C., Zhu, L. and Lu, D. 2016. Neuroprotective effect of lutein on NMDA-induced retinal ganglion cell injury in rat retina. Cell Molecular Neurobiology, 36, 531–540.

[45] Yu, H., Wark, L., Ji, H., Willard, L., Jaing, Y., Han, J., He, H., Ortiz, E., Zhang, Y., Medeiros, D.M., et al. 2013. Dietary wolfberry upregulates carotenoid metabolic genes and enhances mitochondrial biogenesis in the retina of db/db diabetic mice. Molecular Nutrition & Food Research, 57, 1158–1169.

[46] Evans, J.R. and Lawrenson, J.G. 2017. Antioxidant vitamin and mineral supplements for slowing the progression of age-related macular degeneration. Cochrane Database of Systematic Reviews, 7, CD000254.

[47] Choi, E.H., Daruwalla, A., Suh, S., Leinonen, H. and Palczewski, K. 2021. Retinoids in the visual cycle: Role of the retinal G protein-coupled receptor. Journal of Lipid Research, 62, 100040.

[48] Moeller, S.M., Jacques, P.F. and Blumberg, J.B. 2000. The potential role of dietary xanthophylls in cataract and age-related macular degeneration. Journal of the American College of Nutrition, 19(sup5), 522S–527S.

[49] Carvalho-Silva, M., Gomes, L.M., Gomes, M.L., Ferreira, B.K., Schuck, P.F., Ferreira, G.C., Dal-Pizzol, F., De Oliveira, J., Scaini, G. and Streck, E.L. 2019. Omega-3 fatty acid supplementation can prevent changes in mitochondrial energy metabolism and oxidative stress caused by chronic administration of L-tyrosine in the brain of rats. Metabolic Brain Disease, 34, 1207–1219.

[50] Bazan, N.G. 2009. Cellular and molecular events mediated by docosahexaenoic acid-derived neuroprotectin D1 signaling in photoreceptor cell survival and brain protection. Prostaglandins, Leukotrienes and Essential Fatty Acids, 81(2–3), 205–211.

[51] Nguyen, C.T., Bui, B.V., Sinclair, A.J. and Vingrys, A.J. 2007. Dietary omega 3 fatty acids decrease intraocular pressure with age by increasing aqueous outflow. Investigative Ophthalmology & Visual Science, 48(2), 756–762.

[52] Tuo, J., Ross, R.J., Herzlich, A.A., Shen, D., Ding, X., Zhou, M., Coon, S.L., Hussein, N., Salem, N., Jr and Chan, C.C. 2009. A high omega-3 fatty acid diet reduces retinal lesions in a murine model of macular degeneration. The American Journal of Pathology , 175, 799–807.

[53] Prokopiou, E., Kolovos, P., Kalogerou, M., Neokleous, A., Nicolaou, O., Sokratous, K., Kyriacou, K. and Georgiou, T. 2018. Omega-3 fatty acids supplementation: Therapeutic potential in a mouse model of Stargardt disease. Investigative Ophthalmology & Visual Science, 59(7), 2757–2767.

[54] Prokopiou, E., Kolovos, P., Georgiou, C., Kalogerou, M., Potamiti, L., Sokratous, K., Kyriacou, K. and Georgiou, T. 2019. Omega-3 fatty acids supplementation protects the retina from age-associated degeneration in aged C57BL/6J mice. BMJ Open Ophthalmology, 4(1), e000326.

[55] Merle, B.M., Benlian, P., Puche, N., Bassols, A., Delcourt, C. and Souied, E.H. 2014. Circulating omega-3 fatty acids and neovascular age-related macular degeneration. Investigative Ophthalmology & Visual Science, 55(3), 2010–2019.

[56] Ichi, T., Higashimura, Y., Katayama, T., Koda, T., Shimizu, T. and Tada, M. 1995. Analysis of crocetin derivatives from gardenia fruits. Nippon Shokuhin Kagaku Kogaku Kaishi, 42, 776–783.

[57] Li, N., Lin, G., Kwan, Y.W. and Min, Z.D. 1999. Simultaneous quantification of five major biologically active ingredients of saffron by high-performance liquid chromatography. Journal of Chromatography A, 849, 349–355.

[58] Yamauchi, M., Tsuruma, K., Imai, S., Nakanishi, T., Umigai, N., Shimazawa, M. and Hara, H. 2011. Crocetin prevents retinal degeneration induced by oxidative and endoplasmic reticulum stresses via inhibition of caspase activity. European Journal Pharmacology, 650, 110–119.

[59] Ishizuka, F., Shimazawa, M., Umigai, N., Ogishima, H., Nakamura, S., Tsuruma, K. and Hara, H. 2013. Crocetin, a carotenoid derivative, inhibits retinal ischemic damage in mice. European Journal Pharmacology, 703, 1–10.

[60] Fernandez-Albarral, J.A., Ramirez, A.I., De Hoz, R., Lopez-Villarin, N., Salobrar-Garcia, E., Lopez-Cuenca, I., Licastro, E., Inarejos-Garcia, A.M., Almodovar, P., Pinazo-Duran, M.D., et al. 2019. Neuroprotective and anti-inflammatory effects of a hydrophilic saffron extract in a model of glaucoma. International Journal of Molecular Sciences, 20, 4110.

[61] Heitmar, R., Brown, J. and Kyrou, I. 2019. Saffron (Crocus sativus L.) in ocular diseases: A narrative review of the existing evidence from clinical studies. Nutrients, 11, 649.

[62] Maheshgauri, R.D., Paaranjpe, R.R., Gahlot, A., Gohil, A. and Pote, S. 2016. Prevalence of vitamin-A deficiency & refractive errors in primary school-going children. National Journal of Medical Research, 6(01), 23–27.

[63] Ozen, S., Ozer, M.A. and Akdemir, M.O. 2017. Vitamin B12 deficiency evaluation and treatment in severe dry eye disease with neuropathic ocular pain. Graefe's Archive for Clinical and Experimental Ophthalmology, 255, 1173–1177.

[64] Macri, A., Scanarotti, C., Bassi, A.M., Giuffrida, S., Sangalli, G., Traverso, C.E. and Iester, M. 2015. Evaluation of oxidative stress levels in the conjunctival epithelium of patients with or without dry eye, and dry eye patients treated with preservative-free hyaluronic acid 0.15% and vitamin B12 eye drops. Graefe's Archive for Clinical and Experimental Ophthalmology, 253, 425–430.

[65]  Tokuda, K., Zorumski, C.F. and Izumi, Y. 2007. Effects of ascorbic acid on UV light-mediated photoreceptor damage in isolated rat retina. Experimental Eye Research, 84(3), 537–543.

[66]  Peponis, V., Papathanasiou, M., Kapranou, A., Magkou, C., Tyligada, A., Melidonis, A., Drosos, T. and Sitaras, N.M. 2002. Protective role of oral antioxidant supplementation in ocular surface of diabetic patients. British Journal of Ophthalmology, 86(12), 1369–1373.

[67]  Flohé, L. 2005. Selenium, selenoproteins and vision. Nutrition and the Eye, 38, 89–102.

Ebrahim Alinia-Ahandani*, Zahra Alizadeh-Tarpoei, Farjad Rafeie,
Zeliha Selamoglu, Seyed Sara Heidari-Bazardehy,
Chonoor Mohammadi, Elifsena Canan Alp Arici

# Chapter 15
# Medicinal plants and pregnancy sickness

**Abstract:** Medicinal plants is today accepted as an undeniable part of the food basket
of the people of the world, including their safe use as medicine for promoting health
and preventing number of ailments. In the current chapter, some of the potent antioxidant herbs that play a protective role for the pregnancy system of the fetus and mothers are assessed. We have tried to gather related articles and information that will
provide an overview of this subject. Some aspects of the herbal usage by various studies are shown as graphs. Some effects of medicinal plants in nausea and vomiting in
pregnant cases were assessed, like: *Citrus limon, Citrus medica L., Malus domestica
Borkh*, etc. In another section, several specific health risks have been emphasized,
which demonstrate mutagenic and transgenic signs like *Mentha piperita L., Ruta cha-
lepensis / Ruta graveolens*, etc. as also the effective abortifacient influences related to
the use of herbs in pregnancy that are expressed in the ongoing literature. We conclude with an overview of some of the effects of herbs on women' periods in pregnancy and body cycle, with some recommendations.

# 1 Introduction

Complementary medicine is increasingly added in the healthcare of consumers. A financial assessment of storage to a health maintenance organization, where pharmaceuticals are replaced by medicinal plants has been appropriately carried out. This
center spent nearly US$ 1 million annually on selective serotonin reuptake inhibitors.
Researchers concluded that if *Hypericum* (herb in the family of Hypericaceae) was

*Corresponding author: Ebrahim Alinia-Ahandani**, Deputy of Food and Drug, Guilan University of
Medical Sciences, Rasht, Iran, email: dr.ebrahim.alinia@gmail.com
**Zahra Alizadeh-Tarpoei, Seyed Sara Heidari-Bazardehy,** Department of Biology, Faculty of Basic
Sciences, University of Guilan, Iran
**Farjad Rafeie,** Department of Biotechnology, University of Guilan, Rasht, Iran
**Zeliha Selamoglu,** Department of Medical Biology, Medicine Faculty, Nigde Omer Halisdemir Univer-
sity, Nigde, Türkiye; Western Caspian University, Baku, Azerbaijan; Khoja Akhmet Yassawi International
Kazakh-Turkish University, Faculty of Sciences, Department of Biology, Central Campus, Turkestan,
Kazakhstan
**Chonoor Mohammadi,** Biology Department, university of Soran-Kurdistan, Iraq
**Elifsena Canan Alp Arici,** Department of Obstetrics and Gynecology, Faculty of Medicine, Necmettin
Erbakan, University of Meram, Konya, Türkiye

https://doi.org/10.1515/9783111317601-015

meaningful for 25% of these cases, it was predicted there would be a significant saving – approximately, $250,000 in a year. In the United States of America, the assessed use in 1997 adult cases was 42% [1]. Herbal drugs are part of many healthcare protocols around the world. The WHO (World Health Organization) reported estimated that about 80% of the world's people rely on medicinal plants for their primary healthcare necessities [2, 3]. The WHO suggested that historic and cultural aspects of usage must be used along with new scientific research whenever the quality, security, and capability of herbal products are assessed. Research into herbs has stopped, since herbs cannot be patented. Herbal research beneficiaries are the medical and drug companies. In some countries like Germany, however, herbal research is basically supported by grants and contracts to colleges. Most herbal therapies have been used for long in different regions. European countries are considering herbs as effective cure in many cases [4–6]. Although the application of medicinal plants has quadrupled in the last 20 years, reports of toxicity caused by herbs or medicinal plants have remained at low levels. It should be noted that among the most popular and widely known medicinal plants, mint (96.7%), orange spring (89.6%), and cinnamon (85.7%) have been among the most significant. The most important reason for the use and spread of these plants is their effects during pregnancy as well as their safety with no complications (80.5%). Based on the mentioned statistics, half of the women (48%) had used at least one of these medicinal plants during their pregnancy, mint (31%), orange blossom (28.5%) and *Sisymbrium Sophia* (22%). They were among the most popular and widely used herbs. The use of medicinal plants was suggested to a majority of women through other people, including their mothers (37.8%). It should be noted that most women (95.7%) had a positive attitude toward the use of medicinal plants during their pregnancy. Within the definition of complementary and alternative medicine (CAM), herbal drugs are useful and could be used without any pre-assessment. Herbal medicines have the ability to elicit similar kinds of reactions as synthetic medicine since they include the total extracts or, more usually, expressed organs of plants (roots, rhizomes, leaves, and flowering heads) that include many active molecules [7]. In many countries, medicinal plants are sold as supplements without any license [8–11]. In a Canadian interview research containing 27 selected pregnant women, it was seen that the females considered medicinal plants safer than pharmaceuticals [12, 13]. An Italian survey of 1420 females demonstrated that phototherapy was increasing and that females focused on natural products; they felt that natural products were more beneficial, useful, and appropriate to control illness as they are seen as generally safer than usual medicines [14]. So, the best reason for herbal usage looks to be its security aspect – medicinal plants are considered to be safer than pharmaceuticals. It is clear that while pregnant females consider the possible risks of using medicine, they do not find that natural materials, if taken indirectly, can also have toxicity. This derives from the implicit belief that herbal therapies, from the point of view of being traditional, are totally safe; information of the potential risk aspects of many natural drugs in pregnancy is limited [15], and a few natural usages might be teratogenic in the models of human and animals. So, the Medicines and Healthcare Products Regulatory Agency (MHRA) focuses on the illnesses of pregnant women. Based on this knowledge, it is suggested that natural products be avoided during pregnancy period. In spite of this, many women use herbal products during pregnancy. It is important to understand why data on the extent of the use of natural medicines during pregnancy is scanty in some regions. We cannot explore many studies showing the medical consumption in pregnant females in some specific regions like Guilan and Shahrekord in Iran [16–19]. At least one medicinal plant (licensed or unlicensed ones) was used by around 79.6% of the cases in diet or in some form. The use of herbal drugs was remarkably varied between the higher-educated and older females. Some like the ginger herb, thyme herb, mint, and rosemary herb were the most used licensed medicinal plants. The most frequently unlicensed herbal remedies were mint, frankincense, and olive oil which are cultivated in many fields. This chapter explores the potential and the adaptability of some effective herbs during pregnancy as a source of nutrition as also their location in this subject along with an overview of their positive or negative effects, and suggesting its safe use.

# 2 Use of medicinal plants in pregnancy

As we know, herbal exercise with midwifery solutions include ancient links; medicinal plants have been applied and consumed for long to help in pregnancy and some child healthcare issues. Questionnaires of midwives and plans of midwifery in the US [20] have shown suggestions of medicinal plants to pregnant females; around half of them that answered used natural medicine in their diet – the most used supplements were blue cohosh (64%), black cohosh (45%), and raspberry leaf (63%). It seemed 64% of midwives studying planning reported receiving organized training of natural products. Here, 92% of the educational organizations showed the application of medicinal plants. The medicinal plants most regarded were blue and black cohosh, and red raspberry leaf. The harmful effects related to use of blue and black cohosh included nausea, risen meconium-stained fluid, and transient fetal tachycardia; no meaningful effects were seen from the use of raspberry leaves [20,21]. A questionnaire in the North Carolina region center showed that of the 57.3% who suggested complementary and alternative drugs, 73.2% suggestive natural therapies [22, 23]. Effects included nausea and vomiting, work stimulating, unrelaxed formation, lactation problems, and so on. In a subsequent report 37% of the midwives used medicinal plant remedies before using allopathic pharmaceutics, and 30% expressed belief in their safety compared to conventional medicine. Medicinal plants are a fact and healthcare professionals are recommending herbal products [24]. Pregnant females use natural products and healthcare agent would suggest herbal drugs, thinking that everything is natural and hence would be healthier and safer. The reproductive health of individuals may be harmed by using herbal applications. A simple method to guide us on this issue is to read data on usage along with observation – a lot of herbs have already been used by people, pregnant or not, Ethical effects and trends exist to show the safe usage of medicinal plants [25].

# 3 Medicinal herbs in pregnancy sickness

Pregnancy sickness, i.e., nausea and vomiting during pregnancy, is one of the most common issues for pregnant females. Prevalence of nausea is reported to be 50–80% while vomiting and retardation during pregnancy is estimated to be over 50% as per many sources [26]. The exact cause is unknown due to the involvement of multiple factors involving the rapid release of body hormones such as estrogen and human chorionic gonadotropin (HCG). Pregnancy sickness normally appears in 6–8 weeks of gestation and reaches its peak in the 9th week, and is rarely seen after the 20th week. Generally, black women, old-aged and less educated females, and those with lower social status are at high risk of pregnancy sickness. Severe cases of nausea and vomiting during pregnancy may be associated with maternal issues, unwanted effects on fetus, low birth weight, and maternal malnourishment. It may also affect the quality of life of pregnant females in term of family relationship, social behavior, as well as general health status [27]. Many antiemetic drugs currently in practice to treat pregnancy sickness include vitamin $B_6$, anti-dopaminergic drugs, serotonin antagonists, antihistamines, anticholinergic drugs, and others [28]. With the increasing safety concerns with antiemetic drugs, large number of consumers is shifting toward the use of alternative therapies, including medicinal herbs [29]. Some of the medicinal herbs that are reported to possess beneficial effects in relieving nausea and vomiting in pregnancy are discussed below. They include *Citrus limon* (L.) Osbeck, *Cydonia oblonga* Mill, *Elletaria cardamomum* (L.) Maton, *Mentha piperta* L., *Zingiber officinale* Rosc., and *Punica granatum* L.

## 3.1 *Citrus limon* (L.) Osbeck

*Citrus limon* (L.) Osbeck (lemon) belongs to the Rutaceae family. It is rich in volatile oils and bioactive components such as citric acid and flavonoids that make the plant capable of fighting certain ailments, while improving the general health of the host. A number of studies have demonstrated the beneficial effects of lemon plant in pregnancy-associated nausea and vomiting. The expectant mothers were supplemented with *C. lemon* aromatherapy or acupuncture, and the results showed significant improvement in nausea and vomiting in the aromatherapy group, suggesting *C. lemon* as one of the potential non-pharmacological interventions in pregnancy sickness [30]. In a randomized clinical trial, Yavari Kia et al. assessed the biological effects of lemon-based inhalation aromatherapy on nausea and vomiting in pregnancy, where

100 pregnant women are randomized to receive either lemon essential oils or placebo for 4 days. Significant difference was observed in the mean nausea and vomiting score in aromatherapy group on day 2nd and 4th ($p = 0.017$ and $p = 0.039$, respectively) [31]. In addition, lemon plant is considered safe in the scientific literature for use in expecting mothers and/or pregnancy.

## 3.2 *Cydonia oblonga* Mill

*Cydonia oblonga* Mill (quince) belongs to family Rosaceae, which is known as quite rich in volatile and phenolic compounds. In traditional medicine, quince has been used as one of the effective therapies in pregnancy-associated nauseating conditions. A human study demonstrated considerably decreased nausea and vomiting in subjects receiving quince syrup, as compared to the control group. While determining the comparative efficacy of quince and vitamin $B_6$ in pregnant females with mild-to-moderate nausea/vomiting, Jafari-Dehkordi et al. reported that quince syrup could be a more suitable choice in the management of pregnancy sickness, as the reduction of nausea and vomiting scores (from baseline throughout the treatment period) and change in symptoms were more marked in the quince group [32]. In addition, no adverse effects were observed with the quince syrup. Another study investigating the comparative efficacy of quince (10 mg after meals) and ranitidine (150 mg twice daily) syrups on gastroesophageal reflux disease (GERD) in pregnancy showed a similar efficacy for both treatments in the management of pregnancy-related GERD [33]. In four weeks of treatment, no significant differences were detected in both groups for the mean general symptom and mean major symptom scores, though the mean general symptom score was significantly lowered in the quince treated group after two weeks of treatment.

## 3.3 *Elletaria cardamom* (L.) Maton

*Elletaria cardamom* (L.) Maton (cardamom) medicinal plant is a very important member of the Zingiberaceae family, which is rich in terpenes, esters, and flavonoids. The essential oil of this plant has a long history in the management of gastrointestinal symptoms and stomachache. The combination of three essential oils (*Zingiber officinale, Elletaria cardamomum,* and *Artemisia dracunculus*) is reported to effectively reduce the risk of post-surgery nausea and vomiting, with up to 75% improvement in the condition [34]. Aromatherapy, based on inhalation of cardamom oils, is effective to relieve nausea associated with chemotherapy in cancer patients [35]. Ozgholy et al. demonstrated a significant reduction in the frequency and duration of nausea and vomiting in pregnant females in the randomized clinical trial, with the administration of *E. cardamom* (as compared to placebo ($p < 0.0001$) [36].

### 3.4 *Mentha piperta* L.

*Mentha piperta* L. (peppermint) belongs to the Lamiaceae family. It is a well-known medicinal herb used in relieving post-surgical nausea as well as analgesic, antiseptic, and anticlotting properties. One of the proposed mechanisms for its anti-emetic and spasmolytic effects is the inhibition of serotonin-induced muscle contractions in the gastrointestinal tract. It is also reported to have anesthetic actions on the gut wall that may stop nausea and vomiting. However, human studies demonstrated no significant effects of peppermint aromatherapy in pregnancy-associated nausea and vomiting. Pasha et al. reported no significant reduction in nausea and vomiting in pregnant subjects, with mint oil, as compared to the control group [37]. Similarly, Joulaeerad et al. found no significant difference between the anti-emetic effects of peppermint aromatherapy and almond oil (used as placebo) in women with nausea and vomiting [38].

### 3.5 *Zingiber officinale* Rosc

*Zingiber officinale* Rosc. (ginger) belongs to family Zingiberaceae, whose rhizome is commonly used as a spice in kitchen and as a folk medicine. It is one of the medicinal herbs used widely to relieve nausea and vomiting associated with pregnancy, post-surgery, or chemotherapy, though the exact anti-emetic mechanism is yet to known. A study showed no significant difference between the effectiveness of vitamin $B_6$ and ginger in pregnancy sickness management, and it was concluded that ginger was as effective as vitamin $B_6$ in mitigating nausea and/or vomiting in pregnancy [39]. Another study demonstrated greater effects with ginger capsules in the reduction of pregnancy sickness as compared to the control group [40]. Ozgoli et al., in a randomized clinical trial, demonstrated that daily ginger intake (1,000 mg/day) was significantly more effective than placebo in reducing pregnancy sickness (85% vs. 58%; $p <$ 0.01) [41]. Similarly, Basirat et al. found ginger biscuits more effective in the reduction of nausea and vomiting as compared to placebo in females with early pregnancy in a randomized clinical trial [42]. Fischer-Rasmussen et al. found significant higher relief scores for nausea and vomiting in women with hyperemesis gravidarum, with ginger treatment, in a double-blind randomized cross-over trial [43]. While comparing the effectiveness of ginger with other herbal therapies in nausea and vomiting, Modares et al. concluded that chamomile oral capsules are more effective in alleviating pregnancy sickness symptoms as compared to ginger and placebo [44]. Interestingly, a study reported similar effects (with no significant difference) in treatment with ginger and dimenhydrinate [45] and with ginger and doxylamine plus vitamin $B_6$ [46] in the reduction of nausea and vomiting episodes in pregnant women.

### 3.6 *Punica granatum* L

*Punica granatum* L. (pomegranate) belongs to the Lythraceae family, the seeds of which are using in traditional medicines in pregnant females for the management of nausea and vomiting. A single study showed a significant reduction of the PUQE-24 scores (Pregnancy-Unique Quantification of Emesis and Nausea) in subjects treated with a combination of pomegranate, spearmint, and vitamin $B_6$ as compared to vitamin $B_6$ alone [47]. However, no significant difference was found between the two groups for the duration of nausea and frequency of vomiting and retching.

# 4 Safety concerns

Despite originating from the natural (including edible) sources, medicinal herbs are not always safe to use. The bioactive components they contain may also carry some potential health risks for the host, including in cases of pregnancy and lactation. Adulterated or overdosed plants may possess potential teratogenic and mutagenic risks such as *Mentha piperita* L., *Ruta chalepensis/Ruta graveolens, Stryphnodendron polyphyllum, Lantana camara, Echinodorus macrophyllus* (Kunt) Mich., *Vernonia condensate, Tripterygium wilfordii, Rhazya stricta, Senecio latifolius,* and *Tabebuia* species [48–53]. Plants with abortifacient potential should be used cautiously and they may include *Salvia fruticosa* Mill., *Senna alexandrina* Mill., *Rhamnus purshiana* DC, *Rhamnus catharticus, Mentha pulegium* L., *Mentha arvensis* L., *Chenopodium ambrosioides* L., and *Artemisia absinthium* L. Some plants could also possess Caution should be exercised in using certain plants, such as *Allium sativum* L., *Rhamnus purshiana* DC., *Rhamnus catharticus, Senna alexandrina* Mill., in lactating mothers because of their potential risk to the neonates or infants – they can cause cramps and diarrhea in infants while some others like *Glycine max* (L.) Merr., *Maytenus aquifolium* Mart., and *Maytenus ilicifolia* Mart. ex Reissek may decrease milk production [54–58].

# 5 Summary

In conclusion, herbal medicines such as lemon, quince, cardamom, ginger, and pomegranate provide effective and safe medical alternatives for treating pregnant women with mild-to-moderate pregnancy sickness. These herbs not only provide solution for effective treatment of nausea and vomiting but also reduce the burden of conventional medications; they are likely to reduce the incidence of adverse effects. Results suggested that ginger is the most effective herbal medicine as compared to other alternative therapies; it shows high efficacy in treating pregnancy-associated nausea and vomiting as compared to vitamin $B_6$ and some other conventional anti-emetic drugs.

More importantly, the safety profile and potential health risks should always be considered before initiating therapies with medicinal herbs in pregnant females, as some of the herbs may carry serious health issues either to the female or to the newborn baby. Hence, it is always suggested to limit the self-usage of medications (including medicinal herbs) and seek the advice from healthcare practitioners before their intake. Moreover, preclinical and clinical studies of medicinal herbs using pregnancy experimental models or pregnant females are still very scarce and it could be suggested to expand the literature by conducting human studies using more robust randomized clinical trials on a larger scale before considering them for clinical use.

**Acknowledgment:** I would like to thank Miss. Pourhashem for her kind cooperation in preparation and support of editing the article.

# References

[1]    Adams, C. and Cannell, S. 2001. Women's beliefs about "natural" hormones and natural hormone replacement therapy. Menopause, 8(6), 433–440.

[2]    Akerele, O. 1993. Summary of WHO guidelines for the assessment of herbal medicines. Herbal Gram, 28(13), 13–19.

[3]    Alinia-Ahandani, E. and Sheydaei, M. 2020. Overview of the introduction to the new coronavirus (Covid19): A Review. Journal of Biomedical Science and Research, 6(2), 14–20.

[4]    Alinia-Ahandani, E., Malekirad, A.A., Nazem, H., Fazilati, M., Salavati, H. and Rezaei, M. 2021. Assessment of some toxic metals in Ziziphora (Ziziphora persica) obtained from local market in Lahijan, Northern Iran. Annals of Military and Health Sciences Research, 19(4), e119991.

[5]    Alinia-Ahandani, E. 2018. Medicinal plants and their usages in cancer. Journal of Pharmaceutical Sciences and Research, 10, 2.

[6]    Alinia-Ahandani, E. 2018. Medicinal plants effective on pregnancy, infections during pregnancy, and fetal infections. Journal of Pharmaceutical Sciences and Research, 10, 3.

[7]    Ernst, E. 2001. Complementary medicine: Its hidden risks. Diabetes Care, 24(8), 1486–1488.

[8]    Alinia-Ahandani, E., Boghozian, A. and Alizadeh, Z. 2017. New approaches of some herbs used for reproductive issues in the world: short review. Hospital, 17(1), 10.

[9]    Alinia-Ahandani, E., Fazilati, M., Boghozian, A. and Alinia-Ahandani, M. 2019. Effect of ultraviolet (UV) radiation bonds on growth and chlorophyll content of Dracocephalummoldavica L herb. Journal of Biomolecular Research & Therapeutics, 8(1), 1–4.

[10]    Alinia-Ahandani, E., Kafshdar-Jalali, H. and Mohammadi, C. 2019. Opened approaches on treatment and Herbs' location in Iran. American Journal of Biomedical Science and Research, 5(5), 394–397.

[11]    Dass, S., Ramawat, K.G. and Mathur, M. 2009. The chemical diversity of bioactive molecules and therapeutic potential of medicinal plants. In Ramawat, K.G. (Ed.), Herbal Drugs: Ethnomedicine to Modern Medicine. Berlin, Heidelberg: Springer, pp. 7–32.

[12]    Alinia-Ahandani, E., Nazem, H., Malekirad, A.A. and Fazilati, M. 2022. The safety evaluation of toxic elements in medicinal plants: A systematic review. Journal of Human Environment and Health Promotion, 8(2), 62–68.

[13]    Westfall, R.E. 2003. Herbal healing in pregnancy: women's experiences. Journal of Herbal Pharmacotherapy, 3(4), 17–39.

[14]   Bacchini, M., Cuzzolin, L., Camerlengo, T., Velo, G. and Benoni, G. 2008. Phytotherapic compounds. Drug Safety, 31(5), 424–427.

[15]   Lacroix, I., Damase-Michel, C., Lapeyre-Mestre, M. and Montastruc, J.L. 2000. Prescription of drugs during pregnancy in France. The Lancet, 356(9243), 1735–1736.

[16]   Alinia-Ahandani, E., Fazilati, M., Alizadeh, Z. and Boghozian, A. 2018. The introduction of some mushrooms as an effective source of medicines in Iran Northern. Biology and Medicine, 10(5), 1–5.

[17]   Esmaeilzadeh, M. and Moradi, B. 2017. Medicinal herbs with side effects during pregnancy-an evidence-based review article. The Iranian Journal of Obstetrics, Gynecology and Infertility, 20, 9–25.

[18]   Pakrashi, A. and Bhattacharya, N. 1977. Abortifacient principle of Achyranthesaspera Linn. Indian Journal of Experimental Biology, 15(10), 856–858.

[19]   Sereshti, M., Azari, P., Rafieian-Kopaei, M. and Kheiri, S. 2006. Use of herbal medicines by pregnant women in Shahr-e-Kord. Journal of Reproduction & Infertility, 7(2), 125–131.

[20]   McFarlin, B.L., Gibson, M.H., O'Rear, J. and Harman, P. 1999. A national survey of herbal preparation use by nurse-midwives for labor stimulation: review of the literature and recommendations for practice. Journal of Nurse-Midwifery, 44(3), 205–216.

[21]   Maymunah, A.O., Kehinde, O., Abidoye, G. and Oluwatosin, A. 2014. Hypercholesterolaemia in pregnancy as a predictor of adverse pregnancy outcome. African Health Sciences, 14(4), 967–973.

[22]   Allaire, A.D., Moos, M.K. and Wells, S.R. 2000. Complementary and alternative medicine in pregnancy: a survey of North Carolina certified nurse-midwives. Obstetrics & Gynecology, 95(1), 19–23.

[23]   Eisenberg, D.M., Davis, R.B., Ettner, S.L., Appel, S., Wilkey, S., Van Rompay, M. and Kessler, R.C. 1998. Trends in alternative medicine use in the United States, 1990–1997: results of a follow-up national survey. The Journal of the American Medical Association, 280(18), 1569–1575.

[24]   Gallo, M., Sarkar, M., Au, W., Pietrzak, K., Comas, B., Smith, M., Jaeger, T.V., Einarson, A. and Koren, G. 2000. Pregnancy outcome following gestational exposure to echinacea: A prospective controlled study. Archives of Internal Medicine, 160(20), 3141–3143.

[25]   Rodríguez, F.M., Mourelle, J.F. and Gutiérrez, Z.P. 1996. Actividad espasmolítica del extracto fluido de Matricaria recutita (manzanilla) en órganos aislados. Revista Cubana de Plantas Medicinales, 1(1), 19–24.

[26]   Matthews, A., Haas, D.M., O'Mathúna, D.P. and Dowswell, T. 2015. Interventions for nausea and vomiting in early pregnancy. Cochrane Database of Systematic Reviews, 2015(9), CD007575.

[27]   Khorasani, F., Aryan, H., Sobhi, A., Aryan, R., Abavi-Sani, A., Ghazanfarpour, M., Saeidi, M. and Rajab Dizavandi, F. 2020. A systematic review of the efficacy of alternative medicine in the treatment of nausea and vomiting of pregnancy. Journal of Obstetrics and Gynaecology, 40(1), 10–19.

[28]   Thomas, B., Valappila, P. and Rouf, A. 2015. Medication used in nausea and vomiting of pregnancy – a review of safety and efficacy. GynecolObstet (Sunnyvale), 5(270), 2161–2932.

[29]   Anderka, M., Mitchell, A.A., Louik, C., Werler, M.M., Hernández-Diaz, S. and Rasmussen, S.A. 2012. National Birth Defects Prevention Study. Medications used to treat nausea and vomiting of pregnancy and the risk of selected birth defects. Birth Defects Research Part A: Clinical and Molecular Teratology, 94(1), 22–30.

[30]   Nahdiana, N., Cholifah, S., Purwanti, Y. and Widowati, H. 2023. The Role of Citrus Lemon Aromatherapy in Alleviating Pregnancy-Induced Nausea and Vomiting: A Food Science Perspective. IOP Conference Series: Earth and Environmental Science, 1242(1), 12025.

[31]   Yavari Kia, P., Safajou, F., Shahnazi, M. and Nazemiyeh, H. 2014. The effect of lemon inhalation aromatherapy on nausea and vomiting of pregnancy: A double-blinded, randomized, controlled clinical trial. Iranian Red Crescent Medical Journal, 16(3), e14360.

[32]   Jafari-Dehkordi, E., Hashem-Dabaghian, F., Aliasl, F., Aliasl, J., Taghavi-Shirazi, M., Sadeghpour, O., Sohrabvand, F., Minaei, B. and Ghods, R. 2017. Comparison of quince with vitamin B6 for treatment

of nausea and vomiting in pregnancy: a randomised clinical trial. Journal of Obstetrics and Gynaecology, 37(8), 1048–1052.

[33] Shakeri, A., Hashempur, M.H., Mojibian, M., Aliasl, F., Bioos, S. and Nejatbakhsh, F. 2018. A comparative study of ranitidine and quince (*Cydonia oblonga* mill) sauce on gastroesophageal reflux disease (GERD) in pregnancy: a randomised, open-label, active-controlled clinical trial. Journal of Obstetrics and Gynaecology, 38(7), 899–905.

[34] De Pradier, E. 2006. A trial of a mixture of three essential oils in the treatment of postoperative nausea and vomiting. International Journal of Aromatherapy, 16(1), 15–20.

[35] Khalili, Z., Khatiban, M., Faradmal, J., Abbasi, M., Zeraati, F. and Khazaei, A. 2014. Effect of cardamom aromas on the chemotherapy-induced nausea and vomiting in cancer patients. Avicenna Journal of Nursing and Midwifery Care, 22(3), 64–73.

[36] Ozgoli, G., Gharayagh Zandi, M., Nazem Ekbatani, N., Allavi, H. and Moattar, F. 2015. Cardamom powder effect on nausea and vomiting during pregnancy. Complementary Medicine Journal, 14, 1056–1076.

[37] Pasha, H., Behmanesh, F., Mohsenzadeh, F., Hajahmadi, M. and Moghadamnia, A.A. 2012. Study of the effect of mint oil on nausea and vomiting during pregnancy. Iranian Red Crescent Medical Journal, 14(11), 727.

[38] Joulaeerad, N., Ozgoli, G., Hajimehdipoor, H., Ghasemi, E. and Salehimoghaddam, F. 2018. Effect of aromatherapy with peppermint oil on the severity of nausea and vomiting in pregnancy: a single-blind, randomized, placebo-controlled trial. Journal of Reproduction & Infertility, 19(1), 32.

[39] Firouzbakht, M., Nikpour, M., Jamali, B. and Omidvar, S. 2014. Comparison of ginger with vitamin B6 in relieving nausea and vomiting during pregnancy. Ayu, 35(3), 289.

[40] Saberi, F., Sadat, Z., Abedzadeh-Kalahroudi, M. and Taebi, M. 2014. Effect of ginger on relieving nausea and vomiting in pregnancy: a randomized, placebo-controlled trial. Nursing and Midwifery Studies, 3(1), e11841.

[41] Ozgoli, G., Goli, M. and Simbar, M. 2009. Effects of ginger capsules on pregnancy, nausea, and vomiting. The Journal of Alternative and Complementary Medicine, 15(3), 243–246.

[42] Basirat, Z., Moghadamnia, A., Kashifard, M. and Sarifi-Razavi, A. 2009. The effect of ginger biscuit on nausea and vomiting in early pregnancy. Acta Medica Iranica, 4(1), 51–56.

[43] Fischer-Rasmussen, W., Kjær, S.K., Dahl, C. and Asping, U. 1991. Ginger treatment of hyperemesis gravidarum. European Journal of Obstetrics & Gynecology and Reproductive Biology, 38(1), 19–24.

[44] Modares, M., Besharat, S. and Mahmoudi, M. 2012. Effect of Ginger and Chamomile capsules on nausea and vomiting in pregnancy. Journal of Gorgan University of Medical Sciences, *14*(1), 46–51.

[45] Pongrojpaw, D., Somprasit, C. and Chanthasenanont, A. 2007. A randomized comparison of ginger and dimenhydrinate in the treatment of nausea and vomiting in pregnancy. Journal-Medical Association of Thailand, 90(9), 1703.

[46] Biswas, S.C., Dey, R., Kamliya, G.S., Bal, R., Hazra, A. and Tripathi, S.K. 2011. A single-masked, randomized, controlled trial of ginger extract in the treatment of nausea and vomiting of pregnancy. Journal International Medical Sciences Academy, 24, 167–169.

[47] Abdolhosseini, S., Hashem-Dabaghian, F., Mokaberinejad, R., Sadeghpour, O. and Mehrabani, M. 2017. Effects of pomegranate and spearmint syrup on nausea and vomiting during pregnancy: a randomized controlled clinical trial. Iranian Red Crescent Medical Journal, 19(10), e13542.

[48] Almeida, F.C. and Lemonica, I.P. 2000. The toxic effects of Coleus barbatus B. on the different periods of pregnancy in rats. Journal of Ethnopharmacology, 73(1–2), 53–60.

[49] Edraki, M., Moghaddampour, I.M., Alinia-Ahandani, E., Keivani, M.B. and Sheydaei, M. 2021. Ginger intercalated sodium montmorillonite nano clay: assembly, characterization, and investigation antimicrobial properties. Chemical Review and Letters, 4(2), 120–129.

[50] Shewamene, Z., Dune, T. and Smith, C.A. 2017. The use of traditional medicine in maternity care among African women in Africa and the diaspora: a systematic review. BMC Complementary and Alternative Medicine, 17(1), 1–16.

[51] Sheydaei, M. and Alinia-Ahandani, E. 2021. Breast cancer and the role of polymer-carriers in treatment. Biomedical Journal of Scientific & Technical Research, 34(5), 27057–27061.

[52] Sheydaei, M. and Alinia-Ahandani, E. 2020. Cancer and the role of polymeric-carriers in diagnosis and treatment. Journal of Fasa University of Medical Sciences, 10(3), 2408–2421.

[53] Yazdi, N., Salehi, A., Vojoud, M., Sharifi, M.H. and Hoseinkhani, A. 2019. Use of complementary and alternative medicine in pregnant women: a cross-sectional survey in the south of Iran. Journal of Integrative Medicine, 17(6), 392–395.

[54] Alinia-Ahandani, E., Sheydaei, M., Shirani-Bidabadi, B. and Alizadeh-Terepoei, Z. 2020. Some effective medicinal plants on cardiovascular diseaes in Iran-a review. Journal of Global Trends in Pharmaceutical Sciences, 11(3), 8021–8033.

[55] Ali-Shtayeh, M.S., Jamous, R.M. and Jamous, R.M. 2015. Plants used during pregnancy, childbirth, postpartum and infant healthcare in Palestine. Complementary Therapies in Clinical Practice, 21(2), 84–93.

[56] Costa, K.C.D.S., Bezerra, S.B., Norte, C.M., Nunes, L.M.N. and Olinda, T.M.D. 2012. Medicinal plants with teratogenic potential: current considerations. Brazilian Journal of Pharmaceutical Sciences, 48, 427–433.

[57] Dahar, B., Khavasi, N., Kamali, K. and Rashtchi, V. 2022. Comparison of the Effect of Cydonia oblonga and PhyllanthusEmblica on Gastric Residual Volume and Pulmonary Aspiration in Patients under Mechanical Ventilation in Mousavi Hospital ICU of Zanjan in 2020. Journal of Advances in Medical and Biomedical Research, 30(141), 357–364.

[58] Orief, Y.I., Farghaly, N.F. and Ibrahim, M.I.A. 2014. Use of herbal medicines among pregnant women attending family health centers in Alexandria. Middle East Fertility Society Journal, 19(1), 42–50.

Paula Barciela,[†] Ana Perez-Vazquez,[†] Sepidar Seyyedi-Mansour,
Pauline Donn, Franklin Chamorro, Maria Carpena, Maria Fraga-Corral*,
Jesus Simal-Gandara* and Miguel A. Prieto*

# Chapter 16
# Toxicity and safety of nutraceuticals

**Abstract:** Currently, consumers are more conscious about the importance of having a good diet and lifestyle; so the demand for products with potential health benefits has increased. Nutraceuticals can be classified according to different criteria. Based on their chemical structure, they can be divided into phenolic compounds, fatty acids and structural lipids, carbohydrates and amino-acids derivatives, minerals, and terpenoids derivatives. All these compounds are linked to different health benefits, such as antioxidant, anticancer and anti-inflammatory activity, prevention of neurodegenerative and cardiovascular diseases, or health bones maintenance. However, the potential positive impact of these compounds when incorporated in the human body is limited because of their

[†]These authors contributed equally to the publication.

*Corresponding author: Jesus Simal-Gandara, Nutrition and Bromatology Group, Department of Analytical Chemistry and Food Science, Instituto de Agroecoloxia e Alimentacion (IAA) CITEXVI, Universidade de Vigo, 36310 Vigo, Spain, e-mail: jsimal@uvigo.es
*Corresponding author: Maria Fraga-Corral, Nutrition and Bromatology Group, Department of Analytical Chemistry and Food Science, Instituto de Agroecoloxia e Alimentacion (IAA) CITEXVI, Universidade de Vigo, 36310 Vigo, Spain, e-mail: mfraga@uvigo.es
*Corresponding author: Miguel A. Prieto, Nutrition and Bromatology Group, Department of Analytical Chemistry and Food Science, Instituto de Agroecoloxia e Alimentacion (IAA) CITEXVI, Universidade de Vigo, 36310 Vigo, Spain, e-mail: mprieto@uvigo.es
Paula Barciela, Nutrition and Bromatology Group, Department of Analytical Chemistry and Food Science, Instituto de Agroecoloxia e Alimentacion (IAA) CITEXVI, Universidade de Vigo, 36310 Vigo, Spain, e-mail: paula.barciela.alvarez@alumnos.uvigo.es
Ana Perez-Vazquez, Nutrition and Bromatology Group, Department of Analytical Chemistry and Food Science, Instituto de Agroecoloxia e Alimentacion (IAA) CITEXVI, Universidade de Vigo, 36310 Vigo, Spain, e-mail: anaperezvaz@alumnos.uvigo.es
Sepidar Seyyedi-Mansour, Nutrition and Bromatology Group, Department of Analytical Chemistry and Food Science, Instituto de Agroecoloxia e Alimentacion (IAA) CITEXVI, Universidade de Vigo, 36310 Vigo, Spain, e-mail: sepidar.seyyedi@uvigo.es
Pauline Donn, Nutrition and Bromatology Group, Department of Analytical Chemistry and Food Science, Instituto de Agroecoloxia e Alimentacion (IAA) CITEXVI, Universidade de Vigo, 36310 Vigo, Spain, e-mail: donn.pauline@uvigo.es
Franklin Chamorro, Nutrition and Bromatology Group, Department of Analytical Chemistry and Food Science, Instituto de Agroecoloxia e Alimentacion (IAA) CITEXVI, Universidade de Vigo, 36310 Vigo, Spain, e-mail: franklin.noel.chamorro@uvigo.es
Maria Carpena, Nutrition and Bromatology Group, Department of Analytical Chemistry and Food Science, Instituto de Agroecoloxia e Alimentacion (IAA) CITEXVI, Universidade de Vigo, 36310 Vigo, Spain, e-mail: mcarpena@uvigo.es

The original version of this chapter was revised. Unfortunately, the names and institutional addresses of several authors were incorrect in the original publication. This has been corrected. We apologize for the mistake.
https://doi.org/10.1515/9783111317601-016

bioavailability, which is mainly affected because of the hydrophobic properties of these compounds. In this way, different delivery systems were developed as suitable pathways for the bioavailability improvement of nutraceuticals. The most successful delivery systems to incorporate nutraceuticals include micro and nanoemulsions, micelles, nanoparticles, hydrogels, and solid-lipid nanoparticles. They achieved several improvements such as higher bioavailability, bioactivity maintenance, shelf-life improvement, or incorporation of the nutraceutical in a target system/organ. In this way, and considering the increasing demand for these products, the development of regulation is needed so that consumers' safety is assured, although any country has developed it yet. Thus, considering the increasing demand for these compounds, the aim of this study was to compile the available current data of nutraceuticals and also their safety concerns.

# 1 Introduction: bioavailability and delivery systems

The bioactivity of nutraceuticals has raised the interest of the scientific community for preventive treatments as also as health booster. Indeed, they are considered endogenous, meaning they have the ability to increase activity in healthy people [1]. Several health benefits have been related to nutraceuticals, including eye, bone, cardiovascular and mental health, whereas their deficiency may lead to different health issues such as neurological diseases or cardiovascular and inflammatory disorders [2]. Among the multiple compounds providing health benefits, consumption of soy proteins leads to lower lipoprotein and lipid concentration in plasma, reducing the risk of coronary diseases, while soy isoflavones are used for bone health maintenance. Polyunsaturated fatty acids can also help with balance bone density. Sesame oil is also a suitable source of polyunsaturated fatty acids, which are linked to antidiabetic, anti-inflammatory and hypercholesterolemic activities. For eye health maintenance, carotenoid consumption is key. In this way, lutein and zeaxanthin play an important role, being present in spinach, parsley and kale plants.

However, the health benefits of nutraceuticals are limited by their hydrophobicity, low permeability, pre-systemic metabolism, and low *in vivo* stability, leading to low bioavailability. Hence, their direct incorporation into food products is difficult [2–4], even when they are easily soluble in water, such as in the case of tannins and flavonoids that show difficulties to be absorbed [1], and so their health benefits become limited. Indeed, the bioavailability of nutraceuticals can be compromised by different factors, such as the chemical characteristics of the molecule/s, the gastric residence time, the low permeability or solubility in the gastrointestinal (GI) tract, or the stability during food processing and storage [3, 4]. In general terms, bioavailability refers to the fraction of the food component that, once ingested, is transported through blood following the ingestion process. Mathematically, bioavailability is defined as eq. (1):

$$F = F_B \times F_T \times F_M \qquad (1)$$

$F_B$ being the bio-accessible fraction released into the intestinal fluids; $F_T$, the transported fraction through the intestinal epithelium; and $F_M$, the unmetabolized fraction that is not broken down before reaching the blood [4]. The effectiveness of nutraceuticals depends on the bioavailability preservation; so, the therapeutic or physiologic benefit expected for its consumption is appropriate [3]. In this way, different delivery systems have been introduced so that the bioavailability of these compounds could be increased. There are several parameters affecting the effect of delivery systems on bioavailability, such as particle size and shape, vehicle composition, crystallinity, and food matrix composition of the surrounding area. Large compounds inhibit intestinal absorptions; so, a low particle size is required to increase the surface area of the bioactive compound and consequently, a narrow particle distribution is achieved [5, 6]. In this way, micro and nanoscale particles have demonstrated the improvement of oral dose efficiency of compounds. Micro and nanoemulsions, solid lipid nanoparticles (SLNs), and nano dispersions are usually selected as suitable delivery systems because of the size limitation of gastric absorption [5]. Vehicle composition may be a key factor affecting bio-accessibility and bioavailability of the bioactive compounds delivered [4]. In this way, the wall type material has been demonstrated as a factor influencing the *in vitro* bio-accessibility results of hydrophobic compounds. Moreover, the bioavailability of the compounds inside the delivery system is more protected from degradation during processing, storage, and digestion. Besides, by using delivery systems, unpleasant organoleptic changes and interactions between other food components can be avoided. For instance, the bioavailability increment was proved in a *in vivo* study where β-lactoglobulin (β-Lg) was incorporated as a vehicle for epigallocatechin-3-gallate (EGCG). Results showed the positive effect of β-Lg wall since EGCG was protected from oxidation at neutral pH during digestion [7]. The physical state of the compounds is one parameter affecting bioavailability. A crystalline solid at body temperature shows less bioavailability than an amorphous solid, as the liquid form is the most bioavailable due to the solubility phenomena [8]. Furthermore, the physical state of the delivery system has been related to toxicity. Thus, $TiO_2$ nanoparticles (NPs) in the rutile phase induced DNA damage, lipid peroxidation, and micronuclei production, while anatase phase did not lead to toxic effects [9]. In a study run by Clifford et al., the influence of food matrix composition in tea polyphenols bioavailability was proved [10]. Therefore, to guarantee the nutraceutical bioavailability, it is important to assess all the factors involved in its modulation. Moreover, when a delivery system is designed, both the structure and base component need to be considered. Thus, delivery systems can be classified in two groups: lipid or surfactant-based systems and biopolymer-based systems. In Figure 1, a delivery system's structure is represented, considering this classification. Emulsions can be formed by oil in water, water in oil, liquid in liquid or liquid in solid phase. Both micro and nanoemulsions are more stable than large-scale emulsions. Moreover, these structures can incorporate various nutraceuticals at once [3]. The bioavailability of micro and nanoemulsions vary, depending on the particle size, the emulsifier used, and the carrier oil composition [4]. Liposomes are

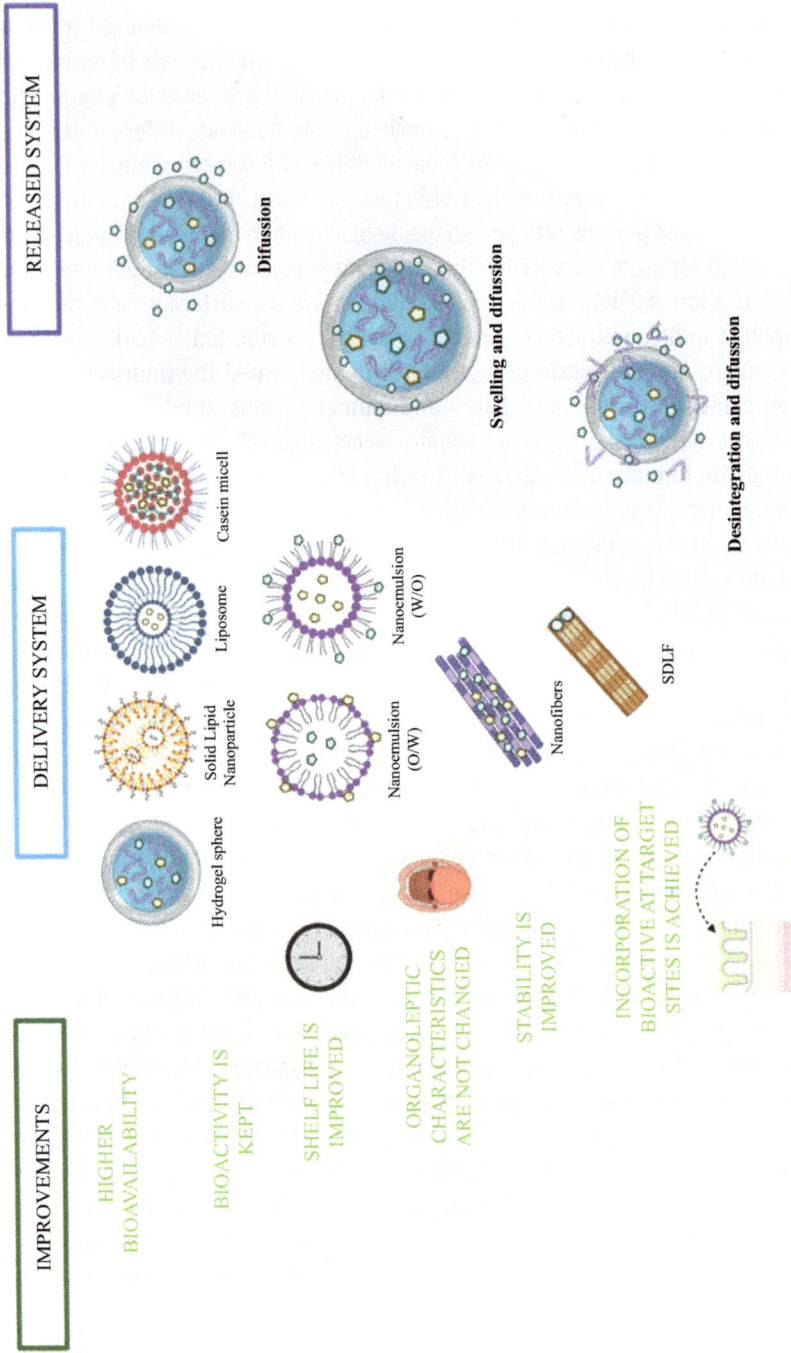

**Figure 1:** Improvements achieved by using delivery systems for the incorporation of nutraceuticals and the released mechanisms associated. O: oil; W: water; SDLF self-dispersing lipid formulations).

sphere-shaped vesicles formed by unilamellar or multilamellar bilayers of lipids containing an aqueous phase at the core, being capable of incorporating both hydrophilic and lipophilic compounds [3, 5]. Moreover, these structures have shown high biocompatibility, biodegradability, and low toxicity [3]. The encapsulation of nanofibers by using electrospin nanofibers has also been studied, being a suitable method to produce a delivery system for nutraceuticals [2]. Nanofibers are one-dimensional structured materials at the nanoscale level, characterized by an interconnected pore structure. These structures form amorphous solid dispersions, leading to an increment in the solubility and stability of bioactive compounds; so their application in nutraceuticals is promising. Hydrogels are hydrophilic networks made of biopolymers (proteins or polysaccharides), formed by both covalent and hydrogen bonds, van der Waals interactions, or physical entanglements [3, 11]. Moreover, hydrogels are interesting for the nutraceuticals' encapsulation, since they are structures with high specific release capacity.

The most common incorporated components as nutraceuticals are polyphenols and vitamins, with the main purpose of food fortification. However, and as it is compiled in Table 1, several compounds have been studied to be incorporated in food products, using different delivery system structures to increase bioavailability, inhibit negative organoleptic effects, and enhance the bioactive compounds' health benefits. For instance, the incorporation of vitamin D as a nutraceutical using micelles as delivery systems leads to cardiovascular and autoimmune improvement, as well as an anticancer function and in bone health improvement [12, 13]. Similarly, the encapsulation of vitamin E in a nanoemulsion as a delivery system was also linked to anticancer and cardiovascular protective activity [14].

**Table 1:** Nutraceutical's delivery systems, health benefits, and food applications.

| Nutraceutical | Delivery system | Agent carrier | Health benefits | Food application | Ref. |
|---|---|---|---|---|---|
| Quercetin | SDLF | Monoglycerides | Potent antioxidant and anti-inflammatory agent | Controlled release of aroma, Creation of flavors, Structure of food products | [3] |
| EGCG | NPs | β-lactoglobulin | Prevention of the metabolic syndrome, cancer, and neurodegenerative disease | Food fortification | [7] |
| Vitamin D3 | Micelles | Casein | Improvement of bone health and prevention of cancer, cardiovascular, and autoimmune diseases | Food fortification | [12] |
| Vitamin D | Micelles | Protein | Cardiovascular health, cancer prevention, immune regulation, and prevention of autoimmune disease | Food fortification | [13] |

**Table 1** (continued)

| Nutraceutical | Delivery system | Agent carrier | Health benefits | Food application | Ref. |
|---|---|---|---|---|---|
| Vitamin E | Nanoemulsion | Medium chain triacylglycerols | Antioxidant and non-antioxidant biological activities | Food fortification | [14] |
| Iron | NPs | Lactoferrin | Essential mineral | Food fortification | [15] |
| Rutin | NPs | Mushroom and date β-glucans | Anti-obesity, antioxidant, and antimicrobial activities | Food fortification | [16] |
| β-carotene | Nanoemulsion | Low chain triacylglycerols | Prevention of cancer, cardiovascular disease, and age-related macular degeneration; prevention of cataract | Food fortification | [17] |
| (+)-catechin | Liposome | Elastic liposomes | Antioxidant and neuroprotective activities | Food fortification | [18] |
| Vitamin A | Liposome | Lecithin and cholesterol | Potent antioxidant activity | Control of undesirable effects in flavor and odor | [19] |

SDLF, self-dispersing lipid formulations; NPs, nanoparticles; EGCG, epigallocatechin gallate.

Therefore, considering the current available data, the use of nutraceuticals may be a promising pathway for the good maintenance of human health since they can be incorporated in human diet, preventing different diseases. Besides, their inclusion into delivery systems is highly recommended to protect the nutraceuticals against adverse external factors by correctly incorporating them into food matrices; controlling their release, maximizing their functionality, and avoiding negative organoleptic changes in the final products. Nevertheless, it is crucial to keep in mind that the structure of the delivery systems is a key factor to consider for ensuring high bioavailability of nutraceuticals.

# 2 Safety and legal status of nutraceuticals

## 2.1 Safety assessment

Food regulation of nutraceuticals is perceived as challenging because of the lack of authoritative controls [20, 21]. On the one hand, stakeholders generally fail to view food regulations in the same manner as the regulator does and consumers frequently are uncertain of their enforceability due to unfamiliarity [22]. Nutraceuticals are apparently exempt from some preclinical efficacy and toxicity testing assessments by regulatory agencies [21]. As the companies are not obliged to carry out clinical trials on the efficacy and safety of the products, the responsibility in USA primarily falls on the Food and Drug Administration (FDA), which checks their effectiveness, and authorizes them to be marketed [23]. Legal characterization of nutraceuticals in the European Union is carried out according to the body's effects on the organism. Where a product helps to support the maintenance of healthy tissues or organs, it can be classified as a food ingredient; however, if it is capable of altering a physiological process, it will probably be considered a medicinal substance [24]. Yet, European legislation, overseen by the European Food Safety Authority (EFSA), does not make any reference to or officially recognize the concept of nutraceuticals. By way of example, the updated EU Novel Foods Regulation EU No. 2015/2283 does not distinguish food supplements from nutraceuticals when applying for a health claim for new food products. This regulation just refers to food supplements, which are defined as concentrated sources of nutrients (vitamins and minerals) or other substances with nutritional or physiological effect [25]. By 2017, EFSA established the tolerable upper intake level (UL) for some nutrients (vitamins and minerals), which means their maximum level of total chronic intake is judged to be unlikely to pose a risk of adverse health effects in humans [26].

From a normative viewpoint, nutraceuticals are not visibly classified as either food or pharmaceuticals, but rather occupy an in-between gray zone [24, 27]. Nevertheless, they are being marketed under the vague and dissimilar definition of "food-pharmaceuticals" and advertising about their ability to prevent disorders is misleading, which increases ambiguity and thus consumer confusion [28]. Indeed, apart from the safety facts, curiously, surveys point to the fact that consumers are willing to spend a premium price to acquire these products in the markets [29]. In this sense, health claims for nutraceuticals must be in line with the same regulatory requirements (efficacy, security, and quality testing) established for other medicinal products [20, 23], to avoid the confusion of consumers between nutraceuticals and food supplements, which are available as over-the-counter (OTC) products [30]. For many cases, the fact that nutraceuticals are OTC could be a possible risk to be taken into account, as the statement "all natural", regarding naturally occurring substances in a food-derived product, does not necessarily imply "safe" [23]. At the same time, nutraceuticals frequently base their health benefits on being sourced from food or a part of it and, thus, are in many cases, recognized as safe or Generally Recognized as Safe (GRAS) [22]. And, even though the Rapid Alert System for Food and Feed (RASFF) is also at the disposal of the European Union to handle the monitoring, detection, and reaction to any risk in these and other foods [27], it is possible that adverse effects on the health of consumers are not ruled out even though nutraceuticals are well-known for their safety, being perceived as lower risk in the likelihood of side effects [21, 31].

On the other hand, the category of nutraceuticals is defined as a range of products, from single ingredients, nutrients and food components, to processed foods, probiotics or prebiotics, functional or herbal products and, above all, dietary supplements [32]. Thus, nutraceuticals and functional foods or supplements are made up of active substances extracted from foods of plant or animal sources that may compensate and/or have positive health effects. Unlike the other two terms, nutraceuticals are concentrated and delivered in the proper dosage form, and must have medicinal efficacy as well as nutritional value [33]. It is possible to draw a compulsory common characteristic for drugs, functional foods, and nutraceuticals, which is their preventative rather than curative nature for a variety of illnesses, even if it is currently a top priority to draw a distinctive boundary between each of them [34]. Evidence of the potential health benefits of nutraceuticals is mostly supported by epidemiological studies, which, based on their findings, reveal linkages between nutraceuticals and health benefits [35]. For instance, evidence has been reported that they might work as cellular and functional modulators, supporting the homeostasis of physiological pathways [36]. There are also certain nutraceuticals in foods and plants that can manifest immune modulating activities [37]. The most used nutraceuticals are compounds extracted from fruits and vegetables with antioxidant or anti-inflammatory properties, which protect against chronic diseases, diabetes, cancer, and so on [38]. Among the most popular nutraceuticals are flavonoids, plant pigments such as anthocyanins from fruits as berries, flavonols from dark chocolate, polyphenols such as resveratrol from red grapes, and catechins from tea and quercetin. Besides, numerous studies have related the significant preventive and healing ability of nutraceuticals obtained from spices such as garlic, coriander, cloves, onion, ginger, pepper, or turmeric against several chronic illnesses by directly acting through the chronic inflammatory pathways [39]. Some of the commercialized nutraceutical products include soylife, which promotes the development of strong bones, xangold that helps to maintain healthy vision, betatene, which contributes to boost immunity, or cholestaid and oatwell, which reduce the levels of blood cholesterol [40]. Sampaio et al. assessed the impact of different nutraceutical formulations that combine *Limosilacto bacillus fermentum,* quercetin, or resveratrol in the presence of various gut bacterial populations, microbial metabolite production, and antioxidant capacity. Results evidenced the beneficial impact on the capacity of the assayed nutraceutical formulations to positively modulate the metabolite composition and production of the human intestinal gut microbiota [32]. Nenseth et al. carried out a randomized, controlled study on the efficacy of a novel nutraceutical formulation of essential vitamins, minerals, and trace elements. Results revealed that regular intake of the nutraceutical for three months produced a significantly increased serum levels of vitamins A, C, D, E, $B_1$, $B_2$, $B_6$, $B_{12}$, and calcium, iron, IGF-1, and FT-3. Therefore, it may be a promising and beneficial supplement for people suffering from nutritional deficiencies [41].

Shrestha et al. studied the prescribing pattern, affordability, knowledge, and attitude of patients toward nutraceuticals. They observed a high prevalence of consumption of prescribed nutraceuticals due to information suggested by health professionals. Also, the majority exhibited a moderately positive attitude, though about half of the population studied showed poor knowledge about nutraceuticals [42]. The overall goals of regulating nutraceuticals as food have attempted to focus on safety and labeling, with a lesser emphasis, compared to pharmaceuticals, on product declarations and intended use [28]. Pharmaceuticals are generally made up of a single substance, in contrast to nutraceuticals, which are composed of a set of substances [23]. Also, nutraceuticals ought to be intrinsically softer when compared to drug products, which in some cases, could hamper the screening for statistically significant effects [35].

Major drawbacks to the efficient and effective utilization of nutraceuticals in preventive and therapeutic purposes are mainly associated with their limited oral bioavailability and the unavailability of clinical data to support their efficacy [43]. Additionally, as will be explained below, legislation concerning nutraceutical products is quite dissimilar at the international level, which is another drawback due to certain gaps in technical analysis; worldwide sales via the Internet in turn make cross-border purchases convenient, and consequently, result in increased public health risks and economic burden [24].

Even though healthcare authorities have recommended nutraceuticals as a helpful approach to maintaining optimal health and to provide a curative effect against acute and chronic diseases due to nutrition, such products are still under development in clinical practice and further research has yet to tackle crucial clinical and pharmaceutical issues [40]. For example, it is recommended to be careful when a nutraceutical is administered as an adjuvant or therapeutic agent, particularly in pre-surgery patients (e.g., products that contain gingko biloba, valerian, ginseng, echinacea or St. John's wort extracts) because they might induce undesirable side effects [23]. The evaluation of the optimum use conditions of nutraceuticals ought to be guided, to complement the data about their safety, bioavailability, and bioaccessibility, in order to propose them to prevent and heal some pathological conditions in individuals, who, for example, are not suitable for standard pharmacological therapy [22]. Additionally, as the trend toward personalized nutrition and lifestyle changes grows, nutraceutical ingredients may mean progress in bridging the gap between medicine and food [44].

## 2.2 International regulation and status of nutraceuticals

Despite the disparity in the regulatory framework of nutraceuticals in different countries or countries unions, the consumer safety concern is at the center of their international regulatory status. In the United State of America, nutraceutical was defined as dietary supplements in 1994 by the FDA, under the Dietary Supplement Health and Education Act (DSHEA). In 1997, a modernization act of the FDA has been laid out. In that document, in the section concerning the food labeling, it is mentioned that health claims on the label should be done after the investigation and authorization of the Academy of Science or a federal authority. Then, they must notify the FDA and wait for 4 months, before introducing a nutraceutical or dietary supplement with the mentioned health claims benefits in the market [45]. Thus, FDA has the authority to act against any unsafe nutraceutical product in the market.

According to European Union (EU), there is no clear distinction in the regulation of food supplements and functional foods or nutraceuticals, even though their authorization is conditioned by a certified clinical trial confirming there efficacity and safety. Moreover, the basis of food safety is presented in the Regulation (EC No. 178/2002) of the EU General Food Law [46]. The European Parliament Council of 28 January 2002 has established the EFSA to be in charge of the authorizations or approvals of health claims for any food products, supplements, or nutraceuticals [45].

In Australia, the regulation of nutraceuticals, considered as "Therapeutic Goods", has been established under the Therapeutic Goods Acts of 1989. Recently, in 2018, a regulatory guideline has been published for complementary medicines, a category to which nutraceuticals belong. Under this regulation, it is mentioned that products with low risk must be listed in the Australian Register of Therapeutic Goods (ARTG) and prior to their listing, even though they do not need an evaluation, they must require a certification before entering the market. On the other side, concerning products with high risk, the regulation framework imposes their registration in the ARTG. So, for high-risk products, an assessment on quality, efficacy, and safety is required. The Australian regulation framework of the registered complementary medicines tends to be close to some international regulatory requirements and procedures such as USFDA and EFSA [24, 47].

In Japan, the regulation of food stuff and food supplements with health benefits includes nutraceuticals under the Foods for Specified Health Use (FOSHU). This legislation settled in 1991, has been incorporated in the Health promotion law; the latest being in 2003 [45,48]. According the FOSHU, the approval of beneficial heath claims is given even without proven scientific evidence. The health claims are accepted if it has been established for one of the ingredients. The FOSHU requirements in Japan are mostly focused on the nutritional ingredients' content and safety. In 2015, Japan introduced "Foods with Function Claims (FFC)" as a category, including nutraceuticals, under which the approval is testified by the Secretary-General of the Consumer Affairs Agency (CAA). The government's intention was to differentiate and categorize the dietary supplements that are intended to prevent and promote heath. But, in that case, the approvals of new products falling into this category were not conditioned by an evaluation of the efficacy of the claimed function and neither by a safety assessment [47].

# 3 Industrial and market status of nutraceuticals

Nutraceuticals contain a variety of health-promoting compounds; they are also undergoing research as a natural remedy. An increasing demand for these supplements is resulting in a growing number of nutraceuticals being marketed. Hence, in parallel to the food and pharmaceutical industries, the global nutraceutical market is expanding rapidly. Since nutraceuticals are still underutilized for treating life-threatening diseases, their development may yield better results [49].

The term "probiotic" was analyzed and finally established by the International Association for Scientific Prebiotics and Probiotics (ISAPP) as "live microorganisms that, when administered in adequate amounts, confer a health benefit on the host" [50]. The probiotics market, which is worth hundreds of million dollars, is increasing daily, owing to their many proven benefits and the growing awareness of healthy foods and lifestyle in people [51, 52]. However, bringing a probiotic product to the market presents some drawbacks and so it may be time-consuming due to the microbiological and regulatory issues involved. The beneficial properties of probiotics include the enhancement of GI tract microbial balance, when ingested (such as *Bifidobacterium* and *Lactobacillus*) [53].

As discussed before for other nutraceuticals, encapsulation is also an efficient technique to preserve and protect probiotic bacteria from the stomach's harshness. In fact, the use of prebiotics as encapsulating agents has been demonstrated to create synergies that enhance the bioactivity of both components in the digestive tract and, consequently, further health benefits [54]. The study of real-time cell imaging techniques and nutrigenomics is increasingly used to develop new nutritional products by encapsulating the essential components in a capsule. For this purpose, it is important to isolate and purify nutraceuticals [35]. The use of encapsulation techniques is key to control the release since probiotic bacteria that can get broken down into beneficial compounds can be absorbed in the large intestine, which may not be digested or absorbed in the early part of the digestive tract. In this sense, even though many natural and synthetic prebiotics are available on the market, more functional natural prebiotic products obtained from natural sources such as fruits and vegetables are being developed by biotechnological sciences [51].

The increasing population and resource depletion have led to a dramatic increase in the use of microbial enzymes. Food and dairy, pharmaceuticals, and nutraceuticals are all industries in which enzymes have significant potential. There is no doubt that multiple industries widely use microbial enzymes to produce high-quality products that respond quickly, and are less prone to contamination [55,56]. Several strains of fungus are GRAS-eligible for food production and can secrete large quantities of extracellular enzymes and, similarly, various yeasts. Therefore, enzymes have a significant share of the market – industrial enzymes are used in the food industry. Indeed, to obtain functional food products, it has become essential to utilize enzymes, since they remove factors that are not nutritional, make nutraceuticals digestible and bioavailable, and facilitate the extraction of bioactive compounds from different sources. Even though molecular mechanisms have enabled food companies to use enzymes more extensively in recent years, enzymes must be evaluated to unlock their potential for biotransformation of critical metabolites and food products. Indeed, due to their sensitivity to harsh industrial environments, enzymes have been reluctantly adopted by many chemical industry sectors over the past decades. The optimal exploitation of enzyme applications in industrial processes requires finding new ways to overcome their weaknesses. Therefore, to prompt the use of biocatalytic processes, different approaches may be established. For this purpose, new enzymes may be sought from unknown natural sources, existing catalytic properties may be improved, enzymes may be re-designed to perform new functions, enzyme preparation formulations may be optimized, or new biocatalysts may be developed [57]. And so, today, bioprocessing relies heavily on enzymes that are difficult to produce, purify, and use in industrial processes. There are also challenges related to the storage and modification of enzymes [58–60]. In this field, drawbacks have been turned into advantages. Enzymes that affected nutraceuticals and metabolites were tuned to help industries when exposed to harsh conditions. In fact, it is currently possible to obtain enzymes from known microorganisms in laboratories. Screening unculturable enzymes can lead to discovering new ones with unusual activity and stability. With advances in genome engineering tools, enzymes can now be used to create functional foods and nutraceuticals, and so it has become a common way to boost productivity and yield since it is an excellent alternative to traditional methods of food production. For instance, metagenomics may be used to identify enzymes that can survive harsh environmental conditions and be active [61], or they can be further modified using genetic engineering methods to withstand these extreme industrial conditions, thereby improving their use in food processing [62]. Regarding the market context, most enzymatic processes and products are developed in China, the United States, Japan, France, and South Korea [63]. On the global market, enzymes have experienced a rapid increase in demand in recent years. For the next few years, carbohydrase enzymes will continue to lead in market segments such as food and beverage, animal feed, and detergent. Growing per capita incomes will lead to greater access to healthcare in developing regions, which will increase pharmaceutical enzyme applications. The Asia/Pacific region is predicted to overtake Western Europe as the largest consumer of enzymes in the coming years since the Asia/Pacific region, China, and India account for the largest demand due to their large economic size and strength. Nevertheless, Western Europe is gaining field and may dominate enzyme production for the next ten years if Asia/Pacific companies, particularly Japanese and Chinese, do not gain a competitive advantage to succeed.

Host pharmacokinetic studies and preclinical and clinical evaluations of active peptides are necessary to confirm biological efficacy, observe their mode of action, and confirm their effects on healthcare [64]. The need for more innovation and detailed research interest for new product development are among the problems ahead in this industry, along with a focus on conventional food ingredients and traditional methods of adding capabilities to food by adding vitamins, minerals, or plant extracts. In this industry, new products will emerge, based on innovation, technology, and research. Using biotechnology tools such as genetically modified food has the potential to expand this industry with more efficacy and safety data, supported by scientific evidence. Personalized food systems for populations with defined risk factors, ageing populations, healthy lifestyle awareness in developing countries, and a movement to promote living in harmony with nature will ensure that nutraceutical industries will grow in the future [65]. As nutraceuticals are derived mainly from natural ingredients, their demand and value are increasing due to new research; so the future market for nutraceuticals will be shaped by natural products. Vitamins, nutraceutical supplements, functional beverages, and functional foods made up about $209 billion of the global nutraceutical market in 2017 [53]. According to various estimates, USA holds a 75.9-billion-dollar market share for nutritional supplements in 2018, with a 5.3% CAGR. Additionally, nutraceuticals were valued at $382 and $412 billion in 2019 and 2020, respectively [66]. Consumer interest in functional foods and beverages has caused the nutraceutical market to overgrow. Approximately US$ 275 billion was expected to be generated by 2021. Cannabidiol (CBD) was the most significant innovation in the nutraceutical industry, due to the growing interest in marijuana's medicinal properties. Market analysts project a 16-billion-dollar growth in USA CBD market between now and 2025. In the future, powerful countries like the USA, Germany, China, and the UK anticipate high profits from nutraceutical crops because they are investing more in research and development. USA has seen the use of nutraceuticals grow to more than 50% of its population in recent years [53]. A multi-billion-dollar industry has developed around nutraceuticals and functional foods, and Canada's nutraceuticals market is estimated to grow to US$ 50 billion by 2025. With a steady average growth rate of 9.6% per year, Japan's food industry is the second largest in the world after USA [67]. Table 2 shows some of the nutritional products available in the world market.

Commercially, nutraceuticals such as enzymes, probiotics, prebiotics, and fortified foods are substances that significantly impact human health, but they must successfully pass strict regulation procedures [63]. Nutraceuticals play a primary role in the development of therapeutics in the context of self-medication in the present day. However, they will only succeed if they maintain quality, purity, stability, safety, and efficacy [70].

Regulatory issues related to probiotic product marketing and classification will be simplified and improved by government regulatory bodies collaborating with the industry and academia. Consumers will benefit from these improvements through more effective and more accurate labeling of products [71]. Nutraceuticals must be regulated as the market grows to protect consumer rights and health, and manufacturers must comply with regulatory requirements and ensure that their products are safe for consumers [72], since, as explained before, several scenarios expose consumers to significant risks and cases of adverse reactions. Therefore, before products are placed on the market, strict regulations are required to ensure their quality and safety [73]. By 1962, the Food and Agriculture Organization (FAO) and World Health Organization (WHO) developed the Codex Alimentarius notification to harmonize global regulatory aspects for food and nutraceuticals and several countries regulated

**Table 2:** Commercialized and functional nutraceutical products [53,67,68,69].

| Brand product | Functionality | Substances |
|---|---|---|
| Aciforce, Bacilac, Bactisubtil, Bififlor, Proflora | Probiotics | Freeze-dried probiotics, bacteria |
| DanActive® | Probiotics | *Lactobacillus casei* DN114001 |
| DanimalsR | Probiotics to prevent infant diarrhea | *Lactobacillus rhamnosus GG* |
| Luna bar | Prebiotics | Fiber |
| Helios nutrition's, organic kefir | Prebiotics | Bifidogenesis, calcium absorption |
| ZonePerfect shakes | Prebiotics | Fiber |
| Ensure fiber | Prebiotics | Fiber, digestive health |
| Builder's bar | Prebiotics | Fiber |
| Pure red reishi | Antidepressant | *Ganoderma lucidum* |
| Appetite Intercept ™ | Appetite suppressant | Caffeine, tyrosine, and phenylalanine |
| Metabolife ultra caffeine free ™ | Appetite suppressant | B-vitamins |
| Lumatol AC | Appetite suppressant | Cacti |
| Biovinca ™ | Neurotonic | Vinpocetine |
| PNer plus ™ | Neuropathic pain supplement | Vitamin and other natural supplements |
| Olivenol ™ | Dietary supplement | Natural antioxidant, hydroxytyrosol |
| GRD® | Nutritional supplement | Proteins, vitamins, minerals, and carbohydrates |
| Revital® | Daily health supplement | Vitamins and minerals |
| Threptin® Diskettes | Protein supplements | Proteins and vitamin B |
| ProteineX® | Protein supplements | Carbohydrates, minerals, predigested, protein, and vitamins |
| Coral calcium | Calcium supplement | Calcium and trace minerals |
| Calcirol D-3® | Calcium supplement | Calcium and vitamins |
| Celestial healthtone | Immune booster and immunomodulator | Dry fruit extract |
| Betafactor® capsules | Immune supplement | Beta-glucan |
| Omega women | Immune supplement | Antioxidants, vitamins, and phytochemicals (*e.g.,* Lycopene) |
| Mushroom optimizer ™ | Immune supplement | Mushrooms polysaccharides and folic acid |

nutraceutical products differently [53]. In this scenario, industry-related companies need to be more innovative, since there is no regulation, no uniformity in rules, and no health claims can be gotten for nutraceuticals due to the complex regulatory issues. Several aspects of the development and marketing of nutraceuticals have been challenging in the global market because of the ambiguity of information, including clinical studies, validation of biological markers, definition of effective dosage based on a dose-response curve, and information about adverse effects. Most nutraceuticals are not unique and are not regulated by patents because they are not inimitable and can be used to make similar products by any company that belongs to the same industry. Companies that strive to achieve health claims lose their competitive advantage as a result. There is a lot of potential in this industry despite many problems. It is becoming increasingly clear to consumers and food companies that nutraceuticals can positively impact people's health, when used in conjunction with medical, behavioral, and nutritional interventions [65].

It is currently possible to purchase a wide range of nutraceutical formulations, including vitamin and herbal formulations on the global market. In recent years, there has been a significant increase in the number of nutraceuticals, based on supporting research. Manufacturing companies innovate new products to meet consumer demands in the global market, and are supported by advertising and web media to reach most consumers. As a result of ageing population, increasing consumer awareness, expanding distribution channels, and rising healthcare costs, the global food industry market is growing. Boosting immunity, reducing stress and anxiety, and seeking other wellness products were among the products that consumers sought out because of the Corona pandemic. In addition, consumers in USA are increasingly interested in nutraceuticals that promote gastrointestinal health and beauty, as evidenced by the increasing number of products that contain prebiotics, probiotics, and collagen [53].

# 4 Summary

In recent decades, consumers have been more health-conscious, and are increasingly looking for foods and food products with potential health benefits. Moreover, since aging population is increasing rapidly, the use of nutraceuticals may be an interesting pathway to guarantee human wellness even in advanced ages. In this way, nutraceuticals are playing an important role in the pharmacological and food industry, since they can not only be directly consumed in capsules but can be incorporated into food matrices, leading to food fortification. Nutraceuticals consumption is related to several health benefits, such as antioxidant and anticancer activities or cardiovascular protection effect. However, most of the bioactive compounds, known as nutraceuticals, have low water solubility and, therefore, their bioavailability is limited. Thus, different delivery systems have been proposed. Nanoemulsions, micelles, and nanoparticles or hydrogels have been studied as suitable delivery systems, showing several advantages such as the increment of intestinal absorption, thanks to the lower particle size, the maintenance of the compounds' bioactivities, or the controlled release. These positive characteristics of nutraceuticals lead to an increased demand for these supplements, and therefore a growing number of nutraceuticals are commercialized, resulting in an expansion of the global market. As the global nutraceutical market is growing, consumers' awareness about these supplements has also increased, and so specific regulation must be considered. Thus, the commercialization of nutraceuticals seems to open an interesting new pathway for both food and pharmacological industry. In this way, the development of a specific regulation as also more studies are required to guarantee consumers' safety.

**Acknowledgements:** The research leading to these results was supported by MICINN supporting the Ramón y Cajal grant for M.A. Prieto (RYC-2017-22,891); by Xunta de Galicia for supporting the program EXCELENCIA-ED431F 2020/12, the post-doctoral grant of M. Fraga-Corral (ED481B-2019/096), and the pre-doctoral grant of M. Carpena (ED481A 2021/313). Authors are grateful to Ibero-American Program on Science and Technology (CYTED – AQUA-CIBUS, P317RT0003), and to the Bio Based Industries Joint Undertaking (JU) under grant agreement No 888003 UP4HEALTH Project (H2020-BBI-JTI-2019). The JU receives support from the European Union's Horizon 2020 research and innovation program and the Bio Based Industries Consortium. The project SYSTEMIC Knowledge hub on Nutrition and Food Security, has received funding from national research funding parties in Belgium (FWO), France (INRA), Germany (BLE), Italy (MIPAAF), Latvia (IZM), Norway (RCN), Portugal (FCT), and Spain (AEI) in a joint action of JPI HDHL, JPI-OCEANS and FACCE-JPI launched in 2019 under the ERA-NET ERA-HDHL (n° 696295).

# References

[1]  Manocha, S., Dhiman, S., Grewal, A.S. and Guarve, K. 2022. Nanotechnology: An approach to overcome bioavailability challenges of nutraceuticals. The Journal of Drug Delivery Science and Technology, 72, 103418.

[2]  Mishra, A., Pradhan, D., Biswasroy, P., et al 2021. Recent advances in colloidal technology for the improved bioavailability of the nutraceuticals. The Journal of Drug Delivery Science and Technology, 65, 102693.

[3]  Gonçalves, R.F.S., Martins, J.T., Duarte, C.M.M., et al 2018. Advances in nutraceutical delivery systems: From formulation design for bioavailability enhancement to efficacy and safety evaluation. Trends in Food Science and Technology, 78, 270–291.

[4]  Abuhassira-Cohen, Y. and Livney, Y.D. 2022. Enhancing the bioavailability of encapsulated hydrophobic nutraceuticals: Insights from in vitro, in vivo, and clinical studies. Current Opinion in Food Science, 45, 100832.

[5]  Ting, Y., Jiang, Y., Ho, C.T. and Huang, Q. 2014. Common delivery systems for enhancing in vivo bioavailability and biological efficacy of nutraceuticals. Journal of Functional Food, 7, 112–128.

[6]     Delfanian, M. and Sahari, M.A. 2020. Improving functionality, bioavailability, nutraceutical and sensory attributes of fortified foods using phenolics-loaded nanocarriers as natural ingredients. Food Research International, 137, 109555.

[7]     Zagury, Y., Kazir, M. and Livney, Y.D. 2019. Improved antioxidant activity, bioaccessibility and bioavailability of EGCG by delivery in β-lactoglobulin particles. Journal of Functional food, 52, 121–130.

[8]     McClements, D.J. 2015. Enhancing nutraceutical bioavailability through food matrix design. Current Opinion in Food Science, 4, 1–6.

[9]     Gorantla, S., Wadhwa, G., Jain, S., et al 2022. Recent advances in nanocarriers for nutrient delivery. Drug Delivery and Translational Research, 12, 2359–2384.

[10]    Clifford, M.N., Der Hooft JJJ, V. and Crozier, A. 2013. Human studies on the absorption, distribution, metabolism, and excretion of tea polyphenols. The American Journal of Clinical Nutrition, 98(6 Suppl), 1619S–1630S.

[11]    McClements, D.J. 2017. Recent progress in hydrogel delivery systems for improving nutraceutical bioavailability. Food Hydrocolloids, 68, 238–245.

[12]    Cohen, Y., Ish-Shalom, S., Segal, E., et al 2017. The bioavailability of vitamin D3, a model hydrophobic nutraceutical, in casein micelles, as model protein nanoparticles: Human clinical trial results. Journal of Functional Food, 30, 321–325.

[13]    Haham, M., Ish-Shalom, S., Nodelman, M., et al 2012. Stability and bioavailability of vitamin D nanoencapsulated in casein micelles. Food & Function, 3, 737–744.

[14]    Yang, Y. and McClements, D.J. 2013. Vitamin E bioaccessibility: Influence of carrier oil type on digestion and release of emulsified α-tocopherol acetate. Food Chemistry, 141, 473–481.

[15]    Martins, J.T., Santos, S.F., Bourbon, A.I., et al 2016. Lactoferrin-based nanoparticles as a vehicle for iron in food applications – Development and release profile. Food Research International, 90, 16–24.

[16]    Shah, A., Ul ashraf, Z., Gani, A., et al 2022. β-Glucan from mushrooms and dates as a wall material for targeted delivery of model bioactive compound: Nutraceutical profiling and bioavailability. Ultrason Sonochem, 82, 105884.

[17]    Qian, C., Decker, E.A., Xiao, H. and McClements, D.J. 2012. Nanoemulsion delivery systems: Influence of carrier oil on β-carotene bioaccessibility. Food Chemistry, 135, 1440–1447.

[18]    Bin, H.Y., Tsai, M.J., Wu, P.C., et al 2011. Elastic liposomes as carriers for oral delivery and the brain distribution of (+)-catechin. Journal of Drug Targeting, 19, 709–718.

[19]    Pezeshky, A., Ghanbarzadeh, B., Hamishehkar, H., et al 2016. Vitamin A palmitate-bearing nanoliposomes: Preparation and characterization. Food Bioscience Volume, 13, 49–55.

[20]    Chopra, A.S., Lordan, R., Horbańczuk, O.K., et al 2022. The current use and evolving landscape of nutraceuticals. Pharmacological Research, 175, 106001.

[21]    Siddiqui, R.A. and Moghadasian, M.H. 2020. Nutraceuticals and nutrition supplements: Challenges and opportunities. Nutrients, 12, 10–13.

[22]    Santini, A. and Novellino, E. 2017. To nutraceuticals and back: Rethinking a concept. Foods, 6, 6–8.

[23]    Daliu, P., Santini, A. and Novellino, E. 2019. From pharmaceuticals to nutraceuticals: bridging disease prevention and management. Expert Review of Clinical Pharmacology, 12, 1–7.

[24]    Komala, M.G., Ong, S.G., Qadri, M.U., et al 2023. Investigating the regulatory process, safety, efficacy and product transparency for nutraceuticals in the USA, Europe and Australia. Foods, 12, 13.

[25]    European Parliament and of the Council. 2015. Regulation (EU) 2015/2283 of the European Parliament and of the Council of 25 November 2015 on novel foods, amending Regulation (EU) No 1169/2011 of the European Parliament and of the Council and repealing Regulation (EC) No 258/97 of the European Parliam. Official Journal of the European Communities Legislation, 327, 1–22.

[26]    EFSA. 2017. Overview on Tolerable Upper Intake Levels as derived by the Scientific Committee on Food (SCF) and the EFSA Panel on Dietetic Products, Nutrition and Allergies (NDA). https://www.

efsa.europa.eu/sites/default/files/2023-11/ul_summary_tables-version-8.pdf (accessed on 7 June 2023).

[27] Helal, N.A., Eassa, H.A., Amer, A.M., et al 2019. Nutraceuticals' novel formulations: The good, the bad, the unknown and patents involved. Recent Patents on Drug Delivery & Formulation, 13, 105–156.

[28] Chandra, S., Saklani, S., Kumar, P., et al 2022. Nutraceuticals: Pharmacologically active potent dietary supplements. BioMed Research International, Hindawi, 2022, 2051017.

[29] Baker, M.T., Lu, P., Parrella, J.A. and Leggette, H.R. 2022. Consumer acceptance toward functional foods: A scoping review. International Journal of Environmental Research and Public Health, 19(3), 1217.

[30] Puri, V., Nagpal, M., Singh, I., et al 2022. A comprehensive review on nutraceuticals: therapy support and formulation challenges. Nutrients, 14, 4637.

[31] Taroncher, M., Vila-Donat, P., Tolosa, J., et al 2021. Biological activity and toxicity of plant nutraceuticals: an overview. Current Opinion in Food Science, 42, 113–118.

[32] Sampaio, K.B., De Brito Alves, J.L., Do Nascimento, Y.M., et al 2022. Nutraceutical formulations combining Limosilactobacillus fermentum, quercetin, and or resveratrol with beneficial impacts on the abundance of intestinal bacterial populations, metabolite production, and antioxidant capacity during colonic fermentation. Food Research International, 161, 111800.

[33] Santini, A., Tenore, G.C. and Novellino, E. 2017. Nutraceuticals: A paradigm of proactive medicine. European Journal of Pharmaceutical Sciences, 96, 53–61.

[34] Ramirez, D., Abellán-Victorio, A., Beretta, V., et al 2020. Functional ingredients from brassicaceae species: Overview and perspectives. International Journal of Molecular Sciences, 21(6), 1998.

[35] Sauer, S. and Plauth, A. 2017. Health-beneficial nutraceuticals – myth or reality? Applied Microbiology and Biotechnology, 101, 951–961.

[36] Alba, G., Dakhaoui, H., Santa-Maria, C., et al 2023. Nutraceuticals as Potential Therapeutic Modulators in Immunometabolism. Nutrients, 15, 411.

[37] Petrarca, C. and Viola, D. 2023. Redox remodeling by nutraceuticals for prevention and treatment of acute and chronic inflammation. Antioxidants, 12, 132.

[38] Ronis, M.J.J., Pedersen, K.B. and Watt, J. 2018. Adverse effects of nutraceuticals and dietary supplements. Annual Review of Pharmacology and Toxicology, 58, 583–601.

[39] Kunnumakkara, A.B., Sailo, B.L., Banik, K., et al 2018. Chronic diseases, inflammation, and spices: How are they linked? Journal of Translational Medicine, 16, 1–25.

[40] Khalaf, A.T., Wei, Y., Alneamah, S.J.A., et al 2021. What is new in the preventive and therapeutic role of dairy products as nutraceuticals and functional foods? BioMed Research International, Hindawi, 2021, 8823222.

[41] Nenseth, H.Z., Sahu, A., Saatcioglu, F. and Osguthorpe, S. 2021. A nutraceutical formula is effective in raising the circulating vitamin and mineral levels in healthy subjects: A randomized trial. Frontiers in Nutrition, 8, 703394.

[42] Shrestha, R., Shrestha, S., Badri, K.C. and Shrestha, S. 2021. Evaluation of nutritional supplements prescribed, its associated cost and patients knowledge, attitude and practice towards nutraceuticals: A hospital based cross-sectional study in Kavrepalanchok, Nepal. PLoS One, 16, 1–15.

[43] Calvani, M., Pasha, A. and Favre, C. 2020. Nutraceutical boom in cancer: Inside the labyrinth of reactive oxygen species. International Journal of Molecular Sciences, 21(6), 1936.

[44] Leena, M.M., Silvia, M.G., Vinitha, K., et al 2020. Synergistic potential of nutraceuticals: Mechanisms and prospects for futuristic medicine. Food & Function, 11, 9317–9337.

[45] Santini, A., Cammarata, S.M., Capone, G., et al 2018. Nutraceuticals: opening the debate for a regulatory framework. British Journal of Clinical Pharmacology, 84, 659–672.

[46] Vettorazzi, A., De Cerain, A.L., Sanz-Serrano, J., et al 2020. European regulatory framework and safety assessment of food-related bioactive compounds. Nutrients, 12, 1–16.

[47] Blaze, J. 2021. A comparison of current regulatory frameworks for nutraceuticals in Australia, Canada, Japan, and the United States. Innovation in Pharmacy, 12, 8.

[48] Yamada, K., Sato-Mito, N., Nagata, J. and Umegaki, K. 2008. Health claim evidence requirements in Japan. Journal of Nutrition, 138(6), 1192S–8S.

[49] Varia, A. and Soni, R. 2022. Biomolecules of Fungi: Nutraceutical, Pharmaceutical, Industrial and Environmental Protection. In: *Multiple Roles of Fungal Organic Molecules (Nutraceuticals) in Different Fields*. pp 225–238

[50] Hill, C., Guarner, F., Reid, G., et al 2014. The International Scientific Association for probiotics and prebiotics consensus statement on the scope and appropriate use of the term probiotic. Nature Reviews Gastroenterology & Hepatology, 11, 506–514.

[51] Thammarutwasik, P., Hongpattarakere, T., Chantachum, S., et al 2009. Prebiotics – a review. Songklanakarin Journal of Science and Technology, 4, 401–408.

[52] Douglas, L.C. and Sanders, M.E. 2008. Probiotics and prebiotics in dietetics practice. Journal of the American Dietetic Association, 108, 510–521.

[53] Amalraj, A., Kuttappan, S., K, V.A.C. and Matharu, A. 2022. *Herbs, Spices and Their Roles in Nutraceuticals and Functional Foods*. Elsevier.

[54] Reque, P.M. and Brandelli, A. 2021. Encapsulation of probiotics and nutraceuticals: Applications in functional food industry. Trends in Food Science and Technology, 114, 1–10.

[55] Tahergorabi, R., Beamer, S.K., Matak, K.E. and Jaczynski, J. 2012. Functional food products made from fish protein isolate recovered with isoelectric solubilization/precipitation. LWT – Food Science and Technology, 48, 89–95.

[56] Ottens, M. and Chilamkurthi, S. 2010. Advances in process chromatography and applications in the food, beverage and nutraceutical industries. In *Separation, Extraction and Concentration Processes in the Food, Beverage and Nutraceutical Industries*. Woodhead Publishing Limited, pp. 109–147.

[57] Li, S., Yang, X., Yang, S., et al 2012. Technology prospecting on enzymes: Application, marketing and engineering. Computational and Structural Biotechnology Journal, 2, e201209017.

[58] Fernández-Lucas, J., Castañeda, D. and Hormigo, D. 2017. New trends for a classical enzyme: Papain, a biotechnological success story in the food industry. Trends in Food Science and Technology, 68, 91–101.

[59] Speranza, P., Lopes, D.B. and Martins, I.M. 2019. Development of functional food from enzyme technology: A review. In *Enzymes in Food Biotechnology: Production, Applications, and Future Prospects*. Elsevier Inc, pp. 263–286.

[60] Da silva, R.R., Da Conceição, P.J.P., De Menezes, C.L.A., et al 2019. Biochemical characteristics and potential application of a novel ethanol and glucose-tolerant β-glucosidase secreted by Pichia guilliermondii G1.2. Journal of Biotechnology, 294, 73–80.

[61] Boddu, R.S. and Divakar, K. 2018. Metagenomic insights into environmental microbiome and their application in food/ pharmaceutical industry. In *Microbial Biotechnology*. Springer Singapore, pp. 23–38.

[62] Zhang, Y., Geary, T. and Simpson, B.K. 2019. Genetically modified food enzymes: A review. Current Opinion in Food Science, 25, 14–18.

[63] Rai, A.K., Sirohi, R., De Souza Vandenberghe, L.P. and Binod, P. 2023. Microbial enzymes in production of functional foods and nutraceuticals. In Microb Enzym Prod Funct Foods Nutraceuticals. pp. 1–301.

[64] Singh, A.R., Desu, P.K., Nakkala, R.K., et al 2021. Nanotechnology-based approaches applied to nutraceuticals. Drug Deliveryand Translational Research, 12, 485–499.

[65] Daliri, E.B.-M. and Lee, B.H. 2015. Current trends and future perspectives on functional foods and nutraceuticals. In *Beneficial Microorganisms in Food and Nutraceuticals*. Springer, pp. 221–244.

[66] Mohd, S. and Kaur, J. 2020. A review on list of herbal nutraceuticals having health benefits. European Journal of Molecular & Clinical Medicine, 7.

[67]   Keservani, R.K., Kesharwani, R.K., Vyas, N., et al 2010. Nutraceutical and functional food as future food: A review. Der Pharmacia Lettre, 2, 106–116.

[68]   Bele, A.A. and Khale, A. 2013. An approach to a nutraceutical abstract. Research Journal of Pharmacy and Technology, 6, 1161–1164.

[69]   Srivastava, S., Sharma, P.K. and Guru, S.K. 2015. Nutraceuticals : A review. Journal Chronopharmaceutical Drug Delivery, 6, 1–10.

[70]   Khan, R.A., Elhassan, G.O. and Qureshi, K.A. 2014. Nutraceuticals : In the treatment & prevention of diseases – an overview. Journal of Pharmaceutical Innovation, 3, 47–50.

[71]   Grumet, L., Tromp, Y. and Stiegelbauer, V. 2020. The development of high-quality multispecies probiotic formulations: From bench to market. Nutrients, 12, 1–19.

[72]   Jha, S.K., Roy, P. and Chakrabarty, S. 2021. Nutraceuticals with pharmaceutics : Its importance and their applications. International Journal of Drug Development and Research, 13, 002.

[73]   Mishra, S.S., Behera, P.K., Kar, B. and Ray, R.C. 2018. Advances in probiotics, prebiotics and nutraceuticals. In *Innovations in Technologies for Fermented Food and Beverage Industries*. Springer, pp. 121–141.

# Index

www.ingramcontent.com/pod-product-compliance
Lightning Source LLC
Chambersburg PA
CBHW080710220326
41598CB00033B/5364